Advances in Experimental Medicine and Biology

Volume 781

For further volumes:
http://www.springer.com/series/5584

Advances in Experimental Medicine and Biology

Recent Volumes in this Series

Volume 782
PROGRESS IN MOTOR CONTROL
Michael A. Riley, Michael J. Richardson and Kevin D. Shockley

Volume 783
THE NEW PARADIGM OF IMMUNITY TO TUBERCULOSIS
Maziar Divangahi

Volume 784
KISSPEPTIN SIGNAING IN REPRODUCTIVE BIOLOGY
Alexander S. Kauffman and Jeremy T. Smith

Volume 785
CROSSROADS BETWEEN INNATE AND ADAPTIVE IMMUNITY IV
Bali Pulendran, Peter D. Katsikis and Stephen P. Schoenberger

Volume 786
TRANSCRIPTIONAL AND TRANSLATIONAL REGULATION OF STEM CELLS
Gary Hime and Helen Abud

Volume 787
BASIC ASPECTS OF HEARING
Brian CJ Moore

Volume 788
NEUROBIOLOGY OF RESPIRATION
Mieczyslaw Pokorski

Volume 789
OXYGEN TRANSPORT TO TISSUE XXXV
Sabine Van Huffel, Gunnar Nualaers, Alexander Caicedo, Duane F. Bruley and David K. Harrison

Volume 790
VIRAL ENTRY INTO HOST CELLS
Stefan Pöhlmann and Graham Simmons

A Continuation Order Plan is available for this series. A continuation order will bring delivery of each new volume immediately upon publication. Volumes are billed only upon actual shipment. For further information please contact the publisher.

Christian R. Landry • Nadia Aubin-Horth
Editors

Ecological Genomics

Ecology and the Evolution of Genes and Genomes

Springer

Editors
Christian R. Landry
Département de Biologie
Local 3110 Institut de Biologie
Intégrative et des Systèmes
Université Laval
Québec, QC
Canada

Nadia Aubin-Horth
Département de Biologie
Local 1241 Institut de Biologie
Intégrative et des Systèmes
Université Laval
Québec, QC
Canada

ISSN 0065-2598 ISSN 2214-8019 (electronic)
ISBN 978-94-007-7346-2 ISBN 978-94-007-7347-9 (eBook)
DOI 10.1007/978-94-007-7347-9
Springer Dordrecht Heidelberg New York London

Library of Congress Control Number: 2013954790

Printed on acid-free paper

Springer is part of Springer Science+Business Media (www.springer.com)

Contents

Recent Advances in Ecological Genomics: From Phenotypic Plasticity to Convergent and Adaptive Evolution and Speciation

1

Christian R. Landry and Nadia Aubin-Horth

Abstract

Biological diversity emerges from the interaction between genomes and their environment. Recent conceptual and technological developments allow dissecting these interactions over short and long time-scales. The 16 contributions to this book by leaders in the field cover major recent progresses in the field of Ecological Genomics. Altogether, they illustrate the interplay between the life-history and genomic architecture of organisms, how the interaction of the environment and the genome is shaping phenotypic variation through phenotypic plasticity, how the process of adaptation may be constrained and fueled by internal and external features of organisms and finally, how species formation is the result of intricate interactions between genomes and the ecological conditions. These contributions also show how fundamental questions in biology transcend the boundaries of kingdoms, species and environments and illustrate how integrative approaches are powerful means to answer the most important and challenging questions in ecology and evolution.

Keywords

Phenotypic plasticity • Ecological genomics • Life history • Speciation • Adaptation

1.1 Introduction to Ecological Genomics

One of the major challenges in biology is to connect genotypes to phenotypes and to identify the ecological and demographic parameters that have shaped genotype frequencies in natural populations (Pavey et al. 2012). Meeting this challenge requires the use of integrative approaches, as different techniques and combination of disciplines are needed to understand processes that act at different levels of organizations, from ecosystems to genes. These integrative approaches have given rise to the field of ecological genomics (Feder and Mitchell-Olds 2003). The chapters contributed to this book reflect the most recent and exciting progress in this young field of research.

C.R. Landry (✉) • N. Aubin-Horth
Département de Biologie, Institut de Biologie Intégrative et des Systèmes, Université Laval, 1030 Avenue de la Médecine, Québec, QC G1V 0A6, Canada
e-mail: Christian.Landry@bio.ulaval.ca

C.R. Landry and N. Aubin-Horth (eds.), *Ecological Genomics: Ecology and the Evolution of Genes and Genomes*, Advances in Experimental Medicine and Biology 781,
DOI 10.1007/978-94-007-7347-9_1, © Springer Science+Business Media Dordrecht 2014

1.2 The New Synthesis

As a second year graduate student and a new postdoc at Harvard University, we registered to a Gordon Research Conference in 2003 that was taking place in nearby New Hampshire. The conference was the inaugural edition of the "Evolutionary & Ecological Functional Genomics" meeting that takes place biyearly ever since. Because of the limited access to genomic resources for most species, major questions asked at the time were "what are good species in ecological genomics"? and "how can we balance the tradeoff between the ability to perform genomics experiments and the interest of the ecological problem to be investigated for a particular species"? This book demonstrates that we are moving ahead from the preoccupations of what species to choose in ecological genomics. Questions have reached another level, where available genomics tools can be applied to virtually any species, and the range of questions one can ask is virtually unlimited. This book is organized around 16 chapters contributed by leaders in the field and covers three major subjects of prime interest in ecological genomics.

The first three chapters illustrate how life-history and mating systems can impact or can be influenced by patterns of genomic variation. First, Vincent Castric and colleagues use the diversity of plant mating systems to illustrate how life-history transitions can impinge on plant genomic architecture and diversity and how genomic architecture and physiological features may in turn affect these transitions. This field of research is a particularly rich one in ecological genomics as it is based on strong theoretical predictions, to which are combined very well documented molecular underpinnings of reductive mechanisms. In the third chapter, Paul Magwene examines the role of life-history traits on patterns of genomic variation in budding yeast on a much shorter time scale. The budding yeast is one of the most studied organisms in the laboratory and it has recently become an excellent model as well in ecological genomics (Landry et al. 2006). Its mode of reproduction has intrigued geneticists for several years because although there is ample opportunity for inbreeding by mating between mother and daughter cells, natural isolates of the budding yeast are often heterozygous at many molecular markers. Here, Magwene reviews these concepts and the main hypotheses that have been put forward to explain these observations, and shows how genome-wide analysis of genetic variation can be used to test these models. Finally, Jean-Baptiste Leducq examines how life-history styles modulate or correlate with the architecture of genomes, using fungi as models. Because most fungi have genomes of limited sizes, a large number of fungal genomes have been fully sequenced or are underway, such that they offer unique opportunities to test hypotheses regarding the relationship between genome organization and ecological traits. Leducq reviews these studies and also many others that have examined the evolutionary ecological genomics of fungi on shorter time scales, such as during experimental adaptation and speciation.

The second part of the book focuses on one aspect of biological diversity that can hardly be explained from the patterns of genomic variation alone, the diversity that emerges from the interaction between a given genome and its environment. In his book "The triple helix", Richard Lewontin (2000) argues that while we can learn a lot about an organism from the analysis of its genomic sequence, our ability to predict what the phenotype of an organism will be based solely on this knowledge is very limited, because phenotypes emerge from the interaction between genes and the environment, which itself is greatly influenced by the organism. With sequencing capabilities that have exploded in the recent years, one would have hoped that the complexity of genotype-to-phenotype mapping would be reduced. However, the major issues pointed by Lewontin are even more critical at this point in time and understanding the role of the environment on the genome in shaping phenotypic variation is all the more important. One type of phenotypic variation that is particularly challenging to grasp is the result of phenotypic plasticity.

Phenotypic plasticity is the ability of a genotype (genomic sequence) to give rise to different phenotypes in different environmental conditions. While for a long time phenotypic plasticity has been studied under statistical terms in quantitative genetics (Stearns 1992), the emergence of genomics approaches in the recent past has allowed to directly probe what the genomics programs of alternative phenotypes are (Aubin-Horth and Renn 2009). In Chap. 5, Matthew Morris and Sean Rogers provide a broad overview of phenotypic plasticity, focusing on its evolution in terms of adaptation and maladaptation and in turn on its impact on evolutionary processes, especially in novel environments and in link with speciation. Morris and Rogers discuss how ecological genomics allow to study these questions further and in novel ways and how the genomic architecture of plastic phenotypes will eventually be uncovered using new approaches. In Chap. 6, Ehab Abouheif and colleagues present the state of the emerging field of eco-evo-devo, which combines the study of development of an organism with the measurement of the environmental factors that affect it and of its evolution. Abouheif and colleagues present several examples of recent advances and outstanding questions about the evolution of development in an ecological context and how the framework of ecological genomics has recently allowed this field to move in giant steps. In Chap. 7, Armin Moczek and colleagues present a model system with ideal characteristics to dissect the genomic mechanisms underlying developmental plasticity of morphology among individuals, among populations and among species: the horned beetles. The presence of a horn has evolved independently in several groups of insects, and in many species, some males of the same population develop a horn while others do not, as a result of developmental plasticity. Moczek et al. present how understanding the molecular pathways that underlie horn development is central to understanding the evolution of plasticity in this trait. They also propose a statistical approach to test the effects of several factors of interest on gene expression, allowing to test hypotheses with ecological genomics data in the most efficient way.

While several studies such as the ones featuring horned beetles have focused on plasticity in morphology, it has also been recently recognized that plasticity in behavior, resulting both from the environment encountered during development and in adult individuals, can play a major role in the success of organisms in variable abiotic and biotic conditions (Aubin-Horth and Renn 2009). Rayna Harris and Hans Hofmann (Chap. 8) present the state of the field of neurogenomics, which aims, among other things, at understanding the molecular mechanisms underlying behavioral plasticity. Harris and Hofmann review the different temporal scales at which this plasticity can be studied from a genomics perspective. They also give an overview of the wide array of systems that have been studied, from bees to fish, birds and mammals, and of the diversity of behaviors, showing the power of a comparative approach in ecological genomics. Finally, they also advocate the use of reverse genomics to complement the most current approach of going from a phenotype to the molecular level.

One of the environmental conditions that can widely affect an individual's phenotype is the presence of parasites. A fascinating effects of infection by parasites is the alteration of its host's behavior, often resulting in a higher success rate of transmission to the parasite's final host where it can reproduce. This plastic change in behavior has been proposed to be an adaptation of the parasite, although this is still an unsettled debate (Poulin 2010). In Chap. 9, François-Olivier Hébert and Nadia Aubin-Horth argue that in order to determine if a parasite actively manipulates its host behavior or, alternatively, if this behavior alteration is merely a "side-effect" of the infection, it is essential to uncover the mechanistic basis of how parasites and their hosts' genomes interact. They propose that diverse molecular mechanisms could be involved, and that the study of less well known levels of biological organization, such as the interactome, the phosphorylome and the microRNAome, in both the host and the parasite, and an integrative view of the

"altered phenotype" in the host will lead to a better understanding of this genome-genome interaction. The final chapter of this part of the book investigates what is thought to be an important mechanism underlying plasticity: epigenetics. Kilvitis and colleagues (Chap. 10) discuss how epigenetic modifications of the genome can affect ecologically important traits such as floral or growth traits and potentially the success of individuals facing different environmental conditions. They also discuss how the use of molecular approaches transferred to ecologically and evolutionary model species is gaining ground and will help understand this new layer of complexity in the determination of phenotypic variation.

In the third part of the book are contributions to the field of the study of adaptation and speciation. The genomic mechanisms of speciation and adaptation are central to ecological genomics (Pavey et al. 2012), since among the major challenges of the field is the identification of the genes and gene networks involved in these processes. One of the evidence to support evolution by natural selection is the evolution of the same phenotype repeatedly in similar environments (Arendt and Reznick 2008). While there are documented cases of repeated adaptation in the wild, we still have a poor understanding of the likelihood of repeated adaptations to occur and whether they result from similar selective forces or from the fact that there are only a limited number of genetic solutions to any required adaptation. In Chap. 11, Achaz and colleagues examine models dealing with evolutionary convergence and provide examples from the literature where adaptive landscapes have been exhaustively dissected. They demonstrate how experimental evolution allows to strictly control the genotypes studied. Their work has broad impact as they conclude on the need to consider different types of convergent evolution and to integrate them, for instance convergent evolution at the gene level or at the phenotypic level. In addition, they illustrate how experimental evolution, i.e., where the ecology of a population is strictly controlled, can inform us on the genomics bases of adaptive evolution. The issues of convergent evolution at the molecular level are also very important in natural contexts, as these issues are central to Chaps. 2, 13, 16, and 17, confirming that this aspect of adaptation is a major question in ecological genomics (Pavey et al. 2012).

The genomics study of adaptive divergence offers remarkable insights into the molecular mechanisms underlying adaptive changes. One spectacular example comes from the adaptation of *Drosophila* to different host plants. In Chap. 12, Luciano Matzkin examines the genomics changes that accompany shifts in host cacti species by *Drosophila mojavensis*. Among the challenges these shifts pose are different nutritional composition and toxic compounds. Matzkin shows how the combination of genomics tools has allowed to point towards the molecular pathways involved in these host shifts. The *Drosophila* example offers an excellent illustration of how metabolic changes accompany adaptive evolution. Adaptive evolution can also be driven by morphological and color changes. One of the best model systems to study adaptive radiation are *Heliconius* butterflies, which builds on several decades of ecological and behavioral observations. In Chap. 13, Megan Supple and colleagues review the recent work in the genomics of speciation in this rich group of species by highlighting studies connecting phenotypes to genotypes and the role of introgression in adaptation. Another excellent model for the study of speciation and adaptation are species of the genus that comprise tomatoes and chili. This group presents a large diversity of floral variation and divergence in addition of representing a group of species of economic interest. In Chap. 14, David Haak and colleagues present new avenues for the dissection of adaptive evolution and speciation in this clade. They discuss how reproductive traits should be jointly studied in terms of adaptation and speciation, as those are key determinants of fitness. They also identify crucial environmental factors driving adaptive evolution in these species and discuss how genomics approaches will allow to identify the genes involved.

In Chap. 15, Jun Kitano and colleagues emphasize the central role of hormonal systems in the regulation of the development of adaptive

traits. They show how the field of evolutionary endocrinology studies the architecture of hormonal signaling pathways, the role of hormones in integrating external and internal signals, in leading to the observed phenotype and to the presence of correlation between traits, as well as how all these are modified during evolution. They also show how the study of the wide-ranging effects of hormone systems is especially improved by large-scale approaches of ecological genomics. In Chap. 16, Andrew Whitehead presents how the rapid changes imposed by human-altered environments are affecting the evolution of traits. The field of evolutionary ecotoxicogenomics is moving fast and Whitehead presents how the large-scale data obtained using ecological genomics approaches is revealing the genomic architecture of the multi-dimensional phenotypes that evolve to enable individuals to face pollution. Finally, in Chap. 17, Jesse Shapiro presents approaches to survey and understand microbial genome diversity and the evolutionary forces shaping it. He presents how ecological speciation is central in bacteria and how ecological genomics approaches can be used to test hypotheses about the genomics of adaptation in microbes, their ecological functions and the importance of evolutionary convergence.

One the most exciting presentation of the 2003 first "Evolutionary and Ecological Functional Genomics" Gordon conference was that of Leroy Hood, director of the Institute for Systems Biology in Seattle. Hood introduced the audience to the then burgeoning field of Systems Biology and to how cellular circuitry could be analyzed and modeled in a comprehensive manner by the precise measurement of molecules and their interactions in the cell. Once again, one needed a fertile imagination to envision how these approaches could be extended to non-model species and thus serve to understand how ecological factors are shaping gene and protein networks. However, readers of this book will realize that the field is poised for such integrative approaches in ecological genomics and that the field is moving from a description of how genes are affected by environmental factors in the short (physiological) and long terms (evolutionary) to a better understanding of how the relationships among these genes and their products are also changing. Once again, the field is ready for harnessing the most powerful approaches of the life sciences in order to address the most challenging questions in ecology and evolution.

References

Arendt J, Reznick D (2008) Convergence and parallelism reconsidered: what have we learned about the genetics of adaptation? Trends Ecol Evol 23:26–32

Aubin-Horth N, Renn SC (2009) Genomic reaction norms: using integrative biology to understand molecular mechanisms of phenotypic plasticity. Mol Ecol 18(18):3763–3780

Feder ME, Mitchell-Olds T (2003) Evolutionary and ecological functional genomics. Nat Rev Genet 4(8):651–657

Landry CR, Townsend JP, Hartl DL, Cavalieri D (2006) Ecological and evolutionary genomics of Saccharomyces cerevisiae. Mol Ecol 15(3):575–591

Lewontin RC (2000) The triple helix: gene, organism, and environment. Harvard University Press, Cambridge

Pavey SA, Bernatchez L, Aubin-Horth N, Landry CR (2012) What is needed for next-generation ecological and evolutionary genomics? Trends Ecol Evol 27(12):673–678

Poulin R (2010) Parasite manipulation of host behavior: An update and frequently asked questions. In: Brockmann J (ed) Advances in the study of behavior. Elsevier, Burlington, pp 151–186

Stearns SC (1992) The evolution of life histories. Oxford University Press, London, 249p

Trait Transitions in Explicit Ecological and Genomic Contexts: Plant Mating Systems as Case Studies

2

Vincent Castric, Sylvain Billiard, and Xavier Vekemans

Abstract

Plants are astonishingly diverse in how they reproduce sexually, and the study of plant mating systems provides some of the most compelling cases of parallel and independent evolutionary transitions. In this chapter, we review how the massive amount of genomic data being produced is allowing long-standing predictions from ecological and evolutionary theory to be put to test. After a review of theoretical predictions about the importance of considering the genomic architecture of the mating system, we focus on a set of recent discoveries on how the mating system is controlled in a variety of model and non-model species. In parallel, genomic approaches have revealed the complex interaction between the evolution of genes controlling mating systems and genome evolution, both genome-wide and in the mating system control region. In several cases, major transitions in the mating system can be clearly associated with important ecological changes, hence illuminating an important interplay between ecological and genomic approaches. We also list a number of major unsolved questions that remain for the field, and highlight foreseeable conceptual developments that are likely to play a major role in our understanding of how plant mating systems evolve in Nature.

Keywords

Selfing • Outcrossing • Self-incompatibility • Hermaphroditism • Dioecy • Gynodioecy • Androdioecy • Convergent evolution

2.1 Plant Mating Systems as Models to Study Evolutionary Transitions

Evolutionary transitions are discrete changes to biological traits that spread to replace ancestral conditions. A fascinating issue in evolutionary

V. Castric (✉) • S. Billiard • X. Vekemans
Laboratoire de Génétique et Evolution des Populations Végétales (GEPV), UMR 8198; CNRS, Université Lille 1, Sciences et Technologies, Cité Scientifique, Villeneuve d'Ascq, France
e-mail: Vincent.Castric@univ-lille1.fr

C.R. Landry and N. Aubin-Horth (eds.), *Ecological Genomics: Ecology and the Evolution of Genes and Genomes*, Advances in Experimental Medicine and Biology 781,
DOI 10.1007/978-94-007-7347-9_2,

biology is to understand why replicated character state transitions occur among unrelated lineages and whether this may indicate either similar selective mechanisms and functional convergence or similar genomic constraints. A powerful way to address these questions is to study model systems in which the nature of selective processes can be clearly identified. The study of mating systems (the suite of traits affecting mating patterns among sexually reproducing individuals) is a particularly relevant field of investigation to address this issue because of their astonishing diversity, especially in plants, and the many parallel and independent transitions that have been described. This general definition of mating systems collectively refers to all mechanisms that affect who in a species is having sex with whom (Billiard et al. 2011), and includes the prevalence of *selfing* versus *outcrossing*, the occurrence of distinct classes of interbreeding individuals such as genders (male, females and hermaphrodites), mating types in fungi and algae and *self-incompatibility* phenotypes. Mating systems crucially matter from a genetic perspective because they control the way genes are transmitted from one generation to the next (mechanism of inheritance) and hence determine genome diversity and organization, as well as the potential for adaptive or non-adaptive evolution. Mating systems also matter from an ecological perspective because they determine the quantity and quality of propagules available for dispersal and hence metapopulation dynamics, including the capacity to colonize new habitat patches. They also directly impact key interspecific interactions such as the types of resources transferred between plants and their pollinators. Putting aside transitions from sexual to asexual reproduction (reviewed in Glémin and Galtier 2012), our review will focus on two specific types of transitions: (1) that between allogamy (where fertilization occurs between distinct individuals) and autogamy (where offspring are produced by self-fertilization) and (2) that between *hermaphroditism* (where all individuals in a species belong to the same (co)sexual phenotype) and dioecy (where individuals belong to two separate sexual phenotypes, males and females) and their possible intermediate steps.

Phylogenetic mapping of mating system variation across Angiosperms has revealed that the different transitions occur at strikingly contrasted rates. For instance, heterostyly, which is a common form of heteromorphic *self-incompatibility*, has evolved at least 23 times independently within Angiosperm families (Lloyd and Webb 1992), and some authors have even suggested that heterostyly may have evolved as a derived trait within some genera (Graham and Barrett 2004). The rate of loss of heterostyly has not been quantitatively estimated, but homostylous species derived from heterostylous ancestors are common, suggesting that this reverse transition is also frequent (Barrett and Shore 2008). Homomorphic *self-incompatibility* is associated with no obvious morphological differences among *self-incompatibility* (SI) phenotypes, and so data on its phylogenetic distribution are more difficult to obtain. Still, homomorphic SI has been documented in at least 94 different families, and the detailed molecular analyses that have been conducted in five families (Brassicaceae, Papaveraceae, Rosaceae, Plantaginaceae, and Solanaceae) revealed three completely distinct molecular modes of action (Takayama and Isogai 2005), strongly suggesting that they represent as many independent emergences. The reverse transition, i.e. the loss of SI and ensuing shift to autogamy, has been considered as the most prevalent transition in plant evolution (Stebbins 1974). In Solanaceae, Goldberg et al. (2010) estimated that the rate of transition to *selfing* was as high as 0.55 transitions per lineage per million years, and Igic et al. (2008) estimated that this rate of loss was nearly 70 times higher than the rate of gain. Even more drastically, the data suggest that the asymmetry is so strong that the loss can be considered as irreversible within this family. *Dioecy* occurs in about 5–10 % of all species, but is present in half of Angiosperm families and probably appeared from hermaphroditic ancestors many times independently (Renner and Ricklefs 1995).

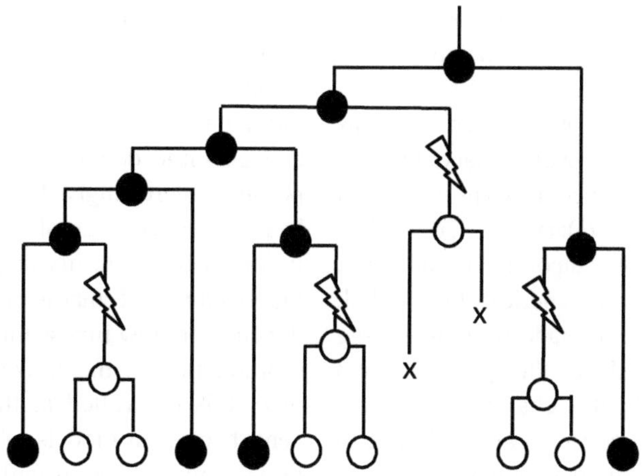

Fig. 2.1 The relative abundance of plant mating systems depends on the transition rates among mating systems as well as on their impact on rates of species diversification. The exhaustive phylogeny of a clade is represented, including extinct taxa, together with current and ancestral states of a trait related to the mating system (e.g. nodes with *black dots* correspond to self-incompatible, SI, species and those with *white dots* correspond to self-compatible, SC, species). Transitions between mating system traits are indicated with the symbol ⚡ (here only transitions from SI to SC were observed). In this clade, the currently most abundant mating system is SC, which is due to a high rate of transition to SC and a high rate of speciation of SC lineages. However, most ancestral nodes are SI, which may suggest that SC taxa are short-lived and have higher rates of extinction (see text and see Goldberg et al. 2010)

Within the Asteraceae, *dioecy* seems to have been derived from *hermaphroditism* at least 5–9 times independently, while the reverse transition from *dioecy* back to *hermaphroditism* seems to have occurred only 0–2 times independently (Torices et al. 2011). With an estimated 133 transitions between *hermaphroditism* and *dioecy*, Bryophytes exhibit remarkable lability in sexual systems, and again the transition rate from *hermaphroditism* to *dioecy* was twice as high as the reverse transition (McDaniel et al. 2013). Overall, beside the great disparity in the rates of transitions, a striking feature is that most of them seem to be asymmetrical or even unidirectional.

This observation raises the question of the ecological and evolutionary factors influencing the rates of transitions between mating systems. Beside extrinsic (adaptive) causes such as Fisher's transmission advantage (Fisher 1941), the strength, genomic architecture and lability of *inbreeding depression* and ecological correlates associated with shifts of the mating system, intrinsic causes have also been invoked such as molecular constraints, possibly making some transition routes more likely than others. In addition, the distribution of plant mating systems in extant species also depends on their impact on the relative evolutionary success of plant lineages. A key step in quantifying relative evolutionary success has been brought recently by methodological innovation in phylogenetic methods allowing to quantify the relative net diversification rate of lineages with contrasting life history traits, in particular allowing estimation of the relative rates of speciation and extinction (Maddison et al. 2007; FitzJohn 2010; Stadler 2011, Fig. 2.1). These approaches have been used for instance to determine the relative success of self-incompatible versus self-compatible lineages in Solanaceae (Goldberg et al. 2010; Goldberg and Igic 2012), of hermaphrodite versus dioecious mosses (McDaniel et al. 2013) or diploid versus polyploid lineages (Mayrose et al. 2010). Getting quantitative estimates, for instance for the relative rates of extinction of lineages with different mating systems, stimulates further theoretical studies to understand the causes of such differences. One important challenge is to develop models that link demographic processes

(species extinction) with population genetic features (e.g. *inbreeding depression*, probability of fixation of adaptive mutations).

Plant mating systems have been a focus of intensive research in evolutionary biology for nearly 150 years, since Darwin's seminal contribution on orchids (Darwin 1876, 1877), including aspects of theoretical analysis, phylogeny, ecology, and genetics. In parallel, major progress have been made on how mating systems are controlled at the genomic and physiological levels. In this chapter, we review how these two fronts of advances are beginning to merge, highlighting in particular how a detailed understanding of the genomic architecture of plant mating systems is providing important insight into the relative importance of extrinsic vs. intrinsic causes of plant mating system transitions and how these transitions affect the evolutionary success of plant lineages. We first focus on theoretical predictions regarding the consequences of several aspects of the genomic architecture of the mating system itself and that of *inbreeding depression* on the rates of transition between mating systems. We then review a set of recent empirical studies that identified mutations having caused mating system transitions in model and non-model species, and confront the nature of these causal mutations with theoretical predictions. Finally, we review another set of recent studies that demonstrate that mating system transitions are associated with major shifts in the patterns of genome organization and evolution, both genome-wide and in the regions involved in mating system determination.

2.2 Genomic Architecture and Plant Mating Systems Transitions: What Does Theory Predict?

In 1977, Lloyd suggested that the end product of evolution generally does not depend on how traits are determined genetically, and thus proposed that phenotypic models should be largely sufficient to investigate trait evolution in a given ecological context. He recognized, however, that phenotypic models are not sufficient if one wants to predict the speed of evolution or the influence of other evolutionary forces, especially drift and mutations, whence taking into account genetic and genomic details into models might be important. Interestingly, Lloyd (1975, 1977) relied heavily on the example of mating system evolution to illustrate his purpose, and argued that the outcome of mating system evolution only depends on how fitness through male and female reproduction is maximized, without regard for how it is controlled at the genomic level. Recent theoretical models of mating system transitions in plants have strongly challenged this view, showing that the genetic and genomic details are indeed important, in two ways: the genetic determination of the mating system itself and the genetic architecture of *inbreeding depression*. In this section, we will assess the importance of these two points from a theoretical point of view, especially regarding the issue of variability of transitions rates in plant mating systems.

2.2.1 Transitions Between Outcrossing and Selfing: The Example of Homomorphic Gametophytic *Self-Incompatibility* in Angiosperms

The genetic determination model for homomorphic SI in Angiosperms is generally assumed to be bipartite: two linked non-homologous genes at a single locus, the S-locus, respectively coding for the pollen and pistil parts of the SI response (Fig. 2.2; note that alternative models with more than one locus are more appropriate in some groups such as the Poaceae). This explicit genetic architecture has dramatic consequences regarding theoretical expectations of when and how SI should be gained and lost.

Loss of SI. Whether a mutation making individuals self-compatible invades and eventually goes to fixation, leading to a loss of SI, mainly depends on five parameters: allelic diversity at the S-locus, the level of *inbreeding depression*, the extent of

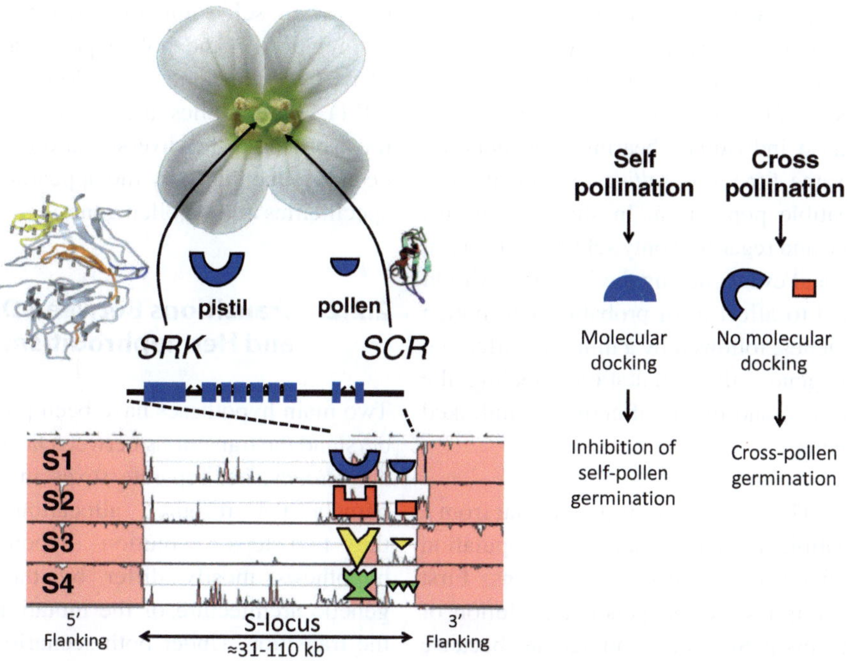

Fig. 2.2 Self-incompatibility in the Brassicaceae at a glance. SI specificity in the Brassicaceae is controlled by two tightly linked genes, *SCR* and *SRK* expressed in anther tapetum and in stigmatic papilla, respectively, that function as a molecular lock-and-key mechanism. When produced by the same S-haplotype (following either self-pollination or cross-pollination by an individual expressing the same S-haplotype), the cognate SCR and SRK proteins physically interact and activate a downstream signalling cascade (including the *ARC1* gene and several others) that disrupts proper pollen hydration, ultimately preventing fertilization. Upon cross-pollination by an individual expressing a different S-haplotype, the two proteins do not physically interact, allowing fertilization to proceed. The genomic region containing the various lock-and-key combinations of *SCR* and *SRK* (represented by distinct shapes and colours) is characterized by extreme sequence divergence, as shown by the complete lack of sequence identity in a region of ca. 31–110 kb (conservation plots at the *bottom*; size varies across the different S-haplotypes, Goubet et al. 2012) (3D models of SCR and the extracellular domain of SRK are reprinted from Chookajorn et al. (2003) and Naithani et al. (2007) respectively, published with kind permission of © National Academy of Sciences, U.S.A. 2003 and 2007. All Rights Reserved)

self-pollination, the level of *pollen limitation* and that of *pollen discounting*. In brief, the invasion of a self-compatible mutation is favoured when *selfing* is high and efficient (i.e. if *inbreeding depression* is low and *pollen limitation* is high), and if the pollen received through outcrossing is also efficient (i.e. if allelic diversity and *pollen discounting* are low). Four different mutational events can be thought of, that all would result in a self-compatible mutant: the mutation can affect the pollen part, the pistil part, both parts at once, or a modifier locus unlinked to the S-locus (e.g. a gene involved in the biochemical processes underlying the SI response). For a given set of parameter values, the conditions under which the mutation invades and becomes fixed were shown to depend on the type of mutations (Charlesworth and Charlesworth 1979; Uyenoyama et al. 2001; Porcher and Lande 2005b). First, the conditions for the invasion of a mutation affecting both pollen and pistil parts are less limited than for the other mutation types, especially regarding *inbreeding depression*. This is because such a mutation benefits from the transmission through both male and female functions. Second, if *pollen limitation* is relatively low, a mutation affecting the pollen part only is more likely to invade than if affecting the pistil part only, because self-compatible pollen benefits both from *selfing* and from outcrossing with all individuals in

the population, while a self-compatible pistil is favoured only by *selfing*. Last, a mutation affecting an unlinked modifier gene is less likely to invade, especially regarding *inbreeding depression*, because individuals bearing this mutation have the same fitness as *selfing* individuals in a self-compatible population. In short, all things being equal and regarding only selection strength, the mutational event causing the loss of SI would be expected to affect most probably both pollen and pistil genes, followed by a mutation affecting the pollen gene only, a mutation affecting the pistil part only, and finally affecting an unlinked modifier gene.

Gain of SI. The reverse transition, i.e. that from a self-compatible to a self-incompatible population seems difficult for at least two reasons. First because it implies the independent evolution of pollen and pistil SI genes, and second because it requires a very high level of *inbreeding depression*. Yet, the very existence of the many independent transitions observed in Angiosperms (Takayama and Isogai 2005) demonstrates that conditions should exist under which such events can occur. The appearance of a functional SI system can be decomposed in two evolutionary steps. First, two genes should appear at the S-locus, expressed in pollen and pistil, respectively, which recognize each other and prevent fertilization when recognition occurs. Second, the S-locus diversifies and becomes fully functional, excluding self-compatible genotypes. Little is known theoretically about the first step (appearance of a SI haplotype), but the conditions under which a self-compatible genotype can be maintained in polymorphism with functional SI alleles have been studied. The maintenance of polymorphism is important since it is necessary for the evolution of new SI haplotypes and thus the evolution towards a fully functional SI system (Uyenoyama et al. 2001). Several models have shown that a self-compatible haplotype can be maintained in polymorphism with SI alleles, but under restricted conditions, except when the number of SI alleles is low (Charlesworth and Charlesworth 1979; Porcher and Lande 2005b). Interestingly, polymorphism is easier to maintain when self-compatible haplotypes are due to mutation in the pollen part than in the pistil part (Uyenoyama et al. 2001; Gervais et al. 2011). This implies the simple prediction that most probably the diversification of the S-locus occurs sequentially by the appearance of new SI specificities at the pollen gene.

2.2.2 Transitions Between Dioecy and Hermaphroditism

Two main hypotheses have been proposed to understand the transitions between *hermaphroditism* and *dioecy*, i.e. separate sexes in a population, namely the resource allocation theory and the two-steps evolution hypothesis. These hypothesis mostly differ by the nature and genetic architecture of the mutations that cause the transition. Under both scenarios, *inbreeding depression* plays an important role since it decreases the fitness of hermaphrodites and thus makes more difficult the invasion of hermaphrodites into populations that already have separate sexes, or conversely facilitates the invasion of male or female individuals into hermaphroditic populations.

The resource allocation theory (Charnov 1982). This theory states that fitness through female and male reproduction can vary and thus differential resource allocation between the sexes can be favoured or disfavoured by natural selection. It is based on two key assumptions. First, there must be a fitness trade-off between male and female functions, i.e. a good pollen producer cannot also be a good ovule provider, and vice-versa. No clear empirical evidence for the occurrence of such trade-off has been demonstrated, although there are several reports of negative genetic correlations between investment in male and female functions in hermaphrodites (reviewed in Ashman 2003). Second, because it is based on models of evolutionary stable strategies (ESS), this theory explicitly assumes that the relative investment in male and female functions varies as the result of mutations that each has a small phenotypic effect. To the best of our knowledge, no empirical

study has tried to identify the genetic architecture of this trait, which would allow testing this basic prediction in the flowering plants.

The two steps hypothesis. It is generally considered that it is impossible for a single mutation to cause the evolution of *dioecy* (Charlesworth and Charlesworth 2010a) because a single mutation entirely abolishing both the male and the female functions at once would lead to fully sterile individuals, and because it is difficult to imagine a single mutation that would sterilize the male function in some individuals and the female function in some others. Hence, the "two steps hypothesis" posits that the transitions from *hermaphroditism* to *dioecy* are caused by two successive evolutionary steps: the invasion and spread of a male-sterility mutation followed by a female-sterility mutation, in either order (Charlesworth and Charlesworth 2010a). The backward transition from *dioecy* to *hermaphroditism* can also be caused by two evolutionary steps restoring male and female fertilities, respectively. Since it is generally thought that the ancestral mating system in Angiosperms is *hermaphroditism*, we will focus on the transition towards *dioecy* (analogous predictions can easily be made for the backward transition). Two pathways can be considered, depending on whether the first mutation causes male or female sterility. If the first mutation causes male sterility, then transient populations are composed of hermaphroditic and female individuals, a mating system that is called *gynodioecy*. If the first mutation causes female sterility, then hermaphroditic and male individuals will coexist in transient populations, a mating system that is called *androdioecy*. Following this scenario, *androdioecy* and *gynodioecy* are seen as necessary intermediates. A male or female sterility mutation can invade a population if it increases female or male fitness, respectively, of the unisexual individuals, i.e. a female sterility mutation must increase male fitness while a male sterility mutation must increase female fitness. The advantage of the sterility mutation to one sex must more than compensate the loss in fitness to the other sex, which is possible only if plants are able to reallocate resources.

The genetic architecture of male and female sterility mutations critically matters for which pathway is most likely to occur. In most Angiosperms, mitochondria and chloroplasts are transmitted by female gametes only, while the nuclear genome is typically transmitted via both male and female gametes, hence creating a sexual asymmetry in the transmission of the two genomes. Accordingly, *gynodioecy* can be controlled by male sterility mutations that are either nuclear or cytoplasmic (Dufaÿ and Billard 2012), although the relative proportion of species with nuclear or cytoplasmic *gynodioecy* is not known. In contrast, only nuclear androdioecy is known so far. The conditions for the invasion and maintenance of nuclear mutations are identical regardless of whether they affect the male or female functions: the fitness of the sex that remains fertile must be at least twice that of the hermaphrodites. The case of cytoplasmic mutations, however, is strikingly different and involves a classical example of genomic conflict. Because cytoplasmic mutations affecting the male function would not suffer from any reduction in fitness (the cytoplasm is maternally transmitted), cytoplasmic *gynodioecy* can arise as soon as the fitness of females gets somewhat higher than that of hermaphrodites, which is a less stringent condition than that expected for nuclear *gynodioecy* (Charlesworth and Charlesworth 2010a). The evolutionary stability of cytoplasmic *gynodioecy*, however, is more difficult to account for. Indeed, fixation of a male sterility mutation would lead to a population composed of female individuals only, which would go extinct by lack of pollen for fertilization. Hence, the sex-ratio bias would create strong selective advantage for any mutation in the nuclear genome able to restore male fertility. If both the cytoplasmic male sterility mutation and the nuclear restorers of male fertility become fixed in the population, the population goes back to *hermaphroditism*. In contrast, the two mutations can stably (or cyclically) segregate in the population, leading to evolutionary stable *gynodioecy* (Gouyon et al.

1991). Overall, because the general conditions for the emergence and maintenance are generally less stringent for *gynodioecy* than for *androdioecy*, theoretical models predict that under the two-steps hypothesis, the first step leading to *dioecy* should be through *gynodioecy*, most likely through a male sterility mutation occurring in the cytoplasmic genome.

Once *gynodioecy* or *androdioecy* has evolved, *dioecy* may then result from the fixation of mutations that decrease resource allocation to the still fertile sex in hermaphrodites. Interestingly, the latter mutations are only expected to invade the population if they are in complete linkage disequilibrium with the initial mutation causing either *gynodioecy* or *androdioecy*. In fact, recombination would produce offspring with both sterility mutations in their genome, which would therefore be fully sterile. Recombination is therefore expected to be extremely low between male and female sterility mutations, involving genes that were either initially adjacent or have become adjacent through translocation. From a genomic point of view, we might therefore expect the evolution of genomic regions determining maleness and femaleness, that may eventually evolve into *bona fide* sexual chromosomes.

Other routes: via monoecy, subdioecy or distyly. Few models have investigated the evolution from *hermaphroditism* to *dioecy* by these alternative evolutionary pathways, even though they are supported by empirical evidence (reviewed in Pannell and Verdu 2006). *Monoecious* populations are composed of hermaphroditic individuals but with separate male and female flowers. It has been suggested that a transition from monoecy to *dioecy* might be possible simply by the successive fixation of small effects mutations changing the ratio between male and female flowers within individuals (Charlesworth and Charlesworth 1978; Lloyd 1980). This pathway is related to the resource allocation theory (see above) but here the genes underlying the transition are suggested to be the ones controlling the number of flowers of a given sex. Subdioecy refers to populations that regularly contain imperfectly sexually differentiated individuals of either or both sexes, in addition to strictly unisexual individuals. Ross (1982) analyzed a population genetic model assuming two genes affecting pollen and ovules production, and assuming a resource allocation trade-off. The model makes the important prediction that evolution towards subdioecy is easier when the underlying genes are genetically linked, which makes a readily testable hypothesis for genomic investigation. Distyly has been suggested as another possible pathway for the transition between *hermaphroditism* to *dioecy*. Specifically, if the two morphs differ in how efficiently they export pollen, for instance because physical contact of the pollinator body with anthers is more important in one morph than in the other, mutations further increasing male reproduction at the cost of female reproduction in the morph that is already better at exporting pollen would become fixed. In response, mutations increasing female reproduction at the cost of male reproduction in the other morph would also become fixed, eventually leading to fully specialized male and female individuals derived from the two morphotypes (Lloyd 1979). A prediction of this model is that such mutations are expected to appear in linkage with the locus determining the morphotypes.

2.2.3 Importance of the Genomic Architecture of Inbreeding Depression

Inbreeding depression is believed to be one the most important evolutionary forces in the evolution of mating systems for at least two reasons. First, it decreases Fisher's "automatic" transmission advantage, whereby *selfing* genes are transmitted 50 % more efficiently to the progeny than genes preventing *selfing* (Fisher 1941). Second, *inbreeding depression* potentially decreases the fitness of hermaphrodite individuals, which is the only category of individuals able to self, leaving male or female individuals relatively unaffected (the highest possible level of inbreeding is obtained when a hermaphrodite reproduces by *selfing*). Accordingly, most models of mating system evolution take *inbreeding depression* into account. Although the issue was

highly controversial in the 1970s, it is now widely accepted that *inbreeding depression* is mainly due to the expression of mutations that are deleterious and at least partially recessive: in inbred individuals, such mutations are more often homozygous and thus expressed than in outbred individuals (Charlesworth and Charlesworth 1987). A difficulty then appears: deleterious mutations, once expressed in homozygous individuals, can be eliminated by natural selection, a process that is called purging. The purging of deleterious mutations is more efficient in inbred, especially *selfing*, populations. The consequence of purging is that *inbreeding depression* is expected to be lower in *selfing* than in outcrossing populations (Roze and Rousset 2004). Theoretical investigations that explicitly modeled *inbreeding depression* and purging showed that the maintenance of outcrossing is possible only under stringent conditions (e.g. Porcher and Lande 2005a), the maintenance of SI especially being difficult (Porcher and Lande 2005b). In fact, purging is efficient in selfers mostly when selection is strong (deleterious mutations have large effect and are not too recessive), while *inbreeding depression* can remain high in the case of mildly deleterious and recessive mutations (Charlesworth and Willis 2009), especially if deleterious mutations appear at multiple loci and if there is epistasis. It has therefore been proposed that during the transition from outcrossing to *selfing*, deleterious mutations causing *inbreeding depression* should be purged, making the reverse transition (back to outcrossing) less likely to occur. Based on the dynamics of *inbreeding depression* caused by recessive deleterious mutations, theory therefore predicts that transitions from outcrossing to *selfing* should be unidirectional.

Two other processes can however permit the maintenance of high *inbreeding depression* even in *selfing* populations. The first process is overdominance, whereby heterozygotes are more fit than homozygotes, allowing polymorphism (and thus *inbreeding depression*) to be maintained. The second process is associative overdominance where there is strong linkage between deleterious mutations, in particular in genomic regions with low recombination. In such case, some haplotypes can accumulate a different suite of linked deleterious mutations from that accumulated by other haplotypes. It has been suggested that non-recombining mating system control regions, such as the S-locus or heteromorphic sex chromosomes, can accumulate such different suites of linked deleterious mutations, which have been called the "sheltered load" (Uyenoyama 2003). Few models have analyzed the importance of such phenomena on the evolution of mating systems. Charlesworth and Charlesworth (1990) showed that outcrossing may be more easily maintained when *inbreeding depression* is due to loci at which heterozygotes are advantaged. Porcher and Lande (2005b) showed that the existence of a sheltered load can facilitate the maintenance of SI. To our knowledge, no model has investigated the case of the transition between *hermaphroditism* and *dioecy*. While this clearly highlights the central importance of *inbreeding depression*, very little is known empirically about its key parameters, i.e. the distribution of the effects of deleterious mutations and of their dominance coefficient (see Sect. 2.5).

2.2.4 Long-Term Evolutionary Consequences of Mating System Transitions

Why and when should selfing be an evolutionary dead-end? A long-standing debate regarding the transition between *selfing* and outcrossing is whether the mating system affects the diversification rates of species, i.e. whether extinction and speciation rates depend on the mating system of a species. In fact, based on the distribution of *selfing* species in phylogenies, which are more frequent at the leaves of phylogenetic trees (Fig. 2.1), it has been suggested that *selfing* should be an evolutionary dead-end (Stebbins 1974). Two evolutionary processes have been proposed (Takebayashi and Morrell 2001). First, *selfing* and outcrossing species may differ in the rate at which they accumulate deleterious mutations. Indeed, the effective population size of *selfing*

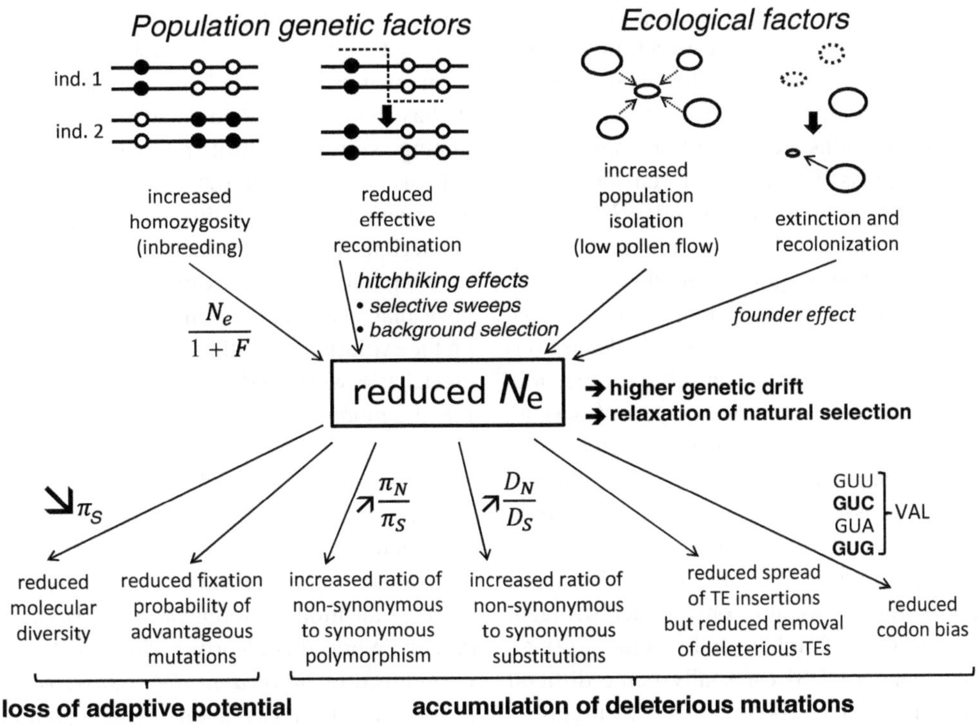

Fig. 2.3 Genomic effects and evolutionary consequences of a transition from outcrossing to selfing in plant populations. Population genetic factors and ecological factors combine to increase the importance of genetic drift as opposed to natural selection. These population genetic processes impact genome features and thereby may induce a loss of adaptive potential and an accumulation of deleterious mutations. F, inbreeding coefficient; π_N and π_S, non-synonymous and synonymous nucleotide diversity, respectively; D_N and D_S, among-species non-synonymous and synonymous substitutions, respectively

populations is expected to be considerably lower than that of outcrossing populations (Fig. 2.3) because the number of independent sampling of gametes is reduced (Nordborg 2000) and because effective recombination rate is decreased, which increases the effect of hitchhiking, leading to large selective sweeps (Kaplan et al. 1989) and intense background selection (e.g. Charlesworth et al. 1995). The decrease of both effective population size and effective recombination rate decreases the efficacy of selection and thus the fixation of advantageous mutations as well as the elimination of deleterious mutations. Second, *selfing* and outcrossing species may also differ in how they adapt to changing environments. Two opposing processes should be taken into account, which makes it difficult to determine whether *selfing* or outcrossing is the most favourable (Glémin and Ronfort 2013). On the one hand, the rate at which advantageous mutations enter the population is high when there is much standing variation, which is expected to be the case in outcrossers. On the other hand, the rate at which advantageous mutations are fixed is expected to be higher in selfers. This is however modulated by the genetic architecture of the beneficial mutations, i.e. whether beneficial mutations are dominant or recessive, and by the speed and strength of the environmental changes. Finally, even though the accumulation of deleterious mutations can lead to extinction of populations in the case of obligate selfers (Lynch et al. 1995b), a slight outcrossing rate is generally sufficient to purge deleterious mutations and keep population viability unaffected (Charlesworth et al. 1993). Overall, the outcome of these opposing forces may be that *selfing* species would have a lower adaptive potential rather than a higher accumu-

lation of deleterious mutations than outcrossers (Glémin and Ronfort 2013).

Dioecy and genomic changes: the evolution of sexual chromosomes. As explained above, the transition from *hermaphroditism* to *dioecy* is expected to occur by the successive fixation of male-sterility and female-sterility mutations. If the male-sterility mutation is recessive and the female-sterility is dominant, the heterozygous individuals at these loci will be males, while the homozygous individuals will be females. This would lead to the evolution of a proto-X and proto-Y genomic region determining the sex of individuals. Any mutation that has a sexually antagonistic effect, i.e. a positive effect on female (or male) and a negative effect on male (or female) fitness will become fixed if it is closely linked to the proto-X and proto-Y regions. Such a genomic region with no recombination is expected to evolve because of several processes. Deleterious mutations can accumulate because of Muller's Ratchet, due to hitchhiking by beneficial mutations and because of a reduced effective size of sexual chromosomes (Charlesworth 2002a). This process will lead to genetic degeneration and decrease in gene content of the Y chromosome (Bachtrog 2008). Finally, it has recently been suggested that there can be a turn-over of sexual chromosomes: the male determination on the Y chromosome moves to an autosomal chromosome which in turn evolves into a neo-Y chromosome (van Doorn and Kirkpatrick 2007). Hence, the evolution towards separate sexes is expected to have major consequences on the evolution of a whole chromosome and involves large genomic reorganisation.

2.3 Causes of Mating System Transitions: What Have We Learned from Molecular Genomic Approaches?

A series of recent advances in molecular genetic analyses have pinpointed the causal mutations for major shifts in the mating system of several plant species and provided key insight about the mutational constraints involved.

2.3.1 Mutations to the Pollen Component of SI and the Extent of Pollen Limitation

Causal mutations for the transition from outcrossing to *selfing* have been identified in a handful of species. SI in the Brassicaceae is controlled by a molecular lock-and-key mechanism involving a transmembrane receptor protein deposited on the stigma surface (SRK, the "pistil part") and a small ligand protein deposited on the pollen coat (SCR, the "pollen part", Fig. 2.2). These two genes largely differ in size, SCR being a much smaller protein than SRK (respectively 83 vs. 432 aa on average in *Arabidopsis*, Goubet et al. 2012). Taking these figures at face value, the size of the mutational target would be predicted to differ between them, with more opportunities for the *SRK* gene to be knocked down by random mutation than *SCR*. In line with this simple expectation, the breakdown of SI in several domesticated *Brassica* cultivars was caused by female – disabling mutations (Tsuchimatsu et al. 2012). In sharp contrast, Tsuchimatsu et al. (2010) demonstrated that mutations in the pollen component (*SCR*) were responsible for the loss of SI in several natural *A. thaliana* accessions where *SRK* was intact and fully functional, as well as in *A. kamchatia* (Tsuchimatsu et al. 2012). Similarly, self-compatibility in *Capsella* also maps to the S-locus (Nasrallah et al. 2007; Slotte et al. 2012), and the fact that *SRK* appears to have retained a full-length coding region in some *C. rubella* accessions suggests that a mutation in the male component might have driven the loss of SI in this species (Guo et al. 2009). In *Leavenworthia alabamica*, controlled crosses and molecular analyses demonstrated that the self-compatibility observed in some populations also probably originated from the loss of function of the male component, *SCR* (Busch et al. 2011; Chantha et al. 2013). Hence, currently

available empirical evidence suggests that evolution of self-compatibility in wild Brassicaceae species tends to be driven by mutations in the male rather than in the female component. This contrast between natural and artificial selection is enlightening, because population genetics theory (Uyenoyama et al. 2001) predicts that natural selection should favour more strongly mutations disabling the pollen gene than those disabling the stigma gene (see Sect. 2.2.1). Indeed, this asymmetry in the direction of evolution rests on the key assumption that only male reproduction is limited by the SI mechanism, all ovules being ultimately fertilized, i.e. the amount of pollen available for fertilization is considered infinite and *pollen limitation* is absent. Hence, while *pollen limitation* seems to be frequent in natural populations (Busch and Schoen 2008), the fact that natural selection mostly favours male mutations tells us that the intensity of *pollen limitation* is weak in many ecological situations, or at least not strong enough to oppose the (apparently) higher rate of spontaneous mutation of the female component due to a wider mutational target. Clearly then, these observations provide a clear link between conclusions that can be made from genomic analyses and the underlying ecological conditions.

Interestingly, the genetic architecture of pollen SI specificity differs across species. In particular, Kubo et al. (2010) reported that in some Solanaceae species, pollen specificity for a given S-haplotype is determined by a series of tandemly duplicated pollen genes acting in a coordinated manner, each with a specific spectrum of antitoxin activity against the full repertoire of S-RNAse proteins they may encounter in pistils from other individuals of the species. Beside the mystery of how such an exquisitely co-evolved mechanism may have arisen and diversified, the multiplicity of pollen-S genes may potentially increase the size of the mutational target, leading to the prediction that these Solanaceae species may be even more prone to mating system shifts through the pollen component than species in which the pollen component is encoded by a single gene.

2.3.2 Transition to *Selfing*: Coupled to or Uncoupled from Speciation?

Genomic analyses of polymorphism have provided some important insights into the demographic context of the transition from outcrossing to *selfing* in two model species (Fig. 2.4). In *A. thaliana*, transgenic complementation experiments showed that a pair of functional *SRK-SCR* genes from *A. lyrata* was sufficient to largely rescue (albeit not fully) the SI phenotype. Hence, the signalling cascade downstream of *SCR* and *SRK* has remained largely intact in several accessions (Nasrallah et al. 2002, 2004; Boggs et al. 2009; Tsuchimatsu et al. 2010, but see Indriolo et al. 2012). This observation has two major implications: (1) the transition to *selfing* must have been recent, otherwise random mutations would have disrupted this signalling cascade if it has no other function and (2) it must have involved mutations at either *SCR* or *SRK* or both. Bechsgaard et al. (2006) identified in *A. lyrata* and *A. halleri* functional orthologs of the three *SRK* haplogroups that are still segregating in *A. thaliana*, but found no evidence for an acceleration of non-synonymous evolution along *A. thaliana* branches. Since *SRK* is currently pseudo-geneized in most *A. thaliana* accessions, this suggests that the loss of function of *SRK* must have been sufficiently recent that the relaxation of functional constraint could not be detected. Following this line of reasoning, Bechsgaard et al. (2006) concluded that the loss of function of *SRK* could not have occurred earlier than 413,000 years ago, while *A. thaliana* became separated from its close relatives *A. lyrata* and *A. halleri* about 5,000,000 years ago (Al-Shehbaz and O'Kane 2002 but see Beilstein et al. 2010). In other words, the currently highly *selfing A. thaliana* would have outcrossed for most of its evolutionary history. A second set of results, however, pointed to a more ancient transition to *selfing*. First, population genetics simulations showed that the genome-wide pattern of linkage disequilibrium (LD) decay with physical distance in *A. thaliana* is not compatible with such a

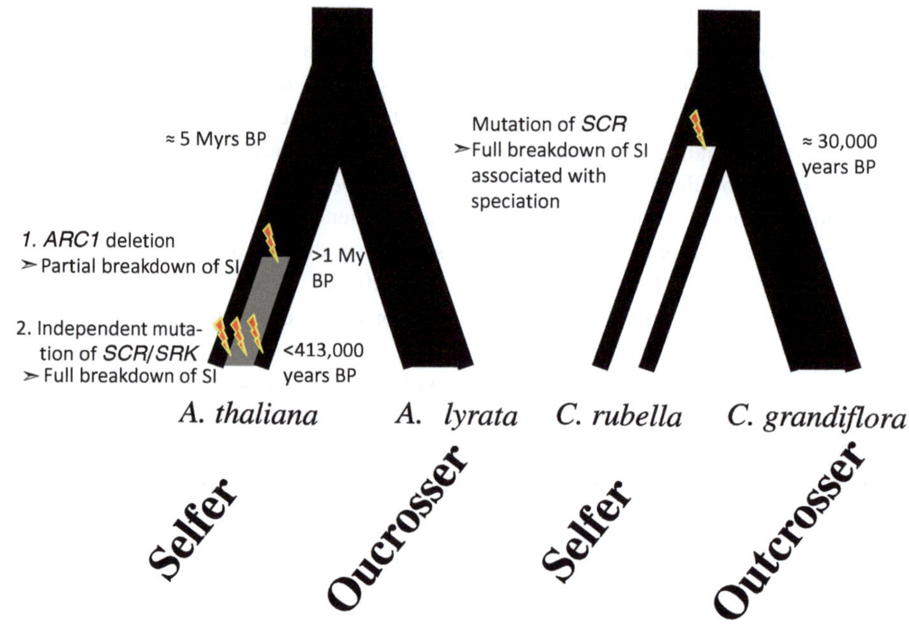

Fig. 2.4 Transition to selfing: coupled to or uncoupled from speciation? In *Arabidopsis*, the shift to autogamy in *A. thaliana* seems to be much more recent than speciation from the outcrosser *A. lyrata* (which occurred about 5,000,000 years ago). In *A. thaliana*, *SRK* became a pseudogene recently, no earlier than 413,000 years ago (Bechsgaard et al. 2006), possibly through three distinct causal mutations (Shimizu et al. 2008; Boggs et al. 2009). However, the pattern of LD decay along chromosomes suggests that *A. thaliana* became a selfer much more anciently, possibly 1,000,000 years ago. A possible solution to this paradox has been suggested by the recent discovery by Indriolo et al. (2012) that the downstream signalling pathway is actually not fully intact in *A. thaliana*, lacking the *ARC1* gene. Fixation of this deletion may have caused an intermediate period of mixed mating system (in *grey*), followed by more recent full breakdown of SI and ensuing shift to autogamy (in *white*). In *Capsella* in contrast, the breakdown of SI is due to the recent fixation of a single S-haplotype whose pollen part is non-functional. Speciation is also extremely recent (30–50,000 years ago, Guo et al. 2009), suggesting that the two events were coupled. Note the sharply different time scales of speciation in the two genera

recent transition to *selfing*, but more closely resembles that predicted at equilibrium in an ancient selfer, which would have selfed for at least 1,000,000 years (Tang et al. 2007). Second, Indriolo et al. (2012) recently showed that *A. thaliana* was fixed for a deletion in one of the genes of the downstream signalling cascade of the SI response (*ARC1*). Hence, contrary to previous claims, this cascade is not fully intact in *A. thaliana*. When inactivated in *A. lyrata*, the *ARC1* gene leads to partial breakdown of SI, possibly suggesting that fixation of this deletion in *A. thaliana* may have led to a mixed mating system for some time, before *SRK* became a pseudogene. Interestingly, Tantikanjana et al. (2009) suggested that at least in one mutant *A. thaliana* background *SRK* has a dual role, jointly controlling both the SI response and pistil exertion, hence possibly accounting for the delay with which *SRK* became a pseudogene even after its downstream SI signalling cascade started to decay. Overall, the data in *A. thaliana* therefore seem to be consistent with a two-steps scenario (Fig. 2.4), whereby an early (>1 Myrs) mutation in *ARC1* slightly decreased the strength of SI, while the SI system was fully inactivated only recently as *SRK* became a pseudogene, possibly through independent causal mutations (Shimizu et al. 2008; Boggs et al. 2009). In *Capsella* in contrast, Guo et al. (2009), reported that the time at which *SRK* became a pseudogene in *C. rubella* was largely consistent with the time of the split from the closely related outcrosser *C. grandiflora* (Foxe et al. 2009; St Onge et al. 2011), suggesting that, contrary to *Arabidopsis*, the transition to *selfing* in *Capsella* was coupled with speciation (Fig. 2.4). Interestingly, a recent

phylogenetic analysis (Goldberg and Igic 2012) suggests that the loss of SI in Solanaceae seems to occur largely by the cladogenetic mode (whereby the shift to *selfing* is associated with speciation, like in *Capsella*) rather than by the alternative anagenetic mode (whereby the transition to *selfing* occurs through fixation of self-compatible mutants in a species that is already isolated, like in *Arabidopsis*). It will now be interesting to extend this kind of analysis to other groups and to assess the generality of this conclusion.

2.3.3 The *Selfing* Syndrome: Neutral Degeneration or Selection for More Efficient Selfing?

While the loss of SI, at least partial, is a prerequisite step towards the evolution of *selfing*, it does not necessarily imply that *selfing* will predominate in the self-compatible lineage. In fact, most *selfing* species exhibit a "*selfing* syndrome", a characteristic suite of traits, including reduction in size of floral organs, reduced flower opening, shorter physical distance between anthers and stigmas, and reduced temporal separation between male and female maturity. They also typically produce less pollen, and invest relatively more in ovule production, all features that are collectively believed to promote selfing (Darwin 1876; Ornduff 1969). Fishman et al. (2002) showed that the basis of phenotypic floral divergence between the large-flowered outcrosser *Mimulus guttatus* and the small-flowered selfer *M. nasutus* was highly polygenic, involving a large number of small-effect QTLs. Two recent studies in the species pair formed by *C. rubella* (a selfer) and its close outcrossing relative *C. grandiflora* also reported numerous strong QTLs for floral morphology, some of which overlapped across the traits measured (Sicard et al. 2011; Slotte et al. 2012). In an elegant backcrossing experiment, Sicard et al. (2011) introgressed the non-functional S-locus of the selfer *C. rubella* into the genomic background of the outcrosser *C. grandiflora*, resulting in a plant whose floral morphology closely resembled that of the outcrosser but was fully self-compatible. Interestingly, these plants selfed autonomously only about half as efficiently as native *C. rubella* plants, hence demonstrating experimentally that the traits associated with the selfing syndrome do indeed provide more efficient self-pollination in *C. rubella*. Slotte et al. (2012) provided further evidence for the adaptive significance of the selfing syndrome, by comparing the direction of QTL effects. Most of them (86 %) were in the direction expected from phenotypic differences between the two species, an observation consistent with directional selection having favoured the evolution of the selfing syndrome in *C. rubella*. Moreover, while *C. rubella* has low genomic diversity overall as compared to *C. grandiflora*, the QTL regions exhibit an even more extreme reduction in diversity and an excess of fixed differences relative to shared polymorphism than other genomic regions in *C. rubella*. This observation suggests that QTL regions may have been the targets of recent selective sweeps, hence again supporting the notion that natural selection for efficient self-pollination rather than neutral processes have driven evolution of the *selfing* syndrome.

2.3.4 Genomic Constraints on How SI May Arise in the First Place

While it is relatively straightforward to imagine how gene-disruptive mutations can lead to the loss of SI in a species, the primary evolution of SI in a plant family is a more difficult issue to address empirically. Yet, a recent ground-breaking study (de Graaf et al. 2012) reported that the genes controlling SI in *Papaver* (*PrpS* and *PrsS*, controlling the pollen and pistil specificity, respectively) were able to function normally when transformed into *A. thaliana* and result in pollen inhibition in this species. This result is spectacular because *Papaver* and *Arabidopsis* became separated ca. 140,000,000 years ago, and also because SI in outcrossing *Arabidopsis* species involves completely different molecular mechanisms. This strongly suggests that SI in Papaveraceae functions through a very general signalling pathway that is highly conserved across

Angiosperms rather than through a specific set of genes that would function solely in SI. Hence, the evolution of SI in *Papaver* only had to involve recruitment of the two recognition proteins, with minimal constraint on the pre-existence of the downstream signalling cascade. Accordingly, a meta-analysis by Ferrer and Good (2012) reported that at least 22 Angiosperm families with SI were polymorphic in the type of SI mechanism, hence suggesting the possibility of multiple origins of SI in different species even within a given family, i.e. at a very short phylogenetic scale. While the genomic basis of most of these SI mechanisms is currently unknown (having been elucidated at the molecular level in only five families), it will be exciting to learn whether distinct genes have been recruited to control SI in each of these species, or whether some genes or gene families are more prone to serve as SI genes. A recent study in *Leavenworthia alabamica* does indeed suggest that the SI recognition function may have been transferred secondarily from *SCR-SRK* to a distinct pair of highly linked genes that belong to the same two gene families but are at a different genomic location (Chantha et al. 2013). Another recent study in *Senecio* provided evidence for yet another mechanism contributing to the lability of mating systems. Indeed, Kim et al. (2008) reported that a cluster of regulatory genes promoting flower asymmetry in the inflorescence (the proportion of disc- vs. ray-florets) had been transferred by introgression from the diploid *S. squalidus* into the tetraploid *S. vulgaris*, leading to an increase in the rate of outcrossing. These results highlight how a complex trait such as SI may be regained during evolution either by recruitment of paralogous genes or by horizontal gene transfer.

2.3.5 Gender Transitions

Data on the molecular events that provoked gender transitions (from *hermaphroditism* to *dioecy* for instance) are much more scarce. The best-studied system is that of *gynodioecy*, whereby female individuals are produced in a species as a consequence of the presence of male-sterility mutations. In some species, male-sterility is encoded by cytoplasmic factors, and evolves as the result of a nuclear-cytoplasmic conflict (see Sect. 2.2.2). Male sterility mutations have been characterized at the molecular level in a handful of species only and were found to have different molecular natures, including novel gain-of-function chimeric genes resulting from intra-genomic recombination. Generally speaking however, the details of how these mutations function to ultimately lead to male sterility remain poorly known (Touzet 2012). Genes able to mask the action of male-sterility factors (coined "nuclear restorer genes"), have been cloned in petunia, radish and rice and found to generally belong to the pentatricopeptide (PPR) gene family, which is involved in organelle gene expression (Touzet and Budar 2004).

Across *monoecious* or dioecious plants, arrest of reproductive organs occurs at all stages of development with no apparent preference for a particular stage, although there is a clear tendency for the male and female organ abortions to occur at the same stage, raising the question of the mechanisms by which coordinated regulatory processes between male and female organs have evolved (Diggle et al. 2011). In the few species in which details of the molecular mechanisms controlling the production of unisexual flowers have been investigated, they were found to proceed through various alterations of developmental processes. In dioecious *Spinacia oleraceae* for example, where flowers are unisexual from inception, two floral organ identity homeotic genes of class B are differentially expressed between the sexes throughout flower development, and knocking them down in males results in the development of a functional gynoecium, suggesting that they act as masculinizing genes. In maize, where gynoecium abortion occurs later in development of male inflorescences (right after initiation), individual mutations in at least four distinct genes have been shown to cause the development of hermaphroditic instead of unisexual flowers (Acosta et al. 2009; Banks 2008), suggesting the involvement of many different pathways. Similarly in melon where flowers become unisexual after primordia initiation, *andromonoecious* lines

(in which individual plants carry both male and hermaphroditic flowers) are due to a single SNP in an ethylene biosynthesis enzyme expressed after carpel primordia have been initiated (Boualem et al. 2008), while the female flowers observed in some other lines (called gynoecious) result from the insertion of a transposon that caused heritable epigenetic changes in the promoter of a transcription factor expressed early in flower development (Martin et al. 2009). Beside the fact that the detailed mechanisms by which these causal mutations ultimately lead to organ abortion are currently not known, these mutations have been observed in cultivars only and may therefore tell us little of which mutations may actually trigger mating system shifts in natural ecological settings. Yet, these studies do tell us that not many mutational steps are indeed required to shift gender distribution in a species. It will now be especially interesting to determine whether similar molecular mechanisms are also encountered in natural populations, and whether a small number of large-effect mutations (as in melon) or rather a large number of small-effect mutations are typically involved, thus providing insight into which of the proposed scenarios for this transition (see Sect. 2.2.2) is most frequent. While the small number of species in which detailed genetic and molecular mechanisms have been uncovered remain too small to make general statements at this point, it seems that species in which sex determination occurs early in development (before primordia initiation) could use the same specific organ identity homeotic genes in a convergent manner, whereas species in which sex determination occurs later in development (after primordia initiation) could use a broader diversity of general developmental processes to achieve abortion of sexual organs (Diggle et al. 2011). It will be interesting to assess the generality of this prediction.

Two recent studies have suggested that the evolution of genders may interfere with other features of the mating system that are generally considered independently from one another. In a recent landmark paper, Saumitou-Laprade et al. (2010) demonstrated that the maintenance of *androdioecy* in *Phillyrea* can be explained by the existence of a novel as yet undescribed SI system that restricts mate availability of hermaphrodites but does not function in males, hence offsetting the reproductive disadvantage that males face due to the loss of female function. Similarly, Ehlers and Schierup (2008) showed that breakdown of SI is more likely to occur in gynodioecious species, whereas in turn the breakdown of SI tends to promote stability of gynodioecious populations. These two studies clearly demonstrate the fact that different aspects of the mating system do interact strongly (*androdioecy* or *gynodioecy* and SI), and therefore highlight the importance of considering them jointly.

2.4 Genomic Consequences of Mating Systems Transitions: Genome-Wide Effects and Local Effects on the Mating System Control Region

While the data required to quantify the population genomic effects of mating system transitions has long remained a technical challenge, the increasing availability of polymorphism data at the whole genome level within and between closely related species with contrasted mating systems now allows more accurate estimates of how mating system transitions impact genome organization and evolution. In this section, we review the recent literature and explore the genomic consequences of mating system transitions at two levels, first looking at genome-wide patterns (Table 2.1); second looking at the patterns of molecular evolution in the genomic regions involved in mating system determination.

2.4.1 Genome-Wide Effects of Transitions from Outcrossing to *Selfing*

Several population genetic and ecological factors combine to cause a strong expected reduction in effective population size in association with the transition from outcrossing to *selfing*

Table 2.1 Empirical tests of predictions about the genomic consequences of a transition from outcrossing to selfing in plant populations. Studies performed using genome wide data are underlined

Predicted consequence of transition from outcrossing to selfing	Prediction confirmed	Prediction tested but not confirmed
Reduced level of molecular diversity	Roselius et al. (2005) – *Solanum* Modliszewski and Willis (2012) – *Mimulus* Pettengil and Moeller (2011) – *Clarkia* Ross-Ibarra et al. (2008)/Nordborg et al. (2005) – *Arabidopsis* St Onge et al. (2011) – *Capsella* Ness et al. (2010) – *Eichhornia*	
Reduced ratio of non-synonymous to synonymous polymorphism within species	Glémin et al. (2006) – meta-analysis Slotte et al. (2010) – *Arabidopsis* Slotte et al. (2013) – *Capsella*	
Reduced ratio of non-synonymous to synonymous substitution among species		Wright et al. (2002) – *Arabidopsis* Haudry et al. (2007) – Triticeae Escobar et al. (2010) – Triticeae
Reduced bias in synonymous codon usage	Qiu et al. (2011) – *Arabidopsis* and *Capsella* Haudry et al. (2008) – Triticeae	
Reduced level of transposition of transposable elements	Morgan (2001) – meta-analysis Tam et al. (2007) – Solanum de la Chaux et al. (2012) –*Arabidopsis*	
Reduced selection efficacy against transposed elements	Wright et al. (2001) – *Arabidopsis* Lockton and Gaut (2010) – *Arabidopsis*	

(summarized in Fig. 2.3, Sect. 2.2.4). Briefly, this reduction is expected to cause a decrease in genome-wide levels of polymorphism within *selfing* taxa, but is also expected to impact genome-wide patterns of molecular evolution (e.g. relative abundance of non-synonymous versus synonymous polymorphisms or substitutions; patterns of codon usage; dynamics of selfish elements) as a consequence of a reduction in the efficacy of positive (adaptive) and negative (purifying) directional selection in *selfing* taxa (reviewed by Wright et al. 2008; Glémin and Galtier 2012).

Levels of molecular diversity. Levels of neutral nucleotide diversity are expected to depend on the product of the effective population size times the neutral mutation rate. If we assume that mutation rates are not impacted by the mating system, we expect a twofold or higher decrease in neutral polymorphisms (expressed with statistics such as π_S, the nucleotide diversity at synonymous sites) in *selfing* taxa (Charlesworth 2003). In a large survey comprising 105 Angiosperm species, Glémin et al. (2006) found overall evidence for higher values of π_S (measured at the species level) in outcrossing as compared to *selfing* species with a twofold difference. Similar results but with more striking differences in π_S were generally obtained in studies focusing on narrower phylogenetic groups but with larger genome sampling: 12-fold difference between three outcrossing and two *selfing* species of tomato wild relatives (14 nuclear genes, Roselius et al. 2005); sixfold difference between the outcrosser *Mimulus guttatus* and the selfer *M. nasutus* (six genes, Modliszewski and Willis 2012); fivefold difference between the outcrosser *Clarkia xantiana* ssp *xantiana* and the selfer *C. xantiana* ssp *parviflora* (eight loci, Pettengill and Moeller 2011); threefold difference between the outcrosser *Arabidopsis lyrata* (77 genes, Ross-Ibarra et al. 2008) and the selfer *A. thaliana* (876 genes, Nordborg et al. 2005); but only twofold difference between the outcrosser *Capsella grandiflora* and the selfer *C. rubella* (16 genes, St Onge et al. 2011), and between outcrossing and *selfing* populations of *Eichhornia paniculata* (ten loci, Ness et al. 2010). Overall these results indicate that population genetic factors (inbreeding,

genetic hitchhiking, background selection) are probably not sufficient to explain the observed reduction in neutral diversity in some *selfing* taxa. This discrepancy is thought to be related to additional differences in ecology or demographic history between selfers and outcrosser. Selfing taxa may indeed experience strong bottlenecks at the time of speciation when evolving from outcrossing ancestors (Modliszewski and Willis 2012) and/or frequent founder effects in relation to their higher colonization potential (Baker 1955; Schoen and Brown 1991), and these events will not be compensated by genetic exchange among populations or closely related species as gene flow through pollen is highly reduced in *selfing* taxa (Ingvarsson 2002).

Ratio of non-synonymous to synonymous polymorphism within species. Although neutral diversity is indeed lower in *selfing* taxa, an increase in the relative occurrence of non-synonymous (π_N) vs. synonymous (π_S) polymorphisms is expected as compared to outcrossing taxa (Charlesworth and Wright 2001). This is because selfers are less efficient in eliminating weakly deleterious mutations, corresponding to a large fraction of non-synonymous mutations, owing to their reduced effective population size (Glémin 2007). In the meta-analysis of Glémin et al. (2006), a significantly higher value of π_N/π_S was indeed observed for *selfing* taxa. With a dataset on 257 loci, Slotte et al. (2010) found evidence for a larger proportion of slightly deleterious non-synonymous mutations occurring in populations of the selfer *A. thaliana*, as compared to the outcrosser *Capsella grandiflora*. Hence, clear evidence for relaxed selection on weakly deleterious mutations in *selfing* species has been brought by recent genomic data. In addition, using whole genome data on 80 accessions from different populations of *A. thaliana*, Cao et al. (2011) showed that populations with lower effective population size experience greater relaxation of selection, suggesting strong heterogeneity among populations in the efficacy of selection in *selfing* species, which could have important ecological and evolutionary implications.

Ratio of non-synonymous to synonymous substitution among species. Differences in the efficacy of selection between *selfing* and outcrossing taxa should lead to higher rates of fixation of weakly deleterious alleles in selfers (Glémin 2007). This could be detected by comparing estimates of the ratio ω of non-synonymous to synonymous substitutions occurring in *selfing* versus outcrossing lineages in specific groups. In a comparison between the outcrosser *A. lyrata* and the selfer *A. thaliana* based on 23 genes, Wright et al. (2002) did not find evidence for higher fixation rates of non-synonymous mutations. Similar negative results were obtained by Haudry et al. (2008) in an analysis of two *selfing* and two outcrossing species of the tribe Triticeae (Poaceae) using data from 46 genes, and by Escobar et al. (2010) on 19 species of Triticeae using data from 27 genes. Although the datasets used in these studies are limited, the results suggest that the expected relaxation of selection in selfers is not so apparent in the long term and may not greatly influence substitution rates and genome evolution. However, a possible explanation for the discrepancy between results on polymorphisms and substitutions could be that the *selfing* taxa investigated have recently evolved (e.g. Bechsgaard et al. 2006), such that the time elapsed may have been too short to allow detecting significant effects on the fixation of mutations.

Codon usage. The relaxation of selection in selfers in association with reduced effective population size is also expected to alter patterns of codon usage in protein coding genes, as mutations towards non-optimal synonymous codons can be considered as slightly deleterious mutations subject to purifying selection (Marais et al. 2004). Using large datasets from two *selfing* species (*A. thaliana* and *C. rubella*) and their closely related outcrossing species (*A. lyrata* and *C. grandiflora*, respectively), Qiu et al. (2011) found clear evidence for a relaxed selection on synonymous codons in selfers as compared to their outcrossing relatives. Similar results, but with weaker statistical support, have been obtained by comparing two *selfing* with two outcrossing species in Triticeae (Haudry et al.

2008). A potential pitfall in these analyses is the confounding effect of GC-biased gene conversion (gBGC), which is a neutral process occurring during double-strand break recombination repair and leading to a bias towards G and C alleles, mimicking the effect of selection for optimal codons (Marais 2003). Indeed, gBGC is associated with recombination at heterozygous sites, but because of the low polymorphism found in *selfing* taxa, it is believed to be ineffective in selfers, while potentially important in outcrossers (Marais et al. 2004). In the study on two Brassicaceae genera (*Arabidopsis* and *Capsella*, that both contain *selfing* and outcrossing species), the effect of gBGC was tested using intron sequence data but no difference was detected between selfers and outcrossers (Qiu et al. 2011). In contrast, in the study on Triticeae, higher levels of gBGC were detected in outcrossing as compared to *selfing* taxa, and this potentially can impede the detection of relaxed selection on codon bias (Haudry et al. 2008).

Transposable elements dynamics. In equilibrium models of TE evolution, the number of copies in a given TE family results from a balance between the process of transposition leading to genomic spread of the family at the population level, and the selective removal of copies with deleterious effects. A transition from outcrossing to *selfing* is expected to have opposite effects on the processes of transposition and purging, so the net outcome is not straightforward to predict (Wright and Schoen 1999; Morgan 2001). Indeed, the high homozygosity occurring in selfers, and low rates of genetic exchange among individuals will cause a reduced spread of new TE insertions throughout the population, whereas the selective removal of weakly deleterious insertions will be either more or less effective in selfers depending on the level of dominance and nature of the deleterious effects (disruption of gene function vs. mediation of ectopic exchange between distinct chromosomes). Empirical results are still limited. Morgan (2001) reviewed studies on Ty1 copia-like elements for a number of species pairs and concluded that copy number is reduced overall in selfers as compared to outcrossers. Tam et al. (2007), observed lower frequencies of insertion of copia-like elements in *selfing* as compared to outcrossing species of tomato wild relatives, but no differences in current copy numbers. In *Arabidopsis*, copy number for a vast majority of TE families was found to be lower in the selfer *A. thaliana* than in the outcrosser *A. lyrata* (de la Chaux et al. 2012), and the results suggest that the rate of transposition has recently decreased in *A. thaliana*, probably after the shift in mating system. Regarding the issue of selective removal of TE insertions, two studies reported that TE insertions segregate at higher frequencies in *A. thaliana*, as compared to *A. lyrata*, suggesting relaxation of selection against weakly deleterious mutations in the *selfing* lineage (Wright et al. 2001; Lockton and Gaut 2010).

Evolutionary consequences of genome-wide effects: is selfing an evolutionary dead end? Although the number of detailed empirical studies of the genome-wide effects of a transition to *selfing* is still scarce, and some results are controversial, a consensus can be found for two major consequences: a decrease in nucleotide polymorphism genome-wide, and a relative increase in frequency of mildly deleterious alleles in *selfing* populations. These empirical evidence are related to the two classical genetic threats associated with *selfing*, namely the loss of adaptive potential caused by a lack of standing variation (Stebbins 1957; Glémin and Ronfort 2013), and the accumulation of deleterious mutations, potentially leading to a "mutational meltdown" whereby deleterious alleles become fixed and contribute to population extinction (Lynch et al. 1995a; see Sect. 2.2.4 and Fig. 2.3). Altogether, these processes could cause an increase in the rate of extinction of *selfing* clades as compared to outcrossing clades, but this still needs to be demonstrated empirically. In this context, phylogenetic approaches involving estimation of differences in diversification rates in relation to mating system traits, or comparative analyses of rates of diversification in *selfing* versus outcrossing clades are interesting approaches to estimate indirectly the evolutionary consequences of mating system

transitions. Such approaches have recently shown very clear evidence for a reduction in evolutionary success of self-compatible clades in Angiosperms (as measured by the net diversification rate, i.e. speciation minus extinction rates; Goldberg et al. 2010; Goldberg and Igic 2012; Ferrer and Good 2012).

2.4.2 Genome-Wide Effects of Transitions from Hermaphroditism to Gynodioecy and/or Dioecy

In contrast to the evolution of *selfing*, a transition from *hermaphroditism* to *gynodioecy* and/or *dioecy* is not expected to cause major changes in population genetics processes affecting effective population size. If the hermaphroditic ancestor was partially reproducing by *selfing*, however, one could predict that the transition would lead to an increase in outcrossing level (Charlesworth and Charlesworth 1978), and thus to a (moderate) increase in effective population size in gynodioecious or dioecious taxa. In contrast, two types of ecological factors associated to the transition to *dioecy* could alter metapopulation dynamics, thereby causing a reduction in effective population size (Kafer et al. 2013): the "seed-shadow handicap", i.e. the overall reduction in seed dispersal efficiency due to a lower proportion of seed-producing individuals (Heilbuth et al. 2001) and a higher sensitivity to variation in pollinator abundance in animal-pollinated dioecious species as a consequence of sexual selection for increased pollinator attraction in males (Vamosi and Otto 2002). In contrast to studies on the effect of *selfing*, the empirical literature on genomic effects of the transition to *gynodioecy* or *dioecy* is strikingly scarce. In agreement with the predictions based on ecological factors, Kafer et al. (2013) found evidence for a reduction in the efficacy of selection (measured on patterns of nucleotide substitutions with the ratio ω) in dioecious species of the section Melandrium in the genus Silene (e.g. *S. latifolia*), as compared to a gynodioecious close relative (*S. vulgaris*). However they found no difference in a second comparison involving the dioecious *S. otites* and the gynodioecious *S. nutans*. They suggest that the discrepancy could be explained by differences in the timescale of the two comparisons, with a much more recent evolution of *dioecy* in *S. otites* as compared to the Melandrium section, so that the time is too short for the evolution of genomic differences (Kafer et al. 2013). This suggestion is confirmed by another study that showed the occurrence of shared polymorphism at nuclear and mitochondrial genes between *S. otites* and *S. nutans*, and which observed similar levels of genetic diversity in both species (Lahiani et al. 2013). Finally, a comparison of transposable elements accumulation between the dioecious *S. latifolia* and the gynodioecious *S. vulgaris* showed a very striking accumulation of a *gypsy* retroelement in the former that could partially account for its much larger genome (Cegan et al. 2012).

2.4.3 Local Effects of Mating System Transitions on Genomic Regions Involved in Mating System Determination

Regions involved in mating system determination often show very different patterns of molecular evolution as compared to unlinked control regions. These are due to two major properties of such regions (Charlesworth 2006). First, they are generally subject to strong negative frequency-dependent selection on sex ratio or on mating type frequencies, due to the transmission advantage of the rarest sex or fecundity advantage of the rarest mating type/SI allele. Second, the recombination rate is frequently highly reduced in such regions. As a consequence, the different functional haplotypes at these regions are expected to show very strong nucleotide divergence, some of it being due to positive (diversifying) selection on the functional genes involved in mating, and some being due to a drift process intensified by long-term maintenance of the polymorphism and low effective population size of sets of gene copies of a given functional haplotype (Vekemans and Slatkin 1994; Charlesworth et al. 2005). The lack of recom-

bination in these regions is also expected to reduce the efficacy of selection and drive genetic degeneration over time causing accumulation of repetitive DNA and decrease in gene content (Charlesworth and Charlesworth 2000; Bachtrog 2008), as commonly found in non-recombining regions of Y chromosomes (Wang et al. 2012; Bergero et al. 2008). Transitions in mating systems may potentially have a large impact on patterns of molecular evolution in regions involved in mating system determination, as for instance the transition from *hermaphroditism* to *dioecy*, which causes the formation of neo sex chromosomes (Ming et al. 2011) and the breakdown of *self-incompatibility*, which may cause a loss of diversity at the S-locus and restoration of recombination (Guo et al. 2009). The transition that occurred within Chlorophyceae between isogamous (no differences in gamete sizes between mating types, as in *Chlamydomonas*) and anisogamous (e.g. *Volvox*) mating systems constitutes another striking example of a transition that had a major impact on the mating type locus region. Indeed, the mating type locus in the anisogamous *Volvox* was found to be homologous to that in the isogamous *Chlamydomonas*, but with a fivefold increase in size, due to inclusion of a higher number of genes (including some genes controlling gamete size) in association with a large inversion preventing intra-locus recombination (Ferris et al. 2010; Charlesworth and Charlesworth 2010b).

Patterns of molecular evolution at the self-incompatibility locus before and after a breakdown of incompatibility. In analogy to the differences between sex chromosomes, comparative surveys of full sequences of the S-locus region in self-incompatible species of the *Arabidopsis* genus have revealed the following patterns: high variability in size and gene organization among functional haplotypes, a complete absence of sequence similarity in intergenic sequences, and strong accumulation of transposable elements (Guo et al. 2011; Goubet et al. 2012). Similar patterns have been observed at the S-locus of other species of Brassicaceae (*Brassica rapa*, Fukai et al. 2003; *B. oleracea*, Fujimoto et al. 2006), or in other multiallelic SI systems (*Prunus mume*, Entani et al. 2003). In a sporophytic SI system, these features were found to be associated with the dominance level of the S-locus haplotypes, with for instance dominant S-locus haplotypes accumulating more transposable elements than recessive ones (Goubet et al. 2012), because of a reduction in the efficacy of selection in the former due to lower recombination in dominant haplotypes (Castric et al. 2010), and/or because of their lower population frequencies (Schierup et al. 1997). It was suggested that recessive S-locus haplotypes (which can recombine in individuals homozygous at the S-locus) are analogous to X-chromosomes (recombining in females), while dominant haplotypes (which are always present in S-locus heterozygotes) are analogous to Y-chromosomes (Goubet et al. 2012). After a transition from SI to a selfing mating system, two different outcomes have been observed in the S-locus region, probably depending on whether *selfing* arose in association with a speciation event. The breakdown of SI was found to be associated with a species-wide fixation of a self-compatible S-locus haplotype in the selfer *Capsella rubella*, where all S-locus haplotypes are highly similar and non-functional (Guo et al. 2009, Fig. 2.4, see Sect. 2.3.2), as well as in one of the *selfing* races within the species *Leavenworthia alabamica* (Busch et al. 2011). In contrast, maintenance of several divergent non-functional haplotypes at the S-locus region has been reported in *Arabidopsis thaliana*, although with strong geographic differences in their relative frequencies (Boggs et al. 2009; Shimizu et al. 2004, 2008; Tsuchimatsu et al. 2010). Similarly in *A. kamchatica*, a recent allopolyploid *selfing* species originating from two self-incompatible species (*A. lyrata* and *A. halleri*), five divergent non-functional haplotypes were found, three of them originating from the parent *A. halleri*, and the last two originating from *A. lyrata* (Tsuchimatsu et al. 2012). These differences are expected to have a major impact on patterns of molecular evolution in the S-locus region in these *selfing* taxa, as the presence or absence of haplotype divergence at the S-locus is

expected to influence levels of recombination in that region.

Evolution of Sex Chromosomes in Dioecious Species. A transition from *hermaphroditism* to *dioecy* is often accompanied by gradual build-up of a set of sex chromosomes carrying sex determination genes, with a progressive reduction in recombination in the heterogametic genotype and genetic degradation of the Y chromosome (Ming et al. 2011). In the first stage the male and female sterility loci are still recombining, allowing the formation of hermaphrodite individuals as in *Fragaria virginiana* (Spigler et al. 2008). The following stages correspond to suppression of recombination between the two sex-determining loci and progressively larger neighbouring regions, allowing an increasing number of Y-linked genes to degenerate and form a male-specific region on the nascent Y chromosome (Ming et al. 2011). This gradual process is believed to involve several subsequent inversions, as suggested by an observed correlation between synonymous divergence between the X and Y sequences of *Silene latifolia* and *S. diclinis* and genetic distance to the pseudo-autosomal region (Nicolas et al. 2005; Bergero et al. 2007). This pattern has been observed in mammalian sex chromosomes and has been termed "evolutionary strata" (Lahn and Page 1999). Although the two sex chromosomes are homomorphic in the first stages (e.g. *Carica papaya* Liu et al. 2004), genetic degradation associated with the extension of the non-recombining region causes the accumulation of transposable elements and duplicated segments in the male-specific region causing significant expansion of the Y chromosome, as found in *S. latifolia* (Bergero et al. 2008), which generates heteromorphic sex chromosomes. Another feature associated with genetic degradation is an overall trend of reduced expression of Y-linked alleles, as observed in *S. latifolia* (Muyle et al. 2012). Interestingly, the genes showing reduced expression of the Y-linked allele were found to have higher expression of the X-linked allele in males than in females, a phenomenon known as dosage compensation, and which has been found in other male heterogametic systems in animals (Muyle et al. 2012). A next step in evolution of sex chromosomes is believed to be a progressive loss of non-functional sequences across the Y accompanied by shrinking of the Y chromosome (Ming et al. 2011), such as been found in *Cycas revoluta* (Segawa et al. 1971).

Evolution of mitochondrial genomes in gynodioecious species with cytoplasmic male sterility factors. Because of the complete linkage disequilibrium within the mitochondrial genome, as well as between the chloroplastic and mitochondrial genomes in most Angiosperms, selection on cytoplasmic male sterility factors in gynodioecious species is expected to affect both cytoplasmic genomes throughout their total length (Touzet 2012). Two alternative selective scenarios have been proposed for the maintenance of gender polymorphism in gynodioecious species with cytoplasmic male sterility: (1) the balancing selection scenario, where two or more mitochondrial haplotypes (either several functionally distinct male sterile haplotypes, or a mixture of male fertile and male sterile haplotypes) are maintained over the long term and experience cycles of frequency changes over time (Gouyon et al. 1991; Dufay et al. 2009); and (2) the epidemic dynamics scenario, where new male-sterilizing haplotypes are continually arising by mutation and sweep through the population (Frank 1989). As noted by Charlesworth (2002b), the two alternative scenarios will have different effects on cytoplasmic neutral-locus diversity. In the case of haplotypes being maintained over a long period of time through balancing selection, diversity is expected to be high because different neutral mutations would accumulate over time in different haplotypes (Stadler and Delph 2002), while under the epidemic dynamics scenario a lower diversity would be expected as new sterilizing cytoplasms will sweep through populations (Ingvarsson and Taylor 2002). Diversity in *Silene* mitochondrial genomes was found to be higher in gynodioecious than in non-gynodioecious species, suggesting that the balancing selection scenario is more

likely (Touzet and Delph 2009; Charlesworth 2010). Moreover, phylogenetic analyses of three male sterilizing mitochondrial haplotypes and three fertile haplotypes in *Beta maritima* show high divergence between the male sterile and the fertile lineages, suggesting long-term maintenance of the polymorphism (Darracq et al. 2011). The latter study also shows that the male sterile lineage has experienced an increase in mutation rate in the mitochondrial genome, which could suggest either that the sterilizing phenotype is a consequence of the higher mutation rate, or that sterilizing mutations could have caused an increase in the production of cellular reactive oxygen species in the mitochondria causing an elevation of mutation rates (Touzet 2012).

2.5 Perspectives: Open Questions

Convergent and parallel mating system transitions: what is the importance of gene reuse? A first set of open issues highlighted in our review is whether independent mating system transitions have been caused by mutations at the same nucleotides, genes or gene networks, or whether they were caused by mutations on entirely different genes. Along the same line, a major question is whether these mutations were independent (i.e. de novo mutations in genetic "hotspots" of variation and phenotypic evolution), or whether different lineages acquired them by ancestral polymorphism or lateral gene transfer. These questions are central in evolution not only for mating systems transitions, but for phenotypic evolution in general (Martin and Orgogozo 2013) and has been debated for decades. Thanks to the advances of genomics, there are now dozens of documented cases of the same genes being used repeatedly for phenotypic evolution in eukaryotes and prokaryotes (reviewed in Martin and Orgogozo 2013), although these cases are generally limited to convergent evolution between a few lineages only, generally two, except in the case of experimental evolution of simple organisms. As we detailed in this review, the recent literature on plant mating systems reported clear evidence for molecular parallelism between lineages, e.g. the pollen gene involved in the breakdown of SI, paralogous *SCR/SRK*-like genes recruited in *Leavenworthia*, PPR genes recruited as male fertility restorers in gynodioecious species. Since mating system transitions occurred in plants dozens of times independently, we argue that they are excellent biological systems to assess the evolutionary importance of gene reuse. A current limitation is the low number of model systems that have been described down to the molecular level. The intense research aimed at dissecting the putative supergene controlling morphological differences between morphotypes in heterostylous species makes this mating system particularly promising for the years to come, once the supergene will have been identified in different groups.

If gene reuse and genetic hotspots are confirmed in independent cases of mating system transitions in plants in many lineages, then a further question arises: why are some genes used more frequently for transitions than others? Several causes have been hypothesized, including variation among genes in (i) mutation rate, whereby genes that mutate more frequently are more prone to cause phenotypic evolution, (ii) mutation size, whereby genes at which single mutations have the largest selective advantage are favoured and (iii) level of pleiotropic effects, favouring the use of genes having lower cascading effect in the regulatory networks (reviewed in Martin and Orgogozo 2013).

Plant mating systems in a broader context. Our review highlighted several clear cases of correlated evolution between mating systems and other life history traits or specific genomic features. A second set of open questions in the field will thus now be to assess the generality of this phenomenon, explicitly considering the evolution of mating systems in their broader genomic context. First, recent theoretical and empirical advances challenged the notion that the various aspects of plant mating systems can be considered as evolving independently from one another. Specifically, SI and *gynodioecy* mutually tend to decrease the conditions of their maintenance

(Ehlers and Schierup 2008) and the maintenance of *androdioecy* is facilitated by the existence of a functional SI (Saumitou-Laprade et al. 2010; Husse et al. 2013). An open question is therefore whether the existence of SI in plants facilitates the transition from *hermaphroditism* to *dioecy*. The joint evolution of SI and gender differentiation might even be a general feature of plant evolution, since anisogamy in the green Algae *Volvox* evolved by the recruitment of mutations in the genomic region determining SI (Ferris et al. 2010). Second, while studies on the loss of SI in *C. rubella* and *A. thaliana* highlight very different evolutionary scenarios in these two model species, with an apparent co-occurrence of the shift in mating system with a speciation event in the history of *C. rubella* but not in *A. thaliana* (Fig. 2.4), the relative importance of these scenarios among flowering plants remains to be investigated, as well as a possible association between the evolutionary scenario and the outcome in terms of maintenance of polymorphism at the mating system control region. Further, the putative occurrence of a transient period of mixed mating in the scenario that is not associated with a speciation event should be addressed. Third, the extent of *inbreeding depression* and its evolutionary lability constitute key features that affect transitions in plant mating systems. These features of *inbreeding depression* depend on whether the mutations involved are deleterious or under balancing selection, whether they have small or large effects on fitness, whether they are recessive or dominant, and on their organization and genomic location (the relative importance of genome wide load vs. sheltered load). While these issues are crucial in order to better understand mating systems transition, very little is currently known, especially in natural plant populations. We believe that this is a very important priority for research in ecological genomics.

Plant mating systems and community ecology. Finally, while significant progress has been made in the field through genomic approaches, a major frontier will now be to place back mating system evolution in the ecological context of interspecific interactions, especially at the community level. Thirty years ago, Charnov (1982) speculated that the evolution of *dioecy* might be affected by community-level laws, but this speculation can be extended to any transition in plant mating systems. For instance in plant species with different sexual morphs (heterostyly or *dioecy*), differential behaviour of pollinators on the sexual morphs (Case and Ashman 2009) would affect relative fitness of the latter and hence modify the evolution of genes involved in morphological differences between morphs. Furthermore, a decrease in pollinators abundance is expected to affect more drastically the demography of outcrossing than that of *selfing* species, hence favouring the evolution of higher *selfing* rates (Eckert et al. 2009). In turn, highly *selfing* species tend to offer less energy resources to their insect pollinators (such as nectar), which may further decrease pollinators abundance, eventually putting at risk population viability of both outcrossers and pollinators. How plant-pollinators interaction networks influence and are robust to the evolution of plant mating system transitions, and more generally how interspecific interactions at the community level affect the evolution of mating systems, their genomic architecture, and how in turn genomic features of mating systems affect the dynamics, stability and evolution of communities is an entire field open for investigation.

Acknowledgements We thank two anonymous reviewers for helpful comments. Our work on plant mating systems is supported by the French Agence Nationale de la Recherche (ANR-11-BSV7- 013-03 and ANR 11 JSV7 008 01) and by the Région Nord Pas de Calais.

Glossary

Hermaphroditism Hermaphroditism is the simultaneous coexistence of male and female reproductive organs on the same (co-sexual) individual (Fig. 2.1). Hermaphroditism is believed to be the ancestral state in Angiosperms, and the defining organ of Angiosperms (the flower) is itself a hermaphroditic organ, producing both pollen and ovules.

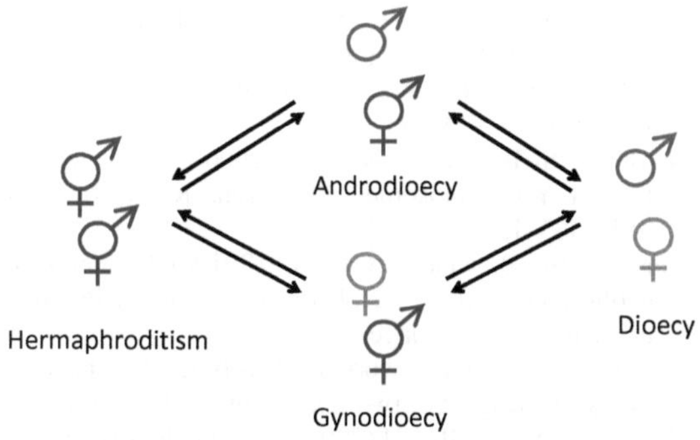

Selfing versus outcrossing A major consequence of hermaphroditism is the potential for self-fertilization, the most extreme form of inbreeding.

Self-incompatibility (SI) A genetic system promoting allogamy in many plants. There are two different types of SI, homomorphic and heteromorphic. In homomorphic SI, the different groups of mating partners differ by the type of recognition proteins they produce, but remain morphologically undistinguishable. In heteromorphic SI (e.g. heterostyly), thc two (distyly) or three (tristyly) self- and within-morph incompatible mating groups typically differ by style length, anther height and pollen size.

Inbreeding depression Inbreeding depression is the decrease of fitness of offspring produced by inbred parents relatively to those produced by unrelated parents.

Pollen limitation A plant is pollen-limited if it does not receive enough pollen to fertilize all its ovules. Pollen limitation thus leads to a reduction in reproductive output through the female function.

Pollen discounting The loss of male reproduction in cross-fertilization due to the decrease of exported pollen, which occurs especially in autogamous species.

Gynodioecy (*resp.* androdioecy) Gynodioecy (*resp.* androdioecy) is a mating system whereby hermaphrodite individuals coexist with female (resp. male) individuals (Fig. 2.1).

Dioecy Dioecy is the separation of sexual functions in specialized (male and female) individuals (Fig. 2.1). Species in which some individuals with incomplete sexual specialization occur along with strictly unisexual individuals are termed **subdioecious.**

Monoecious Monoecious species are composed of hermaphrodites only, but in which male and female flowers are separated on each individual.In **gynomonoecious** species, some individuals produce female-only flowers in variable proportion along with hermaphroditic flowers, while in **andromonoecious** species, some individuals produce male-only flowers along with hermaphroditic flowers.

References

Acosta IF, Laparra H, Romero SP, Schmelz E, Hamberg M, Mottinger JP, Moreno MA, Dellaporta SL (2009) Tasselseed1 is a lipoxygenase affecting jasmonic acid signaling in sex determination of maize. Science 323:262–265

Al-Shehbaz IA, O'Kane SLJ (2002) Taxonomy and phylogeny of *Arabidopsis* (Brassicaceae). In: Somerville CR, Meyerowitz EM (eds) The *Arabidopsis* book. American Society of Plant Biologists, Rockville

Ashman TL (2003) Constraints on the evolution of males and sexual dimorphism: filed estimates of genetic architecture of reproductive traits in three populations of gynodioecious *Fragaria virginiana*. Evolution 57:2012–2025

Bachtrog D (2008) The temporal dynamics of processes underlying Y chromosome degeneration. Genetics 179:1513–1525

Baker HG (1955) Self-compatibility and establishment after "long-distance" dispersal. Evolution 9: 347–348

Banks JA (2008) MicroRNA, sex determination and floral meristem determinacy in maize. Genome Biol 9:204

Barrett SCH, Shore JS (2008) New insights on heterostyly: comparative biology, ecology and genetics. In: Franklin-Tong V (ed) Self-incompatibility in flowering plants: evolution, diversity and mechanisms. Springer-Verlag, Berlin, pp 3–32

Bechsgaard JS, Castric V, Charlesworth D et al (2006) The transition to self-compatibility in *Arabidopsis thaliana* and evolution within S-haplotypes over 10 Myr. Mol Biol Evol 23:1741–1750

Beilstein MA, Nagalingum NS, Clements MD et al (2010) Dated molecular phylogenies indicate a miocene origin for *Arabidopsis thaliana*. Proc Natl Acad Sci 107(43):18724–18728

Bergero R, Forrest A, Kamau E, Charlesworth D (2007) Evolutionary strata on the X chromosomes of the dioecious plant *Silene latifolia*: evidence from new sex-linked genes. Genetics 175:1945–1954

Bergero R, Forrest A, Charlesworth D (2008) Active miniature transposons from a plant genome and its nonrecombining Y chromosome. Genetics 178:1085–1092
Billiard S, López-Villavicencio M et al (2011) Having sex, yes, but with whom? Inferences from fungi on the evolution of anisogamy and mating types. Biol Rev 86:421–442
Boggs NA, Nasrallah JB, Nasrallah ME (2009) Independent S-locus mutations caused self-fertility in *Arabidopsis thaliana*. PLoS Genet 5:e1000426
Boualem A, Fergany M, Fernandez R et al (2008) A conserved mutation in an ethylene biosynthesis enzyme leads to andromonoecy in melons. Science 321(5890):836–838
Busch JW, Schoen DJ (2008) The evolution of self-incompatibility when mates are limiting. Trends Plant Sci 13:128–136
Busch JW, Joly S, Schoen DJ (2011) Demographic signatures accompanying the evolution of selfing in *Leavenworthia alabamica*. Mol Biol Evol 28: 1717–1729
Cao J, Schneeberger K, Ossowski S et al (2011) Whole-genome sequencing of multiple *Arabidopsis thaliana* populations. Nat Genet 43:956–965
Case AL, Ashman TL (2009) Resources and pollinators contribute to population sex ratio bias and pollen limitation in gynodioecious wild strawberry (*Fragaria virginiana*). Oikos 118:1250–1260
Castric V, Bechsgaard J, Grenier S et al (2010) Molecular evolution within and between self-incompatibility specificities. Mol Biol Evol 27:11–20
Cegan R, Vyskot B, Kejnovsky E et al (2012) Genomic diversity in two related plant species with and without sex chromosomes – *Silene latifolia and S. vulgaris*. PLoS ONE 7:e31898
Chantha SC, Herman AC, Platts A, Vekemans X, Schoen DJ (2013) Secondary evolution of self-incompatibility in the mustard genus Leavenworthia. PLoS Biol 11(5):e1001560
Charlesworth D (2002a) Plant sex determination and sex chromosomes. Heredity 88:94–101
Charlesworth D (2002b) What maintains male-sterility factors in plant populations. Heredity 89:408–409
Charlesworth D (2003) Effects of inbreeding on the genetic diversity of populations. Phil Trans R Soc Lond B 358:1051–1070
Charlesworth D (2006) Balancing selection and its effects on sequences in nearby genome regions. PLoS Genet 2:e64
Charlesworth D (2010) Don't forget the ancestral polymorphisms. Heredity 105:509–510
Charlesworth B, Charlesworth D (1978) A model for the evolution of dioecy and gynodioecy. Am Nat 112: 975–997
Charlesworth D, Charlesworth B (1979) Evolution and breakdown of S-allele systems. Heredity 43:41–55
Charlesworth D, Charlesworth B (1987) Inbreeding depression and its evolutionary consequences. Annu Rev Ecol Syst 18:237–268
Charlesworth D, Charlesworth B (1990) Inbreeding depression with heterozygote advantage and its effect on selection for modifiers changing the outcrossing rate. Evolution 44:870–888
Charlesworth B, Charlesworth D (2000) The degeneration of Y chromosomes. Phil Trans R Soc Lond B 355:1563–1572
Charlesworth B, Charlesworth D (2010a) Elements of evolutionary genetics. Roberts and Company Publishers, Greenwood Village
Charlesworth D, Charlesworth B (2010b) Evolutionary biology: the origins of two sexes. Current Biol 20:R520
Charlesworth D, Willis JH (2009) The genetics of inbreeding depression. Nat Rev Genet 10:783–796
Charlesworth D, Wright SI (2001) Breeding systems and genome evolution. Curr Opin Genet Dev 11: 685–690
Charlesworth D, Morgan MT, Charlesworth B (1993) Mutation accumulation in finite outbreeding and inbreeding populations. Genet Res 61:39–56
Charlesworth D, Charlesworth B, Morgan MT (1995) The pattern of neutral molecular variation under the background selection model. Genetics 141: 1619–1632
Charlesworth D, Vekemans X, Castric V, Glémin S (2005) Plant self-incompatibility systems: a molecular evolutionary perspective. New Phytol 168:61–69
Charnov EL (1982) The theory of sex allocation. Princeton University Press, Princeton
Chookajorn T, Kachroo A, Ripoll DR, Clark AG, Nasrallah JB (2003) Specificity determinants and diversification of the Brassica self-incompatibility pollen ligand. Proc Natl Acad Sci USA 101:911–917
Darracq A, Varré JS, Maréchal-Drouard L et al (2011) Structural and content diversity of mitochondrial genome in beet: a comparative genomic analysis. Genome Biol Evol 3:723–736
Darwin C (1876) The effects of cross and self fertilisation in the vegetable kingdom. John Murray, London
Darwin C (1877) The different forms of flowers on plants of the same species. John Murray, London
de Graaf BH, Vatovec S, Juárez-Díaz JA, Chai L, Kooblall K, Wilkins KA, Zou H, Forbes T, Franklin FC, Franklin-Tong VE (2012) The *Papaver* self-incompatibility pollen S-determinant, PrpS, functions in *Arabidopsis thaliana*. Curr Biol 22(2): 154–159
de la Chaux N, Tsuchimatsu T, Shimizu KK, Wagner A (2012) The predominantly selfing plant *Arabidopsis thaliana* experienced a recent reduction in transposable element abundance compared to its outcrossing relative *Arabidopsis lyrata*. Mob DNA 3:2
Diggle PK, Di Stilio VS, Gschwend AR, Golenberg EM, Moore RC, Russell JRW, Sinclair JP (2011) Multiple developmental processes underlie sex differentiation in angiosperms. Trends Genet 27(9):368–376
Dufaÿ M, Billard E (2012) How much better are females? The occurrence of female advantage, its proximal causes and its variation within and among gynodioecious species. Ann Bot 109:505–519

Dufay M, Cuguen J, Arnaud JF, Touzet P (2009) Sex ratio variation among gynodioecious populations of sea beet: can it be explained by frequency-dependent selection? Evolution 63:1483–1497

Eckert CG, Ozimec B, Herlihy CR, Griffin CA, Routley MB (2009) Floral morphology mediates temporal variation in the mating system of a self-compatible plant. Ecology 90(6):1540–1548

Ehlers BK, Schierup MH (2008) When gametophytic self-incompatibility meets gynodioecy. Genet Res Camb 90:27–35

Entani T, Iwano M, Shiba H et al (2003) Comparative analysis of the self-incompatibility (S-) locus region of *Prunus mume*: identification of a pollen-expressed F-box gene with allelic diversity. Genes Cells 8: 203–213

Escobar JS, Cenci A, Bolognini J et al (2010) An integrative test of the dead-end hypothesis of selfing evolution in Triticeae (Poaceae). Evolution 64:2855–2872

Ferrer MM, Good SV (2012) Self-sterility in flowering plants: preventing self-fertilization increases family diversification rates. Ann Bot 110(3):535–553

Ferris P, Olson B, Hoff PD et al (2010) Evolution of an expanded sex-determining locus in *Volvox*. Science 328:351–354

Fisher RA (1941) Average excess and average effect of a gene substitution. Ann Eugen 11:53–63

Fishman L, Kelly AJ, Willis JH (2002) Minor quantitative trait loci underlie floral traits associated with mating system divergence in *Mimulus*. Evolution 56(11):2138–2155

FitzJohn RG (2010) Quantitative traits and diversification. Syst Biol 59:619–633

Foxe JP, Slotte T, Stahl EA, Neuffer B, Hurkad H, Wright SI (2009) Recent speciation associated with the evolution of selfing in *Capsella*. Proc Natl Acad Sci USA 106(13):5241–5245

Frank SA (1989) The evolutionary dynamics of cytoplasmic male-sterility. Am Nat 133:345–376

Fujimoto R, Okazaki K, Fukai E, Kusaba M, Nishio T (2006) Comparison of the genome structure of the self-incompatibility (S) locus in inter-specific pairs of S haplotypes. Genetics 173:1157–1167

Fukai E, Fujimoto R, Nishio T (2003) Genomic organization of the S core region and the S flanking regions of a class-II S haplotype in *Brassica rapa*. Mol Gen Genomics 269:361–369

Gervais CE, Castric V, Ressayre A, Billiard S (2011) Origin and diversification dynamics of self-incompatibility haplotypes. Genetics 188:625–636

Glémin S (2007) Mating systems and the efficacy of selection at the molecular level. Genetics 177: 905–916

Glémin S, Galtier N (2012) Genome evolution in outcrossing versus selfing versus asexual species. Methods Mol Biol 855:311–335

Glémin S, Ronfort J (2013) Adaptation and maladaptation in selfing and outcrossing species: new mutations versus standing variation. Evolution 67:224–240

Glémin S, Bazin E, Charlesworth D (2006) Impact of mating systems on patterns of sequence polymorphism in flowering plants. Proc R Soc Land B 273:3011–3019

Goldberg EE, Igic B (2012) Tempo and mode in plant breeding system evolution. Evolution 66:3701–3709

Goldberg EE, Kohn JR, Lande R et al (2010) Species selection maintains self-incompatibility. Science 330:493–495

Goubet PM, Bergès H, Bellec A et al (2012) Contrasted patterns of molecular evolution in dominant and recessive self-incompatibility haplotypes in *Arabidopsis*. PLoS Genet 8:e1002495

Gouyon PH, Vichot F, Van Damme JMM (1991) Nuclear-cytoplasmic male sterility: single point equilibria versus limit cycles. Am Nat 137:498–514

Graham SW, Barrett SCH (2004) Phylogenetic reconstruction of the evolution of stylar polymorphisms in Narcissus (Amaryllidaceae). Am J Bot 91: 1007–1021

Guo YL, Bechsgaard JS, Slotte T, Neuffer B, Lascoux M, Weigel D, Schierup MH (2009) Recent speciation of *Capsella rubella* from *Capsella grandiflora*, associated with loss of self-incompatibility and an extreme bottleneck. Proc Natl Acad Sci USA 106(13): 5246–5251

Guo Y, Zhao X, Lanz C, Weigel D (2011) Evolution of the S-locus region in *Arabidopsis thaliana* relatives. Plant Physiol 157:937–946

Haudry A, Cenci A, Ravel C, Bataillon T, Brunel D, Poncet C, Hochu I, Poirier S, Santoni S, Glémin S, David J (2007) Grinding up wheat: a massive loss of nucleotide diversity since domestication. Mol Biol Evol 24:1506–1517

Haudry A, Cenci A, Guilhaumon C et al (2008) Mating system and recombination affect molecular evolution in four Triticeae species. Genet Res 90:97–109

Heilbuth JC, Ilves KL, Otto SP (2001) The consequences of dioecy for seed dispersal: modeling the seed-shadow handicap. Evolution 55:880–888

Husse L, Billiard S, Lepart J, Vernet P, Saumitou-Laprade P (2013) A one-locus model of androdioecy in a context with two homomorphic self-incompatibility groups: expected vs. observed male frequencies. J Evol Biol 26(6):1269–1280

Igic B, Lande R, Kohn JR (2008) Loss of self-incompatiblity and its evolutionary consequences. Int J Plant Sci 169:93–104

Indriolo E, Tharmapalan P, Wright SI, Goring DR (2012) The *ARC1* E3 Ligase gene is frequently deleted in self-compatible Brassicaceae species and has a conserved role in *Arabidopsis lyrata* self-pollen rejection. Plant Cell 24:4607–4620

Ingvarsson PK (2002) A metapopulation perspective on genetic diversity and differentiation in partially self-fertilizing plants. Evolution 56:2368–2373

Ingvarsson PK, Taylor DR (2002) Genealogical evidence for epidemics of selfish genes. Proc Natl Acad Sci USA 99:11265–11269

Käfer J, Talianov M, Bigot T et al (2013) Patterns of molecular evolution in dioecious and non-dioecious Silene. J Evol Biol 26:335–346
Kaplan NL, Hudson RR, Langley CH (1989) The hitchhiking effect revisited. Genetics 123:887–899
Kim M, Cui ML, Cubas P, Gillies A, Lee K, Chapman MA, Abbott RJ, Coen E (2008) Regulatory genes control a key morphological and ecological trait transferred between species. Science 322(5904): 1116–1119
Kubo K, Entani T, Takara A, Wang N, Fields AM, Hua Z, Toyoda M, Kawashima S, Ando T, Isogai A, Kao TH, Takayama S (2010) Collaborative non-self recognition system in S-RNase-based self-incompatibility. Science 330(6005):796–799
Lahiani E, Dufaÿ M, Castric V, Le Cadre S, Charlesworth D, Van Rossum F, Touzet P (2013) Disentangling the effects of mating systems and mutation rates on cytoplamic diversity in gynodioecious Silene nutans and dioecious Silene otites. Heredity 111:157–164
Lahn BT, Page DC (1999) Four evolutionary strata on the human X chromosome. Science 286:964–967
Liu Z, Moore PH, Ma H et al (2004) A primitive Y chromosome in papaya marks incipient sex chromosome evolution. Nature 427:348–352
Lloyd DG (1975) The maintenance of gynodioecy and androdioecy in angiosperms. Genetica 45:325–339
Lloyd DG (1977) Genetic and phenotypic models of natural selection. J Theor Biol 69:543–560
Lloyd DG (1979) Evolution towards dioecy in heterostylous populations. Plant Syst Evol 131:71–80
Lloyd DG (1980) The distribution of gender in four angiosperm species illustrating two evolutionary pathways to dioecy. Evolution 34:123–134
Lloyd DG, Webb CJ (1992) The evolution of heterostyly. In: Barrett SCH (ed) Evolution and function of heterostyly. Springer, Berlin/Heidelberg/New York, pp 151–178
Lockton S, Gaut BS (2010) The evolution of transposable elements in natural populations of self-fertilizing *Arabidopsis thaliana* and its outcrossing relative *Arabidopsis* lyrata. BMC Evol Biol 10:10
Lynch M, Conery J, Burger R (1995a) Mutation accumulation and the extinction of small populations. Am Nat 146:489–518
Lynch M, Conery J, Burger R (1995b) Mutational meltdowns in sexual population. Evolution 49: 1067–1080
Maddison WP, Midford PE, Otto SP (2007) Estimating a binary character's effect on speciation and extinction. Syst Biol 56:701–710
Marais G (2003) Biased gene conversion: implications for genome and sex evolution. Trends Genet 19:330–338
Marais G, Charlesworth B, Wright SI (2004) Recombination and base composition: the case of the highly self-fertilizing plant *Arabidopsis thaliana*. Genome Biol 5:R45
Martin A, Orgogozo V (2013) The loci of repeated evolution: a catalog of genetic hotspots of phenotypic variation. Evolution 67(5):1235–1250
Martin A, Troadec C, Boualem A, Rajab M, Fernandez R, Morin H, Pitrat M, Dogimont C, Bendahmane A (2009) A transposon-induced epigenetic change leads to sex determination in melon. Nature 461(7267):1135–1138
Mayrose I, Barker MS, Otto SP (2010) Probabilistic models of chromosome number evolution and the inference of polyploidy. Syst Biol 59:132–144
McDaniel SF, Atwood J, Burleigh JG (2013) Recurrent evolution of dioecy in bryophytes. Evolution 67: 567–572
Ming R, Bendahmane A, Renner SS (2011) Sex chromosomes in land plants. Annu Rev Plant Biol 62: 485–514
Modliszewski JL, Willis JH (2012) Allotetraploid *Mimulus sookensis* are highly interfertile despite independent origins. Mol Ecol 21:5280–5298
Morgan MT (2001) Transposable element number in mixed mating populations. Genet Res 77:261–275
Muyle A, Zemp N, Deschamps C et al (2012) Rapid de novo evolution of X chromosome dosage compensation in *Silene latifolia*, a plant with young sex chromosomes. PLoS Biol 10:e1001308
Naithani S, Chookajorn T, Ripoll DR, Nasrallah JB (2007) Structural modules for receptor dimerization in the S-locus receptor kinase extracellular domain. Proc Natl Acad Sci USA 104:12211–12216
Nasrallah ME, Liu P, Nasrallah JB (2002) Generation of self-incompatible *Arabidopsis thaliana* by transfer of two S locus genes from *A. lyrata*. Science 297: 247–249
Nasrallah M, Liu P, Sherman-Broyles S, Boggs N, Nasrallah J (2004) Natural variation in expression of self-incompatibility in *Arabidopsis thaliana*: implications for the evolution of selfing. Proc Natl Acad Sci USA 101:16070–16074
Nasrallah JB, Liu P, Sherman-Broyles S, Schmidt R, Nasrallah ME (2007) Epigenetic mechanisms for breakdown of self-incompatibility in interspecific hybrids. Genetics 175:1965–1973
Ness RW, Wright SI, Barrett SCH (2010) Mating-system variation, demographic history and patterns of nucleotide diversity in the tristylous plant *Eichhornia paniculata*. Genetics 184:381–392
Nicolas M, Marais G, Hykelova V et al (2005) A gradual process of recombination restriction in the evolutionary history of the sex chromosomes in dioecious plants. PLoS Biol 3:e4
Nordborg M (2000) Linkage disequilibrium, gene trees and selfing: an ancestral recombination graph with partial self-fertilization. Genetics 154:923–929
Nordborg M, Hu TT, Ishino Y et al (2005) The pattern of polymorphism in *Arabidopsis thaliana*. PLoS Biol 3:1289–1299
Ornduff R (1969) Reproductive biology in relation to systematics. Taxon 18:121–133
Pannell JR, Verdu M (2006) The evolution of gender specialization from dimorphic hermaphroditism: paths from heterodichogamy to gynodioecy and androdioecy. Evolution 60:660–673

Pettengill JB, Moeller DA (2011) Tempo and mode of mating system evolution between incipient *Clarkia* species. Evolution 66:1210–1225

Porcher E, Lande R (2005a) The evolution of self-fertilization and inbreeding depression under pollen discounting and pollen limitation. J Evol Biol 18: 497–508

Porcher E, Lande R (2005b) Loss of gametophytic self-incompatibility with evolution of inbreeding depression. Evolution 59:46–60

Qiu S, Zeng K, Slotte T et al (2011) Reduced efficacy of natural selection on codon usage bias in selfing *Arabidopsis* and *Capsella* species. Genome Biol Evol 3:868–880

Renner SS, Ricklefs RE (1995) Dioecy and its correlates in the flowering plants. Am J Bot 82:596–606

Roselius K, Stephan W, Stadler T (2005) The relationship of nucleotide polymorphism, recombination rate and selection in wild tomato species. Genetics 171: 753–763

Ross MD (1982) Five evolutionary pathways to subdioecy. Am Nat 119:297–318

Ross-Ibarra J, Wright SI, Foxe JP et al (2008) Patterns of polymorphism and demographic history in natural populations of *Arabidopsis lyrata*. PLoS ONE 3: e2411

Roze D, Rousset F (2004) Joint effects of self-fertilization and population structure on the mutation load, inbreeding depression and heterosis. Genetics 167: 1001–1015

Saumitou-Laprade P, Vernet P, Vassiliadis C, Hoareau Y, de Magny G, Dommée B, Lepart J (2010) A self-incompatibility system explains high male frequencies in an androdioecious plant. Science 327(5973):1648–1650

Schierup MH, Vekemans X, Christiansen FB (1997) Evolutionary dynamics of sporophytic self-incompatibility alleles in plants. Genetics 147:835–846

Schoen DJ, Brown AHD (1991) Intraspecific variation in population gene diversity and effective population size correlates with the mating system in plants. Proc Natl Acad Sci USA 88:4494–4497

Segawa M, Kishi S, Tatuno S (1971) Sex chromosomes of *Cycas revoluta*. Jpn J Genet 46:33–39

Shimizu KK, Cork JM, Caicedo AL et al (2004) Darwinian selection on a selfing locus. Science 306:2081–2084

Shimizu KK, Shimizu-Inatsugi R, Tsuchimatsu T, Purugganan MD (2008) Independent origins of self-compatibility in *Arabidopsis thaliana*. Mol Ecol 17:704–714

Sicard A, Stacey N, Hermann K, Dessoly J, Neuffer B, Bäurle I, Lenhard M (2011) Genetics, evolution, and adaptive significance of the selfing syndrome in the genus Capsella. Plant Cell 23(9):3156–3171

Slotte T, Foxe JP, Hazzouri KM, Wright SI (2010) Genome-wide evidence for efficient positive and purifying selection in *Capsella grandiflora*, a plant species with a large effective population size. Mol Biol Evol 27:1813–1821

Slotte T, Hazzouri KM, Stern D, Andolfatto P, Wright SI (2012) Genetic architecture and adaptive significance of the selfing syndrome in *Capsella*. Evolution 66(5):1360–1374

Slotte T, Hazzouri KM, Ågren JA, Koenig D, Maumus F, Guo Y, Steige K, Platts AE, Escobar JS, Newman LK, Wang W, Mandakova T, Vello E, Smith SM, Steffen J, Takuno S, Brandvain Y, Coop G, Andolfatto P, Hu TT, Blanchette M, Clark RM, Queseville H, Nordborg M, Gaut BS, Lysak MA, Jenkins J, Grimwood J, Prochnick S, Shu S, Rokhsar D, Schmutz J, Weigel D, Wright SI (2013) The *Capsella rubella* genome and the genomic consequences of rapid mating system evolution. Nat Genet 45:831–835

Spigler RB, Lewers KS, Main DS, Ashman TL (2008) Genetic mapping of sex determination in a wild strawberry, *Fragaria virginiana*, reveals earliest form of sex chromosome. Heredity 101:507–517

St Onge KR, Kallman T, Slotte T et al (2011) Contrasting demographic history and population structure in *Capsella rubella* and *Capsella grandiflora*, two closely related species with different mating systems. Mol Ecol 20:3306–3320

Stadler T (2011) Inferring speciation and extinction processes from extant species data. Proc Natl Acad Sci USA 108:16145–16146

Stadler T, Delph LF (2002) Ancient mitochondrial haplotypes and evidence for intragenic recombination in a gynodioecious plant. Proc Natl Acad Sci USA 99:11730–11735

Stebbins GL (1957) Self fertilization and population variability in higher plants. Am Nat 91:337–354

Stebbins GL (1974) Flowering plants: evolution above the species level. Harvard University Press, Cambridge

Takayama S, Isogai A (2005) Self-incompatibility in plants. Annu Rev Plant Biol 56:467–489

Takebayashi N, Morrell PL (2001) Is self-fertilization an evolutionary deed end? Revisiting an old hypothesis with genetic theories and a macroevolutionary approach. Am J Bot 88:1143–1150

Tam SM, Causse M, Garchery C et al (2007) The distribution of copia-type retrotransposons and the evolutionary history of tomato and related wild species. J Evol Biol 20:1056–1072

Tang C, Toomajian C, Sherman-Broyles S, Plagnol V, Guo YL et al (2007) The evolution of selfing in *Arabidopsis thaliana*. Science 317:1070–1072

Tantikanjana T, Rizvi N, Nasrallah ME, Nasrallah JB (2009) A dual role for the S-locus receptor kinase in self-incompatibility and pistil development revealed by an *Arabidopsis rdr6* mutation. Plant Cell 21(9):2642–2654

Torices R, Méndez M, Gomez MJ (2011) Where do monomorphic sexual systems fit in the evolution of dioecy? Insights from the largest family of angiosperms. New Phytol 190:234–248

Touzet P (2012) Mitochondrial genome evolution and gynodioecy. In: Marechal-Drouard L (ed) Advances in botanical research, vol 63. Academic Press/Elsevier Science, London, pp 71–98

Touzet P, Budar F (2004) Unveiling the molecular arms race between two conflicting genomes in cytoplasmic male sterility? Trends Plant Sci 9:568–570

Touzet P, Delph LF (2009) The effect of breeding system on polymorphism in mitochondrial genes of Silene. Genetics 181:631–644

Tsuchimatsu T, Suwabe K, Shimizu-Inatsugi R et al (2010) Evolution of self-compatibility in *Arabidopsis* by a mutation in the male specificity gene. Nature 464:1342–1346

Tsuchimatsu T, Kaiser P, Yew C-L, Bachelier JB, Shimizu KK (2012) Recent loss of self-incompatibility by degradation of the male component in allotetraploid *Arabidopsis kamchatica*. PLoS Genet 8(7):e1002838

Uyenoyama MK (2003) Genealogy-dependent variation in viability among self-incompatibility genotypes. Theor Popul Biol 63:281–293

Uyenoyama M, Zhang Y, Newbigin E (2001) On the origin of self-incompatibility haplotypes: transition through self-compatible intermediates. Genetics 157:1805–1817

Vamosi JC, Otto SP (2002) When looks can kill: the evolution of sexually dimorphic floral display and the extinction of dioecious plants. Proc R Soc Lond B 269:1187–1194

van Doorn GS, Kirkpatrick M (2007) Turnover of sex chromosomes induced by sexual conflict. Nature 449:909–912

Vekemans X, Slatkin M (1994) Gene and allelic genealogies at a gametophytic self-incompatibility locus. Genetics 137:1157–1165

Wang J, Naa JK, Yub Q et al (2012) Sequencing papaya X and Y chromosomes reveals molecular basis of incipient sex chromosome evolution. Proc Natl Acad Sci USA 109:13710–13715

Wright SI, Schoen DJ (1999) Transposon dynamics and the breeding system. Genetica 107:139–148

Wright SI, Le QH, Schoen DJ, Bureau TE (2001) Population dynamics of an Ac-like transposable element in self- and crosspollinating *Arabidopsis*. Genetics 158:1279–1288

Wright SI, Lauga B, Charlesworth D (2002) Rates and patterns of molecular evolution in inbred and outbred *Arabidopsis*. Mol Biol Evol 19:1407–1420

Wright S, Ness RW, Foxe JP, Barrett SC (2008) Genomic consequences of outcrossing and selfing in plants. Int J Plant Sci 169:105–118

Revisiting Mortimer's Genome Renewal Hypothesis: Heterozygosity, Homothallism, and the Potential for Adaptation in Yeast

3

Paul M. Magwene

Abstract

In diploid organisms, the frequency and nature of sexual cycles have a major impact on genome-wide patterns of heterozygosity. Recent population genomic surveys in the budding yeast, *Saccharomyces cerevisiae*, have revealed surprising levels of genomic heterozygosity in what has been traditionally considered a highly inbred organism. I review evidence and hypotheses regarding the generation, maintenance, and evolutionary consequences of genomic heterozygosity in *S. cerevisiae*. I propose that high levels of heterozygosity in *S. cerevisiae*, arising from population admixture due to human domestication, coupled with selfing during rare sexual cycles, can facilitate rapid adaptation to novel environments.

Keywords

Heterozygosity • Adaptation • Mating systems • Domestication • Admixture

3.1 Introduction

In 1994 the pioneering yeast geneticist Robert Mortimer proposed the "Genome Renewal Hypothesis" to explain patterns of genetic variation observed in the budding yeast *Saccharomyces cerevisiae* (Mortimer et al. 1994; Mortimer 2000). Mortimer and colleagues observed that most yeast strains isolated from vineyards were diploid and heterozygous at one or more loci. The vast majority were also homothallic, meaning that haploid cells produced from these strains were capable of undergoing mating-type switching followed by mother-daughter mating. This process, known as autodiplodization or haploselfing, leads to diploid cells that are homozygous at all but the mating type locus. Mortimer documented a negative correlation between the number of detectable heterozygosities in vineyard isolates and the percentage of viable spores produced; homozygous isolates had nearly 100 % spore viability while heterozygous isolates showed clear evidence for deleterious or sometimes lethal alleles. Finally, isolates that were homozygous were inferred to have

P.M. Magwene (✉)
Department of Biology, Duke University, Box 90338, Durham, NC 27708, USA
e-mail: paul.magwene@duke.edu

C.R. Landry and N. Aubin-Horth (eds.), *Ecological Genomics: Ecology and the Evolution of Genes and Genomes*, Advances in Experimental Medicine and Biology 781,
DOI 10.1007/978-94-007-7347-9_3, © Springer Science+Business Media Dordrecht 2014

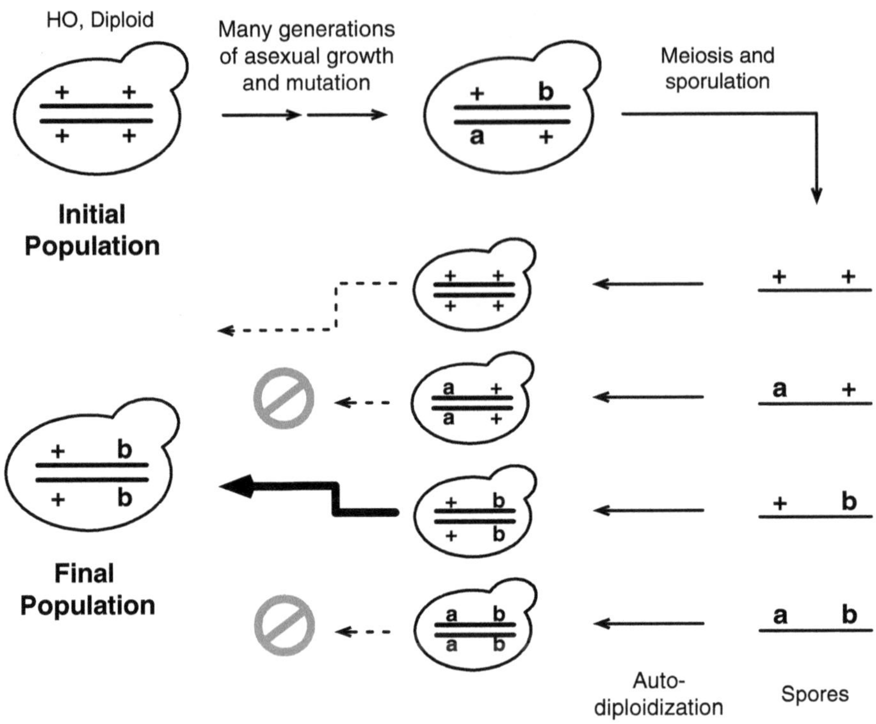

Fig. 3.1 A schematic illustration of Mortimer's Genome Renewal Hypothesis. This figure illustrates the key features of the scenario Mortimer described for Genome Renewal, starting from a homothallic (HO) diploid background. The *pluses* indicate wild-type alleles, while '*a*' and '*b*' indicate recessive alleles arising during periods of asexual propagation. In this example, '*a*' is a deleterious allele, while '*b*' is a beneficial

been derived from heterozygous backgrounds via autodiploidization. Mortimer and colleagues proposed that these observations could be explained by an evolutionary scenario involving long periods of clonal reproduction in which diploid strains accumulated recessive, primarily deleterious alleles in a heterozygous state. They posited that rare sexual cycles involving meiosis followed by mating type switching and autodiploidization would facilitate the loss of deleterious alleles and fix beneficial alleles, thus leading to "Genome Renewal" (Fig. 3.1).

Recently, Masel and Lyttle (2011) developed a mathematical model of evolution under a life history regime like that proposed by Mortimer. This model, which assumed a small number of heterozygous sites and additivity of allelic effects, considered the effect of different mating strategies on heritable genetic variation for different selection regimes and for a range of dominance coefficients. Masel and Lyttle showed that clonal reproduction coupled with rare selfing can lead to an epistatic increase in heritable phenotypic variation per sexual episode, relative to an outcrossing strategy. They concluded that clonal expansion coupled with rare selfing can thus act as a type of 'evolutionary capacitor', allowing cryptic genetic variation to build up during periods of clonality, and exposing that variation to selection following periods of environmental stress that are severe enough to induce sexual cycles (Masel and Lyttle 2011).

The Genome Renewal hypothesis is based on the assumption that the 'ground state' is genomic homozygosity, and the number of heterozygous loci that accumulate between periods of selfing should therefore be modest. However, recent genome sequencing of environmental isolates of *S. cerevisiae* has revealed that many strains harbor abundant polymorphism in the form of thousands of heterozygous sites across the genome (Magwene et al. 2011; Borneman et al. 2011).

In the following pages I re-examine the Genome Renewal hypothesis in light of this discovery, focusing in particular on the implications of extensive heterozygosity coupled with homothallism with respect to adaptation to new niches. I argue that, for highly heterozygous homothallic strains, the adaptive evolutionary landscape has a high degree of "accessibility" because offspring that sample large regions of genotypic and phenotypic space can be generated rapidly from a single founding individual.

3.2 Evidence for the Genome Renewal Hypothesis

Since Mortimer and colleagues first put forth the Genome Renewal Hypothesis, a large number of studies have contributed to an increasingly detailed portrait of population genetic and genomic variation in *S. cerevisiae* (e.g. Fay and Benavides 2005; Gresham et al. 2006; Liti et al. 2009; Schacherer et al. 2009; Skelly et al. 2009). Below I touch on only a fraction of this literature, that which bears most directly on the Genome Renewal Hypothesis. For a more exhaustive overview of yeast population genetics and genomics I refer the reader to several recent reviews (Liti and Schacherer 2011; Sipiczki 2011; Hittinger 2013).

3.2.1 Most Environmental Isolates of *S. cerevisiae* Are Diploid and Homothallic

Saccharomyces cerevisiae has a haplo-diploid life cycle, and can propagate asexually as either haploid or diploid cells. Despite the ability to grow vegetatively in a haploid state, *S. cerevisiae* is predominantly isolated from the environment as diploid cells, though aneuploid and polyploid isolates are not uncommon (Guijo et al. 1997). For example, Cubillos et al. (2009) found that approximately 95 % of the 200+ wine strains they examined had a DNA content consistent with diploidy. Muller and McCusker (2009), based on a diverse sample of 170 isolates, estimated that 70–80 % of their strains were diploid, with the remaining 20–30 % of strains being triploid or tetraploid. Similarly, Sniegowski et al. (2002) show that each of the ten *S. cerevisiae* strains they isolated from an oak forest were diploid.

With respect to homothallism, Mortimer (2000) found that approximately 89 % of the wine isolates he analyzed were homothallic, and Cubillos et al. (2009) similarly found that the majority of the wine strains they characterized were homothallic. All of the oak isolates studied by Sniegowski et al. (2002) were homothallic. The survey by Muller and McCusker (2009), which included strains from a wider variety of environmental contexts, paints a somewhat more complicated picture of homothallism. Twenty-seven of the twenty-eight non-clinical isolates they examined were homothallic, but four of the eight clinical isolates the examined were heterothallic. Muller and McCusker suggested that an increased frequency of heterothallism in clinical isolates might result from indirect selection associated with a selective advantage for heterozygosity in clinical environments.

3.2.2 Patterns of Heterozygosity in *S. cerevisiae*

Mortimer's assessments of heterozygosity in wine isolates was based on segregation of phenotypic traits such as the ability to grow on different carbon sources, growth rate mutations, etc. (Mortimer et al. 1994). Mortimer and colleagues found that roughly 65 % of the more than 200 strains analyzed were heterozygous at one or more loci, as determined by tetrad analysis.

Subsequent studies that have reported data on heterozygosity in environmental isolates of *S. cerevisiae* have primarily focused on molecular genotyping. For example, Fay and Benavides (2005) analysis of 81 strains indicated that approximately 40 % of the strains they characterized were heterozygous for at least one of five loci. Muller and McCusker (2009), based on data from 12 microsatellite markers, found that approximately 80 % of their strains were heterozy-

gous for at least one locus, with clinical isolates having higher average heterozygosity than non-clinical strains. Similarly, Diezmann and Dietrich (2009) undertook a population genetic survey of 103 *S. cerevisiae* strains at five loci and found that between 33 and 88 % of strains from human associated environments (clinical, brewery, fruit) were heterozygous at one or more loci. In contrast all of the soil or bark isolates were homozygous at all loci examined, leading them to conclude that the soil isolates represent lineages that have experienced no or little outcrossing in contrast to the more clearly recombinant strains associated with agricultural and clinical settings.

These studies, based on a modest number of loci, established that heterozygosity is relatively common in *S. cerevisiae* isolates. However, the initial sequencing of the yeast genome (Goffeau et al. 1996; Wei et al. 2007) and population genomic studies (Gresham et al. 2006; Schacherer et al. 2009; Liti et al. 2009) used strains derived from monosporic derivatives, thus making it impossible to characterize heterozygosity on a genome-wide scale. Therefore the genomic extent of heterozygosity wasn't appreciated until the first studies describing the sequencing of unmanipulated diploid genomes were published. The first report of this kind was the characterization of the genome of a diploid strain used in bioethanol production (Argueso et al. 2009). Argueso et al. (2009) arrived at an estimate of $\sim$2 heterozygous SNPs per Kb, corresponding to at least $\sim$24,000 heterozygous sites genome wide. Shortly thereafter the first population genomic surveys of unmanipulated diploids demonstrated that many *S. cerevisiae* strains isolated from both industrial and non-industrial contexts were highly heterozygous, many possessing greater than 30,000 heterozygous sites across the genome (Magwene et al. 2011; Borneman et al. 2011). For example, Magwene et al. (2011), based on whole genome sequencing of 11 diploid isolates plus genotyping of 9 loci in 18 additional strains, concluded that approximately 60 % of strains they examined had modest (>5,000 sites) to extensive (>15,000 sites) genomic heterozygosity. Subsequent genome sequencing of additional environmental isolates has led to similar findings (Akao et al. 2011; Babrzadeh et al. 2012; Hyma and Fay 2013).

Several authors have noted that strains with high levels of heterozygosity are preferentially isolated from human associated environments (Diezmann and Dietrich 2009; Muller and McCusker 2009; Magwene et al. 2011). Clinical and industrial isolates stand out in this regard, but a number of strains isolated from agricultural contexts, such as fruit trees, also have high levels of heterozygosity (Diezmann and Dietrich 2009; Hyma and Fay 2013; Magwene et al. 2011). By contrast, *S. cerevisiae* isolated from relatively undisturbed environments such as oak forests in both North America and Asia have very low levels of heterozygosity (Kuehne et al. 2007; Wang et al. 2012). Similarly, isolates of the undomesticated sister species, *Saccharomyces paradoxus*, also exhibit very little genomic heterozygosity (Tsai et al. 2008). These data suggest that for budding yeast homozygosity may be the rule in the absence of human domestication.

3.2.3 What Generates and Maintains Heterozygosity?

The nature and extent of genomic heterozygosity that sequencing of diploid strains has revealed is surprising given that previous studies (Ruderfer et al. 2006) had suggested that sex in yeast primarily involves inbreeding via intratetrad mating (Tsai et al. 2008). Inbreeding of any type quickly leads to loss of heterozygosity, and mating type switching by halplo-selfing immediately homozygoses the entire genome except at the mating type locus (Kirby 1984). Magwene et al. (2011) concluded that the extensive heterozygosity seen in many strains most likely resulted from outcrossing between genetically diverse lineages. This seems to be at odds with studies that have estimated that outcrossing occurs only about once every 50,000–100,000 mitotic generations in yeast (Ruderfer et al. 2006; Tsai et al. 2008). However, population genomic analyses demonstrate quite clearly that there has

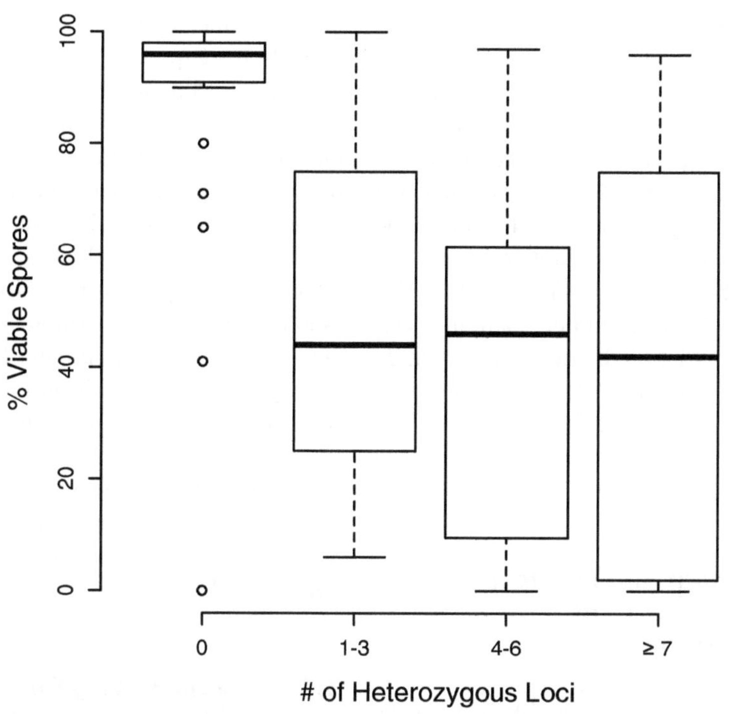

Fig. 3.2 Relationship between the number of heterozygous loci and spore viability for 108 diploid strains, based on data from Muller and McCusker (2009). The rank correlation between heterozygosity and percent spore viability is −0.5

been significant admixture between *S. cerevisiae* lineages (Liti et al. 2009; Schacherer et al. 2009), and a recent study suggests that outcrossing in *S. cerevisiae* may be considerably higher than previous estimates (Kelly et al. 2012). Regardless of the actual rates of outcrossing, the very high levels of heterozygosity observed in clinical, industrial, and many agricultural isolates begs the question of what contributes to the maintenance of heterozygosity? Magwene et al. (2011) suggested that differences between heterozygous and homozygous strains might be the result of alternate life history strategies that favor different frequencies of sexual cycles, and proposed that heterozygous strains represent lineages that are less likely to undergo sexual cycles, thus preserving heterozygosity. Heterothallism should also tend to favor the preservation of heterozygosity; as noted above, clinical isolates of *S. cerevisiae*, which tend to have among the highest levels of genome-wide heterozygosity, show a trend towards a greater frequency of heterothallism (Muller and McCusker 2009).

3.2.4 Heterozygosity and Spore Viability

One of the findings that motivated the Genome Renewal hypothesis was the observation that spore viablity tended to decrease with increased heterozygosity (Mortimer et al. 1994). Mortimer found that strains that were completely homozygous sporulated at high frequency and had spore viabilities near 100 %. By contrast approximately 47 % of strains that were heterozygous at one or more loci were also heterozygous for lethal or deleterious alleles (Mortimer 2000). In Fig. 3.2 I present a reanalysis of data from Muller and McCusker (2009) that supports the pattern suggested by Mortimer; spore viability is negatively correlated with the number of heterozygous loci (Fig. 3.2; Spearman rank correlation $\rho = -0.50$, $p < 0.0001$ by permutation test).

Some of this reduction of spore viability is likely due to recessive deleterious or lethal alleles as suggested by Mortimer, however this may also

reflect incompatible genetic combinations arising under reproductive isolation between the backgrounds that contributed to the formation of the heterozygotes (Cubillos et al. 2011). Such incompatibilities may result from both large scale genomic changes (e.g. aneuploidy) between strain backgrounds or may involve single nucleotide changes, either neutral or adaptive, that arise under reproductive isolation. For example, Demogines et al. (2008) identified naturally segregating allelic variation in two genes, *MLH1* and *PMS1*, involved in DNA mismatch repair. Particular combinations of alleles at these two loci result in a low-fitness mutator phenotype (Demogines et al. 2008).

3.3 Molecular and Phenotypic Consequences of Heterozygosity

High levels of heterozygosity are likely to have a significant impact on molecular interactions. For example, the strain EM93 is the primary ancestor of the standard reference strain S288c and has more than 24,000 heterozygous sites (Magwene et al. 2011; Esberg et al. 2011). Table 3.1 shows the predicted impact of this heterozygosity on the proteome; approximately 37 % of proteins in EM93 are present as two different peptide sequences. While it is hard to know how much of this protein polymorphism has functional effects, this nevertheless represents a very large pool of variation present within a single strain background. There is also abundant heterozygosity in non-coding regions, which has the potential to affect regulatory networks through effects such as allele specific gene expression (Gagneur et al. 2009) and differential protein-DNA interactions (Zheng et al. 2010). At the level of cellular phenotypes, extensive heterozygosity might contribute to cell-to-cell heterogeneity in clonal populations, through mechanisms such as allele specific gene and protein expression (Levy et al. 2012) and differential regulation of epigenetic silencing (Halme et al. 2004).

Table 3.1 Estimated number of heterozygous proteins per chromosome in the genome of the strain EM93

Chr	*#ORFs*	% ORFs
I	44	38
II	101	22
III	60	33
IV	174	21
V	117	36
VI	75	53
VII	233	40
VIII	109	34
IX	119	49
X	192	48
XI	191	55
XII	87	15
XIII	223	44
XIV	211	49
XV	285	48
XVI	191	37
Total	2,412	37

3.4 Evolutionary Consequences of Heterozygosity and Homothallism

Mortimer's Genome Renewal hypothesis and related models (Masel and Lyttle 2011; Sipiczki 2011) primarily consider the case of accumulation of a modest number of heterozygous sites against an otherwise homozygous genomic background. However, as detailed above, many environmental isolates of *S. cerevisiae* are heterozygous at more than 30,000 sites across the genome. How does this degree of heterozygosity modify our view of the Genome Renewal hypothesis?

Consider the following scenario – a single heterozygous individual is introduced into a novel environment and generates a large clonal population. When nutrients become limiting the population undergoes meiosis and sporulation (most heterozygous isolates are slow to sporulate but eventually do so under extended nutrient limitation; Magwene et al. 2011). The meiotic products generated from this clonal population represent a sampling of a very large combinatorial space – the 2^n possible allelic combinations representing the alternative alleles at heterozygous sites in the founder individual (even accounting for linkage disequilibrium, n is likely to be in the hundreds

to thousands). Assuming subsequent germination of those spores, either through reintroduction of nutrients or dispersal to nearby nutrient rich environments (e.g. by insect vectors; Stefanini et al. 2012), haploids will return to diploidy either through mating type switching and autodiploidization or through intra- or intertetrad matings. Due to a high frequency of selfing and other forms of inbreeding, a large number of allelic combinations will be exposed to local selection in a primarily homozygous state, thus increasing the probability that one or more favorable genotypes will become established and thrive in the new environment.

How likely are such introductions and how much potential phenotypic variability can be exposed to selection under such a scenario? The likelihood of introductions is hard to assess given the challenges of studying yeast ecology in a natural setting, but *S. cerevisiae* is commonly isolated during environmental sampling in a wide variety of contexts (Hyma and Fay 2013; McCusker et al. 1994; Naumov et al. 1998; Sweeney et al. 2004) and such introductions are likely to be facilitated, either purposefully or inadvertantly, by human activity (Goddard et al. 2010).

The "phenotypic potential" of heterozygous, homothallic strains can be addressed directly in the laboratory by sporulating such strains, germinating the spores at low enough density to induce autodiploidization, and assessing the phenotypes of the resulting homozygous offspring. Figure 3.3 shows an example of carrying out such an experiment on a highly heterozygous clinical isolate, YJM311 (McCusker et al. 1994). Each of the subfigures represents the distribution of a different phenotype of interest in homozygous offspring of YJM311. There is abundant phenotypic variation for each of the traits and the multivariate phenotypic space represented by these offspring is equally rich (Magwene, unpublished data). Illustrative of the genetic and phenotypic diversity of such strains, my laboratory has recently used such a population to map QTLs for biofilm formation in *S. cerevisiae* (Granek et al. 2013).

Given the very large number of allelic combinations that can be generated from a single individual, and the corresponding phenotypic variability that accompanies this genetic variation, I argue that the evolutionary landscape can be thought of as relatively "accessible" for heterozygous, homothallic strains. For highly heterozygous and homothallic strains the phenotypes of their offpsring can be radically different than their own, and the ability to generate large clonal populations means that large regions of genotypic and phenotypic space can be potentially sampled in a single sexual generation. These properties therefore seem likely to promote invasion of new niches and adaptation to novel environments. The range of phenotypes accessible from a heterozygous founder should generally increase as a function of the number of heterozygous sites and the magnitude of epistatic interactions between loci, but should be negatively correlated with the average degree of dominance (Masel and Lyttle 2011) (Fig. 3.4).

3.5 Admixture, Heterozygosity, and Domestication

Saccharomyces cerevisiae has been at the forefront of studies of yeast population genetics but in recent years several additional species within the *Saccharomyces sensu stricto* complex have begun to garner similar attention (Libkind et al. 2011; Liti et al. 2009). Thus far, *S. cerevisiae* seems unique in the degree of admixture that has occurred between lineages. The amount of heterozygosity, and the genetic structure of geographically distinct populations in species such as *Saccharomyces paradoxus* seems consistent with what one would expect based on selfing followed by slow accumulation of heterozygous mutations during extended periods of asexual growth (Tsai et al. 2008).

It is likely that admixture between divergent *S. cerevisiae* lineages, and the resultant heterozygosity, has been both a consequence of and a contributor to domestication in *S. cerevisiae*. Human activity can bring divergent lineages into contact, facilitating outcrossing that establishes heterozygosity. Subsequently, the adaptive potential of such heterozygotes may have facilitated

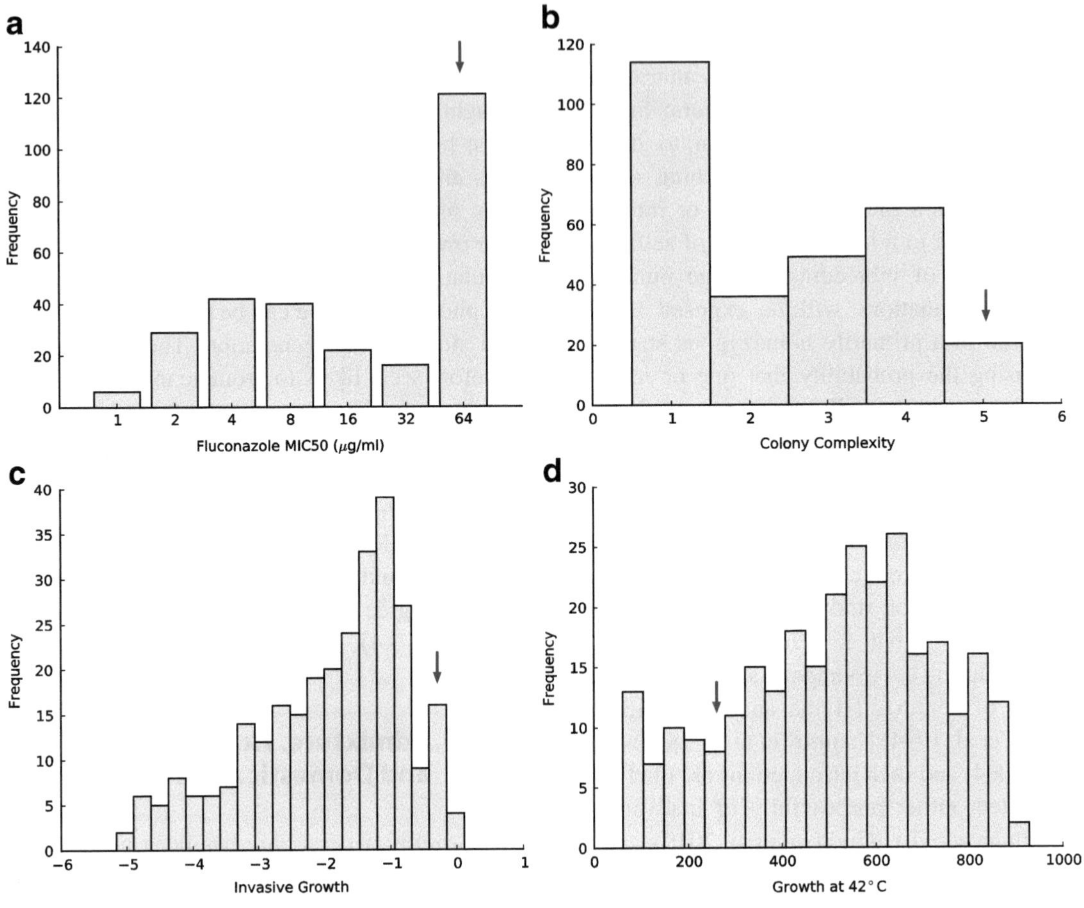

Fig. 3.3 Phenotypic distributions of four traits among offspring of the highly heterozygous clinical isolate YJM311. Two hundred and eighty eight homozygous diploid offspring were generated by sporulation followed by autodiploidization. Traits assessed include: (**a**) resistance to the antifungal drug fluconazole (MIC_{50} as determined in microtiter plates); (**b**) a measure of colony biofilm complexity (Granek et al. 2013); (**c**) a measure of invasive growth on agar substrates (logarithm of the ratio of post-wash to pre-wash colony density; Drees et al. 2005); and (**d**) a measure of growth at high temperature on agar plates (square root of the mean spot density for three replicates of each segregant grown at 42 °C). The *arrow* in each subfigure indicates the phenotype of the parental strain YJM311

selection for traits that are desirable for brewing, baking, and industrial uses. However, these same processes may help to facilitate *S. cerevisiae* adaptation to new, less beneficial (from a human perspective), niches. For example, *S. cerevisiae* is an emerging human pathogen (McCusker 2006), and clinical isolates are frequently highly heterozygous (Magwene et al. 2011).

It is also interesting to consider the extent to which heterozygosity and homothallism may facilitate the establishment of interspecific hybrids in the *Saccharomyces* genus. Prezygotic barriers to hybridzation are relatively weak and hybrids show robust vegetative growth (Morales and Dujon 2012). However, most yeast hybrids are sterile, primarily as a result of chromosomal translocations that distinguish the different species and which lead to incomplete meiosis in hybrids (Delneri et al. 2003; Fischer et al. 2000). However, if rare compatible combinations of alleles from the two species are also homothallic (Greig et al. 2002), then the resulting heterozy-

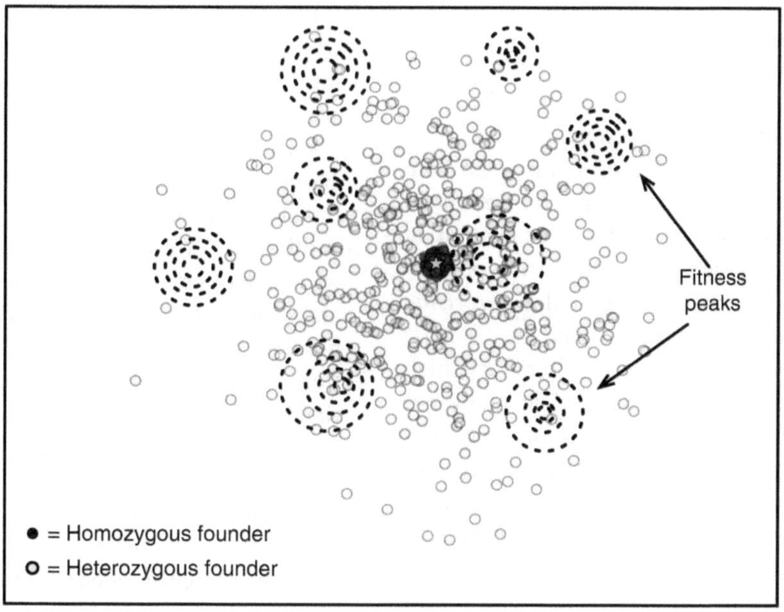

Fig. 3.4 A hypothetical evolutionary landscape for two different populations generated from single founding individuals. The plane of the figure represents a high dimensional phenotypic space, and the concentric dashed circles represent local fitness optima. The light grey circles represent individuals generated from a highly heterozygous founder; the black circles represent derivatives of a homozygous founder. The founder phenotypes are identical for both populations (indicated with a star). The populations are assumed to have been generated by initial clonal growth (with mutation) followed by a single round of meiosis and sporulation coupled with subsequent mating and/or haploselfing

gosity from such interspecies crosses may play a critical role in subsequent adaptation and/or domestication. For example, it has been shown that *Saccharomyces bayanus* is a complex hybrid of three species – *S. eubayanus*, *S. uvarum* and *S. cerevisiae*. The type strain for *S. bayanus*, CBS 380, has significant heterozygosity that reflects the contributions of the different parental species (Libkind et al. 2011).

3.6 Genome Renewal in Other Fungi?

How might a modified model of genome renewal apply more broadly to other fungal lineages? Particularly interesting in this regard is recent work on the fungal pathogen *Candida albicans*. *C. albicans* is primarily isolated as a diploid, and most strains of *C. albicans* have extensive genomic heterozygosity (Jones et al. 2004). However, unlike *S. cerevisiae*, neither a true sexual cycle or haploids had been described in *C. albicans* until recently when Hickman et al. (2013) showed that viable haploid cells can arise from diploids, presumably through a mechanism of concerted chromosome loss. These haploid cells can mate with cells of the opposite mating type as well as autodiploidize. However, in contrast to *S. cerevisiae*, the homozygous diploids examined tended to have reduced fitness (though it should be noted that this observation is based on derivatives of a single diploid background). Like Mortimer, Hickman et al. (2013) propose that a temporary reduction to haploidy should help to purge recessive deleterious alleles.

3.7 Conclusions

The standard version of the Genome Renewal hypothesis is that infrequent sexual cycles, characterized by a high degree of selfing, can help to purge deleterious alleles and fix beneficial alleles, thus helping to facilitate adaptation in yeast. However, recent discoveries from population genomic sequencing of natural *S. cerevisiae* strains has forced us to re-evaluate the Genome Renewal hypothesis to account for very high levels of heterozygosity observed in many environmental isolates. The hypothesis put forth here is that extensive heterozygosity coupled with clonal expansion and homothallism can act as an engine for adaptation by greatly increasing the size of the genotypic and phenotypic spaces that can rapidly be explored in a single sexual generation in a population founded by a single individual. Further studies involving population genomic sequencing, experimental evolution, and microbial ecology will be needed to determine the extent to which this hypothesis holds and whether the patterns observed here are peculiar to *S. cerevisiae* or if they extend to other taxa that operate in a similar ecological and evolutionary milieu of clonal growth, selfing, and admixture.

Acknowledgements I thank Helen Murphy and Cliff Zeyl for discussions and feedback, as well as two anonymous reviewers for their critical comments and suggestions on the manuscript. Ludo Muller and John McCusker kindly provided data on spore viability and heterozygosity. The work was supported in part by awards from NSF (MCB-0614959) and NIH (P50GM081883-01).

Glossary

Admixture Interbreeding between two or more genetically distinct populations.

Autodiploidization Mating between mother and daughter haploid cells following mating type switching. Autodiploidization leads to diploid cells that are completely homozygous across the genome, except at the mating type locus.

Heterothallic Yeast strains that are incapable of undergoing mating type switching as haploid cells are referred as heterothallic. Such strains can be stably propagated as either haploids or diploids. Heterothallism in *S. cerevisiae* is usually a result of loss-of-function mutations at the *HO* locus.

Homothallic Yeast strains that are capable of undergoing mating type switching and autodiploidization are referred to as homothallic. The haploid phase is usually transient in homothallic strains because they rapidly switch mating types and initiate mother-daughter cell matings, leading to diploidy.

Mating type switching In *S. cerevisiae* and related yeast, mating type switching is facilitated by a high-frequency, site specific gene conversion events at the mating type (MAT) locus, induced by a site-specific endonuclease called HO, and involving silenced mating type sequences at 'hidden' MAT loci. in See Haber (1998) for a review of the mechanisms underlying mating type switching.

Tetrad analysis During sporulation, haploid spores are packaged together in a structure called an ascus. The ascus, plus the four haploid spores, are referred to as a tetrad. Tetrads can be teased apart with a micromanipulator following enzymatic digestion of the ascus. Subsequent germination of spores and scoring of phenotypes facilitates genetic analyses such as linkage mapping and distinguishing between Mendelian and non-Mendelian traits.

References

Akao T, Yashiro I, Hosoyama A, Kitagaki H, Horikawa H, Watanabe D, Akada R, Ando Y, Harashima S, Inoue T, Inoue Y, Kajiwara S, Kitamoto K, Kitamoto N, Kobayashi O, Kuhara S, Masubuchi T, Mizoguchi H, Nakao Y, Nakazato A, Namise M, Oba T, Ogata T, Ohta A, Sato M, Shibasaki S, Takatsume Y, Tanimoto S, Tsuboi H, Nishimura A, Yoda K, Ishikawa T, Iwashita K, Fujita N, Shimoi H (2011) Whole-genome sequencing of sake yeast *Saccharomyces cerevisiae* Kyokai no. 7. DNA Res 18(6):423–434

Argueso JL, Carazzolle MF, Mieczkowski PA, Duarte FM, Netto OVC, Missawa SK, Galzerani F, Costa GGL, Vidal RO, Noronha MF, Dominska M, Andrietta MGS, Andrietta SR, Cunha AF, Gomes LH, Tavares FCA, Alcarde AR, Dietrich FS, McCusker JH, Petes TD, Pereira GAG (2009) Genome structure of a *Saccharomyces cerevisiae* strain widely used in bioethanol production. Genome Res 19(12):2258–2270

Babrzadeh F, Jalili R, Wang C, Shokralla S, Pierce S, Robinson-Mosher A, Nyren P, Shafer RW, Basso LC, de Amorim HV, de Oliveira AJ, Davis RW, Ronaghi M, Gharizadeh B, Stambuk BU (2012) Whole-genome sequencing of the efficient industrial fuel-ethanol fermentative *Saccharomyces cerevisiae* strain CAT-1. Mol Genet Genomics 287(6):485–494

Borneman AR, Desany BA, Riches D, Affourtit JP, Forgan AH, Pretorius IS, Egholm M, Chambers PJ (2011) Whole-genome comparison reveals novel genetic elements that characterize the genome of industrial strains of *Saccharomyces cerevisiae*. PLoS Genet 7(2):e1001287

Cubillos FA, Vásquez C, Faugeron S, Ganga A, Martínez C (2009) Self-fertilization is the main sexual reproduction mechanism in native wine yeast populations. FEMS Microbiol Ecol 67(1):162–170

Cubillos FA, Billi E, Zörgö E, Parts L, Fargier P, Omholt S, Blomberg A, Warringer J, Louis EJ, Liti G (2011) Assessing the complex architecture of polygenic traits in diverged yeast populations. Mol Ecol 20(7): 1401–1413

Delneri D, Colson I, Grammenoudi S, Roberts IN, Louis EJ, Oliver SG (2003) Engineering evolution to study speciation in yeasts. Nature 422(6927): 68–72

Demogines A, Wong A, Aquadro C, Alani E (2008) Incompatibilities involving yeast mismatch repair genes: a role for genetic modifiers and implications for disease penetrance and variation in genomic mutation rates. PLoS Genet 4(6):e1000103

Diezmann S, Dietrich FS (2009) *Saccharomyces cerevisiae*: population divergence and resistance to oxidative stress in clinical, domesticated and wild isolates. PLoS One 4(4):e5317

Drees BL, Thorsson V, Carter GW, Rives AW, Raymond MZ, Avila-Campillo I, Shannon P, Galitski T (2005) Derivation of genetic interaction networks from quantitative phenotype data. Genome Biol 6(4):R38

Esberg A, Muller LAH, McCusker JH (2011) Genomic structure of and genome-wide recombination in the *Saccharomyces cerevisiae* S288C progenitor isolate EM93. PLoS One 6(9):e25211

Fay JC, Benavides JA (2005) Evidence for domesticated and wild populations of saccharomyces cerevisiae. PLoS Genet 1(1):66–71

Fischer G, James SA, Roberts IN, Oliver SG, Louis EJ (2000) Chromosomal evolution in *Saccharomyces*. Nature 405(6785):451–454

Gagneur J, Sinha H, Perocchi F, Bourgon R, Huber W, Steinmetz LM (2009) Genome-wide allele- and strand-specific expression profiling. Mol Syst Biol 5:274

Goddard MR, Anfang N, Tang R, Gardner RC, Jun C (2010) A distinct population of *Saccharomyces cerevisiae* in New Zealand: evidence for local dispersal by insects and human-aided global dispersal in oak barrels. Environ Microbiol 12(1):63–73

Goffeau A, Barrell BG, Bussey H, Davis RW, Dujon B, Feldmann H, Galibert F, Hoheisel JD, Jacq C, Johnston M, Louis EJ, Mewes HW, Murakami Y, Philippsen P, Tettelin H, Oliver SG (1996) Life with 6000 genes. Science 274(5287):546, 563–567

Granek JA, Murray D, Kayrkçi O, Magwene PM (2013) The genetic architecture of biofilm formation in a clinical isolate of *Saccharomyces cerevisiae*. Genetics 193(2):587–600

Greig D, Louis EJ, Borts RH, Travisano M (2002) Hybrid speciation in experimental populations of yeast. Science 298(5599):1773–1775

Gresham D, Ruderfer DM, Pratt SC, Schacherer J, Dunham MJ, Botstein D, Kruglyak L (2006) Genome-wide detection of polymorphisms at nucleotide resolution with a single dna microarray. Science 311(5769):1932–1936

Guijo S, Mauricio JC, Salmon JM, Ortega JM (1997) Determination of the relative ploidy in different *Saccharomyces cerevisiae* strains used for fermentation and 'flor' film ageing of dry sherry-type wines. Yeast 13(2):101–117

Haber JE (1998) Mating-type gene switching in *Saccharomyces cerevisiae*. Annu Rev Genet 32:561–599

Halme A, Bumgarner S, Styles C, Fink GR (2004) Genetic and epigenetic regulation of the FLO gene family generates cell-surface variation in yeast. Cell 116(3): 405–415

Hickman MA, Zeng G, Forche A, Hirakawa MP, Abbey D, Harrison BD, Wang Y-M, Su C-H, Bennett RJ, Wang Y, Berman J (2013) The 'obligate diploid' *Candida albicans* forms mating-competent haploids. Nature 494(7435):55–59

Hittinger CT (2013) *Saccharomyces* diversity and evolution: a budding model genus. Trends Genet 29: 309–317

Hyma KE, Fay JC (2013) Mixing of vineyard and oak-tree ecotypes of *Saccharomyces cerevisiae* in North American vineyards. Mol Ecol 22:2917–2930

Jones T, Federspiel NA, Chibana H, Dungan J, Kalman S, Magee BB, Newport G, Thorstenson YR, Agabian N, Magee PT, Davis RW, Scherer S (2004) The diploid genome sequence of *Candida albicans*. Proc Natl Acad Sci U S A 101(19):7329–7334

Kelly AC, Shewmaker FP, Kryndushkin D, Wickner RB (2012) Sex, prions, and plasmids in yeast. Proc Natl Acad Sci U S A 109(40):E2683–E2690

Kirby GC (1984) Breeding systems and heterozygosity in populations of tetrad forming fungi. Heredity 52: 35–41

Kuehne HA, Murphy HA, Francis CA, Sniegowski PD (2007) Allopatric divergence, secondary contact, and

genetic isolation in wild yeast populations. Curr Biol 17(5):407–411

Levy SF, Ziv N, Siegal ML (2012) Bet hedging in yeast by heterogeneous, age-correlated expression of a stress protectant. PLoS Biol 10(5):e1001325

Libkind D, Hittinger CT, Valério E, Gonçalves C, Dover J, Johnston M, Gonçalves P, Sampaio JP (2011) Microbe domestication and the identification of the wild genetic stock of lager-brewing yeast. Proc Natl Acad Sci U S A 108(35):14539–14544

Liti G, Schacherer J (2011) The rise of yeast population genomics. C R Biol 334(8–9):612–619

Liti G, Carter DM, Moses AM, Warringer J, Parts L, James SA, Davey RP, Roberts IN, Burt A, Koufopanou V, Tsai IJ, Bergman CM, Bensasson D, O'Kelly MJT, van Oudenaarden A, Barton DBH, Bailes E, Nguyen AN, Jones M, Quail MA, Goodhead I, Sims S, Smith F, Blomberg A, Durbin R, Louis EJ (2009) Population genomics of domestic and wild yeasts. Nature 458(7236):337–341

Magwene PM, Kayikçi Ö, Granek JA, Reininga JM, Scholl Z, Murray D (2011) Outcrossing, mitotic recombination, and life-history trade-offs shape genome evolution in *Saccharomyces cerevisiae*. Proc Natl Acad Sci U S A 108(5):1987–1992

Masel J, Lyttle DN (2011) The consequences of rare sexual reproduction by means of selfing in an otherwise clonally reproducing species. Theor Popul Biol 80(4):317–322

McCusker JH (2006) *Saccharomyces cerevisiae*: an emerging and model pathogenic fungus. In: Heitman J, Edwards JE, Filler SG, Mitchell AP (eds) Molecular principles of fungal pathogenesis. American Society for Microbiology, Washington, DC

McCusker JH, Clemons KV, Stevens DA, Davis RW (1994) *Saccharomyces cerevisiae* virulence phenotype as determined with CD-1 mice is associated with the ability to grow at 42 °C and form pseudohyphae. Infect Immun 62(12):5447–5455

Morales L, Dujon B (2012) Evolutionary role of interspecies hybridization and genetic exchanges in yeasts. Microbiol Mol Biol Rev 76(4):721–739

Mortimer RK (2000) Evolution and variation of the yeast (*Saccharomyces*) genome. Genome Res 10(4):403–409

Mortimer RK, Romano P, Suzzi G, Polsinelli M (1994) Genome renewal: a new phenomenon revealed from a genetic study of 43 strains of *Saccharomyces cerevisiae* derived from natural fermentation of grape musts. Yeast 10(12):1543–1552

Muller LAH, McCusker JH (2009) Microsatellite analysis of genetic diversity among clinical and nonclinical *Saccharomyces cerevisiae* isolates suggests heterozygote advantage in clinical environments. Mol Ecol 18(13):2779–2786

Naumov GI, Naumova ES, Sniegowski PD (1998) *Saccharomyces paradoxus* and *Saccharomyces cerevisiae* are associated with exudates of North American oaks. Can J Microbiol 44(11):1045–1050

Ruderfer DM, Pratt SC, Seidel HS, Kruglyak L (2006) Population genomic analysis of outcrossing and recombination in yeast. Nat Genet 38(9): 1077–1081

Schacherer J, Shapiro JA, Ruderfer DM, Kruglyak L (2009) Comprehensive polymorphism survey elucidates population structure of *Saccharomyces cerevisiae*. Nature 458(7236):342–345

Sipiczki M (2011) Diversity, variability and fast adaptive evolution of the wine yeast (*Saccharomyces cerevisiae*) genome—a review. Ann Microbiol 61:85–93.

Skelly DA, Ronald J, Connelly CF, Akey JM (2009) Population genomics of intron splicing in 38 Saccharomyces cerevisiae genome sequences. Genome Biol Evol 1:466–478

Sniegowski PD, Dombrowski PG, Fingerman E (2002) *Saccharomyces cerevisiae* and *Saccharomyces paradoxus* coexist in a natural woodland site in North America and display different levels of reproductive isolation from European conspecifics. FEMS Yeast Res 1(4):299–306

Stefanini I, Dapporto L, Legras J-L, Calabretta A, Di Paola M, De Filippo C, Viola R, Capretti P, Polsinelli M, Turillazzi S, Cavalieri D (2012) Role of social wasps in *Saccharomyces cerevisiae* ecology and evolution. Proc Natl Acad Sci U S A 109(33):13398–13403

Sweeney JY, Kuehne HA, Sniegowski PD (2004) Sympatric natural *Saccharomyces cerevisiae* and *S. paradoxus* populations have different thermal growth profiles. FEMS Yeast Res 4(4–5):521–525

Tsai IJ, Bensasson D, Burt A, Koufopanou V (2008) Population genomics of the wild yeast *Saccharomyces paradoxus*: quantifying the life cycle. Proc Natl Acad Sci U S A 105(12):4957–4962

Wang Q-M, Liu W-Q, Liti G, Wang S-A, Bai F-Y (2012) Surprisingly diverged populations of *Saccharomyces cerevisiae* in natural environments remote from human activity. Mol Ecol 21(22):5404–5417

Wei W, McCusker JH, Hyman RW, Jones T, Ning Y, Cao Z, Gu Z, Bruno D, Miranda M, Nguyen M, Wilhelmy J, Komp C, Tamse R, Wang X, Jia P, Luedi P, Oefner PJ, David L, Dietrich FS, Li Y, Davis RW, Steinmetz LM (2007) Genome sequencing and comparative analysis of *Saccharomyces cerevisiae* strain YJM789. Proc Natl Acad Sci U S A 104(31):12825–12830

Zheng W, Zhao H, Mancera E, Steinmetz LM, Snyder M (2010) Genetic analysis of variation in transcription factor binding in yeast. Nature 464(7292): 1187–1191

Ecological Genomics of Adaptation and Speciation in Fungi

4

Jean-Baptiste Leducq

Abstract

Fungi play a central role in both ecosystems and human societies. This is in part because they have adopted a large diversity of life history traits to conquer a wide variety of ecological niches. Here, I review recent fungal genomics studies that explored the molecular origins and the adaptive significance of this diversity. First, macro-ecological genomics studies revealed that fungal genomes were highly remodelled during their evolution. This remodelling, in terms of genome organization and size, occurred through the proliferation of non-coding elements, gene compaction, gene loss and the expansion of large families of adaptive genes. These features vary greatly among fungal clades, and are correlated with different life history traits such as multicellularity, pathogenicity, symbiosis, and sexual reproduction. Second, micro-ecological genomics studies, based on population genomics, experimental evolution and quantitative trait loci approaches, have allowed a deeper exploration of early evolutionary steps of the above adaptations. Fungi, and especially budding yeasts, were used intensively to characterize early mutations and chromosomal rearrangements that underlie the acquisition of new adaptive traits allowing them to conquer new ecological niches and potentially leading to speciation. By uncovering the ecological factors and genomic modifications that underline adaptation, these studies showed that Fungi are powerful models for ecological genomics (eco-genomics), and that this approach, so far mainly developed in a few model species, should be expanded to the whole kingdom.

J.-B. Leducq (✉)
Institut de Biologie Intégrative et des Systèmes, Département de Biologie, PROTEO, Pavillon Charles-Eugène-Marchand, Université Laval, 1030 avenue de la Médecine, Québec, QC G1V 0A6, Canada
e-mail: jean-baptiste.leducq.1@ulaval.ca

C.R. Landry and N. Aubin-Horth (eds.), *Ecological Genomics: Ecology and the Evolution of Genes and Genomes*, Advances in Experimental Medicine and Biology 781,
DOI 10.1007/978-94-007-7347-9_4,

Keywords
Fungi • Genomics • Life history traits • Adaptation • Reproduction • Population genomics • Speciation • Hybridization

4.1 Introduction: The Importance of Fungi in Ecology

Fungi are probably the most underexplored kingdom of Eukaryotic life. Paradoxically, they also contain one of the most studied organisms: the budding yeast *Saccharomyces cerevisiae*. This astonishing contrast stems from two specificities of Fungi. First, unlike most plants and animals, but like many bacteria, the majority of Fungi are either microscopic, non free-living, or dependant on highly specific biotic and abiotic conditions to live. Hence, a substantial part of fungal diversity cannot be observed *in natura*, isolated, or described and studied with the standard methods used in other eukaryotes. Second, fungal diversity has been mischaracterized for a long time, mostly because of the high diversity of sexual and growth forms, whereas these two features have been highly conserved during plant and animal evolution. This has often resulted in several species described for the same organism. For all these reasons, as Hibbet and Taylor (2013) recently pointed out, experimental genetics and evolutionary research on the fungal kingdom often focused on few organisms that, like *S. cerevisiae*, had small genomes and could be easily isolated, described and studied, but that represented a biased view of the actual fungal diversity.

Given their ability to absorb complex carbon sources and minerals from many environments, their resistance to unfavourable conditions and high dispersal abilities of their *spores*, Fungi have adopted various life history traits, and colonized most ecological niches (James et al. 2006; Stajich et al. 2009). Their simple organization – a mass of mostly undifferentiated cells, the thallus – allows them to form complex structures and associations with other organisms. Hence, even in purely mineral habitats, they are able to associate with algae or cyanobacteria to form lichen, enabling them to draw carbon from the air (Nash 2008). In the soil, Fungi form vast networks of rhizomes that recycle organic material, and they can make billions of connections with plant roots – *mycorrhizae* – providing plants with essential minerals in exchange for complex carbon compounds (Read 1991). They are present in the stomach of herbivorous animals where they help degrade fibres in return for a favourable reproductive environment (Nicholson et al. 2010). Many Fungi are also efficient pathogens and parasites of many plants and animals, able to perform part or all of their metabolism and life cycle at the expense of their hosts (San-Blas and Calderone 2008). Finally, Fungi play a central role in human societies with respect to health, industry, agronomy and research (Kendrick 2011).

Most Fungi can reproduce both asexually and sexually. However, the features of sexual reproduction vary greatly among and within species, and all modes of sexual reproduction can be found, from isogamy to anisogamy, from *homothallism* to *heterothallism*, with sometimes thousands of different mating *idiomorphs* (Billiard et al. 2011).

The high diversity and divergence in life history traits and reproductive modes among species has left a profound imprint on the evolution of fungal genomes. Here I review recent progresses in fungal eco-genomics and their contribution to the understanding of the following issues: (1) what are the main genomic features of adaptation to contrasting life history traits and to various modes of sexual reproduction? (2) How does adaptation to contrasting environments

shape fungal genome evolution? And (3) what are the mechanisms underlying speciation at the genomic level?

4.2 Macro-ecological Genomics in Fungi

Fungi are the result of about 1 billion years of evolution, during which genomes of different clades underwent profound molecular changes. Recent advances in molecular biology and genomics have allowed partial reconstruction of the evolutionary history of the kingdom, and improved the taxonomic classification (Fig. 4.1). Here I present recent contributions of fungal genomics to the understanding of how the main clades of Fungi emerged and how their adaptation to life histories shaped their genome organization.

4.2.1 Main Evolutionary Features of Ascomycota and Basidiomycota Genomes

Comparative genomics has been extensively used to infer the evolutionary history of Ascomycota genomes (Fig. 4.2). Evolutionary reconstruction from divergent clades suggests that the ancestral Ascomycota genome was likely small (about 12 Mb) and remained highly conserved in size during the evolution of Ascomycota yeasts, while the diversification of higher Ascomycota (Pezizomycotina) was characterized by a global increase in genome size that stabilized at about 40 Mb (Kelkar and Ochman 2012). Although the phylogenetic organization of Basidiomycota is less well defined, the same observation holds true for this clade, with a generally small genome size (around 15 Mb) in Pucciniomycotina, located at the root of the evolutionary tree, and a

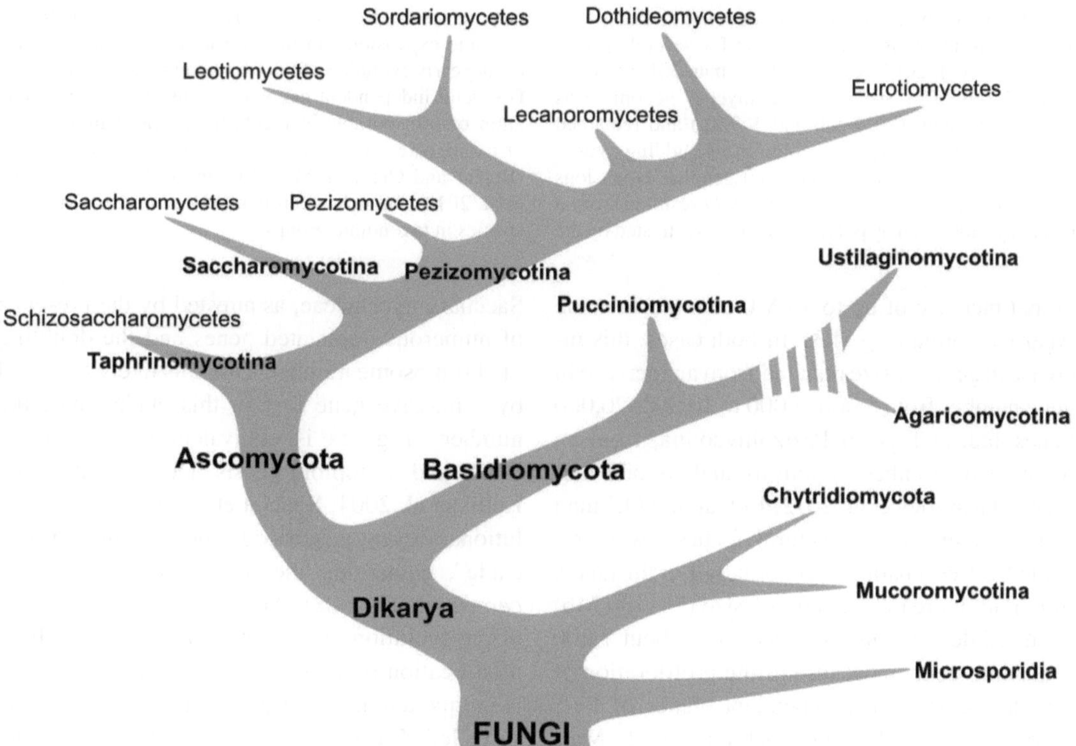

Fig. 4.1 Evolutionary tree of fungi based on whole-genome sequencing. Only main taxonomic groups are represented. Uncertain phylogenetic positions are hatched (Adapted from Kelkar and Ochman (2012); Wang et al. (2009); branch width is unrelated to the number of species in the taxonomic groups)

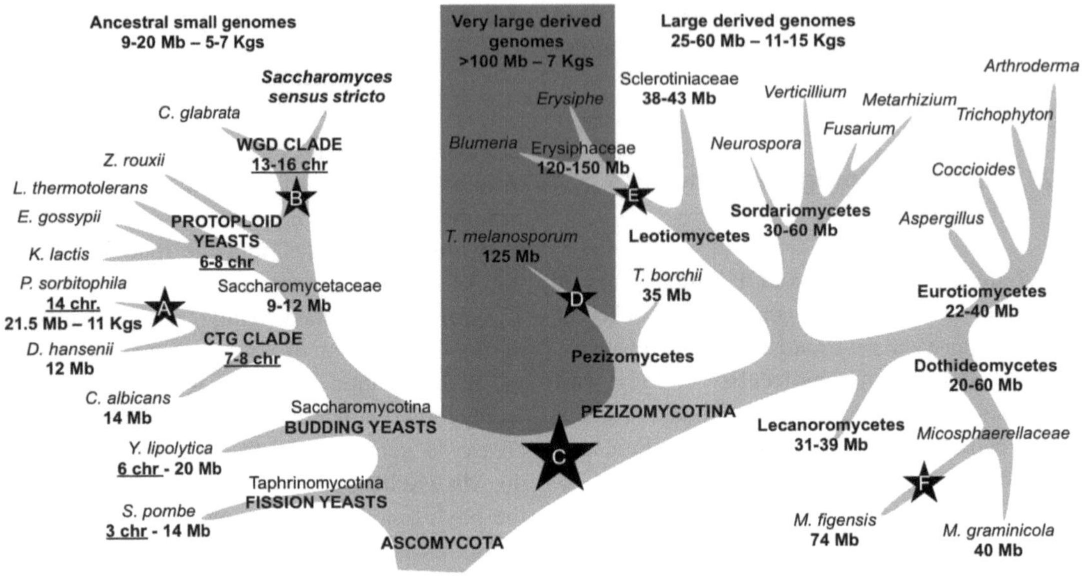

Fig. 4.2 Genome size evolution in Ascomycota based on whole-genome sequencing (Burmester et al. 2011; Dujon 2005; Gao et al. 2011; Kelkar and Ochman 2012; Souciet et al. 2009; Wang et al. 2009). Only representative taxa are shown. Number of chromosomes (chr), genome size (Mb) and number of genes in thousands (kg) are given in intervals when available for several species (Grigoriev et al. 2012; Kelkar and Ochman 2012; Souciet et al. 2009). The ancestral Ascomycota genome was presumably small (9–20 Mb and 5–7 kg) and remained relatively stable during the evolution of budding yeasts. Stars indicate examples of important genome expansions and remodelling. (**a**) *Pichia sorbitophila* resulted from a recent hybridization by polyploidization, as attested by the doubling of genome size and number of genes (Louis et al. 2012). (**b**) Unlike the ancestral and paraphyletic group of protoploid yeasts, yeasts of the WGD clade (including *S. cerevisiae*) resulted from a whole-genome duplication, as attested by the doubling of chromosome number followed by gene losses (Kellis et al. 2004; Souciet et al. 2009). (**c**) Genome expansion and increase in the number of genes during early evolution of Pezizomycotina, followed by (**d**–**f**) several independent genome expansions after proliferation of transposable elements in several symbiotic (**d**) or pathogenic species (**e**–**f**), and (**d**–**e**) massive gene loss (Kelkar and Ochman 2012; Martin et al. 2010; Santana et al. 2012). Branch width is unrelated to the number of species in taxonomic groups

global increase of up to 40 Mb or more in some Agaromycotina (Fig. 4.3). In both cases, this increase in genome size resulted from an increase in gene number from 5,000–7,000 to 10,000–20,000 genes and, at least in Pezizomycotina, from an increase in number of introns and mobile elements (Grigoriev et al. 2012; Kelkar and Ochman 2012). Many other structural features were revealed by comparative genomics. For instance, important increases in genome size (75–150 Mb) occurred despite massive gene loss (about 7,000 remaining genes), because of the proliferation of mobile elements in independent clades of Pezizomycotina (Kelkar and Ochman 2012; Martin et al. 2010; Santana et al. 2012; Fig. 4.2). The evolution of Ascomycota yeasts was marked by a strong conservation in genome size, but a whole-genome duplication (WGD) occurred in Saccharomycetaceae, as attested by the presence of numerous duplicated genes and the doubling of chromosome number. This was likely followed by a massive gene loss in this clade, since the number of genes is equivalent between post-WGD and protoploid yeasts (Dujon et al. 2004; Kellis et al. 2004; Souciet et al. 2009). This evolutionary event gave rise to the so-called "WGD clade", containing the model yeast species *S. cerevisiae* (Fig. 4.2). Another interesting feature of the evolution of Ascomycota yeasts is a slight modification in the standard genetic code, which is unique among eukaryotes and gave rise to the so-called "CTG clade", which includes the model *Candida albicans*. In these yeasts, 97 % of CUG codons are translated into serine and 3 % into leucine, whereas 100 % are translated into leucine in most eukaryotes. The comparison of proteins

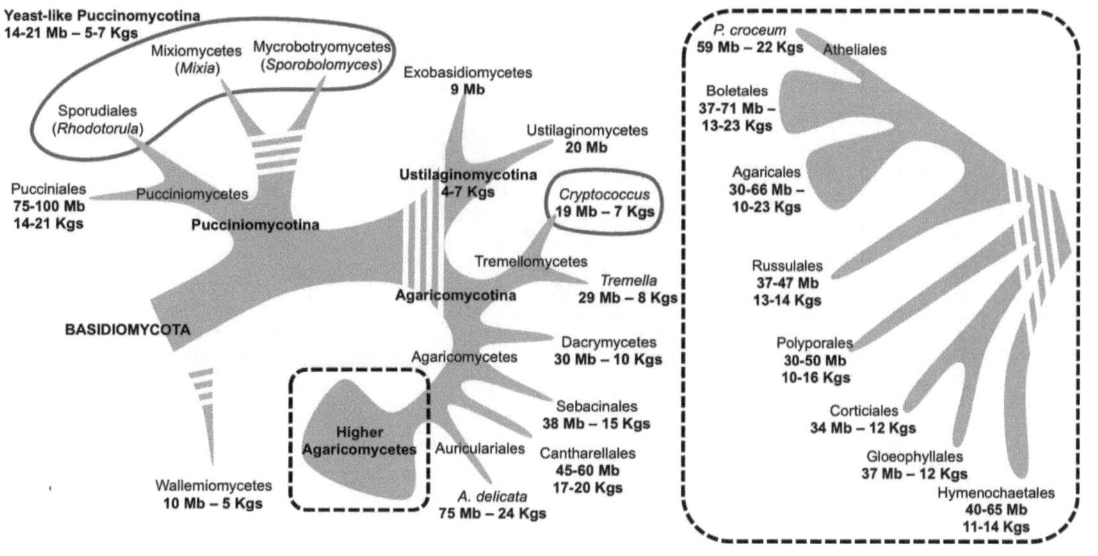

Fig. 4.3 Genome size evolution in Basidiomycota (Aime et al. 2006; Hibbett 2006; Wang et al. 2009). Only main taxonomic groups are represented. Uncertain phylogenetic positions are hatched. Genome size (Mb) and number of thousands of genes (kgs) are given in rank when available for several species (Grigoriev et al. 2012). Yeast-like species, characterized by small genome sizes, are circled in *grey*. The detail of phylogeny in higher Agaricomycetes is represented in the *dotted frame* on the *right*. Branch width is unrelated to the number of species in taxonomic groups

from CTG with their orthologs from non-CTG yeasts showed that the replacement of leucine by serine in the CTG ancestor was not random, but preferentially occurred in regions where it had no effect on protein function (Rocha et al. 2011). Inversely, when serine were experimentally replaced by leucine in *C. albicans*, it had only slight effects on its fitness, like changes in colony morphology (Gomes et al. 2007). Moreover, Gomes et al. (2007) showed that the proportion of CUG codons translated into leucine in *C. albicans* significantly increased in stressful conditions, as temperature increase, pH decrease or oxidative stress. These two studies suggest that flexibilities in the standard genetic code could enhance adaptability of CTG yeasts to changing environments.

4.2.2 Evolution of Life History Traits Modulates Fungal Genomes

Comparative genomics provides insights in the evolutionary processes that remodelled fungal genomes, and is thus tightly linked to the study of life history traits. The large diversity of life forms, reproductive modes and metabolism of Fungi makes them perfect models to investigate the role of adaptation in modelling genomc architecture.

4.2.3 Genome Size Variation in Relation to Lifestyle

The most obvious feature of adaptation to lifestyle in Fungi is the clear correlation between genome size and multicellularity (see Fig. 4.4). Indeed, smaller genomes are usually observed in unicellular or "yeast-like" Fungi, (9–20 Mb), whereas most Fungi producing multicellular thalli have larger genomes (20–125 Mb) and a higher number of genes. This evolutionary trend is independently observed in Ascomycota and Basidiomycota (Fig. 4.4), and broadly observed in multicellular organisms in other branches of life (Koonin 2011). However, many Fungi deviate from this universal pattern because of diverse forms of life, such as parasitism or mycorrhization (Fig. 4.4).

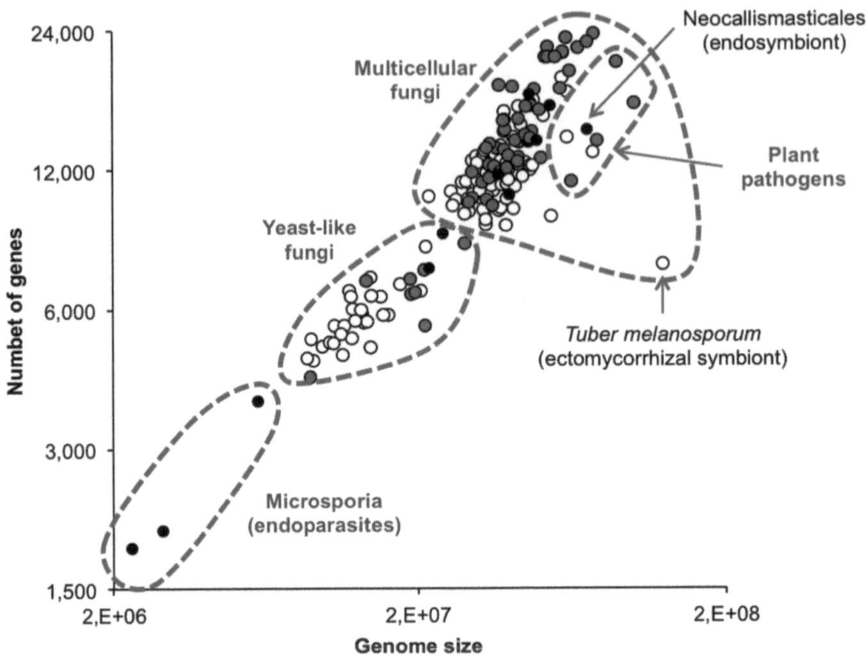

Fig. 4.4 Correlation between genome size and the number of genes reveals footprints of genome evolution on some life history traits of Fungi. *Grey circles* represent Basidiomycota; *white circles* represent Ascomycota; *black circles* represent other sub-phyla of Fungi. All available genomic data from Fungi were compiled here (Grigoriev et al. 2012; Peyretaillade et al. 2011; Souciet et al. 2009)

Microsporidia are the most primitive fungal organisms (Fig. 4.1) and so far have the smallest known eukaryotic genomes. For instance, *Encephalitozoon intestinalis*, a unicellular and obligate intracellular animal parasite, has 1,833 genes contained in a 2.3 Mb genome (Fig. 4.4). Peyretaillade et al. (2011) found that such small genomes are likely derived from a larger genome, and have been probably highly compacted after the reduction of non-coding sequences and the length reduction of protein-coding sequences. The authors suggested that this gene compaction could result from loss of protein domains involved in many protein-protein interactions, since their related functions became non-essential for the parasite. Large genome expansions frequently occurred in phytopathogenic Ascomycota, in spite of a low gene density (Fig. 4.4). These expansion could result from dispensable elements such as small chromosomes in *Mycosphaerella* (Goodwin et al. 2011) or from the proliferation of non-coding sequences, such as introns and mobile elements in *Mycosphaerella* (Santana et al. 2012), *Erysiphe* and *Blumeria* (Kelkar and Ochman 2012). Genome expansion in relation with pathogenicity was also observed in more basal Fungi like Mucoromycotina, whose genome is composed of about 20 % of transposable elements (Ma et al. 2009). In most cases, this accumulation of repeated and mobile elements is found around genes encoding virulence factors such as proteases, suggesting that these elements may be involved in adaptation to pathogenicity (Kelkar and Ochman 2012).

4.2.4 Expansion of Gene Families in Relation to Ecological Specialization

The adaptation of organisms to new ecological niches is the result of the loss and acquisition of specialized functions. These functions are often regulated by families of genes, i.e. genes with similar functions, resulting from repeated

duplications of an ancestral gene, followed by functional divergence (Demuth et al. 2006). The largest fungal genomes so far are found in truffles (Ascomycota), in particular *Tuber melanosporum* (125 Mb), despite its small gene number (7,500). Some of the typical Ascomycota gene families absent in the *T. melanosporum* genome are involved in secondary metabolism. Those became less essential in this ectomycorrhizal species, since the host plant provides secondary metabolites. Comparative genomics in *T. melanosporum* also provided functional evidence for the effect of *symbiosis* on the genome composition, for instance the rise of gene families encoding enzymes involved in degradation of host cellular walls (Martin et al. 2010). Interestingly, genes with identical functions also exist in other symbiotic Fungi like *Laccaria bicolor* (Basidiomycota, Agaricales; Martin et al. 2008), but they arose from independent evolutionary events (Martin et al. 2010). Wood decay Fungi are characterized by efficient lignin depolymerisation, and many functional evidence of this ability were found in fungal genomes. In Agarycomycetes (Basidiomycota), Floudas et al. (2012) used comparative genomics to highlight the early expansion of gene families involved in lignin depolymerisation, such as genes coding for peroxidases. They suggested that wood decay was the ancestral state in this taxon, but that this ability is highly variable among independent lineages where these genes were lost. For instance, in *Phanerochaete chrysosporium* (Basidiomycota, Polyporales), extra-cellular degradation of lignocellulose is permitted by a complex set of multiple gene families, such as cellulases and pyranoseoxidases. However, in the case of *Postia placenta*, a close relative of *P. chrysosporium*, many of these genes were lost, thus making this organism unable to efficiently depolymerize lignin (Martinez et al. 2009). Similarly, in another relative, *Ceriporiopsis subvermispora*, Fernandez-Fueyo et al. (2012) linked decrease in cellulose depolymerisation efficiency with variation in gene composition and expression.

Comparative genomics of pathogenic Fungi provided evidence for the high diversity in genes and molecular pathways underlying pathogenicity, most of which have evolved independently and result from a long-term arms race between hosts and pathogens. For instance, studies carried out in distinct clades revealed a large expansion of gene families encoding proteins involved in pathogenicity, such as secreted proteases, toxins or cell wall degradation enzymes. These increases in copy numbers occurred independently in dermatophytic Fungi such as *Arthroderma* and *Trichophyton* (Ascomycota; Burmester et al. 2011), in insect pathogens such as *Metarhizium* (Ascomycota; Gao et al. 2011), in yeasts of the CTG clade (Butler et al. 2009), in amphibian pathogens such as the Chytrid Fungi *Bratrachochytrium dendrobatidis* (Joneson et al. 2011) and in phytopathogenic species such as those from the *Fusarium* genus (Rep and Kistler 2010). In the human pathogen *Rhizopusoryzae* (Mucoromycotina; Fig. 4.1), the expansion of gene families involved in virulence factors and drug resistance arose after whole genome duplication (Ma et al. 2009). Some exceptions were found, such as in the other phytopathogenic species *Mycosphaerella graminicola*, which showed a surprisingly small number of genes involved in cell wall degradation as compared to other phytopathogens (Fig. 4.5). This feature suggests that *M. graminicola* evolved from an *endophyte* ancestor, and developed a protease activity rather than host cell wall degradation (Goodwin et al. 2011). Recently, Ohm et al. (2012) compiled available genomic data for three clades of Dothideomycetes Fungi (Ascomycota): Pleosporales, containing only phytopathogens, Hysteriales, including only *saprotrophs* (degrading humus) and Carpnodiales, containing *M. graminicola,* but also many other phytopatogens and a saprotrophic species. Based on genomic sequences, they reconstructed and identified major features of the evolutionary history of pathogenicity within this clade (Fig. 4.5). They observed that, despite a great variation in genome size among these Fungi (20–75 Mb), the number of genes was highly conserved (10,000–14,000), confirming that variation in genome size mainly resulted from proliferation of mobile (Kelkar and Ochman 2012; Martin et al. 2010; Santana et al. 2012) and repetitive elements, altogether

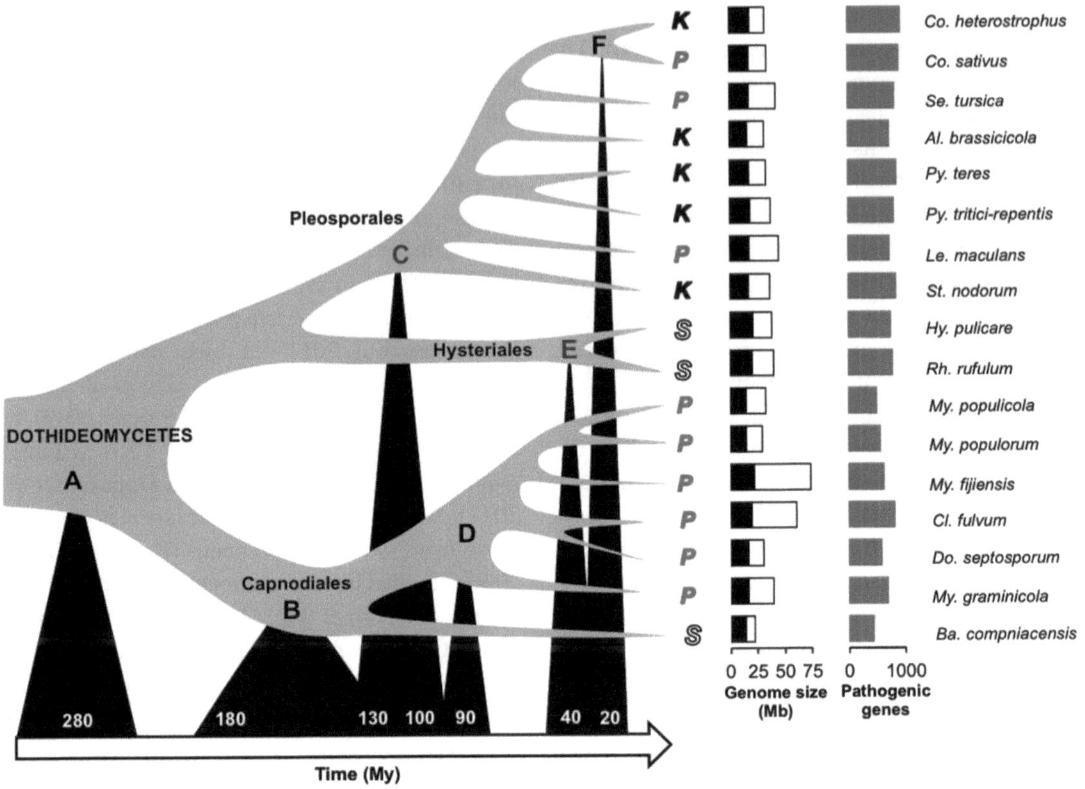

Fig. 4.5 Comparative genomics of 17 genomes allows reconstructing the evolution of pathogenicity in Dothideomycetes Fungi (adapted from Ohm et al. (2012)). The evolutionary tree, based on genomic data, reveals the chronology of the main steps (**a–f**) of Dothideomycetes evolution (*black triangles* on the time line indicate the range for age estimation of each node in million years). The following information is indicated for each species to the *right* of the tree: life style symbolized by a bold letter (*K* killing pathogen, *P* pathogen, *S* saprotroph); coding and non-coding genome size in megabases (*black and white horizontal bars*, respectively); number of pathogenic genes (*grey horizontal bars*); and species name (genus abbreviations: *Co.*: *Cochliobolus*; *Se.*: *Setosphaeria*; *Al.*: *Alternaria*; *Py.*: *Pyrenophora*; *Le.*: *Leptosphaeria; St.*: *Stagonospora*; *Hy.* : *Hysterium*; *Rh.* : *Rhydidhysteron*; *My.*: *Mycosphaerella*; *Cl.*: *Cladosporium*; *Do*: *Dothistroma*; *Ba.*: *Baudoinia*)

comprising about 40 % of the *Mycosphaerella fijiensis* genome (Fig. 4.5). Moreover, these repetitive elements likely resulted from many chromosomal rearrangements occurring during the evolution of this clade, some of them linked to genes involved in pathogenic traits, such as the ability to degrade host cellulose or to lyse host proteins. Interestingly, genomic rearrangements observed in these clades likely enhanced the expansion of pathogenic gene families. These expansions were more important in Pleosporales (700–1,000 pathogenic genes) than in Capnodiales (500–800), in agreement with the more serious pathogenicity of Pleosporales on its plant host, often resulting in cell destruction, compared to Capnodiales, which often leaves the host cell alive until pathogen reproduction (Fig. 4.5). Expansions of pathogenic gene families also occurred in Hysteriales, which have a more recent common ancestor with Pleosporales, suggesting that these *saprotroph* species possibly derived from a phytopathogen ancestor and took advantage of these genes to efficiently degrade fresh plant debris. Conversely, the *saprotroph Baudoinia compniacensis* (Capnodiales) has likely adopted

a distinct strategy to degrade decayed debris, since its small, compact genome (10,000 genes for 20 Mb) only harbours 435 genes related to pathogenicity.

4.2.5 Effect of Sexual Reproduction on Fungal Genomes

Fungi are known to form large asexual colonies, allowing them to efficiently exploit their habitat. However, when ecological conditions are changing, sexual recombination could be necessary to allow the rise of new allelic combinations, which may be advantageous in the new environment. The most obvious evidence of this evolutionary constrain is that stressful conditions are usually used in Fungi to induce sexual reproduction in laboratory conditions. Even apparent asexual species could occasionally perform sex to favour the maintenance of genetic diversity and to colonize new ecological niches (Billiard et al. 2011; Sun and Heitman 2011; Tsui et al. 2013). Hence, sexual reproduction has a great impact on fungal genome evolution and could alter the synteny between genomes of closely related species. For instance, in Dothideomycetes (Ascomycota), genomes of distinct species have similar chromosomes with the same gene composition, but their organization was shuffled by frequent recombination, possibly during meiosis or horizontal transfers between closely related species (Hane et al. 2011).

Sexual recognition in Fungi is governed by one or several mating-type loci (*MAT*), each containing genes coding for complementary sexual *idiomorphs*. When this system is functional, reproduction can only occur between two individuals expressing complementary *MAT idiomorphs* (*heterothallism*). This system of reproduction is widespread and presumably ancestral in most Fungi (Billiard et al. 2011). However, many fungal taxa hijacked or modified this system, resulting in independent and contrasting evolutionary modifications, such as the augmentation of the number of *MAT idiomorphs*, the evolution from *heterothallism* to *homothallism* or the possible loss of sex (Billiard et al. 2011; Lee et al. 2010; Sun and Heitman 2011). With the emergence of genomic data, genomic regions of *MAT* loci have been intensively studied in many Fungi and revealed the high diversity and complexity of function, gene composition and organization underlying the evolution of reproductive systems (Butler et al. 2009; Fraser et al. 2007; Lee et al. 2010; Metin et al. 2010; Tsui et al. 2013). For instance, Butler et al. established that the observed co-occurrence of homothallic and heterothallic species in the CTG clade resulted from multiple losses and reorganizations of *MAT* genes (Butler et al. 2009). In *Neurospora* species (Ascomycota), Wik et al. (2008) found genomic evidence that multiple transitions from *heterothallism* to *homothallism* resulted from independent disruptions of different *MAT* genes in different species. In this particular clade, the genomic region subjected to the effect of *MAT* locus evolution is so large that an entire chromosome was affected. For instance, in the *heterokaryotic* and pseudohomothallic species *N. tetrasperma*, the maintenance of self-fertility is allowed by the co-transmission of two nuclei of opposed *MAT idiomorphs* (Fig. 4.6). Menkis et al. (2008) and Ellison et al. (2011b) showed that this enforced co-transmission is likely to have occurred after inversions of large genomic regions surrounding the *MAT* locus, therefore preventing recombination events between homologous chromosomes containing *MAT idiomorphs*. The non-recombinant region encompasses one fifth of the entire genome (Fig. 4.6). Evidence for similar mechanisms of recombination suppression by complex genomic rearrangements of the *MAT* locus has also been observed in the *Mycrobotrium* (Basidiomycota; Votintseva and Filatov 2009) and *Cryptococcus* (Basidiomycota; Metin et al. 2010) genus. In these species, like in *N. tetrasperma* (Whittle and Johannesson 2011), the absence of homologous recombination between *MAT* chromosomes allows mutations to accumulate without being purged, but also enhances deterioration in preferred codon usage, suggesting a decrease of translation efficiency for genes located in the non-recombinant region

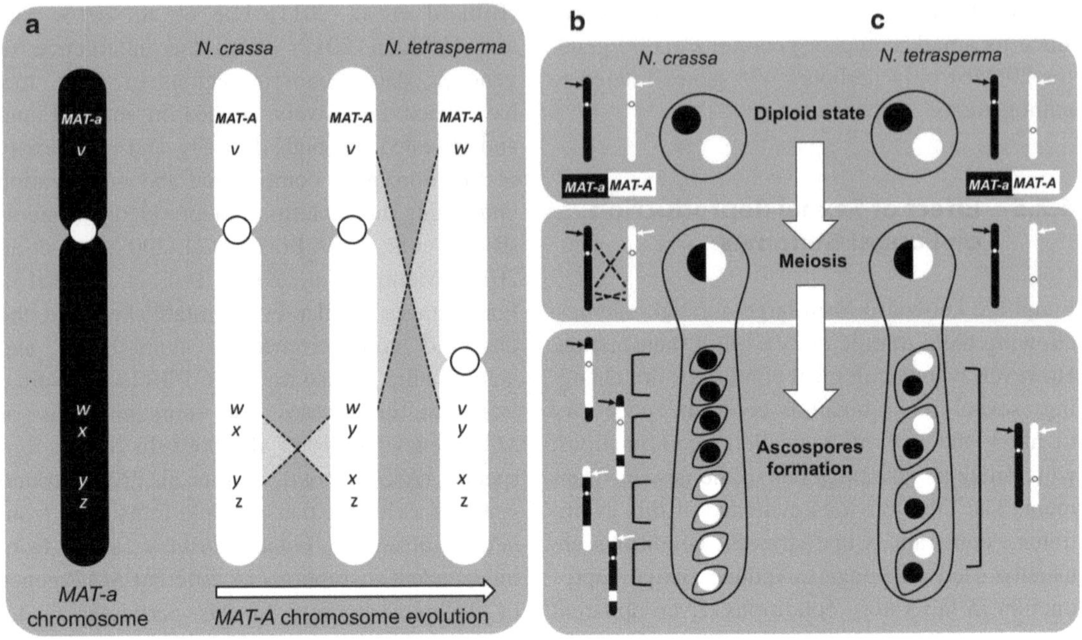

Fig. 4.6 Main genomic features of evolution from self-sterility (*heterothallism*) to self-fertility (pseudo-*homothallism*) in two *Neurospora* species, adapted from Menkis et al. (2008) and Ellison et al. (2011b). (**a**) Evolution of chromosomes including the gene coding for two opposite mating-type determinants (*MAT-a* and *MAT-A*) in *N. tetrasperma*. The *MAT-a* chromosome (*left*) is similar in *N. crassa* and *N. tetrasperma*, and is collinear with the homologous *N. crassa MAT-A* chromosome. In *N. tetrasperma*, the *MAT-A* chromosome underwent two main inversions (*dotted lines*), so that it is not collinear with the *MAT-a* chromosome. Positions of *MAT* locus and five hypothetical loci *v*, *w*, *x*, *y* and *z* are indicated; *circles* indicate the centromere position. (**b**) Segregation of *MAT-a* and *MAT-A idiomorphs* during meiosis and ascospore formation in *N. crassa* (*heterothallism*). A cell with two nuclei is represented on *top*, with the respective karyotype of each nucleus including either the *MAT-a* (*left*; *black*) or the *MAT-A* (*right*; *white*) chromosome. *Arrows* indicate the *MAT* locus position. During the fusion of nuclei and meiosis (*middle*), all chromosomes, including *MAT* chromosomes, undergo crossing over (*dotted lines*). The resulting ascus contains eight *ascospores*, each having a single *MAT* determinant. (**c**) The same steps are indicated for *N. tetrasperma* (pseudo-*homothallism*). Mismatch occurs during meiosis between *MAT-a* and *MAT-A* chromosomes, because they are not collinear and chromosomes are co-transmitted in all of four *ascospores*, all then co-expressing both *MAT idiomorphs*

(Whittle et al. 2011). Such accumulations could possibly lead to chromosome degeneration, which was likely an early step of sexual chromosome evolution in animals (Fraser et al. 2004). Another interesting case of *MAT* chromosome degeneration could be found in Saccharomycetaceae, where *heterothallism* is ancestral but evolved in *homothallism* in higher taxa such as for instance in *C. glabrata* and *S. cerevisiae* (Fig. 4.2). In these species, *homothallism* consists in recombination between homologous regions of active copies of genes coding for *idiomorphs* (either *MAT-a* or *MAT-α*) and silent copies of these genes located somewhere else in the chromosome, allowing recurrent *idiomorph* switching in haploid cells. As a result, a *MAT-a* cell could switch to a *MAT-α* cell and vice versa, allowing mating between any cells. By comparing *MAT* chromosomes of different Saccharomycetaceae species, Gordon et al. (2011) showed that recurrent DNA damages and mis-repairs occurring during *idiomorph* switching progressively resulted in the erosion of the *MAT* chromosome through the loss and transposition of genes flanking the recombination hotspot.

In genomes of several Microsporidia species, Lee et al. (2010) found a sex-related locus, similar to that known in heterothallic Mucoromycotina species. This locus was unrelated to the *MAT* locus and was not found in genomes of higher Fungi (namely Chytridiomycota, Basidiomycota and Ascomycota), suggesting that it is specific to basal lineages. Like in the *MAT* locus, the structure and composition in genes of the sex-related locus greatly vary between Microsporidia and Mucoromycotina species. However, the absence of *idiomorphism* and functional evidence for the sex-related locus in Microsporidia suggests that these species are either homothallic or asexual. Even in higher Fungi, as in some *Candida* species where the *MAT* locus is present but partially lost and inactivated, sex could be maintained, suggesting that more complex and unknown genomic features control for cryptic sex in Fungi (Sun and Heitman 2011). In species reproducing asexually, functionality of *MAT* genes can be maintained, either to perform cryptic metabolism, as the regulation of asexual development in response to light in *Neurospora* (Wang et al. 2012b), or to keep the potential to reproduce sexually, providing the opportunity to increase genetic variability and thus to colonize new ecological niches, as it is the case in Ophiostomatales (Sordariomycetes, Ascomycota), an order of pathogenic Fungi (Tsui et al. 2013).

4.3 Micro-ecological Genomics in Fungi

Large-scale evolutionary changes that shaped fungal genomes were tightly associated with main lifestyle characteristics of Fungi: multicellularity, pathogenicity, *symbiosis*, lignin degradation and reproduction. At a lower evolutionary scale, fungal genomes were also marked by slight modifications such as mutations, translocations and regulatory changes that progressively accumulated to confer advantageous adaptations in changing environments. Here, I review experimental and population studies that have examined the roles of local adaptation and speciation on early steps of fungal genomic evolution and have captured evolution in action.

4.3.1 Genomics of Local Adaptation and Recent Evolution in Fungi

Because genomes are shaped by long-term evolution, it is often impossible to predict the phenotypic response of organisms to a particular environmental variation only from the knowledge of gene function, gene composition, or gene orthology based on inter-species comparisons. The increased genomic data available for closely related species or for several individuals from the same species, and the use of experimental evolution, quantitative trait loci (QTL) approaches and population genomics, now allow the investigation of recent imprints of environmental pressures on fungal genomes.

4.3.2 Using Experimental Evolution to Understand the Early Steps of Genomic Adaptation

Experimental evolution coupled with genomics provides a powerful and developing tool to explore the primary mutations and genomic rearrangements underlying adaptation to different types of environments in eukaryotes. Because of their short generation time and their small genomes, yeast-like Fungi are ideal eukaryotic models for this approach. For instance, Araya et al. (2010) compared the genome of a *S. cerevisiae* strain evolved in sulphate-limited conditions with its ancestor genome. They found single-point mutations responsible for adaptation to this limitation, such as a mutation affecting the regulation of *RRN3*, a gene involved in modulating ribosomal gene expression during nutrient-limiting conditions. Anderson et al. (2010) adopted a similar approach to uncover early mutations underlying high salt concentration and low glucose tolerance in *S. cerevisiae* (see also Sect. 4.4.1). In *C. albicans*, a genome-wide survey of gene expression

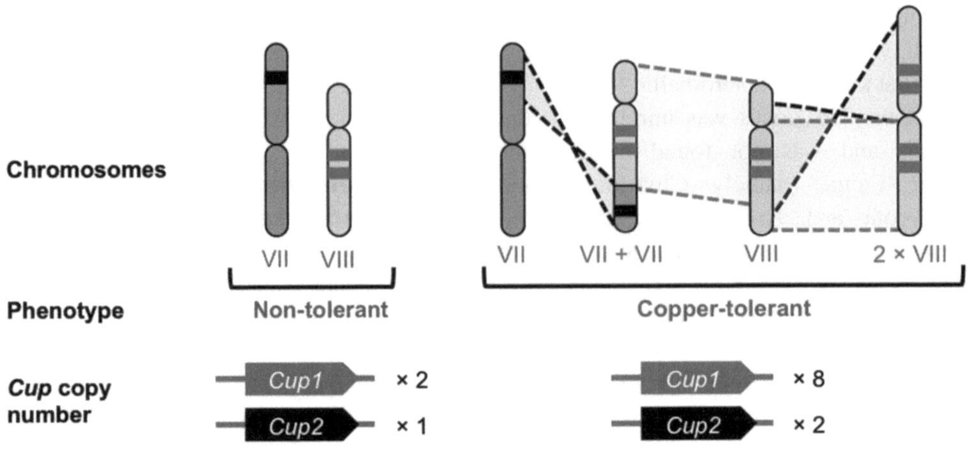

Fig. 4.7 Copper tolerance in natural populations of *Saccharomyces cerevisiae* is allowed by reversible rearrangements (*dotted lines*) of chromosomes VII and VIII fragments. The formation of two additional chimeric chromosomes results in the increase in copy number of *CUP* genes involved in copper regulation in cell (Chang et al. 2013). The position of genes *CUP1* and *CUP2* on chromosomes VIII and VII are indicated by *grey* and *black rectangles*, respectively

conducted in strains adapted to drug resistance suggested that at least two adaptive patterns could occur in response to this stress. These patterns mostly involved nine genes, whose expression levels varied among four independently evolved populations. Hence, one population showed a high expression of a single gene, *CDR2*, controlling drug export, whereas three other populations independently converged on high expression levels of eight other genes, such as *MDR1*, also involved in efflux of drugs, or *YPX98*, *YPR127W*, and *ADH4*, involved in cell protection during oxidative stress (Cowen et al. 2002). Recently, Chang et al. (2013) found that natural strains of *S. cerevisiae* were copper tolerant because of higher expression level of genes *CUP1* and *CUP2* encoding proteins involved in copper regulation. They showed that this increase in expression resulted from an increase in *CUP1* and *CUP2* copy number, allowed by duplications and translocations of large genomic regions encompassing these genes (Fig. 4.7). Interestingly, translocations were reversible when these strains were evolved in less selective environment, suggesting that chromosomal rearrangements could be an important mechanism of rapid adaptation to fluctuating environments.

4.3.3 Population Genomics and Quantitative Trait Loci (QTL) Approaches

An alternative to studying the mutations underlying adaptation is population genomics, which allows detection of natural genomic variations and can associate particular traits to these variations. Because environmental factors are numerous, complex and highly variable within natural populations, the resulting phenotypes can rarely be explained by limited genomic variation, as they can under controlled conditions. The QTL approach allows to experimentally identify genetic variation underlying complex quantitative traits and to measure the influence of selection shaping these traits (Rice and Townsend 2012) by recombining genomes of two phenotypically different individuals and then, by identifying the putative loci involved and their functions. For instance, Liti et al. (2009b) found that telomere length, which is an important factor for buffering DNA loss during replication, varies among *S. paradoxus* populations. Using a QTL approach, they identified two genes likely associated to this variation. Cubillos et al. (2011) crossed individuals from two diverging populations of

S. cerevisiae and grew the progeny in 23 distinct experimental conditions. They found that most of the traits were polygenic, but some, such as copper tolerance or ability to grow on galactose or maltose, were linked with genes whose predicted functions were consistent with the trait, such as genes *CUP1* and *CUP2* involved in copper tolerance (See Sect. 4.3.2), or *GAL3*, involved in galactose metabolism. Using the same approach, Will et al. (2010) showed that the freeze-tolerance of *S. cerevisiae* strains depended on a few mutations in two genes coding for water-transport proteins. Moreover, these genes showed a strong signature of balancing selection. The balanced polymorphism was distributed between two distinct groups of *S. cerevisiae* populations that had contrasting profiles of freeze-tolerance phenotypes.

4.3.4 Population Genomics and Ecological Approaches

When a collection of genomes representing natural variation is available, population genomics can associate the natural environment with natural genomic variations, and then make functional predictions about the genes affected by these variations. By comparing the distribution of synonymous mutations among the genomes of 44 clinical and 44 non-clinical strains of *S. cerevisiae*, Muller et al. (2011) identified a handful of genes likely to be pathogenicity determinants and involved, for instance, in cell wall resistance and cell detoxification, which gives valuable indications about what mechanisms pathogenic strains use to evade the human immune system. Conversely, reverse ecology uses natural genomic variation to predict environmental factors affecting genes, without assumptions about their functions. Ellison et al. (2011a) used a reverse ecology approach to identify factors underlying population structure of *Neurospora crassa* (Ascomycota). Using whole genome sequencing, they found two genetically diverging populations, which also diverged in their ability to grow at low temperature. They looked for genomic distribution of molecular divergence between the two populations, without focusing on particular genes, and found two large genomic regions containing several genes that showed deep divergence between populations. When two of these genes, *MRH4* and *PAC10*, were deleted, strains lost their ability to grow at low temperature, confirming that the divergence between the two genomic regions was responsible for adaptation to cold. When the function of genes and the metabolic pathways underlying a particular trait are known, one can use genome sequences to identify mutations that potentially affect the pathway. This was the case in the yeast *S. kudriavzevii*, where two divergent populations differed in their ability to use galactose as carbon source. Hittinger et al. (2010) identified the underlying mutations affecting genes of the galactose pathway by genome sequencing. They established that, despite frequent gene flow between these two populations, balancing selection maintained inactive copies of these genes in the population unable to metabolise galactose. In this case, the ecological differences between the two *S. kudriavzevii* populations are poorly known. One can only speculate that, because strains were found on different substrates, those provided different sources of carbon: strains able to metabolize galactose were found on barks of trees, whereas strains unable to metabolize it were found in soil.

Population genomics also provided evidence for very recent evolution of pathogenicity in higher Ascomycota at the species level. For instance, Stukenbrock et al. (2011) looked for evidence of selection in genomes of the wheat pathogen species *Mycosphaerella graminicola* and its related non-pathogenic sister species. They found that the number of genes showing positive selection increased after divergence of *M. graminicola* with its closest relative, whereas it remained low when estimated at a higher evolutionary scale. Interestingly, most of these genes encoded pathogen effectors, suggesting that pathogenicity of *M. graminicola* arose recently, likely with the domestication of wheat. In *Verticillium* (Ascomycota) species, de Jonge et al. (2012) identified *AVE1*, a gene

encoding a virulence effector for multiple plant species, which was found to be part of a large genomic region only present in virulent strains. Interestingly, *AVE1* orthologs were also found in other phytopathogen Fungi and bacteria, usually located in a similar large genomic region also containing many transposable elements, suggesting that at least part of *Verticillium* virulence results from horizontal gene transfers.

Altogether, these studies suggest that not only a few genes allow adaptation to new environments. Mutations occurring elsewhere in the genome could also generate variation in regulation of gene expression and lead to new adaptive responses. Rapid and sometimes reversible genomic rearrangements could also lead to new favourable combinations of genes. In extreme cases, as in some pathogenic Fungi, adaptation could occur after horizontal transfer of advantageous genes between bacteria and Fungi. Finally, the opportunity for Fungi to use distinct and alternative molecular pathways involving different sets of genes to reach the same phenotypic response could also instigate early steps of adaptation in a new environment.

4.3.5 Population Genomics of *Saccharomyces cerevisiae* and *Saccharomyces paradoxus*

Population genomics of *Saccharomyces cerevisiae* and *S. paradoxus* has been extensively investigated in the last few years. These sister species have a worldwide distribution and are sympatric – both are found in the wild and are associated with the same tree species (Hyma and Fay 2013; Sniegowski et al. 2002) – but show highly contrasting patterns of population history. On one hand, the population structure of *S. cerevisiae* is highly correlated with its domestication history – association to human pathologies or adaptation to different modes of alcohol fermentation and baking (Hyma and Fay 2013; Liti et al. 2009a; Schacherer et al. 2009) – while almost no geographical signal is observable, except for a few wild and isolated populations in China (Wang et al. 2012a). On the other hand, *S. paradoxus* populations are highly structured according to their geographical locations, with at least three distinct and genetically fixed populations, located in Europe, East Asia and America (Fig. 4.8). This pattern suggests no strong human impact on *S. paradoxus* history and no recent introgression (Hyma and Fay 2013; Liti et al. 2009a), except some evidence of recent hybridization with *S. cerevisiae* in the European *S. paradoxus* lineage (Liti et al. 2006). This structure is emphasized by partial reproductive isolation between the different *S. paradoxus* populations (Kuehne et al. 2007; Liti et al. 2006), although a secondary contact after a recent and likely anthropic immigration event from Europe to America has been reported (Hyma and Fay 2013; Kuehne et al. 2007). Accordingly, Liti et al. (2009a) found that genomic and phenotypic variations were strongly correlated in *S. paradoxus* but not in *S. cerevisiae*. This was in agreement with genomic evidence for frequent introgression events between *S. cerevisiae* lineages, which likely resulted from numerous hybridization during human domestication and acquisition of new and specific genes in industrial strains (Borneman et al. 2011). Additionally, Warringer et al. (2011) found that phenotypic variability was higher in *S. cerevisiae* than in *S. paradoxus*, despite a higher genetic diversity in the latter. Once more, the authors suggested that this paradoxically high phenotypic variation was the result of genetic drift having occurred during multiple and independent domestication events of *S. cerevisiae*. Surprisingly, while selection associated to strong adaptive divergences between species was expected, most of genomic studies failed to detect signal of selection on particular adaptive genes (but see Aa et al. 2006), suggesting either that purifying selection uniformly acted on genomes during *S. cerevisiae* and *S. paradoxus* divergence or that selection was relaxed in *S. cerevisiae* populations after domestication (Liti et al. 2009a). Finally, population genomics studies on these two species quantified how frequently the two species experienced sexual vs. asexual reproduction during their evolution. Based on an estimation of

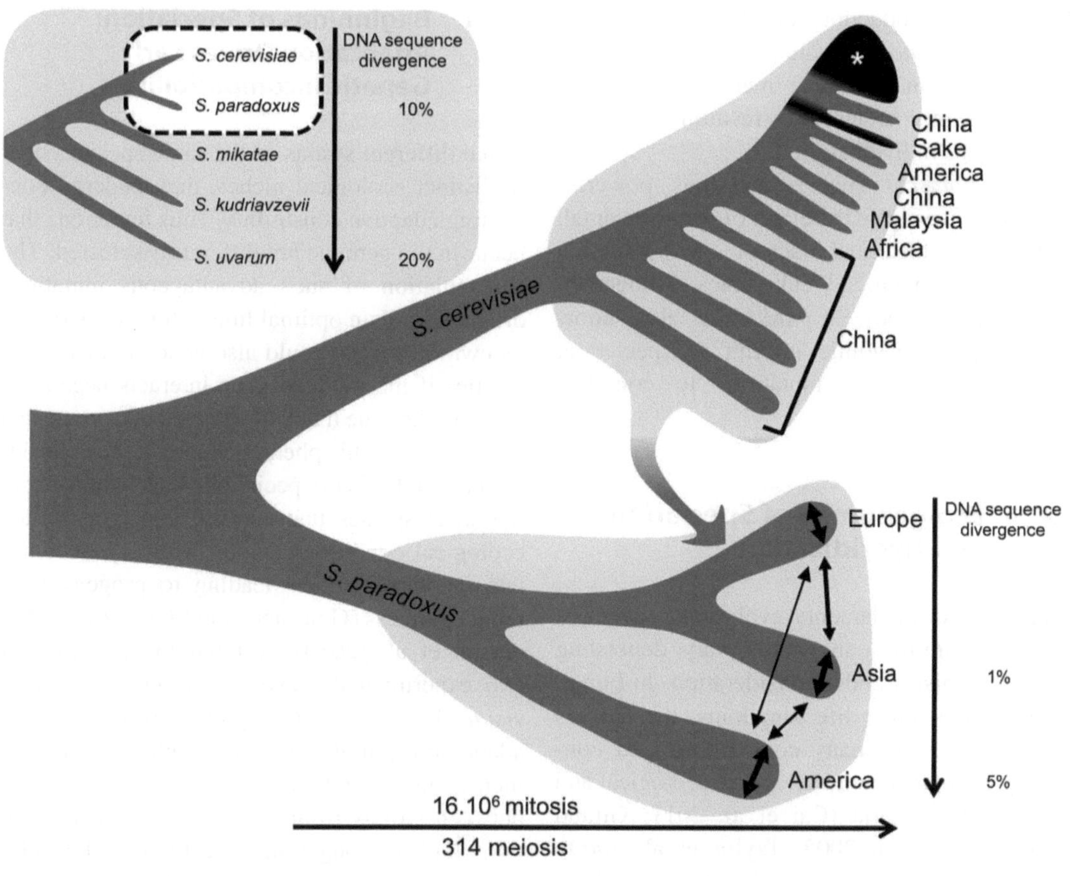

Fig. 4.8 *Saccharomyces cerevisiae* and *S. paradoxus* have contrasting evolutionary histories. *Top left*: evolutionary tree of *Saccharomyces* species indicating the range of genome sequence divergence between *S. cerevisiae* and other species (Dujon 2006). *Dotted frame* indicates the position of *S. cerevisiae* and *S. paradoxus*. Population structure is in agreement with geographical distribution for *S. paradoxus*, whereas *S. cerevisiae*, which underwent human domestication, has no such pattern of structure (Liti et al. 2009a). The *black* area indicates branches in which *S. cerevisiae* domesticated strains are predominant. The *asterisk* indicates the branch containing most of studied strains until recent advances in genomics, and including European, wine, clinical and baking strains. The estimation number of meiosis and mitosis that occurred during species divergence (*arrow on bottom*) was estimated from genomic data (Ruderfer et al. 2006). In *S. paradoxus*, partial reproductive isolation occurred among different lineages. Arrow width is proportional to hybrid progeny survival: 95 % within lineages; 30–70 % among lineages (Kuehne et al. 2007; Liti et al. 2006) and is correlated with sequence divergence among lineages (*vertical axis*) but also with translocations having occurred in some strains (not shown; only results for crosses between collinear genomes are indicated). The *arrow* from *S. cerevisiae* to European *S. paradoxus* indicates evidence for introgression events between the two lineages (Liti et al. 2006)

recombination frequency, Ruderfer et al. (2006) showed that meiosis (i.e. sexual) recombination occurred only once each 50,000 cell divisions in both species, whereas a more recent estimation suggested once per 1,000–3,000 for *S. paradoxus* (Tsai et al. 2008). Both studies support the idea that, despite a fully functional mating system, reproduction in *Saccharomyces* species is mostly clonal (Fig. 4.8). In *S. cerevisiae*, Magwene et al. (2011) proposed an explanation for this phenomenon. They observed that strains with a high proportion of heterozygous sites showed a low capacity to perform meiosis, suggesting that selection favoured asexual reproduction in such strains to maintain advantageous heterozygosity. Similarly, Tsai et al. (2008) estimated that 99 %

of sexual reproduction events that occurred during *S. paradoxus* evolution involved strains originated from the same meiotic events, i.e. from the same parents, thus resulting in highly homozygous individuals.

Saccharomyces species are thus powerful models to investigate the effect of environmental, human and historical factors on population genomics of microbial eukaryotes. Altogether, *S. cerevisiae* and *S. paradoxus* also allow dissecting the genomic imprint of speciation, from early adaptive mutations to complete reproductive isolation.

4.4 Eco-genomics of Speciation and Hybridization

Speciation is a fundamental evolutionary process, but the definition can greatly vary depending on the organisms under consideration. In Fungi, more than in other life kingdoms, the species concept relies on many controversal and conflicting biological, morphological, ecological and phylogenetic criteria (Cai et al. 2011; Giraud et al. 2008; Kohn 2005; Taylor et al. 2000). For instance, depending on the choice of these criteria, one can consider one or three species in *S. paradoxus*, simply because reproductive isolation occurs between different genetic lineages (discussed in Sect. 4.3.5), or four or seven species in *Lentinula* (Basidiomycota), according to morphological or phylogenetic criteria, respectively (Taylor et al. 2006). Similarly in *Neurospora*, most of "genetic" species are reproductively isolated, but exceptions can be found, since some genetically distinct lineages are still able to hybridize (Dettman et al. 2003a, b).

Genomics is the upcoming – but not absolute – criteria to shed light on speciation in Fungi. Here, I review recent studies that investigated a continuum of genomic processes to understand how speciation occurs in Fungi, from the early steps of genetic divergence to the establishment of complete reproductive isolation.

4.4.1 Beginnings of Speciation: Adaptation Drives Early Genetic Incompatibilities

When different strains of the same species occur in distinct ecological niches, they undergo contrasting adaptive constraints. Thus mutations that occur in the genome are differently selected. The accumulation of such advantageous mutations drives toward an optimal fitness for each strain in its own niche, but could also generate incompatibilities if the mutated gene interacts negatively with another one that was selected in a contrasted environment. This phenomenon could be considered as the basis of speciation, since strains from the same species that evolved under contrasted ecological conditions could, in theory, accumulate incompatibilities, leading to progeny with reduced fitness (Gourbiere and Mallet 2010). Anderson et al. (2010) explored the genomes of two experimentally evolved strains of *S. cerevisiae*. They detected early mutations underlying adaptation to high salt and low glucose environments, and found that the progeny of crosses between strains from two experimental populations had a strong fitness decrease in low glucose concentration conditions. They showed that this fitness reduction resulted from strong genetic incompatibility between two mutated alleles, each of them inherited from a distinct parent (Fig. 4.9a). Such within-species incompatibility has also been highlighted in natural populations of *S. cerevisiae* and involved two genes, each present in two allelic states. All combinations between alleles of the two genes could be found in the studied populations, except one. By generating individuals exhibiting the missing allelic combination, Demogines et al. (2008) found that these alleles combined increased the genomic mutation rate, which resulted in long-term accumulation of fitness-defect mutations. These studies suggest that adaptive mutations in a few genes could be sufficient to generate reproductive isolation when selection is strong enough, and eventually lead to speciation. Because yeasts have a

short generation time, one can expect to observe such incompatibilities after a very short evolutionary time. For instance, *S. cerevisiae* and *S. paradoxus* are distinct species that have diverged 0.4–3.5 million years ago. During this time, their genomes have accumulated divergence at 5–10 % of their sites (Fig. 4.8; Liti et al. 2006). This could represent a substantial amount of potential incompatible mutations between species, which however remains to be tested. Large genomic rearrangements, such as chromosomal translocations, could also occur after a very short evolutionary time and generate reproductive isolation. For instance, Liti et al. crossed natural strains of *S. paradoxus* with different chromosome configurations. They found that reproductive isolation, measured as progeny survival, was indeed correlated with the proportion of single nucleotide divergence (Fig. 4.8), but also dramatically decreased when parents had different chromosomal configurations, even if these translocations occurred only a few thousand years ago (Liti et al. 2006).

4.4.2 Genomic Investigation of Inter-species Incompatibilities in Yeasts

When speciation is established after several million years, hybridization is still possible between closely related species, and molecular mechanisms could evolve to maintain and enforce reproductive isolation. For instance, *S. cerevisiae* and *S. paradoxus* produce viable but mostly sterile hybrids, with less than 1 % viability of progeny. Greig et al. (2002) suggested that sterility of hybrids could result from incompatibilities between so-called "speciation genes." To verify this hypothesis, they replaced chromosomes of *S. cerevisiae* by their homologues from *S. paradoxus*. Surprisingly, all tested *S. paradoxus* chromosomes (representing 43 % of the genome) were compatible with the *S. cerevisiae* genome, suggesting that speciation genes were unlikely to play a major role in hybrid sterility (Greig 2007). Two independent genome-wide analyses of progeny from enforced sporulation of interspecies hybrids showed that all expected genomic combinations between *S. paradoxus* and *S. cerevisiae* genomes occurred, even after recombination between parental chromosomes (Kao et al. 2010; Xu and He 2011). Evidence for genetic incompatibilities involving a single pair of speciation genes has been investigated at larger evolutionary scales in *Saccharomyces* species and has so far only been found between the *S. cerevisiae* mitochondrial genome and a nuclear gene of its farthest relative in the group, *S. uvarum* (Lee et al. 2008). Such a lack of evidence for a role of genetic incompatibilities between species is astonishing considering some of the evidence found within *S. cerevisiae* and discussed above, and considering the high genomic sequence divergence (5–20 %) observed between *Saccharomyces* species (Fig. 4.8; Dujon 2006). However, we recently found that high molecular divergence between distant *S. cerevisiae* and *S. kudriavzevii* did not systematically result in functional disorders of essential protein complexes in hybrids, suggesting that vital functions are highly robust to inter-species hybridization (Leducq et al. 2012).

Some other mechanisms that may affect hybrids were also investigated. For instance, genomes of different *Saccharomyces* species are non-collinear, since all of them have undergone independent chromosome translocations. Delneri et al. (2003) demonstrated that these translocations are partly responsible for hybrid sterility between *S. cerevisiae* and *S. mikatae*, since homologous chromosomes could not match perfectly during meiosis, frequently resulting in aneuploid hybrids (Fig. 4.9b). These results confirmed previous findings at the intra-species level (Liti et al. 2006).

All these findings reinforce the hypothesis that no single genetic mechanism is responsible for sterility of inter-species hybrids. Hybrid sterility possibly results from chromosome mismatches during meiosis, combined with multiple complex incompatibilities, which probably involve many genes with individually negligible effects as well as unsuspected underlying molecular processes.

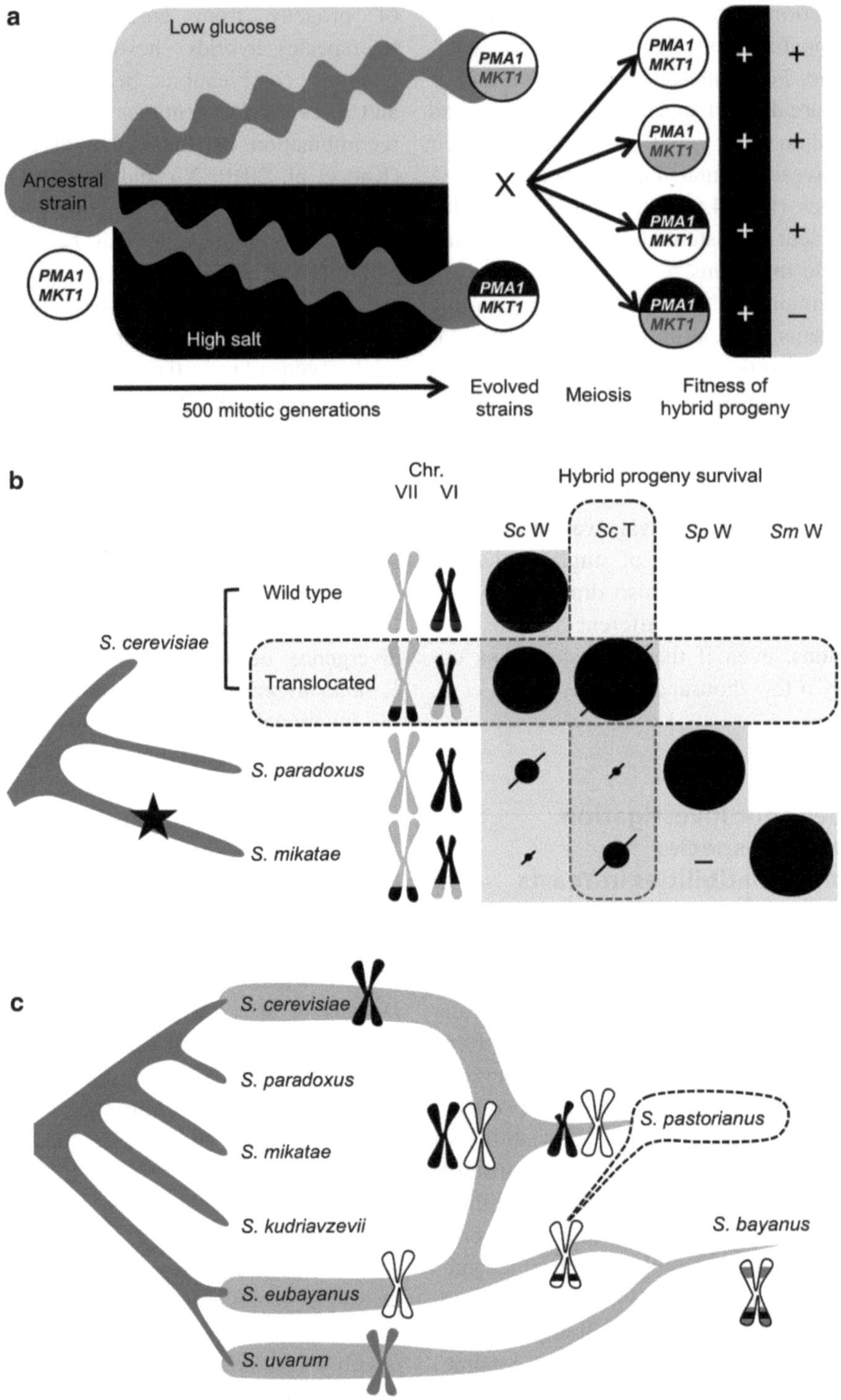

Fig. 4.9 Genomics of speciation and hybridization in *Saccharomyces* yeasts. (**a**) Experimental evolution of *S. cerevisiae* in two contrasting environments (high salt and low glucose concentration, respectively). The analysis of genomes revealed two adaptive mutations affecting genes *PMA1* (high salt; codes for a proton efflux pump) and *MKT1* (low glucose; regulator of mRNA encoding mitochondrial proteins). Hybrid progeny bearing

4.4.3 Genomic Evidence for Speciation by Hybridization

The study of fungal genome evolution provides some evidence that reproductive barriers arising after speciation are not always complete. Indeed, some genomic analyses revealed potential cases of introgression, suggesting that these arose from successful hybridization between two distinct species. Early steps of speciation by hybridization are poorly understood but Dunn et al. (2013) recently showed that rearrangements could experimentally arise among homologous chromosomes of *S. cerevisiae* and *S. uvarum* within their first generation hybrids, when evolved in ammonium-limited conditions. The break points of recombination systematically occurred in the gene *MEP2* coding for an ammonium permease, likely suggesting that chimeric Mep2 proteins confer a higher fitness advantage to the hybrids. However, because interspecific hybrids are often aneuploid and sterile in *Saccharomyces*, this kind of mechanism is likely to lead to an evolutionary dead-end rather than to actual introgression. This is the case of the two sterile brewing yeasts *S. pastorianus* and *S. bayanus*, which resulted from hybridization induced by domestication (Fig. 4.9c). The genome analysis of *S. pastorianus* revealed that it resulted from allotetraploidization between genomes of *S. cerevisiae* and a close relative of *S. uvarum*, *S. eubayanus* (Dunn and Sherlock 2008; Libkind et al. 2011). This hybridization event is likely to have been followed by chromosomal rearrangements resulting in aneuploidy. Similarly, the mosaic genome of *S. bayanus* suggested that it resulted from the fortuitous integration of *S. pastorianus* genomic elements in the *S. eubayanus* genome. In both cases, the genomic hybridization and rearrangements resulted from strong anthropic pressure to select for optimal brewing properties. Similarly, the yeast *Pichia sorbitophila* (CTG clade) is a fortuitous product of industry. Its 14 chromosomes are the results of recent hybridization followed by polyploidization between two unknown but related strains that had only seven chromosomes (Louis et al. 2012) (Fig. 4.2). The evidence of many introgression traces and the absence of entire chromosomes of one of the putative parents was attested by the complete absence of polymorphism between regions of some homologous chromosomes, suggesting that this hybridization was followed by many chromosomal rearrangements and losses. It is interesting to note that the set of genes conferring high osmotic resistance to this species is likely to be the sum of contributions from both parents. For instance, genes enabling metabolism of maltose were inherited from one parent, whereas genes involved in sorbitol metabolism were inherited from both parents.

Finally, in the human pathogenic *Coccioides immitis* (Ascomycota), Neafsey et al. (2010) found evidence for recent genomic introgression from its sister species, especially in geographical areas where both species were sympatric. Introgressed genomic regions contained genes involved in host immune response, once again suggesting that generating new genetic combinations by hybridization could be favoured in selective environments. Hence, fungal genomics provides many examples that it is sometimes advantageous to break reproductive barriers between species in order to generate mosaic

Fig. 4.9 both derived alleles expresses strong fitness decrease (–) in environment with low glucose, suggesting incompatibilities between these derived alleles (Anderson et al. 2010). (**b**) In *S. mikatae* (*Sm* W), a translocation occurred between chromosomes VI and VII (*black star*). Wild-type *S. cerevisiae* (*Sc* W) and *S. paradoxus* (*Sp* W) strains have the ancestral chromosomal configuration. The decrease in *S. cerevisiae* × *S. mikatae* hybrid progeny viability is partly restored when chromosomes of the *Sc*W strains are manipulated so as to be collinear with those of *Sm*W (*Sc* T; *dotted frame*). The *black* disk areas in the cross table are proportional to mean hybrid progeny survival (*bars* indicate standard deviation among replicates; Delneri et al. 2003). (**c**) Multiple anthropic hybridization (*grey*) and *horizontal* transfer (*dotted line*) events between natural *Saccharomyces* species (*dark grey* evolutionary tree) led to the emergence of two brewing species *S. pastorianus* and *S. bayanus*. Evolutionary steps of genome evolution (symbolized by a single duplicated chromosome) were highly simplified. Each color represents the part of the genome from a mother species (Dunn and Sherlock 2008; Libkind et al. 2011)

genomes with new combinations of mutations that evolved independently. Put together, these mutations could bring a fitness advantage in a new ecological niche, but sometimes at the expense of sexual reproduction.

4.5 Conclusion

The advent of genomics in the last 10 years has shed light on the evolution and ecology of Fungi. Budding yeasts, *Neurospora* and other model organisms were yet again the pioneers in exploring the footprint of evolution and ecology on fungal diversity, but unlike classical biological tools, genomics opened this exploration to other diverse forms of Fungi. Using eco-genomics, it is now possible to understand how Fungi conquered such a broad range of ecological niches. First, fungal genomes are the result of 1 billion years of evolution that took place in highly contrasting ecological niches, promoting variable life traits and reproductive modes. During this evolution, genomes were profoundly rearranged through proliferation of mobile elements, whole-genome duplications, gain of adaptive genes, gene compaction and loss of genes linked to metabolisms which became non-essential for Fungi that have developed strong dependency to their host. Second, studies carried out at the lower evolutionary scales investigated early steps of genomic evolution in varying ecological niches. Using experimental evolution, these studies highlighted the role of early mutations in a few key adaptive genes involved in simple traits with strong selection. In natural populations, such adaptive mutations were indeed tightly associated with particular ecological conditions. Adaptive genes could also be gained by horizontal transfers between Fungi and bacteria in some pathogens. In other cases, adaptive traits could be acquired by new, advantageous combinations of genes after chromosomal rearrangements. *Saccharomyces* yeasts are particularly powerful eco-genomics models to investigate all these mechanisms in depth. Finally, local adaptation to contrasting ecological niches could lead to genetic incompatibilities between individuals of the same species, and eventually to reproductive isolation. However, there is little evidence for the occurrence and fixation of such strong incompatibilities in natural populations and for their role in the maintenance of efficient reproductive barriers, even after millions of years of divergence between lineages. Other mechanisms, such as large chromosomal translocations and accumulation of many incompatibilities with smaller effects, are more likely to drive speciation at the genomic level. Moreover, breaking reproductive barriers and reshuffling mutations inherited from different species or divergent strains is sometimes advantageous for organisms to conquer new ecological niches.

Fungi are powerful eco-genomics models to investigate a large range of evolutionary and adaptive mechanisms that drive genome evolution. Of course, the number of studies focusing on a handful of model organisms is still increasing, and these are necessary to understand fundamental mechanisms underlying this evolution. But forthcoming work using eco-genomic approaches will be able to study the remaining underexplored fungal diversity.

Acknowledgments I address special thanks to Mary Thaler, Isabelle Gagnon-Arsenault, Guillaume Charron, Emilie Bernatchez, Marie Filteau, Luca Freschi, Philippe Reignault, Sylvain Billiard, Louis Bernier, Christian R. Landry and two anonymous reviewers for helpful comments. This work was supported by a Natural Sciences and Engineering Research Council of Canada (NSERC) grant and a Human Frontier Science Program (HFSP) RGY0073/2010 grant allocated to Christian R. Landry. I was supported by a fellowship from the Fonds de Recherche en Santé du Québec (FRSQ).

Glossary

Endophyte Organism spending its entire life cycle within a plant.

Heterokaryotic When a cell contains two or more nuclei.

Heterothallism When sexual reproduction can only occur between two phenotypically indistinct individuals from the same species, but expressing different sexual *idiomorphs* (allogamy). Mostly present in algae and Fungi.

Homothallism When sexual reproduction can occur between any individuals from the same species (autogamy), in contrast to heterothallism – **Pseudo-homothallism** derives from heterothallism but the co-transmission of two sexual *idiomorphs* during meiosis allows autogamy.

Idiomorph – or Mating-type Sexual determinants in eukaryotes. Designates compatible sexual partners during reproduction: for instance male and female in plants and animals or *MAT-a* and *MAT-α* in yeasts.

Mycorrhiza Symbiosis between a fungus and roots of a vascular plant. **Ectomycorrhiza** perform the interaction within the host tissues whereas **endomycorrhiza** perform the symbiosis within the host cells.

Quantitative Trait Loci (QTL) Portions of the genome physically co-segregating with an inherited trait, and thus physically linked to at least one gene involved in this trait.

Saprotroph Fungi able to absorb nutrients from dead or decayed organic matter.

Spore In Fungi, a unicellular, resistant reproductive structure formed by meiosis or mitosis, able to produce a new individual after possible dispersal and germination. **Ascospores** are spores produced by Ascomycota.

Symbiosis Close and reciprocally beneficial interaction between two organisms of different species, providing each other with protection, suitable habitat or nutrients.

References

Aa E, Townsend JP, Adams RI, Nielsen KM, Taylor JW (2006) Population structure and gene evolution in *Saccharomyces cerevisiae*. FEMS Yeast Res 6(5): 702–715

Aime MC, Matheny PB, Henk DA, Frieders EM, Nilsson RH, Piepenbring M et al (2006) An overview of the higher level classification of Pucciniomycotina based on combined analyses of nuclear large and small subunit rDNA sequences. Mycologia 98(6):896–905

Anderson JB, Funt J, Thompson DA, Prabhu S, Socha A, Sirjusingh C et al (2010) Determinants of divergent adaptation and Dobzhansky-Muller interaction in experimental yeast populations. Curr Biol 20(15): 1383–1388

Araya CL, Payen C, Dunham MJ, Fields S (2010) Whole-genome sequencing of a laboratory-evolved yeast strain. BMC Genomics 11:88

Billiard S, Lopez-Villavicencio M, Devier B, Hood ME, Fairhead C, Giraud T (2011) Having sex, yes, but with whom? Inferences from fungi on the evolution of anisogamy and mating types. Biol Rev Camb Philos Soc 86(2):421–442

Borneman AR, Desany BA, Riches D, Affourtit JP, Forgan AH, Pretorius IS et al (2011) Whole-genome comparison reveals novel genetic elements that characterize the genome of industrial strains of Saccharomyces cerevisiae. PLoS Genet 7(2):e1001287

Burmester A, Shelest E, Glockner G, Heddergott C, Schindler S, Staib P et al (2011) Comparative and functional genomics provide insights into the pathogenicity of dermatophytic fungi. Genome Biol 12(1):R7

Butler G, Rasmussen MD, Lin MF, Santos MA, Sakthikumar S, Munro CA et al (2009) Evolution of pathogenicity and sexual reproduction in eight Candida genomes. Nature 459(7247):657–662

Cai L, Giraud T, Zhang N, Begerow D, Cai GH, Shivas RG (2011) The evolution of species concepts and species recognition criteria in plant pathogenic fungi. Fungal Divers 50(1):121–133

Chang SL, Lai HY, Tung SY, Leu JY (2013) Dynamic large-scale chromosomal rearrangements fuel rapid adaptation in yeast populations. PLoS Genet 9(1):e1003232

Cowen LE, Nantel A, Whiteway MS, Thomas DY, Tessier DC, Kohn LM et al (2002) Population genomics of drug resistance in *Candida albicans*. Proc Natl Acad Sci USA 99(14):9284–9289

Cubillos FA, Billi E, Zorgo E, Parts L, Fargier P, Omholt S et al (2011) Assessing the complex architecture of polygenic traits in diverged yeast populations. Mol Ecol 20(7):1401–1413

de Jonge R, van Esse HP, Maruthachalam K, Bolton MD, Santhanam P, Saber MK et al (2012) Tomato immune receptor Ve1 recognizes effector of multiple fungal pathogens uncovered by genome and RNA sequencing. Proc Natl Acad Sci USA 109(13): 5110–5115

Delneri D, Colson I, Grammenoudi S, Roberts IN, Louis EJ, Oliver SG (2003) Engineering evolution to study speciation in yeasts. Nature 422(6927): 68–72

Demogines A, Wong A, Aquadro C, Alani E (2008) Incompatibilities involving yeast mismatch repair genes: a role for genetic modifiers and implications for disease penetrance and variation in genomic mutation rates. PLoS Genet 4(6):e1000103

Demuth JP, De Bie T, Stajich JE, Cristianini N, Hahn MW (2006) The evolution of mammalian gene families. PLoS One 1(1):e85

Dettman JR, Jacobson DJ, Taylor JW (2003a) A multilocus genealogical approach to phylogenetic species recognition in the model eukaryote Neurospora. Evolution 57(12):2703–2720

Dettman JR, Jacobson DJ, Turner E, Pringle A, Taylor JW (2003b) Reproductive isolation and phylogenetic divergence in Neurospora: comparing methods of species recognition in a model eukaryote. Evolution 57(12):2721–2741

Dujon B (2005) Hemiascomycetous yeasts at the forefront of comparative genomics. Curr Opin Genet Dev 15(6):614–620

Dujon B (2006) Yeasts illustrate the molecular mechanisms of eukaryotic genome evolution. Trends Genet 22(7):375–387

Dujon B, Sherman D, Fischer G, Durrens P, Casaregola S, Lafontaine I et al (2004) Genome evolution in yeasts. Nature 430(6995):35–44

Dunn B, Sherlock G (2008) Reconstruction of the genome origins and evolution of the hybrid lager yeast Saccharomyces pastorianus. Genome Res 18(10): 1610–1623

Dunn B, Paulish T, Stanbery A, Piotrowski J, Koniges G, Kroll E et al (2013) Recurrent rearrangement during adaptive evolution in an interspecific yeast hybrid suggests a model for rapid introgression. PLoS Genet 9(3):e1003366

Ellison CE, Hall C, Kowbel D, Welch J, Brem RB, Glass NL et al (2011a) Population genomics and local adaptation in wild isolates of a model microbial eukaryote. Proc Natl Acad Sci USA 108(7):2831–2836

Ellison CE, Stajich JE, Jacobson DJ, Natvig DO, Lapidus A, Foster B et al (2011b) Massive changes in genome architecture accompany the transition to self-fertility in the Filamentous Fungus *Neurospora tetrasperma*. Genetics 189(1):55–69, U652

Fernandez-Fueyo E, Ruiz-Duenas FJ, Ferreira P, Floudas D, Hibbett DS, Canessa P et al (2012) Comparative genomics of *Ceriporiopsis subvermispora* and *Phanerochaete chrysosporium* provide insight into selective ligninolysis. Proc Natl Acad Sci USA 109(14):5458–5463

Floudas D, Binder M, Riley R, Barry K, Blanchette RA, Henrissat B et al (2012) The Paleozoic origin of enzymatic lignin decomposition reconstructed from 31 fungal genomes. Science 336(6089):1715–1719

Fraser JA, Diezmann S, Subaran RL, Allen A, Lengeler KB, Dietrich FS et al (2004) Convergent evolution of chromosomal sex-determining regions in the animal and fungal kingdoms. PLoS Biol 2(12):e384

Fraser JA, Stajich JE, Tarcha EJ, Cole GT, Inglis DO, Sil A et al (2007) Evolution of the mating type locus: insights gained from the dimorphic primary fungal pathogens *Histoplasma capsulatum*, Coccidioides immitis, and Coccidioides posadasii. Eukaryot Cell 6(4):622–629

Gao Q, Jin K, Ying SH, Zhang Y, Xiao G, Shang Y et al (2011) Genome sequencing and comparative transcriptomics of the model entomopathogenic fungi *Metarhizium anisopliae* and *M. acridum*. PLoS Genet 7(1):e1001264

Giraud T, Refregier G, Le Gac M, de Vienne DM, Hood ME (2008) Speciation in fungi. Fungal Genet Biol 45(6):791–802

Gomes AC, Miranda I, Silva RM, Moura GR, Thomas B, Akoulitchev A et al (2007) A genetic code alteration generates a proteome of high diversity in the human pathogen *Candida albicans*. Genome Biol 8(10):R206

Goodwin SB, M'Barek SB, Dhillon B, Wittenberg AH, Crane CF, Hane JK et al (2011) Finished genome of the fungal wheat pathogen *Mycosphaerella graminicola* reveals dispensome structure, chromosome plasticity, and stealth pathogenesis. PLoS Genet 7(6):e1002070

Gordon JL, Armisen D, Proux-Wera E, OhEigeartaigh SS, Byrne KP, Wolfe KH (2011) Evolutionary erosion of yeast sex chromosomes by mating-type switching accidents. Proc Natl Acad Sci USA 108(50): 20024–20029

Gourbiere S, Mallet J (2010) Are species real? The shape of the species boundary with exponential failure, reinforcement, and the "missing snowball". Evolution 64(1):1–24

Greig D (2007) A screen for recessive speciation genes expressed in the gametes of F1 hybrid yeast. PLoS Genet 3(2):e21

Greig D, Borts RH, Louis EJ, Travisano M (2002) Epistasis and hybrid sterility in *Saccharomyces*. Proc R Soc B 269(1496):1167–1171

Grigoriev IV, Nordberg H, Shabalov I, Aerts A, Cantor M, Goodstein D et al (2012) The genome portal of the Department of Energy Joint Genome Institute. Nucleic Acids Res 40(Database issue):D26–D32

Hane JK, Rouxel T, Howlett BJ, Kema GHJ, Goodwin SB, Oliver RP (2011) A novel mode of chromosomal evolution peculiar to filamentous Ascomycete fungi. Genome Biol 12(5):R45

Hibbett DS (2006) A phylogenetic overview of the Agaricomycotina. Mycologia 98(6):917–925

Hibbett DS, Taylor JW (2013) Fungal systematics: is a new age of enlightenment at hand? Nat Rev Microbiol 11(2):129–133

Hittinger CT, Goncalves P, Sampaio JP, Dover J, Johnston M, Rokas A (2010) Remarkably ancient balanced polymorphisms in a multi-locus gene network. Nature 464(7285):54–58

Hyma KE, Fay JC (2013) Mixing of vineyard and oak-tree ecotypes of *Saccharomyces cerevisiae* in North American vineyards. Mol Ecol 22:2917

James TY, Kauff F, Schoch CL, Matheny PB, Hofstetter V, Cox CJ et al (2006) Reconstructing the early evolution of fungi using a six-gene phylogeny. Nature 443(7113):818–822

Joneson S, Stajich JE, Shiu SH, Rosenblum EB (2011) Genomic transition to pathogenicity in Chytrid fungi. PLoS Pathog 7(11):e1002338

Kao KC, Schwartz K, Sherlock G (2010) A genome-wide analysis reveals no nuclear Dobzhansky-Muller pairs of determinants of speciation between *S. Cerevisiae* and *S. Paradoxus*, but suggests more complex incompatibilities. PLoS Genet 6(7):e1001038

Kelkar YD, Ochman H (2012) Causes and consequences of genome expansion in fungi. Genome Biol Evol 4(1):13–23

Kellis M, Birren BW, Lander ES (2004) Proof and evolutionary analysis of ancient genome duplication in the yeast *Saccharomyces cerevisiae*. Nature 428(6983):617–624

Kendrick B (2001) Fungi: ecological importance and impact on humans. In: eLS. Wiley & Sons, Ltd: Chichester

Kohn LM (2005) Mechanisms of fungal speciation. Annu Rev Phytopathol 43:279–308

Koonin EV (2011) The logic of chance: the nature and origin of biological evolution. Ft Press, Upper Saddle River

Kuehne HA, Murphy HA, Francis CA, Sniegowski PD (2007) Allopatric divergence, secondary contact, and genetic isolation in wild yeast populations. Curr Biol 17(5):407–411

Leducq JB, Charron G, Diss G, Gagnon-Arsenault I, Dube AK, Landry CR (2012) Evidence for the robustness of protein complexes to inter-species hybridization. PLoS Genet 8(12):e1003161

Lee HY, Chou JY, Cheong L, Chang NH, Yang SY, Leu JY (2008) Incompatibility of nuclear and mitochondrial genomes causes hybrid sterility between two yeast species. Cell 135(6):1065–1073

Lee SC, Corradi N, Doan S, Dietrich FS, Keeling PJ, Heitman J (2010) Evolution of the sex-related locus and genomic features shared in microsporidia and fungi. PLoS One 5(5):e10539

Libkind D, Hittinger CT, Valerio E, Goncalves C, Dover J, Johnston M et al (2011) Microbe domestication and the identification of the wild genetic stock of lager-brewing yeast. Proc Natl Acad Sci USA 108(35):14539–14544

Liti G, Barton DB, Louis EJ (2006) Sequence diversity, reproductive isolation and species concepts in *Saccharomyces*. Genetics 174(2):839–850

Liti G, Carter DM, Moses AM, Warringer J, Parts L, James SA et al (2009a) Population genomics of domestic and wild yeasts. Nature 458(7236): 337–341

Liti G, Haricharan S, Cubillos FA, Tierney AL, Sharp S, Bertuch AA et al (2009b) Segregating YKU80 and TLC1 alleles underlying natural variation in telomere properties in wild yeast. PLoS Genet 5(9):e1000659

Louis VL, Despons L, Friedrich A, Martin T, Durrens P, Casaregola S et al (2012) *Pichia sorbitophila*, an interspecies yeast hybrid, reveals early steps of genome resolution after *polyploidization.* G3 (Bethesda) 2(2):299–311

Ma LJ, Ibrahim AS, Skory C, Grabherr MG, Burger G, Butler M et al (2009) Genomic analysis of the basal lineage fungus *Rhizopus oryzae* reveals a whole-genome duplication. PLoS Genet 5(7):e1000549

Magwene PM, Kayikci O, Granek JA, Reininga JM, Scholl Z, Murray D (2011) Outcrossing, mitotic recombination, and life-history trade-offs shape genome evolution in *Saccharomyces cerevisiae*. Proc Natl Acad Sci USA 108(5):1987–1992

Martin F, Aerts A, Ahren D, Brun A, Danchin EG, Duchaussoy F et al (2008) The genome of *Laccaria bicolor* provides insights into mycorrhizal symbiosis. Nature 452(7183):88–92

Martin F, Kohler A, Murat C, Balestrini R, Coutinho PM, Jaillon O et al (2010) Perigord black truffle genome uncovers evolutionary origins and mechanisms of symbiosis. Nature 464(7291): 1033–1038

Martinez D, Challacombe J, Morgenstern I, Hibbett D, Schmoll M, Kubicek CP et al (2009) Genome, transcriptome, and secretome analysis of wood decay fungus *Postia placenta* supports unique mechanisms of lignocellulose conversion. Proc Natl Acad Sci USA 106(6):1954–1959

Menkis A, Jacobson DJ, Gustafsson T, Johannesson H (2008) The mating-type chromosome in the filamentous ascomycete Neurospora tetrasperma represents a model for early evolution of sex chromosomes. PLoS Genet 4(3):e1000030

Metin B, Findley K, Heitman J (2010) The mating type locus (MAT) and sexual reproduction of *Cryptococcus heveanensis*: insights into the evolution of sex and sex-determining chromosomal regions in fungi. PLoS Genet 6(5):e1000961

Muller LA, Lucas JE, Georgianna DR, McCusker JH (2011) Genome-wide association analysis of clinical vs. nonclinical origin provides insights into *Saccharomyces cerevisiae* pathogenesis. Mol Ecol 20(19):4085–4097

Nash TH (2008) Lichen biology, 2nd edn. Cambridge University Press, Cambridge/New York

Neafsey DE, Barker BM, Sharpton TJ, Stajich JE, Park DJ, Whiston E et al (2010) Population genomic sequencing of Coccidioides fungi reveals recent hybridization and transposon control. Genome Res 20(7):938–946

Nicholson MJ, McSweeney CS, Mackie RI, Brookman JL, Theodorou MK (2010) Diversity of anaerobic gut fungal populations analysed using ribosomal ITS1 sequences in faeces of wild and domesticated herbivores. Anaerobe 16(2):66–73

Ohm RA, Feau N, Henrissat B, Schoch CL, Horwitz BA, Barry KW et al (2012) Diverse lifestyles and strategies of plant pathogenesis encoded in the genomes of eighteen Dothideomycetes fungi. PLoS Pathog 8(12):e1003037

Peyretaillade E, El Alaoui H, Diogon M, Polonais V, Parisot N, Biron DG et al (2011) Extreme reduction and compaction of microsporidian genomes. Res Microbiol 162(6):598–606

Read DJ (1991) Mycorrhizas in ecosystems. Experientia 47(4):376–391

Rep M, Kistler HC (2010) The genomic organization of plant pathogenicity in *Fusarium* species. Curr Opin Plant Biol 13(4):420–426

Rice DP, Townsend JP (2012) A test for selection employing quantitative trait locus and mutation accumulation data. Genetics 190(4):1533–1545

Rocha R, Pereira PJ, Santos MA, Macedo-Ribeiro S (2011) Unveiling the structural basis for translational ambiguity tolerance in a human fungal pathogen. Proc Natl Acad Sci USA 108(34):14091–14096

Ruderfer DM, Pratt SC, Seidel HS, Kruglyak L (2006) Population genomic analysis of outcrossing and recombination in yeast. Nat Genet 38(9):1077–1081

San-Blas G, Calderone RA (2008) Pathogenic fungi: insights in molecular biology. Caister Academic Press, Norfolk

Santana MF, Silva JC, Batista AD, Ribeiro LE, da Silva GF, de Araujo EF et al (2012) Abundance, distribution and potential impact of transposable elements in the genome of *Mycosphaerella fijiensis*. BMC Genomics 13(1):720

Schacherer J, Shapiro JA, Ruderfer DM, Kruglyak L (2009) Comprehensive polymorphism survey elucidates population structure of *Saccharomyces cerevisiae*. Nature 458(7236):342–345

Sniegowski PD, Dombrowski PG, Fingerman E (2002) *Saccharomyces cerevisiae* and *Saccharomyces paradoxus* coexist in a natural woodland site in North America and display different levels of reproductive isolation from European conspecifics. FEMS Yeast Res 1(4):299–306

Souciet JL, Dujon B, Gaillardin C, Johnston M, Baret PV, Cliften P et al (2009) Comparative genomics of protoploid Saccharomycetaceae. Genome Res 19(10):1696–1709

Stajich JE, Berbee ML, Blackwell M, Hibbett DS, James TY, Spatafora JW et al (2009) The fungi. Curr Biol 19(18):R840–R845

Stukenbrock EH, Bataillon T, Dutheil JY, Hansen TT, Li R, Zala M et al (2011) The making of a new pathogen: insights from comparative population genomics of the domesticated wheat pathogen *Mycosphaerella graminicola* and its wild sister species. Genome Res 21(12):2157–2166

Sun S, Heitman J (2011) Is sex necessary? BMC Biol 9(56)

Taylor JW, Jacobson DJ, Kroken S, Kasuga T, Geiser DM, Hibbett DS et al (2000) Phylogenetic species recognition and species concepts in fungi. Fungal Genet Biol 31(1):21–32

Taylor JW, Turner E, Townsend JP, Dettman JR, Jacobson D (2006) Eukaryotic microbes, species recognition and the geographic limits of species: examples from the kingdom fungi. Philos Trans R Soc B 361(1475):1947–1963

Tsai IJ, Bensasson D, Burt A, Koufopanou V (2008) Population genomics of the wild yeast *Saccharomyces paradoxus*: quantifying the life cycle. Proc Natl Acad Sci USA 105(12):4957–4962

Tsui CKM, DiGuistini S, Wang Y, Feau N, Dhillon B, Bohlmann J et al (2013) Unequal recombination and evolution of the Mating-Type (MAT) Loci in the pathogenic fungus *Grosmannia clavigera* and relatives. G3-Genes Genom Genet 3(3):465–480

Votintseva AA, Filatov DA (2009) Evolutionary strata in a small mating-type-specific region of the smut fungus *Microbotryum violaceum*. Genetics 182(4):1391–1396

Wang H, Xu Z, Gao L, Hao B (2009) A fungal phylogeny based on 82 complete genomes using the composition vector method. BMC Evol Biol 9:195

Wang QM, Liu WQ, Liti G, Wang SA, Bai FY (2012a) Surprisingly diverged populations of *Saccharomyces cerevisiae* in natural environments remote from human activity. Mol Ecol 21(22):5404–5417

Wang Z, Kin K, Lopez-Giraldez F, Johannesson H, Townsend JP (2012b) Sex-specific gene expression during asexual development of *Neurospora crassa*. Fungal Genet Biol 49(7):533–543

Warringer J, Zorgo E, Cubillos FA, Zia A, Gjuvsland A, Simpson JT et al (2011) Trait variation in yeast is defined by population history. PLoS Genet 7(6):e1002111

Whittle CA, Johannesson H (2011) Evidence of the accumulation of allele-specific non-synonymous substitutions in the young region of recombination suppression within the mating-type chromosomes of *Neurospora tetrasperma*. Heredity (Edinb) 107(4):305–314

Whittle CA, Sun Y, Johannesson H (2011) Degeneration in codon usage within the region of suppressed recombination in the mating-type chromosomes of *Neurospora tetrasperma*. Eukaryot Cell 10(4):594–603

Wik L, Karlsson M, Johannesson H (2008) The evolutionary trajectory of the mating-type (mat) genes in *Neurospora* relates to reproductive behavior of taxa. BMC Evol Biol 8:109

Will JL, Kim HS, Clarke J, Painter JC, Fay JC, Gasch AP (2010) Incipient balancing selection through adaptive loss of aquaporins in natural *Saccharomyces cerevisiae* populations. PLoS Genet 6(4):e1000893

Xu M, He X (2011) Genetic incompatibility dampens hybrid fertility more than hybrid viability: yeast as a case study. PLoS One 6(4):e18341

5 Integrating Phenotypic Plasticity Within an Ecological Genomics Framework: Recent Insights from the Genomics, Evolution, Ecology, and Fitness of Plasticity

Matthew Morris and Sean M. Rogers

Abstract

E.B. Ford's 1964 book *Ecological Genetics* was a call for biologists to engage in multidisciplinary work in order to elucidate the link between genotype, phenotype, and fitness for ecologically relevant traits. In this review, we argue that the integration of an ecological genomics framework in studies of phenotypic plasticity is a promising approach to elucidate the causal links between genes and the environment, particularly during colonization of novel environments, environmental change, and speciation. This review highlights some of the questions and hypotheses generated from a mechanistic, evolutionary, and ecological perspective, in order to direct the continued and future use of genomic tools in the study of phenotypic plasticity.

Keywords

Community genetics • Speciation • Genetic compensation • Genetic assimilation • Adaptation

5.1 Introduction

E.B. Ford's 1964 book *Ecological Genetics* was a call for biologists to engage in multidisciplinary work in order to elucidate the link between genotype, phenotype, and fitness for ecologically relevant traits. It became rapidly clear that methodologies were the main limiting factor in meeting this goal, but recent next generation sequencing (NGS) technologies have reinvigorated interest in this field (Feder and Mitchell-Olds 2003; Orsini et al. 2013). Ecological genetics has given way to ecological genomics (Fig. 5.1), or the investigation of the entire set of genes that interact to produce the phenotype and shape the evolution of species and communities (Ungerer et al. 2008). Ecological genomics is limited less by technology than by the complexity of statistical tools required to quantify the voluminous data, the interdisciplinary knowledge required to fully understand the production of even a single phenotype, experimental constraints (the space required to perform carefully controlled experiments), and the nature of the organism (challenges to raising

M. Morris (✉) • S.M. Rogers
Department of Biological Sciences, University of Calgary, Calgary, AB, Canada
e-mail: mrjmorri@ucalgary.ca

C.R. Landry and N. Aubin-Horth (eds.), *Ecological Genomics: Ecology and the Evolution of Genes and Genomes*, Advances in Experimental Medicine and Biology 781,
DOI 10.1007/978-94-007-7347-9_5, © Springer Science+Business Media Dordrecht 2014

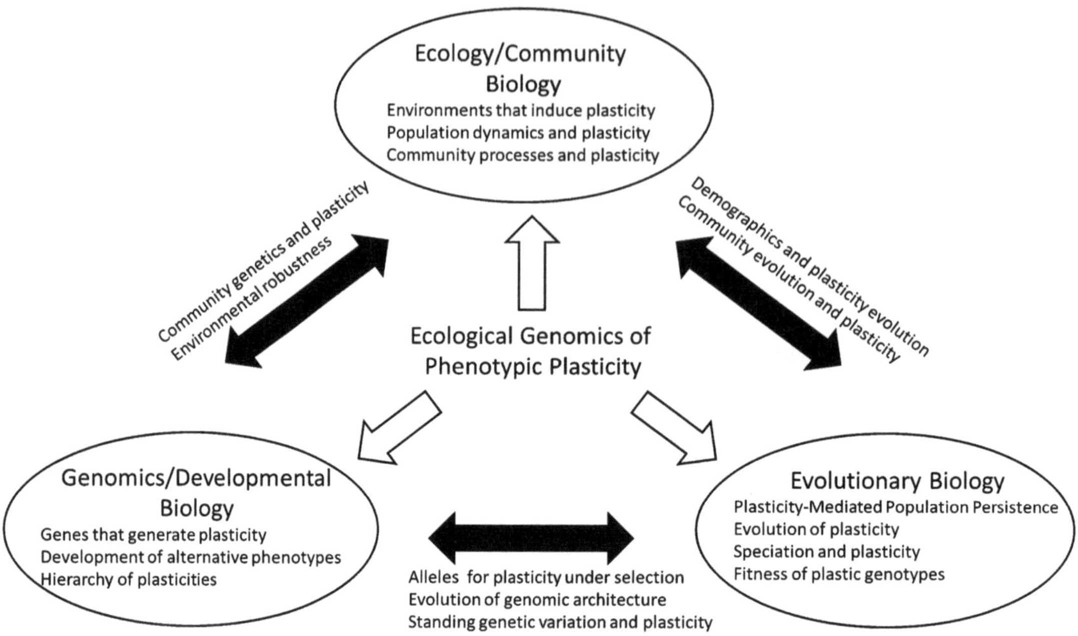

Fig. 5.1 The ecological genomics of phenotypic plasticity involves multidisciplinary work in the interrelated fields of developmental biology/genomics, ecology and community biology, and evolutionary biology. Some of the topics treated in this paper are shown

and breeding the organism, structures that inhibit DNA extraction, etc.), but initial work has been promising (Tollrian and Leese 2010; Whitehead et al. 2012; Andrew et al. 2013).

Ecological genomics is the telling of a complex story about the mechanisms governing the production of a phenotype, but moves beyond functional genomics (e.g., Dalziel et al. 2009) to ask questions concerning the evolution of the phenotype and its role in the greater community (Table 5.1). To date the main focus of ecological genomics has been on the genetic and molecular basis of ecologically relevant traits and their evolutionary consequences. Such an approach cannot capture the full story. Ecological genomics, to be successful, must recognize that genes can only go so far in producing a phenotype – the environment proposes the phenotype in a manner that cannot be separated from the genome (Moczek 2012). Ecological genomic approaches must therefore consider the role of *phenotypic plasticity*.

Phenotypic plasticity, the environmentally sensitive production of alternative phenotypes by a single genotype (DeWitt and Scheiner 2003) (Fig. 5.2), reminds us that individuals can exhibit phenotypic differentiation not only among genotypes, but also across environments. The conceptual framework of phenotypic plasticity broadens the scope of ecological genomics (Table 5.1) by asking questions and generating novel hypotheses that more fully integrate the environment into the production and evolution of the phenotype. Although plasticity has been the topic of numerous reviews (e.g., Bradshaw 1965; Alpert and Simms 2002; West-Eberhard 2003; Ghalambor et al. 2007; Fusco and Minelli 2010; Pfennig et al. 2010; Moczek et al. 2011; Fitzpatrick 2012; Moczek 2012) and theoretical work (e.g., Lande 2009; Thibert-Plante and Hendry 2011; Espinosa-Soto et al. 2011), only recently have researchers been able to focus on the integration of phenotypic plasticity with ecological genomic approaches (examples of

Table 5.1 Phenotypic plasticity increases the scope of questions asked by ecological genomics

Some questions addressed by ecological genomics
What genes underlie a particular phenotype?
What alleles are responsible for phenotypic differences between individuals or populations?
What is the nature of the developmental network that generates the phenotype?
How is phenotype development buffered against genetic or environmental perturbations?
What is the influence of the phenotype on individual success under natural ecological conditions?
How do phenotypically differentiated individuals differ in terms of fitness?
If the same phenotype is found on different phenotypic backgrounds, how does that alter its effect on the organism?
What is the evolutionary history of the phenotype?
How does the phenotype influence or constrain future evolution?
How do populations or species differ in the phenotype of interest?
What generates phenotypic diversity?
How does the phenotype affect population persistence under changing environments?
How does the phenotype affect community processes?
How does the phenotype affect the fitness of conspecifics?
How does the phenotype affect the fitness and evolution of other species?
Further questions raised by phenotypic plasticity
How does an organism sense its environment?
What environmental cues induce phenotypic change?
How reliable are environmental cues?
How do different environmental cues translate into different phenotypes?
How do changes to the internal environment affect developmental trajectories?
What generates non-plasticity?
How do genes shape the environments to which they plastically respond?
What mechanistically constitutes a reaction norm?
How are reaction norms affected by genetic and environmental perturbations?
How do reaction norms work together to produce plastic and non-plastic phenotypes?
What are the costs and limits of plasticity?
Under what circumstances is plasticity expected to evolve?
How does plasticity affect population persistence under changing environments?
How does plasticity drive evolutionary innovation and speciation?
How does plasticity affect the evolution of other species within a community?

initial forays into ecological genomics include Evans and Wheeler 2000; Renn et al. 2008; McCairns and Bernatchez 2010; De Boer et al. 2011; Richards et al. 2012; Schwartz and Bronikowski 2013).

Although plasticity involves a single genotype, genomic tools are essential for understanding the molecular basis for how alternative phenotypes may be produced. Given the goal of linking patterns of phenotypic and genotypic variation to patterns of environmental variation, testing the predicted evolutionary consequences and patterns of plasticity will be necessary to understand community-level processes driven by plasticity, and to provide evidence for the fitness consequences of plastic variation in evolving populations. In this review we identify questions and hypotheses about the role of phenotypic plasticity in adaptive evolution and speciation that can be tested with an ecological genomics framework, illustrating how such integrated approaches will enable a depth of insight into ecological and evolutionary questions we would not have considered asking in the twentieth century.

5.2 Genomics and Plasticity

5.2.1 Why Genomics?

Given that genotypically identical individuals can produce different phenotypes under different environmental conditions, it might seem strange to approach plasticity from a genomic perspective. After all, a phenotype cannot be explained

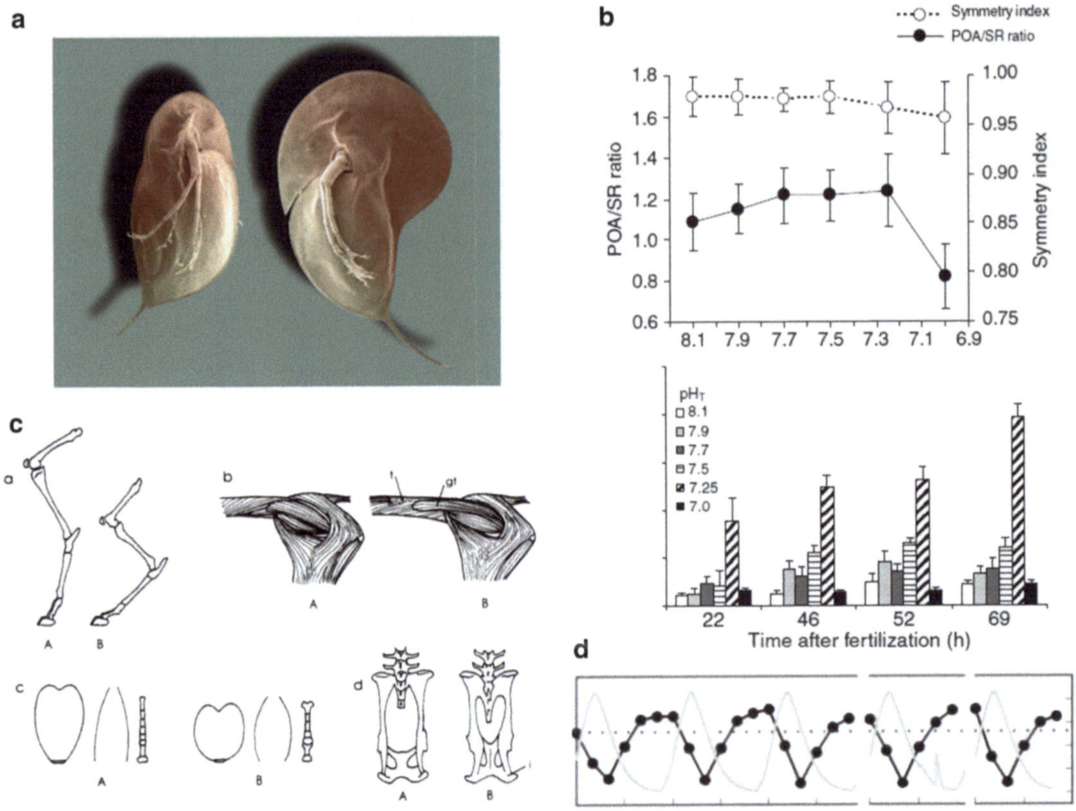

Fig. 5.2 Some examples of phenotypic plasticity. (**a**) Genetically-identical clones of *Daphnia* produce morphological defenses (*right*) in the presence of predator kairomones. (**b**) Sea urchin larvae raised under different pH display morphological plasticity (*top*) at a pH of 7.0, but resist such change at higher pH. This robustness to acidity is due to underlying transcriptional plasticity (*bottom*) that upregulates biomineralization genes at low pH. At a pH of 7, however, this upregulation disappears. (**c**) Slijper's (1942a, b) two-legged goat learned to walk upright on its hindlimbs, resulting in numerous plastic changes to other phenotypes. (**a**) Shows the morphology of a regular goat, (**b**) shows the morphology of the two-legged goat, for (**a**) hindlimb skeletal structure; (**b**) pelvic musculature, showing the elongated gluteal tongue (gt) and tendon reinforcements (t); (**c**) thoracic skeleton, showing a transverse, horizontal, and ventral view (*left* to *right*); (**d**) pelvic bones, showing the kangaroo-like ischium (i) of the two-legged goat. (**d**) Transcriptional plasticity in killifish for the *HMGB1* gene (*black*) with fluctuating temperature (*grey*) (Figure (**a**) Reproduced from Laforsch and Tollrian (2010), image kindly provided by C. Laforsch. Published with kind permission of © Elsevier Inc. 2010. All Rights Reserved) (Figure (**b**) Reproduced from Martin et al. (2011; doi:10.1242/jeb.051169). Published with kind permission of © The Company of Biologists Ltd. 2011. All Rights Reserved) (Figure (**c**) Reproduced from West-Eberhard (2003; Fig. 3.13, p. 53) with kind permission of Oxford University Press after Slijper (1942a, b). Published with kind permission of © Koninklijke Nederlandse Akademie Wetenschappen 1942. All Rights Reserved) (Figure (**d**) Reproduced from Podrabsky and Somero (2004; rightmost box of Fig. 4E). Published with kind permission of © The Company of Biologists Ltd. 2004. All Rights Reserved)

solely through a genetic "blueprint". Along with genes, offspring also inherit *epigenetic* modifications (Hackett et al. 2013), the internal cellular environment of the gamete/embryo (including lipids, polysaccharides, free nucleotides, transcripts, mitochondria, symbionts, and minerals) (West-Eberhard 2003), the external environment of the developing embryo (Refsnider and Janzen 2012), and/or the environment in which juveniles are reared (Dawkins 1976; Rossitter 1996). Each of these components is important in shaping the phenotype. Each of these components also has the capacity to influence an individual's fitness. And each of these may be passed on in

Table 5.2 Some causes of phenotypic variation. Definitions are provided in the glossary

Environmental differences between two habitats	Phenotypes of two populations in two habitats	Two populations are genetically identical clonal lines	Two populations are genetically distinct clonal lines
Distinct and stable	Similar	Environmental robustness OR plastic compensation	Environmental and genetic robustness OR genetic compensation
	Distinct	Plasticity	Adaptive divergence
Similar and stable	Similar	Stochastic robustness	Genetic robustness
	Distinct	Developmental noise	High penetrance
Differentially fluctuating	Similar reaction norms	Environmental robustness of plastic trait	Environmental and genetic robustness of plastic trait OR genetic compensation
	Distinct reaction norms	Lack of environmental robustness	Adaptive divergence for plasticity
Similarly fluctuating	Similar reaction norms	Stochastic robustness of plastic trait	Genetic robustness of plastic trait
	Distinct reaction norms	Developmental noise for plastic trait	Genotype-by-environment interaction

a relatively stable form for several generations (Crews et al. 2012). Furthermore, individuals can shape their own internal and external environments, which can have phenotypic effects (Dawkins 1982). Therefore, how one differentiates between genetic and environmental effects will depend on one's starting point; the relationship between genotype and the environment is more integrated than the term "genomics" implies (West-Eberhard 2003; Moczek 2012). This integration has led to concepts like *phenotypic accommodation*, which questions our ability to discover genes that are "for" certain phenotypes (West-Eberhard 2005).

Of course, this does not imply that the gene is irrelevant, or even equivalent to the actions of the environment, when it comes to the production of phenotypic diversity and its association with fitness. It is the gene that evolves. Selection operates at the level of the phenotype but acts on genetic variation (Lande and Arnold 1983). The environment can produce the effects that it does because gene products are built by selection in such a way as to be so affected. Indeed, genomic tools have established a functional link between *gene expression* and physiological, morphological, and *behavioral plasticity* (Aubin-Horth and Renn 2009). A genomics and developmental perspective of plasticity, therefore, enquires into the mechanistic basis, hierarchical interactions, genetic architecture, and *robustness* of plasticity, while remembering that phenomena other than plasticity exist (Table 5.2). In the context of ecological genomics, the integration of these facets under the predictive framework of the ecological theory of adaptive divergence provides a means to move beyond the notion that plasticity is common in nature and towards actually understanding (and predicting) its role in adaptive evolution (Schluter 2000).

5.2.2 The Mechanisms of Plasticity

There are at least two distinct forms of environmental induction. First, the environment can force the phenotype by virtue of chemical and physical laws (passive induction). For instance, temperature can cause phenotypic changes through enzyme kinetics and diffusion rates, while low nutrient availability can impact growth and morphology. Second, phenotypic change can be wrought through a complex interaction between environmental cues, sensors for the cues, signaling molecules that transfer information about the environment, and all of the machinery involved in phenotypic modification (active induction) (Windig et al. 2003). Understanding

the actual form of induction will therefore be key to understanding the causal link between genotype and phenotype.

The active pathway from cue → receptor → signal → translation of signal → phenotype is being elucidated in a few species (Beldade et al. 2011). For example, predator-secreted cues (kairomones) are known to induce morphological defenses in *Daphnia* species (Fig. 5.2), but little is known about the structure of the receptors associated with these chemical compounds (Peñalva-Arana et al. 2009; Akkas et al. 2010; Miyakawa et al. 2010). Activated kairomone receptors stimulate neural pathways to release hormones into the hemolymph (Barry 2002; Weiss et al. 2012). These hormones, including juvenile and insulin signaling hormones (Miyakawa et al. 2010), target polynucleated cells that control production of the inducible structures (Beaton and Hebert 1997; Barry 2002; Simon et al. 2011), resulting in increased transcriptional activity and post-translational modifications of structural proteins (Schwarzenberger et al. 2009; reviewed in Tollrian and Leese 2010). Plasticity, in turn, comes with a cost to the immune system (Yin et al. 2011). This summary represents decades of research in an easily-reared *model organism* with a sequenced genome, whose plasticity has been known since the early 1900s (Woltereck 1909), and yet the number of genes involved, their function, and their fitness consequences are only beginning to be determined. Even less is known of plasticity in ecologically important *non-model* species, reinforcing the significance of ecological genomics as an approach to understanding the consequences of plasticity.

Overall, phenotypic plasticity is possible because of the environmental sensitivity of gene expression or protein, lipid, and RNA activity, and/or variation in the levels of environmental components that are required for the production of a "normal" phenotype. This environmental sensitivity, in turn, may be driven by epigenetics (Richards et al. 2010), exploratory behavior coupled with intra-individual selection (Frankenhuis and Panchanathan 2011; Snell-Rood 2012), and/or the evolved coordinated response to the stimulation of environmental sensors (Tollrian and Leese 2010). Ecological genomic studies are revealing that the development of alternative phenotypes by a single genotype may be common but is amazingly complex, involving the interplay of numerous plastic and *non-plastic reaction norms* moving through developmental trajectories (Beldade et al. 2011; Sommer and Ogawa 2011; Valena and Moczek 2012; Zhou et al. 2012).

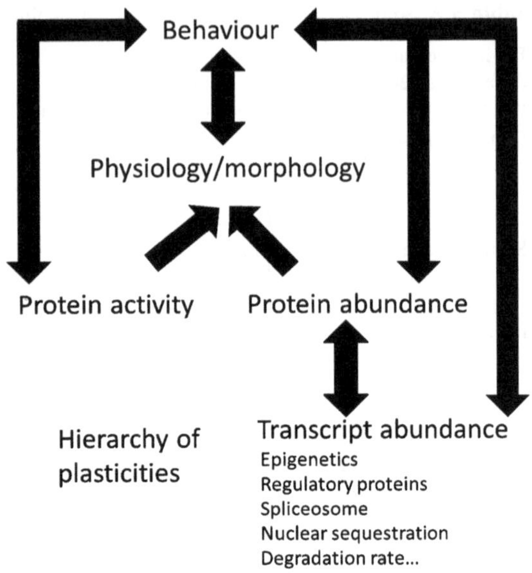

Fig. 5.3 The hierarchy of plasticities (Bradshaw 1965), including transcriptional and proteomic plasticity, protein activity plasticity, physiological and morphological plasticity, and behavioral plasticity. Note that in this hierarchy higher-level reaction norms can affect lower-level reaction norms, and vice versa, as indicated by the two-directional *arrows*. Behavioral plasticity especially can alter the rest of the hierarchy, as behavioral plasticity can bring organisms into new environments

5.2.3 Interactions of Reaction Norms

A major challenge for the ecological genomics of plasticity will be elucidating the relationship between reaction norms at all levels of the phenotype, from *molecular plasticity* to physiological and morphological plasticity, to behavioral plasticity. This interaction is known as the *hierarchy of plasticities* (Bradshaw 1965) (Fig. 5.3).

5.2.3.1 Molecular Plasticity

Gene expression can be measured as a *molecular phenotype* (Ranz and Machado 2006) that responds to the environment (Gracey et al. 2004; Greenberg et al. 2012; Yampolsky et al. 2012). Measuring *transcriptional plasticity* has its advantages: thousands of phenotypes can be measured simultaneously from a small sample, revealing plastic phenotypes that *a priori* predictions may not have anticipated. Since transcripts are gene copies, candidate genes involved in plasticity can be identified (Pavey et al. 2010). For instance, Podrabsky and Somero (2005) subjected killifish to different temperatures and found a tight negative correlation with high mobility group box one protein (*HMGB1*) transcript abundance, identifying *HMGB1* as a putative global temperature sensor on top of its previously described roles as regulatory protein and cytokine (Müller et al. 2001; Vezzoli et al. 2011) (Fig. 5.2). This hypothesis could not have been generated without the integration of transcriptional plasticity.

Proteomic plasticity measures protein abundance for the entire *proteome* under different environmental conditions. Proteomic plasticity has been well-documented in several organisms, although the integration of proteomics with ecological genomics is currently limited (Diz et al. 2012). As with transcriptional plasticity, proteomic plasticity can identify potential candidate genes for plasticity (including some not found in the *transcriptome* nor annotated from the genome; Findlay et al. 2009; Schrimpf et al. 2009), can measure thousands of phenotypes simultaneously, is closely associated with the genome, and can uncover unanticipated plastic phenotypes. Unlike transcript abundance, protein abundance is one step closer to the expression of the *macrophenotype* (Diz et al. 2012).

The importance of transcriptional plasticity for ecological genomics has been questioned in light of advances in proteomics, on biological rather than methodological grounds. For instance, it has been suggested that the control of protein production is more essential than the control of transcript abundance, as it imposes heavier costs (Malakar and Venkatesh 2012). However, estimates of the costs of protein production suggest they are minimal (Stoebel et al. 2008; Shachrai et al. 2010; Eames and Kortemme 2012), but may increase with stress (Vilaprinyo et al. 2010). Furthermore, evidence for selection against long introns in highly expressed genes indicates that transcription is also costly (Castillo-Davis et al. 2002) and can influence energy reserves and fitness (Wagner 2007; Lang et al. 2009). Altogether, studies that aim to understand the fitness consequences of molecular plasticity may shed more light on the adaptive link between transcript and protein abundance.

Developmental noise has also been used to defend a proteomic rather than a transcriptomic perspective. Genes involved in plasticity tend to be transcriptionally noisy; a decoupling between transcript and protein abundance is predicted to evolve as a strategy to reduce the impact of transcriptional noise on the phenotype (Raser and O'Shea 2005; Maier et al. 2011). Indeed, correlations between transcript and protein abundance tend to be low (Diz et al. 2012), although this varies with the type of gene and the type of regulation investigated (Lee et al. 2011; Maier et al. 2011). However, experiments on yeast have demonstrated that plasticity is not as noisy as once thought. There is a negative relationship between how vital the gene is for cellular functions and the amount of noise it generates. This has been achieved through the selection of certain genetic architectures, with greater noise being associated with particular chromatin dynamics and promoter types (Lehner 2010), epigenetic modifications (Viñuelas et al. 2012), and translational efficiencies (Bajić and Poyatos 2012). Overall, noise provides a biologically relevant reason why protein abundance should not be ignored, but this should not preclude efforts to understand the ecological genomics of transcriptional plasticity.

Despite noise, transcript abundance tends to drive protein abundance, linking these two phenotypes together in the hierarchy of plasticities. However, this relationship is in practice difficult to determine. Plasticity in the expression of one gene can have pleiotropic effects on other genes (Zhou et al. 2012), making it difficult

to determine which plastic phenotypes are adaptively responding to environmental change, and which are responding via pleiotropy. Since pleiotropic genes may be less vital and therefore more prone to noise, pleiotropy could mask a positive relationship between transcript and protein abundance for adaptively plastic genes. Furthermore, the causal link between transcript and protein abundance may take several forms, further diminishing our ability to measure their relationship. For instance, increased transcript abundance may maintain protein levels if protein degradation increases, while a lack of transcriptional plasticity may allow proteomic plasticity (Beldade et al. 2011). Protein abundance, in turn, may affect transcript expression in a similar manner (Tomanek and Somero 2002; Tomanek 2008). Even if transcript and protein abundance are not correlated, the fitness consequences of unnecessary plasticity should be of ecological interest (Lang et al. 2009). In short, to understand the hierarchy of plasticities, both transcriptional and proteomic plasticity must be measured for a single gene, and the relationship between these reaction norms ascertained through techniques such as RNA interference, morpholinos, and methylation (Juliano et al. 2005; Zhou et al. 2007; Wang et al. 2012). If, for instance, protein abundance changes across environments, what happens to protein abundance when transcript production is suppressed across environments in adult organisms?

Along these lines, other sources of molecular plasticity such as metabolomics and epigenomics will become increasingly incorporated into ecological genomics studies (Bossdorf et al. 2008; Sardans et al. 2011). The *epigenome* is of special interest as techniques for sequencing methylated regions of DNA have only recently been established (reviewed in Bock 2012). Recent studies on plasticity and the epigenome have shown that epigenetic modifications can produce alternative phenotypes within a single individual (Herrera and Bazaga 2012), can plastically prepare offspring for uncertain future conditions (Angers et al. 2010), and can transfer plastic changes induced in one generation to future generations. The latter is particularly interesting, as plastic modifications to the phenotype in one generation can arise in later generations, even if the later generations never experience the inducing environment (Stern et al. 2012). The effects of the epigenome on plasticity are context-specific, and in some systems have been known to limit the development of alternative phenotypes (Roberts and Gavery 2012). However, even in such cases methylation and histone modifications are induced by the environment, and can be measured as a form of intergenerational plasticity. There are still many questions to answer regarding the relationship between plasticity and the epigenome, but epigenetics does seem to be an important mechanism in at least some forms of plasticity (Richards et al. 2010; Valena and Moczek 2012).

Proteins may have their own reaction norms apart from protein abundance. Protein movement, half-life, and enzyme efficiency are all influenced by the environment. Their degree of plasticity, however, is dependent on their amino acid sequences. Changes to amino acid sequences can alter reaction norms, increasing or decreasing plasticity in protein behavior (Powers and Schulte 1998). Finally, interactions between proteins, genes, non-coding RNA, lipids, etc., can be influenced by the environment and may affect the macrophenotype (Hayward et al. 2007; Tomanek 2008; Deredge et al. 2010).

The integration of molecular plasticity in ecological genomics is driven by several questions: what is the relationship between transcriptional and proteomic plasticity for particular ecologically relevant genes? How does this relationship affect ecologically important traits? Where in the pathway from gene to protein do mutations that alter plasticity lie? Comparing gene and protein sequences for populations with different reaction norms can begin to address these questions. For instance, killifish adapted to cooler waters had an amino acid substitution at site 311 of their lactate dehydrogenase B enzyme that altered the kinetic properties of the enzyme relative to warm-adapted fish (Powers and Schulte 1998). The complex nature of molecular plasticity will be sure to challenge researchers attempting to answer these questions for years to come, but the tools to investigate them are now available.

5.2.3.2 Hierarchy of Plasticities

Transcriptional studies can discover functional relationships between molecular plasticity and physiological, morphological, or behavioral plasticity (e.g. Schwarzenberger et al. 2009; Martin et al. 2011) (Fig. 5.3). What is less appreciated is the relationship between phenotypes that resist environmental change and molecular plasticity.

Non-plastic phenotypes may resist change despite environmental perturbations, and this resistance to the environment may be evolutionarily important. The production of non-plastic traits has been analyzed in some organisms across different environments. For instance, in sea urchin larvae *Paracentrotus lividus*, morphology was relatively insensitive to decreasing pH. This non-plasticity, however, was maintained by transcriptional plasticity for genes involved in biomineralization. At a pH of 7 morphology was disrupted by pH, and this was associated with a breakdown of gene expression regulation (Martin et al. 2011) (Fig. 5.2). This type of study shows the breadth of reaction norm interactions, and reminds us that the environment may influence the phenotype even if plasticity cannot be readily observed (see *plastic compensation*).

The highest rung on the plasticity hierarchy is behavioral plasticity. Although behavioral plasticity is difficult to define (see Glossary), it has long been expected that behavioral plasticity can drive plastic changes in other phenotypes, and may be an important first step in the generation of phenotypic variation (Price 2003; West-Eberhard 2003). For instance, a goat born with congenital limb defects learned to walk on two legs, which sparked numerous plastic changes to its musculature and skeleton (West-Eberhard 2003) (Fig. 5.2), while stickleback ecotypes may have evolved morphological differences via behavioral plasticity in diet acquisition (Wund et al. 2008). One intriguing recent hypothesis suggests that exploratory behavior, for cells and for organisms, is likely an important generator of individual differences in plasticity. Individuals that stochastically sample the environment before developing an appropriate phenotype may plastically respond early in development if the environments they sample are homogeneous, leading to reduced environmental sensitivity during later stages of development. Individuals that stochastically sample an unpredictably heterogeneous environment, however, may maintain a propensity for plasticity in later stages of development (Frankenhuis and Panchanathan 2011). Thus the hierarchy of plasticities cannot be conceived as an inflexible chain, but rather every level of the hierarchy can induce plastic changes at every other level.

5.2.4 Genetic Architecture

Genetic architecture of plasticity is concerned with the number, placement, and effect size of genes involved in the development of alternative phenotypes. Most studies that discuss the genetic architecture of plasticity have yet to address any of these subjects. Experimental work has shown that plasticity can be influenced by single genes of large effect. For instance, *Caenorhabditis briggsae* normally develop into hermaphrodites across all temperatures, but mutations in the *she-1* gene (such as *v49*, which produces an early stop codon, or *vDf2*, which is a 5′ deletion) can lead to the development of XX females at 25°C and XX hermaphrodites at lower temperatures (Guo et al. 2009). Plants ordinarily exhibit density-dependent plasticity in stem length, but the transgenic addition of an oat phytochrome A gene to tobacco induces long stems even under low densities, while *Brassica rapa* mutants for phytochrome B exhibit small stems even under high densities (Schmitt et al. 1995). Other examples could be given (Beldade et al. 2011). However, the genetic architecture of plasticity involves more than comparing phenotypic differences between mutant lines; it involves determining the entire complement of genes involved in plasticity, and these are likely more numerous than mutational studies could ever determine.

Given the complex nature of plastic responses, genes involved in plasticity can include any of the components of a plastic response, from cue reception to signal transduction to phenotype production. This can include protein-coding and RNA-coding genes, such as regulatory genes and

genes involved in epigenetic modifications. Comparative approaches are ill-prepared to identify this diversity of genes. Quantitative Trait Loci (QTL) and expression QTL studies can identify genomic regions associated with divergent plastic phenotypes between genotypes, but cannot capture loci involved in plasticity that lack genetic or phenotypic variation. Gene expression studies have identified thousands of genes induced by a single environmental variable, but it can be difficult to differentiate between transcripts that produce the induced macrophenotype and transcripts that pleiotropically respond to environmental change or plastic changes in other phenotypes (Aubin-Horth and Renn 2009; Fraser 2011). This task is further limited by the lack of ecological annotation for genes that exhibit molecular plasticity (Pavey et al. 2012). In short, standard approaches for quickly ascertaining the number of genes involved in plasticity (e.g., QTL analysis, microarrays, RNA-sequencing) cannot provide basic information regarding the genetic architecture of plasticity, but can identify those loci that lead to divergent plastic responses or those genes whose expression is environmentally sensitive. To provide a complete picture of the genetic architecture of plasticity, gene expression studies need to be extended across multiple tissues and developmental stages under contrasting environments (Beldade et al. 2011). The transcriptome, proteome, metabolome, epigenome, etc., and their interactions, must all be considered, and the relevance of individual genes or gene networks for the plastic response must be ascertained through gene silencing methods (ex. Zhou et al. 2007) or other functional approaches. The focus of genes under selection, or genes producing divergent plastic responses, although important, cannot preclude research on functionally important genes that lack variation, or non-genetic aspects of the organism that are involved in the production of alternative phenotypes. As seen in the example of *Daphnia* given above, this will take a coordinated effort by a multitude of researchers with different areas of specialization, all focused on a single species. Such research is already under way, and the results are promising (some recent examples: Bossdorf et al. 2010; Meister et al. 2011; Greenberg et al. 2012; Srinivasan and Brisson 2012).

5.2.5 Robustness of Plasticity

Robustness (often called *canalization*) tends to be used to describe phenotypes that resist environmental change and are thus non-plastic, but plasticity itself can be robust to stochastic, environmental, and genetic perturbations (Waddington 1953a, b; Gibson and Wagner 2000; Debat and David 2001), as adaptations to maintain a consistent plastic response. *Stochastic robustness* occurs whenever the reaction norm is resistant to developmental noise. Such resistance can occur via alterations to the surrounding genomic structure, or by loose causal links between molecular plasticity and higher levels of plasticity (Raser and O'Shea 2005; Lehner 2010). *Environmental robustness* includes a lack of discontinuous change in reaction norm shape under unnatural extreme environments, or the maintenance of reaction norm shape under one environmental variable when a second environmental variable is introduced. For instance, if temperature-induced plasticity is maintained despite changes in salinity, that reaction norm is robust to salinity. How environmental robustness for plasticity occurs, and how species can evolve such robustness, has never, to our knowledge, been explored. It is important to remember that robust reaction norms at one level of the hierarchy may be driven by non-robust reaction norms at other levels of the hierarchy. Finally, reaction norms are tested against diverse genetic backgrounds. Studies of natural populations have revealed that individual genotypes often have distinct reaction norms, indicating a relative lack of *genetic robustness* (Landry et al. 2006; Bentz et al. 2011). This lack of robustness permits evolution. The alleles generating these changes, however, have rarely been examined, and the relative degree of genetic or environmental robustness for plasticity has not been measured. Going forward, reaction norms induced by a single environmental variable need to be measured when held against other environmental variables or genotypes, and

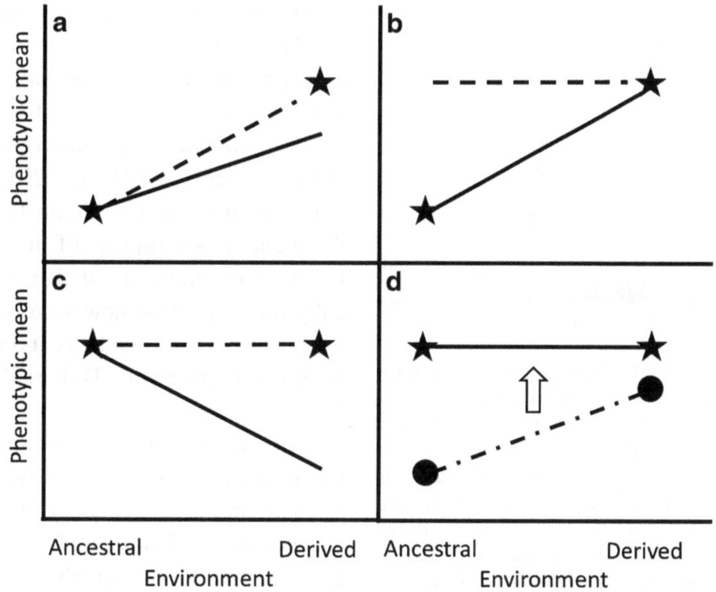

Fig. 5.4 Some important concepts in the evolution of plasticity are shown for simple linear reaction norms. Stars and circles represent phenotypic optima within each environment. Ancestral denotes a stable environment that the population was initially adapted to. Derived denotes a novel stable environment that the population has colonized. Solid lines indicate the reaction norm of the population prior to colonization, and the dashed line indicates the reaction norm of the population after evolving in the derived environment. (**a**) Plasticity-Mediated Population Persistence occurs when plasticity pre-exists and moves the colonizing population towards its new fitness optimum. It may then evolve under directional selection to maximize fitness in both environments (adaptive plasticity). (**b**) If the derived environment is stable, the population may evolve the loss of plasticity (genetic assimilation, *dashed line*), such that a return to the ancestral environment would induce no plastic change. (**c**) Some environments may induce phenotypic changes that move the population away from their phenotypic optimum (maladaptive plasticity, *solid line*). Selection may then work to bring the population back to its optimum (genetic compensation), potentially causing a loss of plasticity (*dashed line*). (**d**) Maladaptive plasticity may not be detected (plastic compensation, *solid line*), if adaptive plasticity in an underlying trait (*dash-dotted line*) counteracts maladaptive plasticity on the affected phenotype (*arrow*). If the underlying trait was not plastic, maladaptive plasticity would be evident in the solid line

the role of molecular plasticity in maintaining a plastic reaction norm against environmental, genetic, and stochastic perturbations needs to be measured (e.g., Lehner 2010).

5.3 Evolution and Plasticity

The ecological genomics of plasticity is concerned not only with the production of ecologically-relevant traits, but also the consequences of plasticity for population differentiation and evolutionary novelty. Figure 5.4 and the glossary define some important terms (*adaptive plasticity, maladaptive plasticity, neutral plasticity*). We favor fitness-based rather than historically based definitions of adaptation, as they avoid unnecessary and often untestable assumptions. Some recent findings of the evolutionary significance of plasticity will be discussed below.

5.3.1 Plasticity and Population Persistence

Baldwin (1896, 1902) hypothesized that adaptive phenotypic plasticity could enable individuals to colonize novel environments (*Plasticity-Mediated Population Persistence* – PMPP, Pavey et al. 2010) (Fig. 5.5). This has recently been supported by theoretical (Ghalambor et al. 2007;

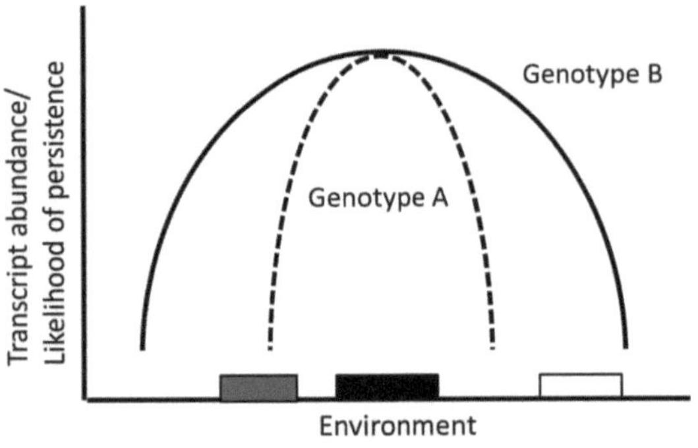

Fig. 5.5 Plasticity-Mediated Population Persistence (PMPP) occurs when plasticity enables colonization of a novel environment. In this diagram, the likelihood of persistence across an environmental range is shown for genotype A and genotype B. This likelihood is directly related to the abundance of transcript induced by the environment. Genotype A (*dashed line*) produces transcript under a narrow set of environments, and so can persist in a narrow set of environments. Genotype B (*solid line*) can produce transcript under a greater range of environments. Both genotypes can persist in the environmental range at which they evolved (*black box*). At the edge of Genotype A's tolerance range (*grey box*), both genotypes could colonize, but Genotype B has an advantage due to its greater level of transcript abundance. Under extreme environments (*open box*) Genotype A cannot successfully colonize, while Genotype B has a small likelihood of survival (PMPP) (Modified from Pavey et al. (2010, Fig. 2) with kind permission of © New York Academy of Sciences 2010. All Rights Reserved)

Thibert-Plante and Hendry 2011) and empirical (Yeh and Price 2004; Hahn et al. 2012) research. For example, *Daphnia lumholtzi* plastically produce head spines in the presence of predators. Its invasive success in North America appears to be mediated by this plasticity: in the presence of native non-plastic *Daphnia pulicaria*, *D. lumholtzi* is an inferior competitor, but when predators are introduced *D. lumholtzi* has a competitive edge (Engel and Tollrian 2009). Numerous other studies have implicated plasticity in invasive success, although it is not always clear if plasticity preexisted or evolved after colonization (Bachmann et al. 2012; Hanshew and Garica 2012; Molina-Montenegro et al. 2012; Mozdzer and Megonigal 2012; Purchase and Moreau 2012; but see Matzek 2012), an important distinction to make when assessing the role of plasticity in population persistence.

There are predictions regarding the likelihood of PMPP. For instance, organisms that adjust their phenotype post-dispersal are more likely to colonize new environments than individuals that adjust their phenotype irreversibly pre-dispersal (Thibert-Plante and Hendry 2011). Post-dispersal plasticity may also facilitate PMPP by reducing the genetic swamping of migrants (migration load), as migrants and their offspring can plastically adjust to their new surroundings, taking on the phenotypes of residents and limiting selection against interbreeding (Thibert-Plante and Hendry 2011).

One underexplored area of PMPP involves the role of *cryptic genetic variation* (CGV), a form of *standing genetic variation* (SGV). Under normal environmental conditions, individuals may exhibit similar phenotypic traits despite genotypic differences, due to the suppression of genetic variation via *phenotypic capacitors* (Levy and Siegal 2008) or the accumulation of neutral mutations in unexposed regions of the reaction norm (Ghalambor et al. 2007). In either case, novel environments may expose CGV in plasticity, increasing heritability for the phenotype and thereby permitting rapid evolution. PMPP will occur for those individuals whose CGV exhibits plasticity in the adaptive direction. This has likely occurred in the colonization of freshwater environments by marine threespine sticklebacks: freshwater salinities exposed CGV in body size, resulting in the rapid parallel evolution of smaller body sizes in freshwater populations (McGuigan et al. 2011). New genomic tools have allowed the mechanisms governing CGV production to be elucidated (Iwasaki et al. 2013), and its evolutionary significance to be tested. For instance, ribozymes selected for their ability to bind to a particular substrate were replicated via mutagenic Polymerase Chain Reaction (PCR) to introduce genetic variation into the ribozyme population. Following ten generations of replication, ribozymes were again selected for their ability to bind to the same substrate, thereby favoring mutations that had no phenotypic effect (CGV). The wild-type and CGV populations were then introduced to a new substrate. The evolution of enzymatic efficiency in

the presence of this new substrate was measured over several generations of moderately-mutagenic PCR, and ribozymes were genotyped each generation. CGV enabled more rapid evolution on this new substrate by "pre-adapting" certain ribozyme genotypes to this new environment (Hayden et al. 2011). Future work is clearly moving away from (albeit important) heritability studies, towards tracking the cryptic alleles responsible for rapid evolution in novel environments.

There are at least six potential consequences of CGV for PMPP under post-dispersal plasticity. First, since selection only favors adaptive plasticity, the colonizing population will have reduced genetic diversity at those loci compared to the ancestral population. This could potentially decrease future evolutionary potential for that phenotype. Second, CGV could increase the likelihood of PMPP relative to small or recently bottlenecked populations that exhibit little CGV (and therefore exhibit plasticity in the same, possibly maladaptive, direction). Third, CGV may increase the likelihood that colonists experience stabilizing rather than directional selection, as the random nature of CGV may produce some individuals with a perfect environment-phenotype match (Ghalambor et al. 2007). Fourth, founder effects and drift could play an important role during PMPP – different colonizing populations from the same ancestral population could have different likelihoods of persistence and be subject to different selection strengths or forms of selection (stabilizing or directional), depending on the subset of CGV present among dispersers. Fifth, CGV could increase the heritability of a trait under new environments, resulting in rapid evolution (Neyfakh and Hartl 1993; Chown et al. 2009; McGuigan et al. 2011). Finally, individuals with different genotypes could produce similar adaptive phenotypes in the novel environment, and be selected together. This could increase their likelihood of reproduction, producing reaction norms comprised of the cryptic alleles from several individuals. This in turn could produce new reaction norms, potentially causing *genetic assimilation* or increased niche breadth. CGV in reaction norms are clearly important for adaptive evolution and must be included in theories of PMPP and adaptive divergence.

Individual-level differences in plasticity that permit the PMPP of certain individuals may exist in the absence of genetic variation. Identical genotypes that experience different levels of environmental heterogeneity early in life may have altered abilities to respond plastically to novel environments later in life (Frankenhuis and Panchanathan 2011). Exploratory behavior is therefore the non-genetic equivalent of CGV. Experiments that actively uncover the alleles generating CGV, or experimentally account for the prior history of the organism, are needed to differentiate between these genetic and non-genetic processes.

The role of maladaptive plasticity in population persistence has also been relatively ignored (Morris and Rogers 2013). Presumably maladaptive plasticity would decrease the possibility of persistence and increase the likelihood of extinction (*plasticity-mediated population extinction*, PMPE). PMPE could occur if the phenotype is forced away from its optimum (Ghalambor et al. 2007), or if the environmental context that favored adaptive plasticity were to change. For instance, freshwater snails *Physella virgata* have evolved adaptive plasticity in shell morphology, such that in the presence of fish predators they can produce crush-resistant rotund shells. These changes come at the cost of reduced fecundity and increased leech predation, and can be induced by non-predatory sunfish. Snails introduced to ponds containing non-predatory sunfish may therefore be less likely to persist because of plasticity (Langerhans and DeWitt 2002). If populations are able to persist despite maladaptive plasticity, this could have some interesting consequences for adaptive divergence (see below).

5.3.2 PMPP and Adaptive Divergence

Ecological speciation results from a combination of colonization of distinct environments and adaptive divergence due to divergent selection. Ironically, theoretical work has shown that post-dispersal plasticity facilitates colonization but

inhibits adaptive divergence (Thibert-Plante and Hendry 2011), as plasticity enables migrants to successfully interbreed with residents. Migrant and resident populations therefore remain phenotypically distinct but genetically homogeneous. Pre-dispersal plasticity, however, is unique in that divergent selection predates genetic divergence, as migrants are selected against when competing with residents. Pre-dispersal plasticity can therefore facilitate adaptive divergence, but it reduces the likelihood of colonization (Thibert-Plante and Hendry 2011).

Divergent selection can occur within a single environment in the absence of migration, if the population experiences a fitness minimum. The theory of adaptive speciation states that population size can alter the fitness landscape. Colonizing populations may first evolve under directional selection, allowing population size to increase over time. As population size increases, the adaptations that occurred under directional selection become less favorable. Directional selection therefore moves the population towards a fitness minimum, at which point individuals on either side of the minimum experience divergent selection and follow different evolutionary trajectories (Dieckmann et al. 2004). Given that plasticity can facilitate colonization to new environments, and that plasticity can occur in response to demographic changes (Svanbäck et al. 2009), the role of PMPP in adaptive speciation needs to be addressed. These sorts of models have opened the door to many exciting theoretical and empirical opportunities for researchers testing predictions about plasticity in cases of ecological or adaptive divergence.

5.3.3 PMPP and Evolutionary Rescue

Evolutionary rescue occurs when populations adapt to stressful environments after a period of population decline, such that population size increases. Theoretical work has shown that plasticity can promote evolutionary rescue by slowing the rate of population decline, permitting time for adaptive changes to occur (Chevin et al. 2013).

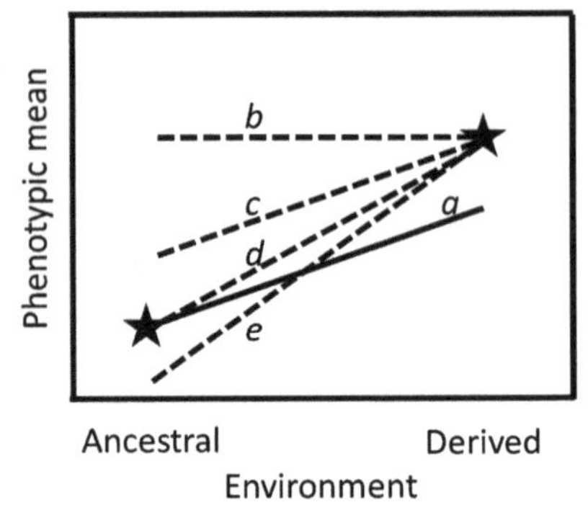

Fig. 5.6 Reaction norms can evolve in height (a → c) or slope (a → b–e). *a* (*solid line*) represents the initially expressed reaction norm upon colonizing a new environment. It does not attain the phenotypic optimum (*star*), and so is subject to directional selection. The reaction norm can then evolve to meet the optima in both environments (*d*) or overshoot the optima in the ancestral environment (*e*). If plasticity is selected against in the derived environment, plasticity could be lost (*d* → *c* → *b*). *d* could also represent a reaction norm under stabilizing selection

5.3.4 Adaptive Plasticity and Adaptive Divergence

Baldwin (1896, 1902) hypothesized that reaction norms could evolve post-colonization. The likelihood and form of such evolution depends on whether plasticity results in stabilizing (Fig. 5.6d) or directional (Fig. 5.6a) selection (Ghalambor et al. 2007). Stabilizing selection occurs when plasticity brings the phenotype to its fitness maximum. It can reduce the likelihood of reaction norm evolution or lead to the loss of plasticity, depending on whether plasticity is expressed or not expressed in the new environment. Directional selection, which occurs when plasticity does not bring the phenotype to its fitness maximum, could result in the evolution of reaction norm height or slope (Fig. 5.6). One would expect an increased slope (Fig. 5.6d, e) if the population routinely migrated between its ancestral and newly-colonized habitat, if the colonized environment fluctuated beyond the conditions experienced in the ancestral environment, or if gene flow between environments was high (Berrigan and Scheiner 2003; Crispo 2008), and a decreased slope (Fig. 5.6b, c) if the maintenance of plasticity was costly and the environment was

stable (see below). Furthermore, polymorphisms in reaction norms may be maintained by fluctuating environments, if different reaction norms produce phenotypes that are optimal in contrasting environments. In short, the nature of environmental fluctuations in the colonized environment, the costs to plasticity, the mutations available to selection, and the strength of selection, will, among others, determine the form that the evolved reaction norm takes. Despite this, there is good evidence that plasticity *does* evolve, leading to adaptive differences between populations (Crispo 2007; McCairns and Bernatchez 2010; Pfennig et al. 2010; Schwander and Leimar 2011; Svanbäck and Schluter 2012).

5.3.5 Genetic Assimilation and Adaptive Divergence

Plasticity can generate dramatic phenotypic divergence, as seen in the case of a two-legged goat whose musculature changed rapidly upon assuming a bipedal form of locomotion (Slijper 1942a, b). West-Eberhard (2003, 2005) proposed that such phenotypic divergence can come under genetic control. That is, genetic changes could occur that result in the loss of adaptive plasticity, canalizing one possible phenotype across environments. This genetic assimilation (Waddington 1953a, b) (Fig. 5.4) is expected to evolve if the environment remains stable, and: (1) plasticity is costly to maintain when it is not required; (2) neutral mutations accumulate in the unexpressed portion of the reaction norm, such that plasticity is lost in other environments; (3) hybridization is permitted due to plasticity, but hybrids incur some fitness cost (genetic assimilation via reinforcement); or (4) selection reduces the environmental threshold required to induce the phenotypic change (Waddington 1956; West-Eberhard 2003). Differentiating between the causes of genetic assimilation has proven difficult, and some possibilities may not even be plausible. For example, could costs to plasticity be reduced rather than plasticity itself (DeWitt et al. 1998)?

Genetic assimilation has important consequences for ecological genomics. The *flexible stem* model of evolution (West-Eberhard 2003; Pfennig et al. 2010), in which a plastic ancestral population births phenotypically divergent non-plastic populations, predicts that population phenotypes may not always be built from the "ground up", but may reflect canalized ends of the same reaction norm. Day et al. (1994) tested plasticity for trophic morphology on benthic and limnetic sticklebacks from Paxton Lake, British Columbia, fed on a "benthic" diet of worms or a "limnetic" diet of plankton. Limnetics, which have a more diverse diet, exhibited significantly greater plasticity in gill raker length than did benthics. Intriguingly, when fish were fed the diet of their contrasting ecotype, plasticity partially moved them in the direction of that ecotype, suggesting that benthic and limnetic individuals were derived from a plastic ancestor. Wund et al. (2008, 2012) complemented this work by comparing diet-induced plasticity in marine, solitary limnetic, and solitary benthic sticklebacks. The marine ancestor was highly plastic, producing a benthic or limnetic morphology depending on the food source (but see Svanbäck and Schluter 2012), but the solitary populations were also plastic. This leads to the intriguing possibility that plasticity was not costly in derived populations, but was reduced in benthic-limnetic species pairs via reinforcement. At the moment this is simply speculation, but the flexible stem model allows such hypotheses to be generated and tested. Phylogenetic studies lend further support to the reality of genetic assimilation (Schwander and Leimar 2011).

5.3.6 Maladaptive Plasticity and Adaptive Divergence

Non-plasticity may evolve for reasons other than genetic assimilation. For instance, adaptive plasticity could become costly if other environmental variables were to change. Populations of *Daphnia melanica* plastically adjust their melanin production with depth as an adaptation to ultraviolet radiation. This plasticity makes *Daphnia* visible to predators at shallow depths. Populations recently

exposed to predators have rapidly evolved the loss of melanin production plasticity (Scoville and Pfrender 2010). Non-plasticity could also evolve if novel environments were to move the phenotype away from its optimum (maladaptive plasticity). Selection can work via mutation to overcome maladaptive plasticity, such that populations that originally exhibited maladaptive plasticity can produce the same phenotype as their ancestors through a novel developmental pathway (Fig. 5.4). This has been called *genetic compensation* (or cryptic evolution) and has been demonstrated in Kokanee salmon (*Oncorhynchus nerka*) (Grether 2005; Fitzpatrick 2012). Genetic compensation may produce a non-plastic reaction norm but it does not need to (Grether 2005), and may explain phenomena like countergradient variation (Conover and Schultz 1995). Genetic compensation can be distinguished from genetic assimilation, in that the pathway of genetic assimilation is phenotypic divergence between populations via adaptive plasticity → genetic divergence → phenotypic divergence via loss of plasticity, while the pathway of genetic compensation is phenotypic divergence between populations via maladaptive plasticity → genetic divergence → phenotypic similarity in divergent environments.

Maladaptive plasticity may be overcome in the absence of novel genetic input. Plastic compensation (Morris and Rogers 2013) (Fig. 5.4), defined as adaptive plasticity overcoming maladaptive plasticity, is likely a common phenomenon that has been underrepresented in discussions of maladaptive plasticity. In plastic compensation, a phenotype that should express maladaptive plasticity does not, or does so transiently, due to an adaptive plastic response in some other phenotype. Plastic compensation may therefore prevent maladaptive plasticity from being identified. Plastic compensation likely comes with a cost. For instance, in the brittlestar *Amphiura filiformis*, the ability to regenerate limbs (the otherwise maladaptively plastic phenotype) was maintained despite decreasing pH. However, this could only be maintained at low pH by digesting muscle tissue for energy (the cost) to presumably fuel increased rates of biomineralization (the adaptively plastic phenotype) (Wood et al. 2008). The pathway of plastic compensation can be described as phenotypic divergence between populations via maladaptive plasticity → phenotypic similarity between populations for the otherwise maladaptively plastic phenotype via adaptive plasticity in some other phenotype → possible genetic divergence to reduce costs. If plastic compensation occurs immediately, the initial step (phenotypic divergence via maladaptive plasticity) may never be observed. A key test of plastic compensation involves the inhibition or deletion of the adaptively plastic phenotype, which should lead to the expression of maladaptive plasticity.

Intriguingly, plastic compensation may be maintained across generations via heritable epigenetic modifications that keep the compensating phenotype induced even in the absence of the environmental inducer (Stern et al. 2012). This could be considered the epigenetic form of genetic assimilation, in which a once-plastic phenotype becomes constitutively produced across environments for several generations, a phenomenon known as epigenetic assimilation (Sollars et al. 2003; Ruden et al. 2005).

5.3.7 Plasticity and Reproductive Isolation

The final component of ecological speciation is reproductive isolation. Fitzpatrick (2012) helpfully reminds us that plasticity may result in reproductive isolation prior to adaptive divergence, if plasticity occurs pre-dispersal. If individuals follow a basic rule such as "breed only with individuals that are phenotypically similar to conspecifics," there should be little reproduction between genetically identical but phenotypically distinct populations. The opposite, however, that plasticity may confound measures of reproductive isolation, has been rarely noted (but see Crispo et al. 2011). For instance, reproductively isolated populations may colonize the same environment, inducing similar plastic changes that reduce the phenotypic differences between them. If reproductive barriers are pre-zygotic, these plastic changes could alter reproductive behavior and increase hybridization. Environmental

disturbance can also affect reproductive isolation. In Lake Victorian haplochromine cichlids, increased turbidity has been shown to reduce reproductive isolation between species, largely because the cues females use to find preferred mates can only be detected under broad spectrum light (Seehausen et al. 1997, 2008). Mate choice behavior is therefore plastic; measures of reproductive isolation in common garden experiments may not reflect actual levels of reproductive isolation. For cichlids, any measure of reproductive isolation performed in clear laboratory waters would overestimate the degree of reproductive isolation experienced under murkier natural conditions.

5.3.8 Summary

Ecological genomics approaches are vital for understanding the multitudinous consequences of phenotypic plasticity for evolutionary biology. Understanding how populations differ in their molecular reaction norms, and how this generates macrophenotypic divergence across environments; ascertaining the direction of plasticity evolution through phylogenetic studies; identifying the alleles that generate reaction norm differences and the alleles that contribute to CGV; testing different molecular strategies for persisting in novel environments; mapping alternative developmental pathways that produce similar phenotypes in different environments; and above all determining how alternative plastic phenotypes are generated, and how these pathways are altered in derived populations and species; all require an ecological genomics framework if we wish to move from conjecture and modelling to testing these ideas in nature.

5.4 Ecology of Plasticity

Phenotypic plasticity involves the environmentally-sensitive production of a phenotype, and therefore the ecological conditions experienced by the organism, both abiotic and biotic (including conspecifics), must be measured and incorporated into the ecological genomics of plasticity. Just as plasticity will not evolve without genetic variation, it also cannot evolve without certain environmental conditions. Demography, which can be influenced by the environment, also has consequences for the evolution of plasticity. Furthermore, plasticity in one organism has ecological and evolutionary consequences for other members within the community, thereby giving a role for plasticity in community and ecosystem processes. Finally, unnatural environments, often eschewed by ecologists, can shed important light on the mechanisms and evolutionary consequences of plasticity. Altogether, current research reveals the importance of integrating ecology into the ecological genomics of plasticity.

5.4.1 Plasticity and the Environment

Plasticity is expected to evolve under predictably varying environments (Berrigan and Scheiner 2003). If the environment does not vary, plasticity is not expected to evolve. If the environment varies in an unpredictable manner, bet-hedging strategies may evolve (Starrfelt and Kokko 2012). Predictable variation alone is not sufficient: for active induction, environments must also produce reliable cues for the environmental change, cues that can be detected by the organism and translated into a phenotypic response (Berrigan and Scheiner 2003). Reliable cues must then permit adequate time between the reception of the cue and the production of the plastic phenotype. If environmental change outpaces phenotypic change, or if the time-lag between cue reception and phenotype production is too long, non-plastic strategies may be favored (Padilla and Adolph 1996). These models show the importance of generating precise measurements of the environments experienced by organisms. This is easier said than done, as motile organisms may reduce the temporal environmental variation that they experience by moving throughout a spatially heterogeneous landscape. Thus environmental metrics taken at a single site may not measure

the environment as experienced by an organism. Basic ecological measures are sorely needed for many species to test the relationship between environmental heterogeneity and plasticity evolution.

5.4.2 Plasticity and Demography

Environmental change can influence plasticity directly, as shown above, but it can also influence plasticity by affecting population migration and abundance. Environments which favor gene flow between subpopulations, for instance, are predicted to favor the evolution of plasticity (Crispo 2008). Population abundance can influence plasticity in at least four ways. (1) Large populations, which have greater opportunities for mutations, are more likely to harbor SGV, which can favor PMPP. (2) Large populations are also more likely to evolve beneficial mutations that positively affect reaction norms and reduce the costs of plasticity (Stern 2010). (3) Small populations produce fewer mutations and therefore are more likely to evolve pleiotropic and costly reaction norms (Stern 2010). (4) Population size can alter fitness landscapes, with individuals experiencing reduced fitness as the population increases (Dieckmann et al. 2004). If populations routinely experience fluctuations in population size, plasticity may evolve to reduce the effect of population size on fitness. For instance, models have shown that population fluctuations due to predator–prey dynamics facilitates the evolution of plasticity rather than adaptive speciation (Svanbäck et al. 2009). The challenge for researchers in the next few years will be to incorporate increasingly more complex population dynamics into their study of the ecological genomics of plasticity.

5.4.3 Plasticity and Community Genetics

Community genetics is an emerging subdiscipline within ecological genomics (Fig. 5.1). It involves studying how certain genotypes affect the distribution, abundance, and evolution of other genotypes within a community, and what genes underlie heritable community traits (Whitham et al. 2008; Hersch-Green et al. 2011). Community genetics of plasticity has received increased interest over the last several years (Rowntree et al. 2011; Tétard-Jones et al. 2011), particularly as plasticity's role in community ecology has been documented (Agrawal 2001; Fordyce 2006) (Fig. 5.7). Plasticity can, among other things, drive selection in one species to overcome plastic changes in another species (ex. induced plant defenses and herbivore tolerance – Mithöfer and Boland 2012); induce plastic changes in another species (ex. parasite modifications of host phenotypes, or behavioral plasticity during competition – Dawkins 1982; Grangier and Lester 2012); cause coevolution of reciprocal plasticity between two or more species (antagonistically or mutualistically – Reimer and Tedengren 1996; Agrawal 2001; Freeman et al. 2009); alter the composition or abundance of other species within the community (ex. irreversible barnacle plasticity affects mussel and algal abundance – Raimondi et al. 2000, Fig. 5.7); determine community composition via *dominant plasticity* (Ashton et al. 2010) or limits to plasticity (ex. homeostatic mechanisms limit species distributions latitudinally – France 1992; Molina-Montenegro and Naya 2012); affect community interactions under changing climates, by altering species in such a way that their interactions are maintained (Cresswell and McCleery 2003; Charmantier et al. 2008) or disrupted (Post and Forchhammer 2008); can place novel selection pressures on species that exist in communities invaded by plastic species (Strauss et al. 2006; Lankau 2012); and can alter fitness landscapes, such that plastic resident organisms reduce the likelihood of successful colonization by plastic invasive species (Peacor et al. 2006).

These multitudinous interactions between plastic genotypes and other members of the community have led to recent studies on plasticity from a community genetics perspective

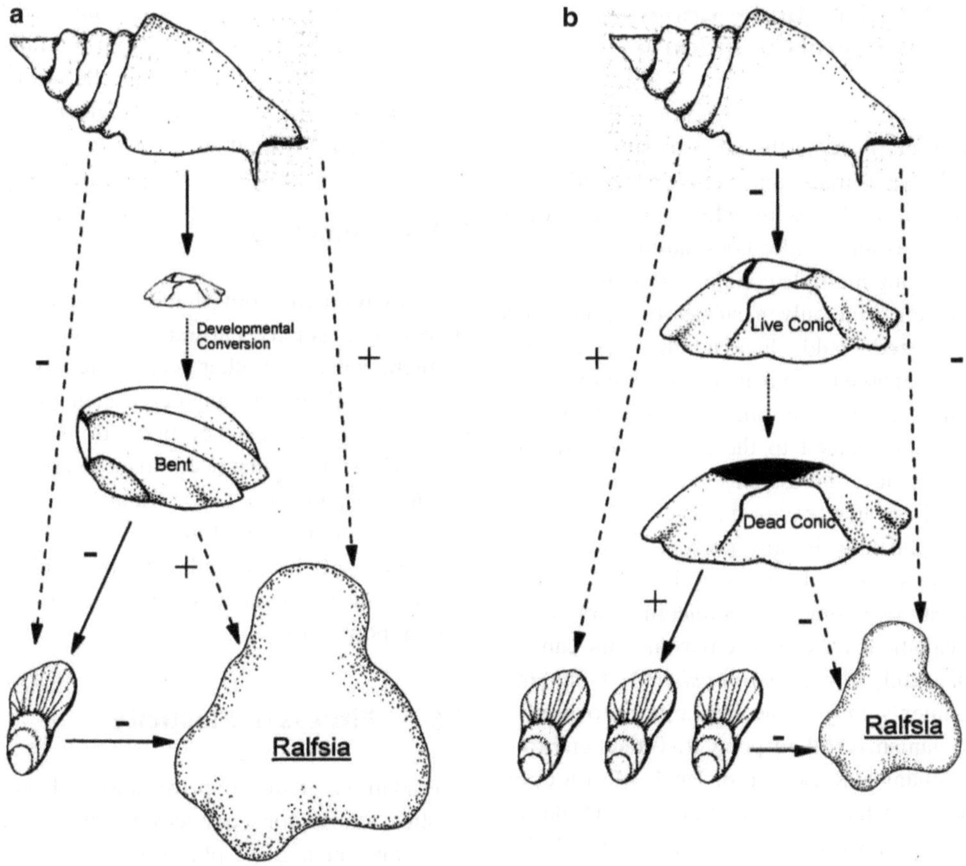

Fig. 5.7 Plasticity can affect the community in numerous ways. This example shows the direct (*solid lines*) and indirect (*dashed lines*) positive and negative interactions between four species of an intertidal community, that are influenced by plasticity in one species. (**a**) Whelks pass over young barnacles that are too small to consume. The barnacles, in response, plastically develop a bent morphology that inhibits whelk predation. Since mussels utilize empty barnacle shells for space, their abundance is negatively affected. Encrusting algae, which compete with mussels for space, increase in density. (**b**) Whelks pass over adult barnacles that developed in the absence of whelks. They consume the barnacle, leaving behind an empty shell that can be colonized by mussels. The mussels outcompete encrusting algae, such that algal density declines while mussel density increases (Reproduced from Raimondi et al. (2000, Fig. 2). Published with kind permission of © John Wiley and Sons 2003. All Rights Reserved)

(Schweitzer et al. 2008; Utsumi 2011). Experimental work is limited, but a recent study likely foreshadows things to come: the interactions between barley, aphids, and rhizobacteria were measured, along with barley and aphid plasticity across rhizobacterial environments. QTL for barley and aphid plasticity were mapped on to the barley genome, thus identifying gene regions in one species that influenced plasticity in another species, as mediated by a third species (Tétard-Jones et al. 2011). Predictions regarding the role of plasticity in community genetics are few, but recent modeling work suggests that plasticity may facilitate community stability in tritrophic systems to a greater extent than genetic variation (Kovach-Orr and Fussmann 2012). Community genetics is ripe for studies on phenotypic plasticity, but requires nuanced predictions and experimental data before patterns can be ascertained.

5.4.4 Plasticity Under Unnatural Ecological Conditions

Although the ecological genomics of adaptive plasticity emphasizes the need to study ecologically relevant traits under natural conditions, there are several reasons why one would want to study plasticity under non-natural conditions. (1) Subjecting multiple populations to an extreme environment that only a subset of populations has colonized could elucidate the mechanisms that have enabled survival in the extreme environment. For instance, one could find that populations not adapted to these extreme environments may nevertheless plastically adjust to survive in these new environments, or that stress-induced plasticity is only present in non-native populations, motivating research into how that stress was overcome. (2) Subjecting organisms to unrealistically extreme environments can allow the study of *symmorphosis*. (3) Unnatural environments may allow researchers to predict how organisms will respond to future environmental change (Reekie et al. 1994; Martin et al. 2011). For instance, exposing congeneric marine species to elevated temperatures revealed that cold-adapted marine organisms live well below their maximum thermal tolerance, while warm adapted marine organisms are negatively affected by very small increases in temperature (Somero 2005, 2010). (4) Studying plasticity using a single manipulated environmental variable is an important first step in elucidating how a particular environmental variable influences the phenotype. Once the production of that phenotype by that variable is understood, multienvironmental variables can be used to determine the influence of environmental interactions on plasticity. In other words, decomposing the environment into its different variables is an important, albeit unnatural, means of learning about plasticity. (5) Subjecting organisms to unnatural environments may allow researchers to uncover how different reaction norms interact to produce the phenotype. Sea urchins subjected to natural pH exhibited plastic compensation in morphology via transcript abundance. Under extremely unnatural acidities, transcriptional activity broke down and morphological plasticity was induced (Martin et al. 2011) (Fig. 5.2). Without the unnatural environment, however, the link between transcript abundance and morphological plasticity would not have been made.

5.4.5 Summary

The environment both shapes plasticity, and is shaped by plasticity, particularly when that environment consists of other genes. The next several years will likely see an increase in the use of genomic tools to test key predictions about the types of mutations that occur in large versus small populations, or the sorts of plastic genes that shape community structure. However, the use of genomics will be limited without precise measures of the ecological conditions experienced by natural populations.

5.5 Fitness of Plasticity

The adaptive or maladaptive nature of plasticity implies fitness consequences for plasticity. Fitness consequences of plasticity can be assessed both indirectly and directly under an ecological genomics framework.

5.5.1 Indirect

Indirect methods for assessing the fitness consequences of plasticity involve everything from modelling the possibility of plasticity evolution to discovering patterns for plasticity among taxa. Plasticity models generally compare plastic and non-plastic phenotypes under distinct stable or fluctuating environments. There are three main types of models (Scheiner 1993): optimality models, which provide cost-benefit analyses of plasticity (Stearns and Koella 1986); quantitative genetic models, which assess the evolution of plasticity given certain selection regimes and genetic variances/covariances for plastic traits (Via and Lande 1985); and gametic models, which assess the consequences of pleiotropy, epistasis, or linkage on the

evolution of plasticity (De Jong 1990). Recent theoretical work has involved the evolution of plasticity given dispersal rates (Scheiner and Holt 2012; Scheiner et al. 2012), colonization events (Thibert-Plante and Hendry 2011), spatial heterogeneity (Chevin and Lande 2011), and an environment that contains genes (Wolf et al. 2003), and the consequences of plasticity for demography (Chevin et al. 2013). Although models are only as good as their assumptions, they have led to several testable predictions about the requirements for plasticity evolution, including the presence of genetic variation, high gene flow, low costs, and a predictable and reliable environmental cue (Crispo 2008).

Studies on natural populations can also provide indirect evidence for the fitness consequences of plasticity. For instance, phylogenetic analyses have shown that plastic traits can arise or become assimilated within a clade (Schwander and Leimar 2011). If this occurs in parallel within a clade, it suggests positive fitness for plasticity or its loss. Finding the ecological relevance of an induced phenotype can also indirectly test the adaptive nature of plasticity, particularly if the induced phenotype is difficult to produce. For instance, the induced defense morphology of *Daphnia* is clearly relevant to the environment that induces it, requiring receptors for predator abundance and the coordinated action of numerous underlying phenotypes. The fact that such a system evolved in association with this environment implies its positive fitness consequences. Finally, comparisons of plasticity that involve multiple populations adapted to different environmental regimes can provide indirect support for the adaptive nature of plasticity or non-plasticity. Three such patterns include: (1) Positive correlations between the degree of plasticity exhibited by a population and the extent of environmental heterogeneity experienced by that population. For instance, the climatic variability hypothesis suggests that plants and ectotherms at high latitudes should exhibit greater temperature-induced plasticity for tolerance and acclimation phenotypes, than populations that reside at lower latitudes. This is because temperature fluctuations are greater at higher latitudes, requiring an increased capacity to maintain homeostasis. Evidence for this hypothesis has been found (Compton et al. 2007; Sunday et al. 2011; Molina-Montenegro and Naya 2012). (2) The parallel evolved loss or gain of plasticity from a known ancestral population. For instance, tiger snakes (*Notechis scutatus*) from mainland Australia feed on relatively small prey, while island colonists consume larger prey. This dietary switch is facilitated by ancestral plasticity in head shape. Head shape plasticity has subsequently been lost in older colonized populations, due to costs associated with the production of smaller heads (Aubret et al. 2004; Aubret and Shine 2009, 2010). (3) Finally, patterns at a genomic level can be assessed. For instance, plasticity was hypothesized to buffer against the effects of selection, resulting in higher rates of evolution for genes whose expression was environmentally sensitive. Leichty et al. (2012) used microarrays to assess genes involved in the production of environmentally-induced morphs of tadpoles. Using 454 sequencing, they then sequenced "biased" (plastic) and "unbiased" (non-plastic) genes, and compared rates of evolution between these genes for multiple plastic and non-plastic amphibian species. Contrary to expectations, they found that plastic genes had higher substitution rates even in non-plastic species, leading to the intriguing hypothesis that non-essential genes in non-plastic species may rapidly accumulate mutations. This rapid evolution then becomes a precondition for the evolution of plasticity, permitting the co-option of these non-essential genes for novel plastic functions under heterogeneous environments (Leichty et al. 2012). Although selection on these plastic genes was not assessed, the repeatability of these patterns in other taxa suggests that plasticity can have positive fitness effects (Hunt et al. 2011).

5.5.2 Direct

Direct methods for assessing the fitness consequences of plasticity involve comparisons of fitness between plastic and non-plastic genotypes. This can be assessed in several ways.

5.5.2.1 Estimates of Fitness in Natural Populations

Long-standing field studies can provide measures of plasticity for individuals, and can measure heritability, fitness, and selection for those plastic phenotypes. Seasonal plasticity in bighorn sheep mass was measured over a 25 year span in both parents and offspring, and was found to have a genetic basis. Selection was measured for summer and winter mass changes, and revealed that plastic individuals had a higher fitness coming out of the winter than less plastic individuals (Pelletier et al. 2007; see also Nussey et al. 2005). Although field studies provide compelling examples of selection under ecologically relevant conditions, true differences in plasticity are difficult to measure due to limited environmental control.

5.5.2.2 Common-Garden/Mesocosm Experiments

Genotypes that differ in their degree of plasticity can be raised in several (often reciprocal) common gardens or mesocosms that manipulate some environmental variable, such that plastic changes are induced between common gardens. The fitness of each plastic and non-plastic genotype can be assessed for each environment (Griffith and Sultan 2012; Matesanz et al. 2012). These genotypes can occur as polymorphisms within a population or between populations, or can be the result of genetic manipulation. For instance, plant genotypes that exhibited plasticity in stem length in response to conspecific density had consistently high fitness at low and high densities, whereas plant genotypes that could produce only long or only short stems had high fitness only at specific densities (Schmitt et al. 1995). Morphologically plastic and non-plastic species of *Daphnia* were raised together and apart in the presence and absence of predators. The plastic species had higher fitness in the presence of predators, but lower fitness in the absence of predators when competing with the non-plastic species. This reduced fitness in the plastic species was not measured when species were raised apart (Engel and Tollrian 2009).

5.5.2.3 Experimental Evolution

Artificial selection experiments, in which plasticity is selected by researchers (Scheiner and Lyman 1991; Scheiner 2002; Kelly et al. 2006), or compared between domestic and wild lineages (Morris et al. 2011; Debes et al. 2012; Solberg et al. 2013), increases our confidence that plasticity is heritable and can evolve. Experimental evolution studies, in which genotypes freely evolve under controlled conditions, have shown that plasticity can evolve when it benefits the organism rather than the researcher (Garland and Kelly 2006). For instance, viruses raised in a combination of single-infection and coinfection conditions experimentally evolved greater plasticity than viruses raised in single-infection conditions alone, and this increased plasticity conferred greater fitness under both environments (Leggett et al. 2013). Plasticity has also been shown to evolve *in silico* for digital organisms (Clune et al. 2007). A genomics approach to the experimental evolution of plasticity would ideally track both the phenotypic and genetic changes that occur under various forms of environmental change to address questions regarding the rules of plasticity evolution, including the types of genes or chromatin structures involved, the importance of pleiotropy, the nature of parallel plasticity evolution, etc. (Bell 2010; Dettman et al. 2012).

5.5.2.4 F2 Selection Experiments

It can be difficult in practice to determine if fitness differences between plastic and non-plastic genotypes are due to plasticity or to other phenotypic differences. For instance, in comparisons between the fitness of invasive plants and their ancestral counterparts, plastic tetraploids had greater fitness than less-plastic diploids, but non-plastic phenotypes affected by tetraploidy could have conferred this fitness benefit (Hahn et al. 2012 – but see their plausible explanation for why this was not the case). If distinct populations can be hybridized, this difficulty can be circumvented. Recombination in the gametes of F1 hybrids results in F2 hybrids with chromosomes that vary in their

distribution of parental alleles. Unless plastic and non-plastic traits are tightly linked, F2 individuals should vary randomly with respect to these phenotypes. F2 individuals could then be measured for their degree of plasticity relative to the parental populations, and fitness assessed in multiple environments. If plasticity does confer a fitness benefit apart from non-plastic traits, then the phenotypic background for the plastic phenotype should not matter. This method relies on a number of practical considerations (plasticity must be measurable in individuals – that is, plasticity must be reversible) and assumptions (no relationship between plastic and non-plastic traits, no linkage). To our knowledge such an F2 experiment has not been employed.

5.5.2.5 Genomics and Fitness

Novel genomic techniques have increased our ability to detect and measure plastic differences between populations. For instance, researchers can now use gene expression tools (quantitative reverse transcriptase PCR, microarrays, RNA-Sequencing, etc.) in association with common garden experiments to associate experimentally-manipulated transcript abundance with fitness (Rest et al. 2013), measure gene expression for fitness-related traits (Zhou et al. 2012), or compare populations for gene expression profiles (Levine et al. 2011). Gene sequencing can identify alleles associated with fitness-related differences in plasticity (Powers and Schulte 1998), while QTL mapping using microsatellites or Single Nucleotide Polymorphisms can identify regions of the genome associated with divergent plasticities (Ungerer et al. 2003; Gerald et al. 2006; Gutteling et al. 2007; Tétard-Jones et al. 2011). F2 selection experiments could provide compelling associations between genotype, phenotype, and fitness, if QTLs for plasticity can be shown to be associated with F2 survival or fecundity in a common garden. Finally, a genomics perspective permits us to ask questions regarding which genes and genomic structures are likely to facilitate the evolution of adaptive plasticity (Leichty et al. 2012).

5.5.3 Summary

One cannot assume a priori that plasticity confers a fitness advantage relative to non-plastic individuals. Indirect evidence supports the adaptive nature of both plasticity and non-plasticity, but direct measures of fitness offer the most compelling results. This direct evidence has revealed that plasticity may increase fitness under all environments, a subset of environments, or can decrease fitness across environments. Ecological knowledge of the study organism is required to make predictions that can differentiate between these alternatives. Just as ecology cannot be ignored, so the underlying genome cannot be ignored, as plasticity will only evolve if genetic variation is present. Future work in the ecological genomics of plasticity will involve identifying alleles that confer fitness advantages in plastic or non-plastic organisms, and determining the genomic architectures that constrain or encourage the evolution of plasticity.

5.6 Conclusion

Phenotypic plasticity research has undergone a renaissance of sorts in developmental, ecological, and evolutionary biology. Integrating these disciplines with novel genomic tools has allowed researchers to test key predictions regarding the production of plasticity (e.g. the importance of transcriptional vs. proteomic plasticity), the evolution of plasticity (e.g. plasticity as a leader or a follower in evolution), the ecological significance of plasticity (e.g. plasticity genes involved in heritable community traits), and the fitness of plasticity (e.g. relaxed selection as a precondition for adaptive plasticity). Furthermore, the ecological genomics of plasticity has emphasized the complexity and difficulty of elucidating the complete developmental, evolutionary, and ecological story for a single phenotype, even in well-known and well-studied species. Theoretical work currently outpaces data, but the tide is changing. Genomic tools are enabling researchers to understand the

significance of plasticity like never before, and the incoming data should stimulate research for years to come.

Acknowledgments MM held a Vanier Canada Graduate Scholarship from the Natural Sciences and Engineering Research Council of Canada (NSERC), as well as a research allowance from Alberta Innovates Technology Futures (AITF). SMR is funded by an NSERC Discovery Grant and an AITF New Faculty Award. The authors wish to thank two reviewers for their insightful suggestions.

Glossary: Some Definitions of Important Terms

Adaptive plasticity The production of alternative phenotypes (continuous or discrete) by the same genotype across some environmental variable, such that there is a better match between the organism and its environment (Beldade et al. 2011). Alleles that confer plasticity are more likely to spread through a population relative to competing alleles that do not confer plasticity.

Behavioural plasticity Environmentally-induced alternative behaviors displayed by a single genotype. Behavioral plasticity is difficult to define. Does an organism display behavioral plasticity if it switches from grazing when there are no predators to predator avoidance when predators are present? Or should behavioral plasticity be restricted to a single behavior type (i.e., different foraging tactics for different foods, or different predator avoidance strategies for different predators)? Or is behavior only plastic if one particular behavioral trait is expressed differently when the same environmental variable in manipulated, such as different foraging strategies for a single food under different light conditions, or different predator avoidance strategies for a single predator under different conspecific densities? One's definition will determine the magnitude of the relationship one finds between behavioral plasticity and other reaction norms.

Canalization See **Robustness**.

Community genetics The study of how genes within a community shape the phenotypes and evolution of other members of the community, and the identification of genes that contribute to heritable community traits.

Cryptic genetic variation The subset of standing genetic variation that exists in a population but does not affect the phenotype or performance under normal environmental conditions. Upon exposure to a novel environment, this genetic variation produces novel phenotypic variation, and may facilitate adaptation.

Developmental noise The production of alternative phenotypes by a single genotype under identical environmental conditions (Raser and O'Shea 2005) due to molecular stochasticity in the birth and death rates of transcripts, the effects of low-abundance regulatory proteins, the stickiness of proteins, and random fluctuations in promoter behavior (Raser and O'Shea 2005; Brettner and Masel 2012; Singh et al. 2012).

Dominant plasticity In niche complementarity, occurs when a superior competitor with high resource use plasticity alters the resources it uses depending on the competitive environment.

Ecological speciation A theory of speciation in which adaptive phenotypic and genetic divergence, contributing to reproductive isolation, is due to divergent selection (Nosil 2012).

Environmental robustness (environmental canalization) The production of a stable reaction norm despite environmental perturbations. Non-plastic reaction norms are canalized against at least one environmental variable. A plastic reaction norm can be environmentally canalized if: (1) the reaction norm does not exhibit discontinuous change under extreme environments, or (2) the reaction norm maintains its height and slope in the presence of a second environmental variable.

Epigenetics The study of environmentally-induced, sometimes heritable modifications to the phenotype, caused by mechanisms other

than changes to the underlying DNA sequence (i.e., DNA methylation, histone modification, etc.).

Epigenome The entire suite of epigenetic modifications that have occurred in a particular cell, tissue, developmental stage, or organism, including the number and placement of methylated sites, the number and nature of histone modifications, etc.

Flexible stem A model of adaptive phenotypic divergence whereby an initially plastic ancestral population diverges into two populations residing in distinct environments, such that each population expresses opposing ends of a reaction norm. These phenotypes become genetically assimilated, such that plasticity is lost and phenotypic divergence is maintained (West-Eberhard 2003; Wund et al. 2008).

Gene expression The context-dependent production of gene product, including pre-mRNA, mRNA, microRNA, and protein. Context can include cell type, tissue type, genotype, developmental stage, time, and environment.

Genetic assimilation The evolved loss of adaptive phenotypic plasticity, such that environmental induction is no longer necessary for the production of the phenotype (Waddington 1953a, b).

Genetic compensation The evolved loss of maladaptive phenotypic plasticity, resulting in phenotypic similarity (cryptic evolution) between populations living in regular and novel environments (Grether 2005).

Genetic robustness (genetic robustness) The production of a stable reaction norm despite different genetic backgrounds (Gibson and Wagner 2000). Genotypically distinct individuals that display the same plastic or non-plastic reaction norm are genetically canalized against the alleles that differentiate them. A lack of genetic robustness can be evidenced by changes to the slope or height of the reaction norm.

Hierarchy of plasticities The production of a plastic or non-plastic macrophenotype due to interactions between numerous underlying reaction norms (Bradshaw 1965).

Macrophenotype The visible manifestation of numerous underlying phenotypes, sometimes referred to as the "end phenotype" (Beldade et al. 2011).

Maladaptive plasticity The production of alternative phenotypes (continuous or discrete) by the same genotype across some environmental variable, such that the match between organism and environment is reduced (Ghalambor et al. 2007).

Model organism Any non-human species that has been readily cultured or raised over many generations in a laboratory setting, for which genomic tools have been developed and applied, and that is used to answer biological questions that can be applied to other species. Examples: *Arabidopsis*, *Drosophila*, *Daphnia*, *Mus*. The ideal model species for ecological genomics has locally adapted populations, characterized phenotypic and genetic variation, a sequenced genome, a known phylogeny, and is studied by a large community of researchers (Feder and Mitchell-Olds 2003).

Molecular phenotype Measures of context-specific gene expression or protein behavior (Ranz and Machado 2006; Pavey et al. 2010).

Molecular plasticity A form of phenotypic plasticity that focuses on environmentally-sensitive gene expression or protein behavior (plasticity in the molecular phenotype).

Neutral plasticity The production of alternative phenotypes (continuous or discrete) by the same genotype across some environmental variable, that does not contribute positively or negatively to fitness.

Non-model organism Any species that has not been readily incorporated into biological research in the last several decades. Basic biological information, including genomic information, is often lacking for these organisms, although genomic tools may be developed and used.

Non-plasticity A reaction norm with a slope of zero. Sometimes referred to as environmental robustness (Gibson and Wagner 2000). Non-plasticity may be adaptive, maladaptive, or neutral, relative to competing alleles that confer plasticity.

Phenotypic accommodation A source of adaptive phenotypic novelty, in which a genetically- or environmentally-induced change to a phenotype during development is accommodated through plastic changes in other phenotypes (West-Eberhard 2005).

Phenotypic capacitor Any phenotype that can suppress phenotypic variation that would otherwise be expressed via developmental noise, microenvironmental variation, and genotypic variation (Queitsch et al. 2002; Levy and Siegal 2008).

Phenotypic plasticity The environmentally sensitive production of alternative phenotypes by a single genotype (DeWitt and Scheiner 2003).

Plastic compensation The production of phenotypic similarity between populations living in regular and novel environments, due to plasticity in some compensating phenotype. Without plasticity in the compensating phenotype, maladaptive plasticity would generate phenotypic divergence between populations. Usually comes with a cost to some other phenotype and may mask the existence of maladaptive plasticity (Morris and Rogers 2013).

Plasticity-mediated population extinction (PMPE) The unsuccessful colonization of a new environment due to phenotypic plasticity induced by the new environment (Morris and Rogers 2013).

Plasticity-mediated population persistence (PMPP) The successful colonization of a new environment due to phenotypic plasticity induced by the new environment (Baldwin 1896; Pavey et al. 2010).

Proteome The full complement of proteins present within a particular context (see gene expression).

Proteomic plasticity A form of molecular plasticity that focuses on environmentally-sensitive protein abundance.

Reaction norm A function of all possible phenotypic states across some environmental gradient.

Robustness (Canalization) The production of a stable reaction norm despite genetic, environmental, or stochastic perturbations. Both plastic and non-plastic reaction norms can display robustness (Waddington 1953a, b).

Symmorphosis The theory that biological structures match their functional requirements, without unnecessarily exceeding those requirements. This includes the idea that the components of a system will not exceed in possible performance their weakest unit.

Standing genetic variation (SGV) Genetic variation that exists at a single locus in natural populations (Barrett and Schluter 2008). One form of SGV, cryptic genetic variation (CGV), is not apparent until exposed by novel environments (Schlichting 2008).

Stochastic robustness (Stochastic robustness) The production of a stable reaction norm despite stochasticity (developmental noise) in transcript abundance, protein activity, etc.

Transcriptional plasticity A form of molecular plasticity that focuses on environmentally-sensitive mRNA transcript abundance.

Transcriptome The full complement of mRNA present within a particular context (see gene expression).

References

Agrawal A (2001) Phenotypic plasticity in the interactions and evolution of species. Science 294:321–326

Akkas S, Kepenek A, Beklioglu M, Severcan F (2010) Molecular approach to the chemical characterization of fish-exuded kairomone: a Fourier transform infrared spectroscopic study. Aquat Sci 72:71–83

Alpert P, Simms E (2002) The relative advantages of plasticity and fixity in different environments: when is it good for a plant to adjust? Evol Ecol 16: 285–297

Andrew RL, Bernatchez L, Bonin A, Buerkle CA, Carstens BC, Emerson BC, Garant D, Giraud T, Kane NC, Rogers SM, Slate J, Smith H, Sork VL, Stone GN, Vines TH, Waits L, Widmer A, Rieseberg LH (2013) A roadmap for molecular ecology. Mol Ecol 22:2605–2626

Angers B, Castonguay E, Massicotte R (2010) Environmentally induced phenotypes and DNA methylation: how to deal with unpredictable conditions until the next generation and after. Mol Ecol 19: 1238–1295

Ashton I, Miller A, Bowman W, Suding K (2010) Niche complementarity due to plasticity in resource use: plant partitioning of chemical N forms. Ecology 91:3252–3260

Aubin-Horth N, Renn S (2009) Genomic reaction norms: using integrative biology to understand molecular mechanisms of phenotypic plasticity. Mol Ecol 18:3763–3780

Aubret F, Shine R (2009) Genetic assimilation and the postcolonisation erosion of phenotypic plasticity in island tiger snakes. Curr Biol 19:1932–1936

Aubret F, Shine R (2010) Fitness costs may explain the post-colonisation erosion of phenotypic plasticity. J Exp Biol 213:735–739

Aubret F, Shine R, Bonnet X (2004) Evolutionary biology: adaptive developmental plasticity in snakes. Nature 431:261–262

Bachmann D, Both S, Bruelheide H, Ding B, Gao M, Härdtle W, Scherer-Lorenzen M, Erfmeier A (2012) Functional trait similarity of native and invasive herb species in subtropical China – environment-specific differences are the key. Environ Exp Bot 83:82–92

Bajić D, Poyatos J (2012) Balancing noise and plasticity in eukaryotic gene expression. BMC Genomics 13:343

Baldwin J (1896) A new factor in evolution. Am Nat 30:441–451

Baldwin J (1902) Development and evolution. MacMillan, New York

Barrett R, Schluter D (2008) Adaptation from standing genetic variation. Trends Ecol Evol 23:38–44

Barry M (2002) Progress toward understanding the neurophysiological basis of predator-induced morphology in *Daphnia pulex*. Physiol Biochem Zool 75:179–186

Beaton M, Hebert P (1997) The cellular basis of divergent head morphologies in *Daphnia*. Limnol Oceanogr 42:346–356

Beldade P, Mateus R, Keller R (2011) Evolution and molecular mechanisms of adaptive developmental plasticity. Mol Ecol 20:1347–1363

Bell G (2010) Experimental genomics of fitness in yeast. Proc R Soc B 277:1459–1467

Bentz B, Ryan B, Mock K, Pfrender M (2011) Genetic architecture and phenotypic plasticity of thermally-regulated traits in an eruptive species, *Dendroctonus ponderosae*. Evol Ecol 25:1269–1288

Berrigan D, Scheiner S (2003) Modeling the evolution of phenotypic plasticity. In: DeWitt T, Scheiner S (eds) Phenotypic plasticity: functional and conceptual approaches. Oxford University Press, Oxford

Bock C (2012) Analysing and interpreting DNA methylation data. Nat Rev Genet 13:705–719

Bossdorf O, Richards C, Pigliucci M (2008) Epigenetics for ecologists. Ecol Lett 11:106–115

Bossdorf O, Arcuri D, Richards C, Pigliucci M (2010) Experimental alteration of DNA methylation affects the phenotypic plasticity of ecologically relevant traits in *Arabidopsis thaliana*. Evol Ecol 24:541–553

Bradshaw A (1965) Evolutionary significance of phenotypic plasticity in plants. Adv Genet 13:115–155

Brettner L, Masel J (2012) Protein stickiness rather than number of functional protein-protein interactions predicts expression noise and plasticity in yeast. BMC Syst Biol 6:128

Castillo-Davis C, Mekhedov S, Hartl D, Koonin E, Kondrashov F (2002) Selection for short introns in highly expressed genes. Nat Genet 31:415–418

Charmantier A, McCleery R, Cole L, Perrins C, Kruuk L, Sheldon B (2008) Adaptive phenotypic plasticity in response to climate change in a wild bird population. Science 320:800–803

Chevin L-M, Lande R (2011) Adaptation to marginal habitats by evolution of increased phenotypic plasticity. J Evol Biol 24:1462–1476

Chevin L-M, Gallet R, Gomulkiewicz R, Holt R, Fellous S (2013) Phenotypic plasticity in evolutionary rescue experiments. Phil Trans R Soc B 368:20120089

Chown S, Jumbam K, Sørensen J, Terblanche J (2009) Phenotypic variance plasticity and heritability estimates of critical thermal limits depend on methodological context. Funct Ecol 23:133–140

Clune J, Ofria C, Pennock R (2007) Investigating the emergence of phenotypic plasticity in evolving digital organisms. In: Almeida e Costa F, Rocha L, Costa E, Harvey I, Coutinho A (eds) Proceedings of the 9th European conference on advances in artificial life. Springer-Verlag, Berlin

Compton T, Rijkenberg M, Drent J, Piersma T (2007) Thermal tolerance ranges and climate variability: a comparison between bivalves from differing climates. J Exp Mar Biol Ecol 352:200–211

Conover D, Schultz E (1995) Phenotypic similarity and the evolutionary significance of countergradient variation. Trends Ecol Evol 10:248–252

Cresswell W, McCleery R (2003) How great tits maintain synchronization of their hatch date with food supply in response to long-term variability in temperature. J Anim Ecol 72:356

Crews D, Gillette R, Scarpino S, Manikkam M, Savenkova M, Skinner M (2012) Epigenetic transgenerational inheritance of altered stress responses. Proc Natl Acad Sci USA 109:9143–9148

Crispo E (2007) The Baldwin Effect and genetic assimilation: revisiting two mechanisms of evolutionary change mediated by phenotypic plasticity. Evolution 61(11):2469–2479

Crispo E (2008) Modifying effects of phenotypic plasticity on interactions among natural selection, adaptation and gene flow. J Evol Biol 21:1460–1469

Crispo E, Moore J, Lee-Yaw J, Gray S, Haller B (2011) Broken barriers: human-induced changes to gene flow and introgression in animals: an examination of the ways in which humans increase genetic exchange among populations and species and the consequences for biodiversity. Bioessays 33:508–518

Dalziel A, Rogers S, Schulte P (2009) Linking genotypes to phenotypes and fitness: how mechanistic biology can inform molecular ecology. Mol Ecol 18: 4997–5017

Dawkins R (1976) The selfish gene. Oxford University Press, Oxford

Dawkins R (1982) The extended phenotype. Oxford University Press, Oxford

Day T, Pritchard J, Schluter D (1994) A comparison of two sticklebacks. Evolution 48:1723–1734
De Boer TE, Birlutiu A, Bochdanovits Z, Timmermans MJTN, Dijkstra TMH, Van Straalen NM, Ylstra B, Roelofs D (2011) Transcriptional plasticity of a soil arthropod across different ecological conditions. Mol Ecol 20:1144–1154
De Jong G (1990) Quantitative genetics of reaction norms. J Evol Biol 3:447–468
Debat V, David P (2001) Mapping phenotypes: canalization, plasticity and developmental stability. Trends Ecol Evol 16:555–561
Debes P, Normandeau E, Fraser D, Bernatchez L, Hutchings J (2012) Differences in transcription levels among wild, domesticated and hybrid Atlantic salmon (*Salmo salar*) from two environments. Mol Ecol 21:2574–2587
Deredge D, Baker J, Datta K, LiCata V (2010) The glutamate effect on DNA binding by Pol I DNA polymerases: osmotic stress and the effective reversal of salt linkage. J Mol Biol 401:223–238
Dettman J, Rodrigue N, Melnyk A, Wong A, Bailey S, Kassen R (2012) Evolutionary insight from whole-genome sequencing of experimentally evolved microbes. Mol Ecol 21:2058–2077
DeWitt T, Scheiner S (2003) Phenotypic variation from single genotypes: a primer. In: DeWitt T, Scheiner S (eds) Phenotypic plasticity: functional and conceptual approaches. Oxford University Press, Oxford
DeWitt T, Sih A, Wilson DS (1998) Costs and limits of phenotypic plasticity. Trends Ecol Evol 13:77–81
Dieckmann U, Doebeli M, Metz J, Tautz D (2004) Adaptive speciation. Cambridge University Press, New York
Diz A, Martínez-Fernández M, Rolán-Alvarez E (2012) Proteomics in evolutionary ecology: linking the genotype with the phenotype. Mol Ecol 21:1060–1080
Eames M, Kortemme T (2012) Cost-benefit tradeoffs in engineered *lac* operons. Science 336:911–915
Engel K, Tollrian R (2009) Inducible defences as key adaptations for the successful invasion of *Daphnia lumholtzi* in North America? Proc R Soc B 276:1865–1873
Espinosa-Soto C, Martin O, Wagner A (2011) Phenotypic plasticity can facilitate adaptive evolution in gene regulatory circuits. BMC Evol Biol 11:5
Evans JD, Wheeler DE (2000) Expression profiles during honeybee caste determination. Genome Biol 2:research0001-research0001.6
Feder M, Mitchell-Olds T (2003) Evolutionary and ecological functional genomics. Nat Rev Genet 4:649–655
Findlay G, MacCoss M, Swanson W (2009) Proteomic discovery of previously unannotated rapidly evolving seminal fluid genes in *Drosophila*. Genome Res 19:886–896
Fitzpatrick B (2012) Underappreciated consequences of phenotypic plasticity for ecological speciation. Int J Ecol 2012:256017
Ford EB (1964) Ecological genetics. Chapman and Hall, London
Fordyce J (2006) The evolutionary consequences of ecological interactions mediated through phenotypic plasticity. J Exp Biol 209:2377–2383
France R (1992) The North American latitudinal gradient in species richness and geographical range of freshwater crayfish and amphipods. Am Nat 139:342–354
Frankenhuis WE, Panchanathan K (2011) Individual differences in developmental plasticity may result from stochastic sampling. Perspect Psychol Sci 6:336–347
Fraser H (2011) Genome-wide approaches to the study of adaptive gene expression evolution. Bioessays 33:469–477
Freeman A, Meszaros J, Byers J (2009) Poor phenotypic integration of blue mussel inducible defenses in environments with multiple predators. Oikos 118:758–766
Fusco G, Minelli A (2010) Phenotypic plasticity in development and evolution: facts and concepts. Phil Trans R Soc B 365:547–556
Garland T Jr, Kelly S (2006) Phenotypic plasticity and experimental evolution. J Exp Biol 209:2344–2361
Gerald J, Lehti-Shiu M, Ingram P, Deak K, Biesiada T, Malamy J (2006) Identification of Quantitative Trait Loci that regulate *Arabidopsis* root system size and plasticity. Genetics 172:485–498
Ghalambor C, McKay J, Carroll S, Reznick D (2007) Adaptive versus non-adaptive phenotypic plasticity and the potential for contemporary adaptation in new environments. Funct Ecol 21:394–407
Gibson G, Wagner G (2000) Canalization in evolutionary genetics: a stabilizing theory? Bioessays 22:372–380
Gracey A, Fraser E, Li W, Fang Y, Taylor R, Rogers J, Brass A, Cossins A (2004) Coping with cold: an integrative multitissue analysis of the transcriptome of a poikilothermic vertebrate. Proc Natl Acad Sci USA 101:16970–16975
Grangier J, Lester P (2012) Behavioral plasticity mediates asymmetric competition between invasive wasps and native ants. Commun Integr Biol 5:127–129
Greenberg J, Xia J, Zhou X, Thatcher S, Gu X, Ament S, Newman T, Green P, Zhang W, Robinson G, Ben-Shahar Y (2012) Behavioral plasticity in honey bees is associated with differences in brain microRNA transcriptome. Genes Brain Behav 11:660–670
Grether G (2005) Environmental change, phenotypic plasticity and genetic compensation. Am Nat 166:E115–E123
Griffith T, Sultan S (2012) Field-based insights to the evolution of specialization: plasticity and fitness across habitats in a specialist/generalist species pair. Ecol Evol 2:778–791
Guo Y, Lang S, Ellis R (2009) Independent recruitment of F Box genes to regulate hermaphrodite development during nematode evolution. Curr Biol 19:1853–1860
Gutteling E, Riksen J, Bakker J, Kammenga J (2007) Mapping phenotypic plasticity and genotype-environment interactions affecting life-history traits in *Caenorhabditis elegans*. Heredity 98:28–37

Hackett J, Sengupta R, Zylicz J, Murakami K, Lee C, Down T, Surani M (2013) Germline DNA demethylation dynamics and imprint erasure through 5-hydroxymethylcytosine. Science 339:448–452

Hahn M, van Kleunen M, Müller-Shärer H (2012) Increased phenotypic plasticity to climate may have boosted the invasion success of polyploid *Centaurea stoebe*. PLoS One 7:e50284

Hanshew B, Garica T (2012) Invasion of the shelter snatchers: behavioural plasticity in invasive red swamp crayfish *Procambarus clarkia*. Freshwater Biol 57:2285–2296

Hayden EJ, Ferrada E, Wagner A (2011) Cryptic genetic variation promotes rapid evolutionary adaptation in an RNA enzyme. Nature 474:92–95

Hayward S, Murray P, Gracey A, Cossins A (2007) Beyond the lipid hypothesis: mechanisms underlying phenotypic plasticity in inducible cold tolerance. Adv Exp Med Biol 594:132–142

Herrera CM, Bazaga P (2012) Epigenetic correlates of plant phenotypic plasticity: DNA methylation differs between prickly and nonprickly leaves in heterophyllous *Ilex aquifolium* (Aquifolaceae) trees. Bot J Linn Soc 171:441–452

Hersch-Green E, Turley N, Johnson M (2011) Community genetics: what have we accomplished and where should we be going? Phil Trans R Soc B 366:1453–1460

Hunt B, Ometto L, Wurm Y, Shoemaker D, Yi S, Keller L, Goodisman M (2011) Relaxed selection is a precursor to the evolution of phenotypic plasticity. Proc Natl Acad Sci USA 108:15936–15941

Iwasaki WM, Tsuda ME, Kawata M (2013) Genetic and environmental factors affecting cryptic variations in gene regulatory networks. BMC Evol Biol 13:91

Juliano RL, Dixit VR, Kang H, Kim TY, Miyamoto Y, Xu D (2005) Epigenetic manipulation of gene expression: a toolkit for cell biologists. J Cell Biol 169: 847–857

Kelly S, Czech P, Wight J, Blank K, Garland T Jr (2006) Experimental evolution and phenotypic plasticity of hindlimb bones in high-activity house mice. J Morphol 267:360–374

Kovach-Orr C, Fussmann G (2012) Evolutionary and plastic rescue in multitrophic model communities. Phil Trans R Soc B 368:20120084

Laforsch C, Tollrian R (2010) Cyclomorphosis and phenotypic changes. In: Likens G (ed) Plankton of inland waters. Academic, San Diego

Lande R (2009) Adaptation to an extraordinary environment by evolution of phenotypic plasticity and genetic assimilation. J Evol Biol 22:1435–1446

Lande R, Arnold S (1983) The measurement of selection on correlated characters. Evolution 37:1210–1226

Landry C, Oh J, Hartl D, Cavalieri D (2006) Genome-wide scan reveals that genetic variation for transcriptional plasticity in yeast is biased towards multi-copy and dispensable genes. Gene 366:343–351

Lang G, Murray A, Botstein D (2009) The cost of gene expression underlies a fitness trade-off in yeast. Proc Natl Acad Sci USA 106:5755–5760

Langerhans R, DeWitt T (2002) Plasticity constrained: over-generalized induction cues cause maladaptive phenotypes. Evol Ecol Res 4:857–870

Lankau R (2012) Coevolution between invasive and native plants driven by chemical competition and soil biota. Proc Natl Acad Sci USA 109:11240–11245

Lee M, Topper S, Hubler S, Hose J, Wenger C, Coon J, Gasch A (2011) A dynamic model of proteome changes reveals new roles for transcript alteration in yeast. Mol Syst Biol 7:514

Leggett H, Benmayor R, Hodgson D, Buckling A (2013) Experimental evolution of adaptive phenotypic plasticity in a parasite. Curr Biol 23:139–142

Lehner B (2010) Conflict between noise and plasticity in yeast. PLoS Genet 6:e1001185

Leichty A, Pfennig D, Jones C, Pfennig K (2012) Relaxed genetic constraint is ancestral to the evolution of phenotypic plasticity. Integr Comp Biol 52:16–30

Levine M, Eckert M, Begun D (2011) Whole-genome expression plasticity across tropical and temperate *Drosophila melanogaster* populations from eastern Australia. Mol Biol Evol 28:249–256

Levy S, Siegal (2008) Network hubs buffer environmental variation in *Saccharomyces cerevisiae*. PLoS Biol 6:e264

Maier T, Schmidt A, Güell M, Kühner S, Gavin A, Aebersold R, Serrano L (2011) Quantification of mRNA and protein and integration with protein turnover in a bacterium. Mol Syst Biol 7:511

Malakar P, Venkatesh K (2012) Effect of substrate and IPTG concentrations on the burden to growth of *Escherichia coli* on glycerol due to the expression of Lac proteins. Appl Micriobiol Biot 93:2543–2549

Martin S, Richier S, Pedrotti M, Dupont S, Castejon C, Gerakis Y, Kerros M, Oberhänsli F, Teyssié J, Jeffree R, Gattuso J (2011) Early development and molecular plasticity in the Mediterranean sea urchin *Paracentrotus lividus* exposed to CO_2-driven acidification. J Exp Biol 214:1357–1368

Matesanz S, Horgan-Kobelski T, Sultan S (2012) Phenotypic plasticity and population differentiation in an ongoing species invasion. PLoS One 7:e44955

Matzek V (2012) Trait values, not trait plasticity, best explains invasive species' performance in a changing environment. PLoS One 7:e48821

McCairns R, Bernatchez L (2010) Adaptive divergence between freshwater and marine sticklebacks: insight into the role of phenotypic plasticity from an integrated analysis of candidate gene expression. Evolution 64(4):1029–1047

McGuigan K, Nishimura N, Currey M, Hurwit D, Cresko W (2011) Cryptic genetic variation and body size evolution in threespine stickleback. Evolution 65(4):1203–1211

Meister P, Schott S, Bedet C, Xiao Y, Rohner S, Bodennec S, Hudry B, Molin L, Solari F, Gasser S, Palladino F (2011) *Caenorhabditis elegans* Heterochromatin protein 1 (HPL-2) links developmental plasticity, longevity and lipid metabolism. Genome Biol 12:R123

Mithöfer A, Boland W (2012) Plant defense against herbivores: chemical aspects. Annu Rev Plant Biol 63: 431–450

Miyakawa H, Imai M, Sugimoto N, Ishikawa Y, Ishikawa A, Ishigaki H, Okada Y, Miyazaki S, Koshikawa S, Cornette R, Miura T (2010) Gene up-regulation in response to predator kairomones in the water flea *Daphnia pulex*. BMC Dev Biol 10:45

Moczek A (2012) The nature of nurture and the future of evodevo: toward a theory of developmental evolution. Integr Comp Biol 52:108–119

Moczek A, Sultan S, Foster S, Ledón-Rettig C, Dworkin I, Nijhout H, Abouheif E, Pfennig D (2011) The role of developmental plasticity in evolutionary innovation. Proc R Soc B 278:2705–2713

Molina-Montenegro M, Naya D (2012) Latitudinal patterns in phenotypic plasticity and fitness-related traits: assessing the climatic variability hypothesis (CVH) with an invasive plant species. PLoS One 7:e47620

Molina-Montenegro M, Peñuelas J, Munné-Bosch S, Sardans J (2012) Higher plasticity in ecophysiological traits enhances the performance and invasion success of *Taraxacum officinale* (dandelion) in alpine environments. Biol Invasions 14:21–33

Morris M, Rogers SM (2013) Overcoming maladaptive plasticity through plastic compensation. Curr Zool 59:526–536

Morris M, Fraser D, Eddington J, Hutchings J (2011) Hybridization effects on phenotypic plasticity: experimental compensatory growth in farmed-wild Atlantic salmon. Evol Appl 4:444–458

Mozdzer T, Megonigal P (2012) Jack-and-master trait responses to elevated CO_2 and N: a comparison of native and introduced *Phragmites australis*. PLoS One 7:e42794

Müller S, Bianchi M, Knapp S (2001) Thermodynamics of HMGB1 interaction with duplex DNA. Biochemistry 40:10254–10261

Neyfakh A, Hartl D (1993) Genetic control of the rate of embryo development: selection for faster development at elevated temperatures. Evolution 47: 1625–1631

Nosil P (2012) Ecological speciation. Oxford University Press, New York

Nussey D, Postma E, Gienapp P, Visser M (2005) Selection on heritable phenotypic plasticity in a wild bird population. Science 310:304–306

Orsini L, Andrew R, Eizaguirre C (2013) Evolutionary ecological genomics. Mol Ecol 22:527–531

Padilla D, Adolph S (1996) Plant inducible morphologies are not always adaptive: the importance of time delays in a stochastic environment. Evo Ecol 10:105–117

Pavey S, Collin H, Nosil P, Rogers S (2010) The role of gene expression in ecological speciation. Ann NY Acad Sci 1206:110–129

Pavey S, Bernatchez L, Aubin-Horth N, Landry CR (2012) What is needed for next-generation ecological and evolutionary genomics? Trends Ecol Evol 27:673–678

Peacor S, Allesina S, Riolo R, Pascual M (2006) Phenotypic plasticity opposes species invasions by altering fitness surface. PLoS Biol 4:e372

Pelletier F, Réale D, Garant D, Coltman D, Festa-Bianchet M (2007) Selection on heritable seasonal phenotypic plasticity of body mass. Evolution 61(8):1969–1979

Peñalva-Arana D, Lynch M, Robertson H (2009) The chemoreceptor genes of the waterflea *Daphnia pulex*: many Grs but no Ors. BMC Evol Biol 9:79

Pfennig D, Wund M, Snell-Rood E, Cruickshank T, Schlichting C, Moczek A (2010) Phenotypic plasticity's impacts on diversification and speciation. Trends Ecol Evol 25:459–467

Podrabsky J, Somero G (2004) Changes in gene expression associated with acclimation to constant temperatures and fluctuating daily temperatures in an annual killifish *Austrofundulus limnaeus*. J Exp Biol 207:2237–2254

Post E, Forchhammer M (2008) Climate change reduces reproductive success of an Arctic herbivore through trophic mismatch. Phil Trans R Soc B 363:2367–2373

Powers D, Schulte P (1998) Evolutionary adaptations of gene structure and expression in natural populations in relation to a changing environment: a multidisciplinary approach to address the million-year saga of a small fish. J Exp Zool 282:71–94

Price T (2003) The role of phenotypic plasticity in driving genetic evolution. Proc R Soc B 270:1433–1440

Purchase C, Moreau D (2012) Stressful environments induce novel phenotypic variation: hierarchical reaction norms for sperm performance of a pervasive invader. Ecol Evol 2:2562–2571

Queitsch C, Sangster T, Lindquist S (2002) Hsp90 as a capacitor of phenotypic variation. Nature 417: 618–624

Raimondi P, Forde S, Delph L, Lively C (2000) Processes structuring communities: evidence for trait-mediated indirect effects through induced polymorphisms. Oikos 91:353–361

Ranz J, Machado C (2006) Uncovering evolutionary patterns of gene expression using microarrays. Trends Ecol Evol 21:29–37

Raser J, O'Shea E (2005) Noise in gene expression: origins, consequences and control. Science 309:2010–2013

Reekie J, Hicklenton P, Reekie E (1994) Effects of elevated CO_2 on time of flowering in four short-day and four long-day species. Can J Bot 72:533–538

Refsnider J, Janzen F (2012) Behavioral plasticity may compensate for climate change in a long-lived reptile with temperature-dependent sex determination. Biol Cons 152:90–95

Reimer O, Tedengren M (1996) Phenotypic improvement of morphological defences in the mussel *Mytilus edulis* induced by exposure to the predator *Asterias rubens*. Oikos 75:383–390

Renn SCP, Aubin-Horth N, Hofmann HA (2008) Fish and chips: functional genomics of social plasticity in an African cichlid fish. J Exp Biol 211:3041–3056

Rest J, Morales C, Waldron J, Opulente D, Fisher J, Moon S, Bullaughey K, Carey L, Dedousis D (2013) Nonlinear fitness consequences of variation in expression level of a eukaryotic gene. Mol Biol Evol 30:448–456

Richards CL, Bossdorf O, Pigliucci M (2010) What role does heritable epigenetic variation play in phenotypic evolution? BioScience 60:232–237

Richards CL, Rosas U, Banta J, Bhambhra N, Purrugganan MD (2012) Genome-wide patterns of *Arabidopsis* gene expression in nature. PLoS Genet 8:e1002662

Roberts SB, Gavery MR (2012) Is there a relationship between DNA methylation and phenotypic plasticity in invertebrates? Front Physiol 2:116

Rossitter M (1996) Incidence and consequences of inherited environmental effects. Annu Rev Ecol Syst 27:451–476

Rowntree J, Shuker D, Preziosi R (2011) Forward from the crossroads of ecology and evolution. Phil Trans R Soc B 366:1322–1328

Ruden DM, Xiao L, Garfinkel MD, Lu X (2005) Hsp90 and environmental impacts on epigenetic states: a model for the trans-generational effects of diethylstibesterol on uterine development and cancer. Hum Mol Gen 14:R149–R155

Sardans J, Peñuelas J, Rivas-Ubach A (2011) Ecological metabolomics: overview of current developments and future challenges. Chemoecology 21:191–225

Scheiner S (1993) Genetics and evolution of phenotypic plasticity. Annu Rev Ecol Syst 24:35–68

Scheiner S (2002) Selection experiments and the study of phenotypic plasticity. J Exp Biol 15:889–898

Scheiner S, Holt R (2012) The genetics of phenotypic plasticity. X. Variation versus uncertainty. Ecol Evol 2:751–767

Scheiner S, Lyman R (1991) The genetics of phenotypic plasticity II. Response to selection. J Evol Biol 4:23–50

Scheiner S, Barfield M, Holt R (2012) The genetics of phenotypic plasticity. XI. Joint evolution of plasticity and dispersal rate. Ecol Evol 8:2027–2039

Schlichting C (2008) Hidden reaction norms cryptic genetic variation and evolvability. Ann NY Acad Sci 1133:187–203

Schluter D (2000) The ecology of adaptive radiation. Oxford University Press, New York

Schmitt J, McCormac A, Smith H (1995) A test of the adaptive plasticity hypothesis using transgenic and mutant plants disabled in phytochrome-mediated elongation responses to neighbors. Am Nat 146:937–953

Schrimpf S, Weiss M, Reiter L, Ahrens C, Jovanovic M, Malmström J, Brunner E, Mohanty S, Lercher M, Hunziker P, Aebersold R, von Mering C, Hengartner M (2009) Comparative functional analysis of the *Caenorhabditis elegans* and *Drosophila melanogaster* proteomes. PLoS Biol 7:616–627

Schwander T, Leimar O (2011) Genes as leaders and followers in evolution. Trends Ecol Evol 26:143–151

Schwartz T, Bronikowski A (2013) Dissecting molecular stress networks: identifying nodes of divergence between life-history phenotypes. Mol Ecol 22:739–756

Schwarzenberger A, Courts C, von Elert E (2009) Target gene approaches: gene expression in *Daphnia magna* exposed to predator-borne kairomones or to microcystin-producing and microcystin-free *Microcystis aeruginosa*. BMC Genomics 10:527

Schweitzer J, Madritch M, Bailey J, LeRoy C, Fischer D, Rehill B, Lindroth R, Hagerman A, Wooley S, Hart S, Whitham T (2008) From genes to ecosystems: the genetic basis of condensed tannins and their role in nutrient regulation in a *Populus* model system. Ecosystems 11:1005–1020

Scoville A, Pfrender M (2010) Phenotypic plasticity facilitates recurrent rapid adaptation to introduced predators. Proc Natl Acad Sci USA 107:4260–4263

Seehausen O, van Alphen J, Witte F (1997) Cichlid fish diversity threatened by eutrophication that curbs sexual selection. Science 277:1808–1811

Seehausen O, Terai Y, Magalhaes I, Carleton K, Mrosso H, Miyagi R, van der Sluijs I, Schneider M, Maan M, Tachida H, Imai H, Okada N (2008) Speciation through sensory drive in cichlid fish. Nature 455: 620–626

Shachrai I, Zaslaver A, Alon U, Dekel E (2010) Cost of unneeded proteins in *E. coli* is reduced after several generations in exponential growth. Mol Cell 38: 758–767

Simon J, Pfrender M, Tollrian R, Tagu D, Colbourne J (2011) Genomics of environmentally induced phenotypes in 2 extremely plastic arthropods. J Hered 102:512–525

Singh A, Razooky B, Dar R, Weinberger L (2012) Dynamics of protein noise can distinguish between alternate sources of gene-expression variability. Mol Syst Biol 8:607

Slijper E (1942a) Biologic-anatomical investigations on the bipedal gait and upright posture in mammals with special reference to a little goat born without forelegs. I. Proceedings of the Koninklijke Nederlandse Akademie Wetenschappen, vol 45, pp 288–295

Slijper E (1942b) Biologic-anatomical investigations on the bipedal gait and upright posture in mammals with special reference to a little goat born without forelegs. II. Proceedings of the Koninklijke Nederlandse Akademie Wetenschappen, vol 45, pp 407–415

Snell-Rood E (2012) Selective processes in development: implications for the costs and benefits of phenotypic plasticity. Integr Comp Biol 52:31–42

Sollars V, Lu X, Xiao L, Wang X, Garfinkel MD, Ruden DM (2003) Evidence for an epigenetic mechanism by which Hsp90 acts as a capacitor for morphological evolution. Nat Genet 33:70–74

Solberg M, Skaala Ø, Nilsen F, Glover K (2013) Does domestication cause changes in growth reaction norms? A study of farmed wild and hybrid Atlantic salmon families exposed to environmental stress. PLoS One 8:e54469

Somero G (2005) Linking biogeography to physiology: evolutionary and acclimatory adjustments of thermal limits. Front Zool 2:1

Somero G (2010) The physiology of climate change: how potentials for acclimatization and genetic adaptation will determine 'winners' and 'losers'. J Exp Biol 213:912–920

Sommer RJ, Ogawa A (2011) Hormone signaling and phenotypic plasticity in nematode development and evolution. Curr Biol 21:R758–R766

Srinivasan DG, Brisson JA (2012) Aphids: a model for polyphenism and epigenetics. Genet Res Int 2012:431531

Starrfelt J, Kokko H (2012) Bet-hedging – a triple trade-off between means variances and correlations. Biol Rev 87(3):742–755

Stearns S, Koella J (1986) The evolution of phenotypic plasticity in life-history traits: predictions of reaction norms for age and size at maturity. Evolution 40: 893–913

Stern D (2010) Evolution, development and the predictable genome. Roberts and Company Publishers, Greenwood Village

Stern S, Fridmann-Sirkis Y, Braun E, Soen Y (2012) Epigenetically heritable alteration of fly development in response to toxic challenge. Cell Rep 1:528–542

Stoebel D, Dean A, Dykhuizen D (2008) The cost of expression of *Escherichia coli lac* operon proteins is in the process, not in the products. Genetics 178:1653–1660

Strauss S, Lau J, Carroll S (2006) Evolutionary responses of natives to introduced species: what do introductions tell us about natural communities? Ecol Lett 9: 357–374

Sunday J, Bates A, Dulvy N (2011) Global analysis of thermal tolerance and latitude in ectotherms. Proc R Soc B 278:1823–1830

Svanbäck R, Schluter D (2012) Niche specialization influences adaptive phenotypic plasticity in the threespine stickleback. Am Nat 180:50–59

Svanbäck R, Pineda-Krch M, Doebeli M (2009) Fluctuating population dynamics promotes the evolution of phenotypic plasticity. Am Nat 174:176–189

Tétard-Jones C, Kertesz M, Preziosi R (2011) Quantitative trait loci mapping of phenotypic plasticity and genotype-environment interactions in plant and insect performance. Philos Trans R Soc Lond B 366:1368–1379

Thibert-Plante X, Hendry A (2011) The consequences of phenotypic plasticity for ecological speciation. J Evol Biol 2:326–342

Tollrian R, Leese F (2010) Ecological genomics: steps towards unraveling the genetic basis of inducible defenses in *Daphnia*. BMC Biol 8:51

Tomanek L (2008) The importance of physiological limits in determining biogeographical range shifts due to global climate change: the heat-shock response. Physiol Biochem Zool 81:709–717

Tomanek L, Somero G (2002) Interspecific- and acclimation-induced variation in levels of heat-shock proteins 70 (hsp70) and 90 (hsp90) and heat-shock transcription factor-1 (HSF1) in congeneric marine snails (genus *Tegula*): implications for regulation of *hsp* gene expression. J Exp Biol 205:677–685

Ungerer M, Halldorsdottir S, Purugganan M, Mackay T (2003) Genotype-environment interactions at quantitative trait loci affecting inflorescence development in *Arabidopsis thaliana*. Genetics 165:353–365

Ungerer M, Johnson L, Herman M (2008) Ecological genomics: understanding gene and genome function in the natural environment. Heredity 100:178–183

Utsumi S (2011) Eco-evolutionary dynamics in herbivorous insect communities mediated by induced plant responses. Popul Ecol 53:23–34

Valena S, Moczek AP (2012) Epigenetic mechanisms underlying developmental plasticity in horned beetles. Genet Res Int 2012:576303

Vezzoli M, Castellani P, Corna G, Castiglioni A, Bosurgi L, Monno A, Brunelli S, Manfredi A, Rubartelli A, Rovere-Querini P (2011) High-mobility group box 1 release and redox regulation accompany regeneration and remodeling of skeletal muscle. Antioxid Redox Signal 15:2161–2174

Via S, Lande R (1985) Genotype-environment interaction and the evolution of phenotypic plasticity. Evolution 39:505–522

Vilaprinyo E, Alves R, Sorribas A (2010) Minimization of biosynthetic costs in adaptive gene expression responses of yeast to environmental changes. PLoS Comput Biol 6:e10000674

Viñuelas J, Kaneko G, Coulon A, Beslon G, Gandrillon O (2012) Towards experimental manipulation of stochasticity in gene expression. Prog Biophys Mol Biol 110:44–53

Waddington C (1953a) Genetic assimilation of an acquired character. Evolution 7:118–126

Waddington C (1953b) The 'Baldwin effect', 'genetic assimilation' and 'homeostasis'. Evolution 7:386–387

Waddington C (1956) Genetic assimilation of the bithorax phenotype. Evolution 10:1–13

Wagner A (2007) Energy costs constrain the evolution of gene expression. J Exp Zool 308B:322–324

Wang Y, Wu L, Wang P, Lv C, Yang Z, Tang X (2012) Manipulation of gene expression in zebrafish using caged circular morpholino oligomers. Nucleic Acids Res 2012:1–8

Weiss L, Kruppert S, Laforsch C, Tollrian R (2012) *Chaoborus* and *Gasterosteus* anti-predator responses in *Daphnia pulex* are mediated by independent cholinergic and gabaergic neuronal signals. PLoS One 7:e36879

West-Eberhard M (2003) Developmental plasticity and evolution. Oxford University Press, Oxford

West-Eberhard M (2005) Phenotypic accommodation: adaptive innovation due to developmental plasticity. J Exp Zool B 304B:610–618

Whitehead A, Roach J, Zhang S, Galvez F (2012) Genomic mechanisms of evolved physiological plasticity in killifish distributed along an environmental salinity gradient. Proc Natl Acad Sci USA 108: 6193–6198

Whitham T, DiFazio S, Schweitzer J, Shuster S, Allan G, Bailey J, Woolbright S (2008) Extending genomics to natural communities and ecosystems. Science 320:492–495

Windig J, De Kovel C, De Jong G (2003) Genetics and mechanics of plasticity. In: DeWitt T, Scheiner S (eds) Phenotypic plasticity: functional and conceptual approaches. Oxford University Press, New York

Wolf J, Brodie E III, Wade M (2003) The genotype-environment interaction and evolution when the environment contains genes. In: DeWitt T, Scheiner S (eds) Phenotypic plasticity: functional and conceptual approaches. Oxford University Press, New York

Woltereck R (1909) Weitere experimentelle untersüchungen über artveränderung speziell über das wesen quantitativer artunterschiede bei daphniden. Verhandlungen de Deutschen Zooligischen Gesellschaft 19: 110–172

Wood H, Spicer J, Widdiecombe S (2008) Ocean acidification may increase calcification rates but at a cost. Proc R Soc B 275:1767–1773

Wund M, Baker J, Clancy B, Golub J, Foster S (2008) A test of the 'flexible stem' model of evolution: ancestral plasticity genetic accommodation and morphological divergence in the threespine stickleback radiation. Am Nat 172:449–462

Wund M, Valena S, Wood S, Baker J (2012) Ancestral plasticity and allometry in threespine stickleback reveal phenotypes associated with derived freshwater ecotypes. Biol J Linn Soc 105:573–583

Yampolsky L, Glazko G, Fry J (2012) Evolution of gene expression and expression plasticity in long-term experimental populations of *Drosophila melanogaster* maintained under constant and variable ethanol stress. Mol Ecol 21:4287–4299

Yeh P, Price T (2004) Adaptive phenotypic plasticity and the successful colonization of a novel environment. Am Nat 164:531–542

Yin M, Laforsch C, Lohr J, Wolinska J (2011) Predator-induced defense makes *Daphnia* more vulnerable to parasites. Evolution 65(5):1482–1488

Zhou X, Tarver MR, Scharf ME (2007) Hexamerin-based regulation of juvenile hormone-dependent gene expression underlies phenotypic plasticity in a social insect. Development 134:601–610

Zhou S, Campbell T, Stone E, Mackay T, Anholt R (2012) Phenotypic plasticity of the *Drosophila* transcriptome. PLoS Genet 8:e1002593

Eco-Evo-Devo: The Time Has Come

6

Ehab Abouheif, Marie-Julie Favé, Ana Sofia Ibarrarán-Viniegra, Maryna P. Lesoway, Ab Matteen Rafiqi, and Rajendhran Rajakumar

Abstract

The major goal of ecological evolutionary developmental biology, also known as "eco-evo-devo," is to uncover the rules that underlie the interactions between an organism's environment, genes, and development and to incorporate these rules into evolutionary theory. In this chapter, we discuss some key and emerging concepts within eco-evo-devo. These concepts show that the environment is a source and inducer of genotypic and phenotypic variation at multiple levels of biological organization, while development acts as a regulator that can mask, release, or create new combinations of variation. Natural selection can subsequently fix this variation, giving rise to novel phenotypes. Combining the approaches of eco-evo-devo and ecological genomics will mutually enrich these fields in a way that will not only enhance our understanding of evolution, but also of the genetic mechanisms underlying the responses of organisms to their natural environments.

Keywords

Evodevo • Evolution • Ecology • Stochastic variation • Robustness • Environmental stress • Developmental recombination • Genetic accommodation • Genetic assimilation • Ancestral developmental potential • Social interactions • Epigenetics • Developmental plasticity • Polyphenism • Ecoevodevo

6.1 Introduction

The time seems to have come when we need to take into account two further aspects of the evolutionary mechanism. In the first place, natural selective pressures impinge not on the hereditary factors themselves, but on the organisms as they develop from fertilized eggs to reproductive adults. We need to bring into the picture not only the genetic system by which hereditary information is passed on from one generation to the next, but also

E. Abouheif (✉) • M.-J. Favé • A.S. Ibarrarán-Viniegra
A.M. Rafiqi • R. Rajakumar
Department of Biology, McGill University, 1205 avenue Docteur Penfield, Montreal, QC H3A 1B1, Canada
e-mail: ehab.abouheif@mcgill.ca

M.P. Lesoway
Department of Biology, McGill University, 1205 avenue Docteur Penfield, Montreal, QC H3A 1B1, Canada

Smithsonian Tropical Research Institute, Apartado Postal 0843-03092, Panamá, República de Panamá

C.R. Landry and N. Aubin-Horth (eds.), *Ecological Genomics: Ecology and the Evolution of Genes and Genomes*, Advances in Experimental Medicine and Biology 781,
DOI 10.1007/978-94-007-7347-9_6, © Springer Science+Business Media Dordrecht 2014

> the "epigenetic system" by which the information contained in the fertilized egg is expanded into the functioning structure of the reproducing individual. Each organism during its lifetime will respond in some manner to the environmental stresses to which it is submitted and in a population there is almost certain to be some genetic variation in the intensity and character in these responses. Natural selection will favor those individuals in which the responses are of most adaptive value. Conrad H. Waddington (1959).

Why should an ecological genomicist be interested in Waddington's (1959) prescient words written almost 60 years ago? The short answer is that development – the processes through which the fertilized egg becomes a reproducing adult – matters. Development mediates interactions between genes and the environment, so if the goal of ecological genomics is to "uncover the genetic mechanisms underlying responses of organisms to their natural environments" (Ungerer et al. 2008), then this goal is unattainable without taking development into account. The long answer is that development mediates these interactions in multiple and complex ways affecting the genetic and phenotypic variation available for natural selection to act upon. Waddington (1959) clearly saw the importance of integrating these interactions into evolutionary theory back in 1959, but the time for this integration seems to have come only now with the recent emergence of the field of ecological evolutionary developmental biology or more simply "eco-evo-devo." This field acknowledges the fact that organisms are continually subject to a changing environment, whether it be changes in nutrition, temperature, predators, competitors or all of these simultaneously. It aims to uncover the rules that underlie the interactions between an organism's environment, genes, and development, and by doing so, aims to expand our current view of how evolution works. As the name of this field implies, eco-evo-devo unites the field of evolutionary developmental biology (evo-devo) with ecology, but also includes the subdisciplines of developmental plasticity, epigenetics, and social evolution, where many recent advances are being made. For example, recent books in development plasticity, such as West-Eberhard's (2003) *Developmental Plasticity and Evolution*, and Gilbert and Epel's (2009) *Ecological Developmental Biology*, have breathed new life in our understanding of how developmental plasticity can promote, rather than retard, evolutionary change. In this explosive cocktail of fields and subdisciplines is a conceptual revolution on the horizon – a potentially new way of thinking about evolution, development, and ecology.

The goal of our chapter is to describe, in a series of sections, some key and emerging concepts from eco-evo-devo. Although the sections cover a wide range of different topics, taken together, they contribute to painting a larger picture of the evolutionary process. First, in Endocrine Signaling (Sect. 6.2), we set the tone by highlighting the genes controlling sensitivity to or titers of hormones during development as the potential locus of gene and environment interactions. Second, through our sections on Ancestral Developmental Potential (Sect. 6.3), Stochastic Variation (Sect. 6.5), Social Interactions (Sect. 6.6), and Ecological Interactions (Sect. 6.7), we discuss the different ways in which the environment can play a dual role – the more familiar role as selective agent and the less familiar role as a source or inducer of phenotypic variation. Third, in Developmental Recombination (Sect. 6.4) and Robustness (Sect. 6.8), we discuss how development can then regulate the interactions between genes and their environment in a number of complex ways such that development can mask, release, or create new combinations of genotypic and phenotypic variation produced by this interaction. Finally, in Genetic Assimilation and Accommodation (Sect. 6.9) and Integrating Levels of Biological Organization (Sect. 6.10), we discuss how variation produced by the environment at all levels of biological organization can be fixed by natural selection to create novel phenotypes. We conclude (Sect. 6.11) with a diagram inspired by Waddington (1959) to summarize the interrelations between all of the following sections.

6.2 Endocrine Signaling: A Locus of Gene and Environment Interactions

Phenotypic traits can be under complete genetic control (e.g. mendelian traits), but most traits are plastic and result from a complex interaction between genetic and environmental inputs (West-Eberhard 2003; Gilbert and Epel 2009). One form of plasticity, polyphenism, is the ability of the same genome to produce two or more alternative phenotypes in a single population in response to an environmental cue, such as temperature or nutrition (Nijhout 1999). Polyphenic traits represent powerful models to understand the effects of environmental inputs on development and provide the opportunity to uncover the physiological and genetic mechanisms underlying the susceptibility to undertake or not particular developmental decisions (Jenner and Wills 2007).

Developmental decisions are often mediated by endocrine signaling (Flatt and Heyland 2011). The decision about which developmental trajectory to undertake depends on several key components of endocrine signaling (Nijhout 1999; Zera et al. 2007): (1) the systemic titer of the hormone (circulating concentration of the hormone); (2) the hormonal receptors in target cells, which can show variation in the degree and timing of sensitivity; and (3) the downstream pathways that are activated upon hormone binding. In insects, the role of hormones in developmental decisions has been extensively studied (Nijhout 1998); juvenile hormone (JH), together with ecdysteroids, orchestrate growth, molts and timing of metamorphosis. For example, in the cricket *Gryllus rubens*, wing-length polyphenism is the result of population density during development (Zera and Tiebel 1988). At low densities, during a critical sensitivity period that spans from mid to penultimate larval instars, JH titer drops below a threshold, whereas ecdysteroid titer exceeds a threshold, which result in the development of macropterous individuals. In contrast, at high densities, JH titer remains above the threshold, whereas ecdysteroid titer is below a threshold during the critical period of sensitivity, which leads to the interruption of wing development and micropterous individuals are formed.

The evolution of environmentally sensitive traits may occur through evolutionary changes in any one of the three key components of endocrine signaling highlighted above (Nijhout 1999; Zera et al. 2007). Mutations in genes controlling hormonal titers can bring individuals to cross a hormonal threshold more readily and therefore express an alternate phenotype. Artificial selection experiments in the tobacco hornworm (*Manduca sexta*) have shown that polyphenic traits can evolve through the modification of hormonal titers. Suzuki and Nijhout (2006) selected for a polyphenic line of *M. sexta*, where individuals express a black pigmentation when raised at a low temperature but develop as green when raised at a high temperature (see Sect. 6.9 for details on the mechanisms). In order to determine if a hormonal titer is involved in controlling the expression of the alternative color phenotypes, they ligated the thorax of the larvae to prevent endogenous JH circulation. They subsequently topically applied a JH analogue, methoprene, from low to high concentrations. They found that regardless of the temperature that they were raised at, individuals develop as black at low methoprene concentrations, while they develop into green individuals at higher methoprene concentrations. These results indicate that this polyphenic switch is mediated largely through differences in hormonal titers rather than changes in hormonal sensitivities or downstream effector molecules.

Alternatively, mutations in the threshold itself (for example, a hormone-binding receptor), may allow for a hormonal titer to cross that threshold more readily, and therefore also express an alternative phenotype (Nijhout 1999; Zera et al. 2007). In dung beetles, *Onthophagus taurus*, males can develop, or not, a pair of horns on their head depending on larval nutrition: large males above a critical threshold body size will develop a pair of horns on their heads while smaller males below the critical body size threshold will not. Introduced populations of *O. taurus* in North Carolina and in Australia have rapidly evolved divergence in the critical body

size threshold that separates alternative morphs. Using hormonal manipulations, Moczek and Nijhout (2002) showed that this divergence has evolved through the modification of the response threshold to JH. North Carolina male beetles develop horns when exposed to a lower methoprene concentration, and during an earlier critical period than male beetles from Australia. These results indicate that divergence of polyphenic traits can occur through changes in the degree and timing of sensitivity to endocrine signaling.

Furthermore, the production of effector molecules downstream of a hormonal switch can also be the target for evolution (Nijhout 1999; Zera et al. 2007). Wing polyphenism in ants is universal and evolved only once: across all ant species, reproductive individuals are winged and worker castes are wingless (Abouheif and Wray 2002). The regulation of expression of the wing patterning gene network is under the control of a JH switch that determines the fate of an egg (queen or worker). This gene network is largely conserved between winged castes and other holometabolous insects (Carroll et al. 2005; Tomoyasu et al. 2009; Shbailat et al. 2010), but is evolutionarily labile across wingless castes of different ant species (Abouheif and Wray 2002; Shbailat and Abouheif 2012). For example, in the ant species *Lasius niger* and *Crematogaster lineolata*, the hormonal switch between queens and workers occurs relatively early during development, and therefore, both species have vestigial imaginal discs of similar size. However, the expression of genes in the wing patterning network in vestigial discs of workers differs between the two species. These results indicate that downstream targets of endocrine signaling can also evolve to generate expression differences in gene regulatory networks.

The underlying genetic basis of polyphenic traits, either in terms of changes in hormone titers or sensitivity, remains to be discovered in most cases, although it is most often assumed to be polygenic (Roff and Fairbairn 1991; Roff 1996; Braendle et al. 2005). Using next-generation sequencing tools may help answer questions such as: (1) is the variation in the response to environmental stimuli between taxa a consequence of changes in a few key genes, or small changes in multiples genes? (2) Which kind of genes and gene networks underlie hormonal thresholds and sensitivities? These examples illustrate how environmental responses can be incorporated into developmental decisions through the action of endocrine signaling. Furthermore, the evolution of any key component of the endocrine signaling pathway may lead to the evolution of new variation and phenotypes.

6.3 Ancestral Developmental Potential

The term atavism refers to the sporadic and spontaneous appearance of ancestral phenotypes in individuals of modern wild populations (Darwin 1868; West-Eberhard 2003). Examples of this include individual whales with hindlimbs, snakes with additional skeletal elements and humans with tails (Dubrow et al. 1988; Hall 2003; Tomić and Meyer-Rochow 2011). Generally, this type of variation is often considered to contribute little, if at all, to the evolutionary process (Levinton 1986; Stiassny 2003). There are also several cases of atavistic traits being induced in the lab (Waddington 1957; Weatherbee et al. 1998; Harris et al. 2006; Chan et al. 2010). Examples of these "experimental atavisms" include chickens with teeth, freshwater stickleback fish with pelvic structures, and flies with hindwings. As is the case with spontaneous atavisms, experimental atavisms are given little weight in understanding the evolutionary process (Levinton 1986) and are more commonly known as "hopeless monsters".

Dollo's Law, which posits that once a complex trait is lost it is unlikely to re-evolve, has been an influential concept in phylogenetic systematics (Goldberg and Igić 2008; Wake et al. 2011). However, so many counter-examples have appeared that Dollo's Law can no longer be considered the rule, but rather the exception (Collin and Miglietta 2008). Several instances have been demonstrated where ancestral traits, that have been lost for millions of years, have subsequently re-evolved in derived lineages such as wings in stick insects (Whiting et al. 2003), teeth in

amphibians (Wiens 2011), digit number in lizards (Kohlsdorf and Wagner 2006; Kohlsdorf et al. 2010) shell coiling and mode of development in marine snails (Collin and Cipriani 2003; Collin 2004; Collin et al. 2007) and herbivore defense in plants (Armbruster et al. 2009). This phylogenetic pattern of reversal, which has been called 'taxic atavism' (Stiassny 2003), may be much more common than originally thought (West-Eberhard 2003; Abouheif 2008; Rajakumar et al. 2012). Although these three forms of atavisms (spontaneous, experimental and taxic) have been described in detail in the literature, little effort has gone into determining whether considering them together would in any way be informative to the further understanding of their occurrence or more generally the evolutionary process.

Recently, Rajakumar et al. (2012) united all three types of atavism in a single study of supersoldier development and evolution in the ant genus *Pheidole*. They demonstrated that the supersoldier caste, known for its complex defensive skills and giant heads (Huang 2010), is actually an ancestral feature that was subsequently lost in most species of the group. The supersoldier subcaste then re-evolved in at least eight species, including one species called *Pheidole obtusospinosa* (taxic atavism). Furthermore, they found in nature several anomalous supersoldier-like individuals in one *Pheidole* species that does not have a supersoldier caste (spontaneous atavism). How did this occur? Using hormonal manipulations, they were able to produce supersoldiers in several species (including the species of which they found the spontaneous atavism) that do not have a supersoldier caste (experimental atavism). This result demonstrates that the potential to produce supersoldiers is ancestral and that there exists an ancestral developmental potential for supersoldiers that can be environmentally induced across the genus *Pheidole*.

How do ancestral developmental potentials persist throughout 35–60 millions years of evolution, such that they can be environmentally induced in extant species? In Rajakumar et al.'s (2012) case, pleiotropy or more specifically hormonal pleiotropy, is proposed to be the mechanism that facilitates the retention of dormant ancestral traits. Specifically, the same hormonal process is involved in the production of both soldier and supersoldier ants. If the underlying process of supersoldier development is compromised, soldier development would be affected as well. This would be disadvantageous as the soldier caste performs functions that are critical to the survival of the entire colony. Therefore, although not phenotypically expressed, the ability to produce a supersoldier is preserved in the genome of all *Pheidole* species indirectly through continued selection for soldier production.

Rajakumar et al. (2012) propose that, if recurrently induced by environmental factors in the wild, what begins as a spontaneous atavism can later evolve into a taxic atavism. Both the underlying developmental process and the eventual evolutionary pattern can be elucidated further with the help of phylogenetics and the induction of experimental atavisms in the lab. Initially, anomalous phenotypes that occur in the wild may not appear to be evolutionarily advantageous. However, the anomaly may be a spontaneous atavism that reflects a historically advantageous trait, which has been evolutionarily preserved by pleiotropy. If it is reactivated and similar selective pressures are present (to that of the ancestor) there is a possibility that this atavism may eventually become fixed in the population. Taken together, anomalies that spontaneously appear in the wild are a source of variation for natural selection to act upon.

6.4 Developmental Recombination: A Source of New Combinations

The modular nature of development has been one of the most important discoveries in developmental and evolutionary biology (Schlosser and Wagner 2004; Gilbert and Epel 2009). The organization of development into modules is largely the emergent consequence of genes being organized into interacting networks capable of responding to discrete morphogen and hormonal thresholds (Schlosser and Wagner 2004; Flatt

et al. 2005). Developmental modules are critical for distinguishing and giving identity to populations of cells within and between tissues (Carroll et al. 2005: Schlosser and Wagner 2004; Davidson 2006). The key implication of the modular nature of development for eco-evo-devo is that developmental modules are quasi-independent, meaning that when these modules are subject to genetic perturbation or environmental stresses they will respond in an almost independent manner from one another. On the one hand, this quasi-independence of developmental modules helps confer robustness during development (see Sect. 6.8) because genetic or environmental perturbations can be confined to specific modules. However, when the genetic perturbation or environmental stress is too great and robustness is compromised, "cryptic variation" is released (Gibson and Dworkin 2004). Examples of the accumulation and the release of cryptic genetic variation include ribozymes, which exhibit higher adaptation rate with accumulated cryptic genetic variation (Hayden et al. 2011) and T cell adaptive immunity where cryptic alleles drive rapid adaptation of activation responses when the cellular population is presented with a novel environment during infection (Whitacre et al. 2012). Several other examples are vulval development in *C. elegans* (Duveau and Félix 2012), feeding strategies in toads (Ledón-Rettig et al. 2010), and wing development in *Drosophila* (Dworkin 2005). While the release of cryptic variation is thought to be instrumental in trait evolution, much less attention has been given to the fact that the variation released may not be completely random, but rather, may reflect variation within and between independent developmental modules. The dissociation or formation of different combinations of modules is called "developmental recombination" which selection can subsequently act upon (West-Eberhard 2003, 2005). For example, stripe patterns in zebras and related equine species appear to be modular in their appearance in nature as described by Darwin (1868). It is possible that particular stripes are controlled independently from one another by different developmental modules and these can be recombined such that they can occur in different combinations in closely-related species.

Rajakumar et al. (2012) provide another example of developmental recombination with the evolution of supersoldiers in the ant genus *Pheidole*. It has previously been shown that the soldier caste in *Pheidole* develops as the result of a pulse of JH that crosses a discrete threshold during a critical time period (Wheeler and Nijhout 1981, 1983; Abouheif and Wray 2002). An additional JH threshold at a second critical period is present in the species *Pheidole obtusospinosa* that regulates the development of an additional 'supersoldier' caste (Rajakumar et al. 2012). When JH is applied during this second critical period, the supersoldier caste is consistently produced. Surprisingly, supersoldiers can also be induced in species that do not have a supersoldier caste due to the activation of an ancestral, but cryptic, JH threshold (Rajakumar et al. 2012). This demonstrates that the second threshold in *Pheidole obtusospinosa* emerged from the re-evolution of a cryptic JH threshold. This has occurred either through the evolution of the threshold itself or the regulation of JH production. The key indicator of supersoldier development is wing imaginal discs, each of which develop as independent modules. Induced supersoldiers exhibited quantitatively and qualitatively more variability in wing imaginal discs as compared to species, like *Pheidole obtusospinosa*, that naturally evolved supersoldiers. They found there was more variation in wing imaginal disc number, size, asymmetry and gene expression (Rajakumar et al. 2012). This novel variation of developmental modules, i.e., wing imaginal discs, generated by developmental recombination can undergo selection. It is likely that this is the type of variation, following the induction of a cryptic hormonal threshold, which was under selection during the course of supersoldier evolution in *Pheidole obtusospinosa*. Through the reorganization of existing developmental modules, the process of developmental recombination may more generally provide a source of variation for selection.

Many of the studies highlighted above were only possible due to recent advancements of molecular and genomic techniques. We are now beginning to appreciate the importance of the

reorganization of developmental modules in generating novel phenotypes, the next step will be to apply this framework to a life-history context in order to make better predictions of the ecological role and adaptive function of this type of variation. Furthermore with the use of comparative genomics and transcriptomics, we can more precisely identify the molecular makeup of different developmental recombinants that arise in nature.

6.5 Stochastic Variation: Molecules and Beyond

Biological systems are far more dynamic and noisy than originally assumed (McAdams and Arkin 1999). Noise at the level of molecular interactions can permeate to higher biological levels leading to stochastic variation in gene expression, and in turn, can contribute to differences in phenotype (Kilfoil et al. 2009). In most cases, developmental systems are robust and will buffer this stochastic variation in gene expression (see Sect. 6.8). However, recent studies have demonstrated that stochastic variation in gene expression has been co-opted during evolution to play an important role in influencing developmental decisions, where an initially stochastic expression of genes is stabilized to determine cell fates (Losick and Desplan 2008). An example of this occurs in the early *Drosophila* embryo, where all cells in proneural clusters initially have the capacity to differentiate into neuroblasts (Heitzler and Simpson 1991, reviewed in Losick and Desplan 2008). One cell in the cluster stochastically expresses more Delta protein than other cells, and as a consequence, it differentiates into a neuroblast while all others become epidermal cells. Cell fate decisions based on the stabilization of stochastic gene expression occur in other multicellular organisms, in different tissue types, and permeates to higher levels of biological organization (Kilfoil et al. 2009). Kilfoil et al. (2009) suggest that stochastic decisions in gene expression may permeate up to the level of individual organisms, such that particular social and behavioral decisions emerge stochastically. For example, reproduction in colonies of the ant species *Harpagnathos saltator* appear to be regulated in a stochastic manner, where an initially stochastic decision is stabilized and made permanent (Hölldobler and Wilson 2008, reviewed in Kilfoil et al. 2009). Queens and workers in *H. saltator* are both capable of reproduction; however, worker reproduction is inhibited by the presence of the queen (Peeters et al. 2000). The removal of the queen results in several antagonistic interactions between workers leading to the emergence of a small group of workers that become reproductively active (Hölldobler and Wilson 2008). These reproductive workers, who are at the top of the colony's dominance hierarchy, emerge because they are the first to acquire a distinct cuticular hydrocarbon profile that signals their fertility to the rest of the low-ranking workers in the colony (Hölldobler and Wilson 2008). Kilfoil et al. (2009) proposed the following model: removal of the queen triggers stochastic variation in the activity levels of enzymes involved in the synthesis of cuticular hydrocarbons. The stochastic expression of these enzymes in workers biases particular individuals towards becoming reproductive workers. Later, positive and antagonistic behavioral interactions between individuals further amplify the differences, resulting in the establishment of a small group of high-ranking reproductive individuals, while the rest of the workers in the colony remain reproductively quiescent. This example suggests that stochastic variation in gene expression may contribute in important ways to phenotypic evolution and behavioral decisions at higher levels of organization. Furthermore, Kilfoil et al. (2009) raise the possibility that phenotypic variation due to stochastic variation in gene expression should be acknowledged formally as a category of phenotypic variation in quantitative genetics (called V_S). Ecological genomics will be critical for defining and quantifying this category of variation, as well as in overcoming the challenges of distinguishing this type of variation from those derived from deterministic processes. Understanding how stochastic variation that is abundant at the molecular level permeates to higher levels of organization is an important arena where ecological genomics and eco-evo-devo meet.

6.6 Social Interactions: Generators of Behavioral and Phenotypic Variation

Social interactions can be thought of as a network and can be not only a source of information that individuals respond to, but also feed information back onto themselves. The way individuals tune their behavior to their social context and how accurate they are at tuning into the social signals affects not only their survival but their overall fitness. Therefore, while traditionally we have considered the importance of interactions between species and at the ecosystem level, we should not disregard the effect of group composition within species. The examples we present below show that social interactions can generate adaptive and novel phenotypic variation upon which natural selection can act upon. Social interactions, for the purpose of this chapter are not limited to social species (Tinbergen 1971), and encompass both an individual's reaction to the presence of at least one other individual of the same species and how their interactions can influence each other.

Social interactions can initiate top-down influences on an individual's phenotype; that is, changes in social interactions can lead to changes in individual behavior that in turn can change gene expression. For example, in the guppy *Poecilia reticulata*, social interactions in the form of mating preferences can affect an individual's behavior. Males choose the social context that will make them more attractive to female guppies based on the appearance of other males (Gasparini et al. 2013). In the African cichlid fish *Astatotilapia burtoni*, social interactions not only affect male behavior but also affect neural gene expression (Renn et al. 2008). Depending on social status, males can show two possible phenotypes: dominant or subordinate. When dominance status changes as a consequence of social interactions, changes in pigmentation and behavior take place within minutes, which is followed by gene expression changes in the brain. This is followed within a couple of weeks by changes in reproductive physiology and the dominant male phenotype is manifested (Renn et al. 2008; Fernald and Maruska 2012).

Social interactions can also influence development. In reptiles, for example, temperature plays a major role in influencing not only sex determination but also developmental timing. McGlashan et al. (2012) have shown in the freshwater turtle *Emydura macquarii* that synchronicity in hatching times are socially driven. Synchronicity in hatching times appears to be influenced by embryo-embryo communication. In this case, the authors suggest that temperature independent developmental timing could be achieved through changes in thyroid hormone production cued to CO_2 concentration in the nest or detection of sibling's heart rate. Interspecific interactions may have ultimately driven the evolution of this socially generated synchronicity because hatching time affects survival in the face of predation. Therefore, social interactions in this case can result in developmental timing changes which permeate to higher levels of organization and result in species success in complex interspecific interactions.

In the previous two examples we discuss how social interactions can generate adaptive phenotypic responses in developing and adult individuals. However, social interactions can also induce novel phenotypic variation during development that is relevant for evolution (see Sects. 6.3, 6.4, and 6.9). Some of the best documented examples come from social insects, where social interactions regulate caste determination (Hölldobler and Wilson 1990, 2008). A good example of social interactions influencing caste determination during development comes from the work by Wheeler and Nijhout (1981, 1983) on the determination of different types of worker subcastes in colonies of the ant genus *Pheidole*. The worker caste in this genus is composed of small 'minor workers' and large, big-headed, 'soldiers,' where minor workers make up 95 % of the colony and perform tasks related to foraging and brood rearing and soldiers make up the other 5 % and specialize mainly in tasks related to defence. Remarkably, colonies can maintain and even slightly adjust this ratio according to changes in their ecological environment (Yang et al. 2004). Passera (1977), Passera et al. 1996 and Wheeler and Nijhout (1984) showed that in

circumstances where the percentage of soldiers in the colony is too high, adult soldiers can inhibit the development of future soldiers using a contact pheromone. This contact pheromone exploits the mechanism normally used by *Pheidole* colonies to determine minor workers and soldiers during development. This developmental mechanism is based on a switch (threshold) that is regulated by the levels of JH, where larvae that produce high levels of JH during a critical period develop into soldiers, and those that produce low levels of JH develop into minor workers. The contact pheromone, which adult soldiers use to inhibit the future development of soldiers when there are too many soldiers in the colony, is thought to reduce the sensitivity of the larva to JH and larvae that would normally develop into soldiers develop instead into minor workers. For example, when Wheeler and Nijhout (1984) treated larva with relatively moderate levels of JH in a colony that has no soldiers, these larvae developed into soldiers, but when they treated larvae in a colony that has 100 % soldiers, they developed into minor workers. The following experiment, however, shows how social interactions can induce phenotypic variation relevant for evolution – when Wheeler and Nijhout (1984) treated larvae with relatively high levels of JH in a colony with 100 % soldiers, the JH treatment was too high for adult soldiers in the colony to completely inhibit the development of these larvae into soldiers. While some larvae still developed into soldiers, some larvae developed into exceptionally large small-headed minor workers that were as large in size as the big-headed soldiers! This shows that social interactions can influence development to produce phenotypic variations not normally observed in the colony. Indeed, the induction of phenotypic variation through social interactions may have played a role in the evolution of the minors and soldiers in this genus. Pie and Traniello (2007) showed that body size is the most variable trait across *Pheidole* species. Ecological genomics can play a critical role in helping to uncover the genes expressed during development of castes in ant species as they socially regulate their colonies to respond to the ecological pressures that surround them. In general, there is great opportunity in these and other systems for ecological genomics to enrich our understanding of how social interactions can generate adaptive and novel phenotypic variation upon which natural selection can act on.

6.7 Ecological Interactions

The role of the environment is twofold: through the action of natural selection, certain phenotypes will be selected in certain environments, but simultaneously, the environment can induce phenotypic variation through plasticity, thereby influencing the ecology of the organism. It is becoming clear that the dual role of the environment may often create a feedback loop that simultaneously influences the evolution of a trait. The fact that selection can act at various stages during the ontogeny of plastic traits can facilitate a rapid reaction and evolution of populations to a changing environment. Plastic phenotypes can subsequently be fixed through genetic assimilation and accommodation (Waddington 1957; West-Eberhard 2003, see Sect. 6.9), therefore facilitating the evolution of adaptive phenotypes.

An example of the environment acting as both a selective force while simultaneously acting as an inducing force in giving rise to different phenotypes can be found in North American Spadefoot toads (Pfennig and Murphy 2000; Ledón-Rettig and Pfennig 2011). These amphibians inhabit xeric habitats and among them all species of *Spea* genus have the ability to produce alternative larval phenotypes: omnivorous larvae that are small and feed on detritus, and carnivorous larvae that are large and feed exclusively on small insects or other anuran larvae. This resource polyphenism is dependant on multiple environmental cues, including nutrition and density in the ephemeral ponds where larvae develop (Pfennig and Murphy 2000; Ledón-Rettig and Pfennig 2011). Spadefoot toads may reach high densities in wetlands, are key for nutrient cycling within ponds, and it has been shown that their larvae can influence the entire trophic structure of these ecosystems (Ghioca-Robrecht and Smith 2010). In allopatry, *S. multiplicata* and *S. bombifrons* both exhibit this resource polyphenism. Interestingly, when they occur in sympatry, the

competitive interaction between larvae of both species have promoted differences in expression of this resource polyphenism: *S. multiplicata* has a tendency to produce omnivorous phenotypes, while *S. bombifrons* has a tendency to produce carnivorous phenotypes in sympatry (Pfennig and Murphy 2000). Therefore, their phenotypic plasticity has permitted the persistence of the two species in sympatry by reducing competitive interactions between them. This example illustrates how the environment acts bi-directionally; both as a selective force in favoring certain phenotypes over others, and simultaneously as an inducing force giving rise to different phenotypes. This eco-evolutionary feedback modifies the ecological interactions among species within communities, and will ultimately affect the evolutionary trajectories of each *Spea* species in these sympatric populations.

Another well-described example where the environment plays a selective role, but may also be playing an inducing role, is the parallel evolution of stickleback phenotypes. Parallel evolution of recurrent phenotypes in similar environments is nearly universal in the natural world and is generally considered an indication that the traits evolved by natural selection (Futuyma 1998; West-Eberhard 2003). These parallel selection pressures can explain in part this outcome, but the additional explanation of the existence of shared ancestral developmental potentials across replicates that respond in a comparable manner when placed into the same environmental conditions has been examined recently (see Sect. 6.3). The parallel evolution of three-spined sticklebacks in freshwater habitats from a marine ancestor is probably one of the most extensively studied systems to uncover the ecological drivers and genetic bases of parallel evolution in wild populations (Schluter 2000). Many freshwater populations show repeated evolution of the same limnetic and benthic ecotypes that differ in several morphological traits and diet. Notably, limnetic ecotypes feed primarily on zooplankton, and have a long and slender mouth whereas benthic ecotypes feed on larger invertebrates and have a short and wide mouth (Bell and Foster 1994). These ecotypes occupy different trophic niches and their respective evolution has therefore affected the ecological interactions within these freshwater communities (Bell and Foster 1994). The role of natural selection in the evolution of these freshwater phenotypes is well-established at both the phenotypic and the genetic level (Schluter 2000; Colosimo et al. 2005), but in comparison, the role of a common developmental potential in their parallel evolution has not received as much attention (West-Eberhard 2005). Wund et al. (2008, 2012) tested the hypothesis that the recurrent evolution of freshwater ecotypes is the result of a plastic developmental potential present in the marine ancestor. They found that when marine sticklebacks were reared on either a "limnetic diet" or a "benthic diet", the phenotypic plasticity of the head and mouth parallels the phenotypic divergence observed among freshwater ecotypes, supporting the role of a developmental potential in the marine ancestor in the recurrent evolution of the limnetic ecotypes, as well as repeated genetic assimilation in this system (see Sects. 6.3 and 6.9).

In the last few years, many organisms have had their genomes published, and exploiting the full potential of these data may reveal insights into the genetic bases of ecological adaptation and recurrent environmental induction of phenotypes. The genome of the three-spined stickleback has recently been published, and quite remarkably, accompanied with 20 additional genome-wide comparisons across populations to detect genomic regions that are repeatedly and consistently associated with the marine-freshwater divergence (Jones et al. 2012). The highly replicated nature of the system, the presence of the ancestral population together with the genomic resources available provides the unique opportunity to identify the genetic bases of this developmental potential and discover the mechanisms underlying its evolution. The presence of fragile sites (specific loci that preferentially exhibit gaps and break on metaphase chromosomes (Durkin and Glover 2007) in the stickleback genome have already been previously identified as targets for repeated evolution of ecologically relevant traits (Chan et al. 2010) and further exciting discoveries are without doubt awaiting to be realized. With genomes becoming available for more and more species, and even for several populations

of the same species, similar approaches for understanding the dualistic role of the environment in selecting and inducing phenotypes may be undertaken in the coming years.

6.8 Robustness: A Regulator of Variation

With the advent of population genetics in the twentieth century followed by molecular population genetics it became evident that natural populations have abundant genetic variation (Lewontin 1974). At the gene expression level, this variation gets further compounded by stochasticity of cellular processes (see Sect. 6.5) (Landry 2009). In addition to these 'internal' sources of variation, organisms must also face variations in their biotic and abiotic environments during development, such as geographical location, seasons, abrupt changes in weather, predatory relations, social interactions, and nutrition (see Sects. 6.6 and 6.7). Some of these variations are predictable but others are often rapid and unpredictable. Given that organisms face variation in both genotype and the environment during development, it is surprising that organisms mostly develop a robust phenotype. Robustness – also known as 'canalization of development' (Waddington 1942) – is the persistence of an organismal trait (organism or organ, gene expression pattern or activity, a cellular process) under different stochastic, environmental and genetic conditions or perturbations (Félix and Wagner 2008). Waddington and Schmalhausen independently characterized robustness in the mid 1900s, although they used different terminology (Waddington 1942; Schmalhausen 1949). When detailed studies of developmental systems at the genetic and molecular levels became available in the last few decades, the problem of robustness was revisited (Gilbert 1991; Eshel and Matessi 1998; Siegal and Bergman 2002; West-Eberhard 2003).

Robustness mostly acts to conceal the underlying variation in the genotype and the responsiveness of the organism to varying environments. In special cases, like that that of polyphenism, robustness allows only a few or specific phenotypic outcomes in response to specific environments. By acting against expression of variation at the phenotypic level, robustness results in accumulation of cryptic variation in the population both in the genotype and in the responsiveness of the organism to the environment. It is only under certain variations or conditions in the genotype or the environment that robustness becomes compromised thus exposing these phenotypes to natural selection.

Genetic and simulation data reveal that regulation of robustness could happen at the gene interaction level in a network involving feedback mechanisms (Crickmore et al. 2009; Holloway et al. 2011). In the same context the degree of robustness of a gene in a network would depend on its additive, dominance, or epistatic relationships with other components of the network (Proulx and Phillips 2005). It has also been shown that gene network hubs contribute to robustness (Levy and Siegal 2008). The genetic or epigenetic regulators that produce robust organisms during development in presence of variation in the genotype and the environment are in the early phases of their exploration (Masel and Siegal 2009). The developmental mechanisms that these regulators employ to achieve robustness still remain as fragmented case examples (Braendle and Félix 2009; Félix 2012; Gursky et al. 2012). One type of regulators of robustness are heat shock proteins that act as capacitors (a term borrowed from electronics), which implies that they accumulate a large amount of variation in an input and transmit it in a controlled manner (Rutherford and Lindquist 1998). This mechanism often involves miRNA (Pal-Bhadra et al. 2004) or piwi RNA pathways (Gangaraju et al. 2011). Redundancy (Wagner 2005) and modularity (von Dassow et al. 2000; Ma et al. 2006) are some other mechanisms that have been proposed. Identifying the molecular genetic mechanisms that are involved in the buffering mechanism of robustness, and their compromise that brings about release of cryptic variation in the genome, are necessary for a complete understanding of how shape, form, and proportion are generated during development and how robustness contributes raw material for

natural selection to act upon. Emerging tools and concepts highlighted in this book allow us to identify genome-wide contributors of robustness and to integrate ecological concepts into specific mechanisms of robustness in the near future.

6.9 Genetic Assimilation and Accommodation: Fixation of Environmentally – Induced Variation

In classical evolutionary genetic models, genetic variation is thought to represent an important source of raw material for evolution (Futuyma 1998; Rockman and Wray 2002; Wray et al. 2003). However, in recent years, phenotypic variation arising from developmental plasticity has been proposed as an equal, if not more, significant source of raw material for evolutionary change (West-Eberhard 2003, 2005). According to this model, developmental systems generally produce robust phenotypes until they become compromised due to the presence of an environmental or genetic perturbation. This results in a systemic response, exposing phenotypic variants to natural selection. Thereafter, through a process called genetic accommodation, natural selection increases the environmental sensitivity of the developmental program such that an environmentally induced trait is always induced when it encounters a recurrent environmental cue (West-Eberhard 2003, 2005). Therefore, genetic accommodation increases phenotypic plasticity often leading to multiple phenotypic outcomes, such as in polyphenism. Alternatively, in a process called genetic assimilation, natural selection decreases environmental sensitivity of the developmental program such that an environmentally induced trait is constitutively expressed in the absence of the recurrent environmental cue (Waddington 1942; Schmalhausen 1949; Waddington 1956). Therefore, genetic assimilation decreases phenotypic plasticity resulting in the evolution of a single phenotypic outcome. Both of these processes alter environmental sensitivity by acting on the genes that control the frequency and form of a trait (genetic accommodation increases while genetic assimilation decreases environmental sensitivity).

In the 1950s, Conrad H Waddington was able to demonstrate the process of genetic assimilation by repeatedly selecting for four-winged flies after an environmental perturbation (Waddington 1957). Gloor (1947) had discovered that environmental perturbation of fruit fly embryos with ether can result in transformation of the third thoracic segment into a duplicate of the second thoracic segment such that these flies develop four wings. A few years later Waddington showed that these four-winged flies could become fixed in the population using artificial selection (Waddington 1957). Waddington repeatedly selected flies with this phenotype after ether shock. This resulted first in an increase in the frequency of the phenotype until after some generations homozygous females consistently produce four-winged individuals without the ether treatment (Waddington 1957). Following up on Waddington's experiments, Gibson and Hogness (1996) demonstrated that this phenotypic response to ether correlated with a loss of expression of *Ultrabithorax* (*UBX*) gene in the imaginal discs of the third thoracic segment.

Elegant work by Suzuki and Nijhout (2006) with the tobacco hornworm moth *M. sexta* provided compelling evidence of the process of genetic accommodation. Wild-type *M. sexta* larvae are green. In some cases, mutant larvae arise that develop a black pigmentation. When these mutants are heat-shocked (42 °C), a spectrum of pigment phenotypes (between green and black) is generated. They established a genetic line for larvae that more readily developed green by selecting for the green variants each generation. After only 13 generations, most larvae would develop green following the heat-shock treatment. Most importantly, when exposed to low temperatures, the larvae developed black but, when exposed to temperatures above 28.5 °C, the larvae would develop green. The response-curve to this temperature continuum was sigmoidal indicating the trait was now polyphenic. It turns out that the black mutant larvae produce very low amounts of JH, whereas the heat-shock induced green larvae

produce significantly more JH (see Sect. 6.2). Therefore, beginning with a single phenotype, selection was able to generate individuals that have an increased plastic response that is mediated by a hormonal threshold (resulting in two phenotypes). Interestingly, genetic assimilation is also possible with this *Manduca* model. In parallel to selecting for a genetic line that could respond to temperature with increased plasticity (genetic accommodation), they also selected for a "monophenic" line, that after seven generations consistently produced black larvae regardless of temperature level due to decreased plasticity (genetic assimilation). Finally, it is known that a sister species of *M. sexta* (*Manduca quinquemaculata*; Hudson 1966) is naturally polyphenic: at low temperatures the larvae develop black and at high temperatures the larvae develop green. Therefore, due to shared developmental modules arising from common ancestry, it is possible that cryptic genetic variation in the genome of *Manduca sexta* includes polyphenic combinations. This evolutionary contribution to cryptic genetic variation is likely due to the presence of an ancestral developmental potential in this group (see Sect. 6.3).

It is not exactly known how genomic loci contribute to fixing of the phenotype via genetic assimilation or genetic accommodation. For example, in Waddington's (1957) artificial selection experiment, when the chromosome providing the four-winged phenotype was brought into the context of a wildtype genome or individual wildtype chromosomes, the phenotype could not be reproduced. This experiment shows that during artificial selection, chromosomal loci scattered throughout the genome contributed to the genetic assimilation of the four-winged phenotype. In this case, and more generally, these contributing loci could come from standing genetic variation or *denovo* mutations in the form of: allelic variants, cis- and trans-regulators, downstream target genes, promoters or coding regions. Advances in ecological genomics make it possible to uncover these loci precisely to understand the mechanistic basis of genetic assimilation and genetic accommodation.

6.10 Integrating Levels of Biological Organization

A major challenge in eco-evo-devo is to uncover the relationships within and among different levels of biological organization; from the level of molecules, to cells, tissues, organs, and organ systems, all of which combine to make up the individual organism. Levels of organization external to the individual organism extend to higher levels, such as groups, populations, and communities. Each level includes multiple members of the same level, which interact to form higher levels of organization in the form of nested hierarchies (MacMahon et al. 1978; Zylstra 1992; Valentine 2003; Jagers op Akkerhuis 2008; Findlay and Thagard 2012). Beyond discussions on the units or levels of selection (Lewontin 1970; Dawkins 1978; see discussion in Pigliucci and Kaplan 2006), many have recognized that organismal complexity is produced from networks of both top-down and bottom-up interactions (Valentine 2003; Longo et al. 2012).

Research in eco-evo-devo over a number of years has shown that evolutionary and developmental changes at some levels of biological organization can either be associated or dissociated with other levels (Abouheif 1997; Wray 1999). For example, across species, the same or homologous phenotypes do not necessarily use the same or homologous genes or gene networks (Wray and Abouheif 1998). To understand how such associations or dissociations evolve, we have to explicitly consider the organization and interaction among hierarchical levels. Gene networks, which can be considered as a distinct level of biological organization from its constituent genes, provides an example (Abouheif 1999). Modularity is a general and well-characterized feature of gene networks that allows for the maintenance of obligate linkages, while at the same time allows some flexibility and redundancy of other linkages in the network (Von Dassow et al. 2000; Ma et al. 2006). Furthermore, network modules as a whole can be co-opted during evolution to function in different processes. A good example of this is

the signal transduction Receptor Tyrosine Kinase (Ras-RTK) pathway, which has been co-opted during evolution to transmit signals between the cell surface and nuclear genes during the development of skin in mammals, eyes in fruit flies, and the female genitalia in nematodes (Gilbert 2010). Therefore, the modularity and co-option of network modules during development and evolution is one of the many ways in which dissociations between genes, gene networks, and phenotypes can evolve (see Sect. 6.4 for further discussion).

Associations and dissociations can also occur between the organ to embryo level. For example, Nijhout and Emlen (1998) show that removal of the hindwing imaginal discs in the butterfly *Precis coenia*, leads to a proportional increase in the final size of the forewings. Similarly, selection for increased or decreased size of horns in male *Onthophagus acuminatus* beetles produces a compensatory (opposite) change in eye size (Nijhout and Emlen 1998). In this case, associations and dissociations can arise because while production of the organ itself (the disc in the butterfly and the horn in the beetle) may be autonomous, it is their interactions that influence the final form of the organism. Furthermore, Abouheif and Wray (2002) showed that although the wingless phenotype in worker castes is evolutionarily conserved across all 15,000 species of ants, the underlying wing organs (wing imaginal discs), as well as the gene network that is responsible for the growth and patterning of these organs, are disrupted in different ways in different species. Because wing development in winged and wingless castes in ants is environmentally determined through the action of hormones, it raises the possibility that the environment may have played a significant role in facilitating the evolution of these associations and dissociations in ants (see Sect. 6.9).

Together, these examples show that associations and dissociations can occur among multiple levels at the same time. Development integrates all of the levels of biological organization outlined above, producing an individual organism that incorporates the interactions, outcomes and variation at multiple levels of organization during its own lifetime and transferring these from one generation into the next (Hall 2013). The tools and concepts of ecological genomics will facilitate the explicit consideration of multiple levels of hierarchical organization into eco-evo-devo studies.

6.11 Conclusion

Indeed "the time seems to have come ..." as Waddington (1959) so eloquently stated, to integrate the complex interactions between environment, genes, and development into our understanding of the evolutionary process. Although each of our individual sections document a particular aspect of this complex interaction, together, our sections tell the larger story of eco-evo-devo and its wider implications on evolutionary theory. We attempt to summarize this larger story in Fig. 6.1, which was actually inspired by the figure in Waddington's (1959) article that attempts to "take into account two further aspects of the evolutionary mechanism."

In Fig. 6.1, the development of an individual in a single generation is represented from top to bottom, from a fertilized egg to a reproducing adult, where arrows represent interactions between environment, genes, and development. Development must buffer the effect of *de novo* and standing genetic variation, as well as stochastic variation, to produce a robust phenotype (Interactions represented by grey arrows). While this type of variation is most often masked or buffered leading to the accumulation of cryptic genetic variation, natural selection can occasionally act upon this variation to produce alternative phenotypes that are stochastically or genetically determined. As a consequence of environmental change (organisms dispersing, constructing, modifying, or suddenly experiencing new environments) developing individuals are subject to environmental stress either directly or indirectly through new ecological and social interactions (represented by black arrows). This environmental stress can either perturb or be buffered by robustness (indicated by a double-headed arrow). If the environmental stress perturbs robustness, it can result in the release of cryptic genetic and phenotypic variation as well as in the induction of

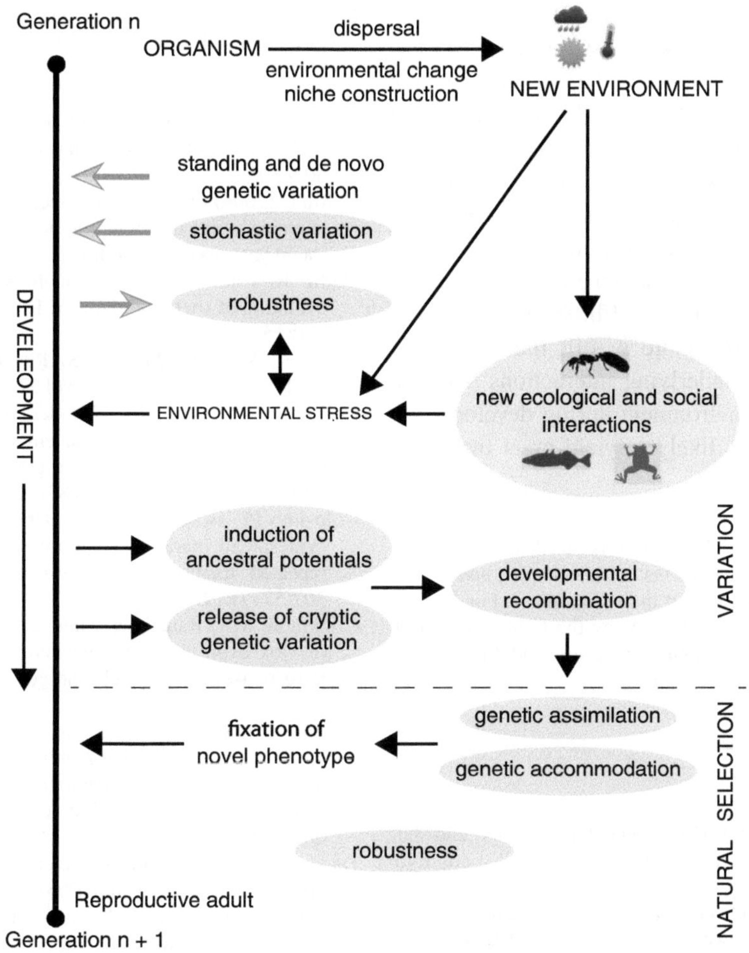

Fig. 6.1 The development of an individual during a single generation (from n to n + 1) is represented by a *thick black line* to the *left* from *top* to *bottom*, beginning with fertilization and ending with a reproducing adult. Shaded *grey ovals* correspond to Sects. 6.2, 6.3, 6.4, 6.5, 6.6, 6.7, 6.8, and 6.9 in the main text (some sections are combined into a *single oval*). *Grey arrows* show the interaction of endogenous sources of variation (standing genetic variation, de novo variation and stochastic variation) with the robustness of the developmental system of the organism. *Black arrows* highlight the developmental and evolutionary consequences following dispersal, niche construction and modification, and sudden changes in the environment of an organism. *Double-headed arrow* highlights the direct interaction between robustness and the environment. The area above the dotted line represents sources of variation, whereas the area below the dotted line represents evolutionary processes where natural selection is acting. Figure 6.1 is inspired by that found in Waddington (1959)

ancestral developmental potentials. The modular nature of gene networks will subsequently cause the variation released by environmental stress to appear in potentially new combinations.

If any of this variation provides an advantage to the reproducing adult, then this variation will be fixed by natural selection through either the process of genetic accommodation or genetic assimilation. The mechanism through which the fixation of environmentally induced phenotypes occurs is most likely to involve standing genetic variation, although de novo mutations and epigenetic mechanisms also contribute. Finally, these fixed phenotypes become robust to further environmental and genetic variation.

This eco-evo-devo view of the evolutionary process takes into account both genetic and epigenetic systems. We will end this chapter by

raising an important quandary for the ecological genomicist – if we view the evolutionary process as one where both genetic and epigenetic systems are important, then where should the focus of the ecological genomicist be: the genes responsible for the adaptive trait itself or the genes underlying the environmental responsiveness of the adaptive trait? This distinction is fundamental and important for understanding the genetic basis of adaptation, yet this has received little attention from biologists. Clearly, there is still much to learn about the rules underlying interactions between genes and the environment during development, the near future is likely to yield great insight in this direction.

Acknowledgements We would like to thank the Editors for their insightful comments on this chapter, and for the opportunity to contribute it in the first place. Funding for this work was provided to E.A. by the Canada Research Chairs Program and Natural Sciences and Engineering Research Council of Canada.

References

Abouheif E (1997) Developmental genetics and homology: a hierarchical approach. Trends Ecol Evol 12:405–408

Abouheif E (1999) Establishing homology criteria for regulatory gene networks: prospects and challenges. In: Homology, vol 222, Novartis foundation symposium. Wiley, Chichester, pp 207–225

Abouheif E (2008) Parallelism as the pattern and process of mesoevolution. Evol Dev 10(1):3–5

Abouheif E, Wray GA (2002) Evolution of the gene network underlying wing polyphenism in ants. Science 297(5579):249–252

Armbruster WS, Lee J, Baldwin BG (2009) Macroevolutionary patterns of defense and pollination in *Dalechampia* vines: adaptation, exaptation, and evolutionary novelty. Proc Natl Acad Sci 106(43): 18085–18090

Bell MA, Foster SA (1994) The evolutionary biology of the threespine stickleback. Oxford University Press, Oxford

Braendle C, Felix MA (2009) The other side of phenotypic plasticity: a developmental system that generates an invariant phenotype despite environmental variation. J Biosci 34(4):543–551

Braendle C, Friebe I, Caillaud MC, Stern DL (2005) Genetic variation for an aphid wing polyphenism is genetically linked to a naturally occurring wing polymorphism. Proc R Soc B 272:657–664

Carroll S, Grenier J, Weatherbee S (2005) From DNA to diversity. Blackwell Science, Oxford, 258 pp

Chan YF, Marks ME, Jones FC et al (2010) Adaptive evolution of pelvic reduction in sticklebacks by recurrent deletion of a Pitx1 enhancer. Science 327(5963): 302–305

Collin R (2004) Phylogenetic effects, the loss of complex characters, and the evolution of development in calyptraeid gastropods. Evolution 58(7):1488–1502

Collin R, Cipriani R (2003) Dollo's law and the re-evolution of shell coiling. Proc R Soc Lond B Biol Sci 270(1533):2551–2555

Collin R, Miglietta MP (2008) Reversing opinions on Dollo's Law. Trends Ecol Evol 23(11):602–609

Collin R, Chaparro OR, Winkler F, Véliz D (2007) Molecular phylogenetic and embryological evidence that feeding larvae have been reacquired in a marine gastropod. Biol Bull 212(2):83–92

Colosimo PF, Hosemann KE, Balabhadra S et al (2005) Widespread parallel evolution in sticklebacks by repeated fixation of Ectodysplasin alleles. Science 307(5717):1928–1933

Crickmore MA, Ranade V, Mann RS (2009) Regulation of Ubx expression by epigenetic enhancer silencing in response to Ubx levels and genetic variation. PLoS Genet 5(9):e1000633

Darwin C (1868) The variation of animals and plants under domestication, vol 2. John Murray, London

Davidson EH (2006) The regulatory genome: gene regulatory networks in development and evolution. Academic Press, London

Dawkins R (1978) Replicator selection and the extended phenotype. Z Tierpsychol 47:61–76

Dubrow TJ, Wackym PA, Lesavoy MA (1988) Detailing the human tail. Ann Plast Surg 20(4):340

Durkin SG, Glover TW (2007) Chromosome fragile sites. Annu Rev Genet 41:169–192

Duveau F, Félix MA (2012) Role of pleiotropy in the evolution of a cryptic developmental variation in *Caenorhabditis elegans*. PLoS Biol 10(1): e1001230

Dworkin I (2005) Towards a genetic architecture of cryptic genetic variation and genetic assimilation: the contribution of KG Bateman. J Genet 84(3):223

Eshel I, Matessi C (1998) Canalization, genetic assimilation and preadaptation: a quantitative genetic model. Genetics 149(4):2119

Félix MA (2012) Evolution in developmental phenotype space. Curr Opin Genet Dev 22(6):593–599

Félix MA, Wagner A (2008) Robustness and evolution: concepts, insights and challenges from a developmental model system. Heredity 100(2): 132–140

Fernald RD, Maruska KP (2012) Social information changes the brain. Proc Natl Acad Sci 109(2): 17194–17199

Findlay SD, Thagard P (2012) How parts make up wholes. Front Physiol 3:1–10

Flatt T, Heyland A (eds) (2011) Mechanisms of life history evolution: the genetics and physiology of life history traits and trade-offs. Oxford University Press, Oxford

Flatt T, Tu MP, Tatar M (2005) Hormonal pleiotropy and the juvenile hormone regulation of *Drosophila* development and life history. Bioessays 27(10): 999–1010

Futuyma DJ (1998) Evolutionary biology, 3rd edn. Sinauer Associates, Sunderland

Gangaraju VK, Yin H, Weiner et al (2011) *Drosophila* Piwi functions in Hsp90-mediated suppression of phenotypic variation. Nat Genet 43(2):153–158

Gasparini C, Serena G, Pilastro A (2013) Do unattractive friends make you look better? Context-dependent male mating preferences in the guppy. Proc R Soc B 280:20123072

Ghioca-Robrecht DM, Smith LM (2010) The role of spadefoot toad tadpoles in wetland trophic structure as influenced by environmental and morphological factors. Can J Zool 89(1):47–59

Gibson G, Dworkin I (2004) Uncovering cryptic genetic variation. Nat Rev Genet 5(9):681–690

Gibson G, Hogness DS (1996) Effect of polymorphism in the *Drosophila* regulatory gene Ultrabithorax on homeotic stability. Science 271(5246):200–203

Gilbert SF (1991) Induction and the origins of developmental genetics. Dev Biol 7:181–206

Gilbert SF (2010) Developmental biology. Sinauer Associates, Sunderland

Gilbert SF, Epel D (2009) Ecological developmental biology. Sinauer Associates, Sunderland

Gloor H (1947) Phänokopie Versuche mit Aether an Drosophila. Rev Suisse Zool 54:637–712

Goldberg EE, Igić B (2008) On phylogenetic tests of irreversible evolution. Evolution 62(11):2727–2741

Gursky VV, Surkova SY, Samsonova MG (2012) Mechanisms of developmental robustness. Biosystems 109(3):329–335

Hall BK (2003) Descent with modification: the unity underlying homology and homoplasy as seen through an analysis of development and evolution. Biol Rev 78(3):409–433

Hall BK (2013) Epigenesis, epigenetics, and the epigenotype: toward an inclusive concept of development and evolution. In: Brian G. Henning, Adam C. Scarfe (eds) Beyond mechanism: putting life back into biology. Lexington Books, Maryland, pp 348–371

Harris MP, Hasso SM, Ferguson MW, Fallon JF (2006) The development of archosaurian first-generation teeth in a chicken mutant. Curr Biol 16(4):371–377

Hayden EJ, Ferrada E, Wagner A (2011) Cryptic genetic variation promotes rapid evolutionary adaptation in an RNA enzyme. Nature 474(7349):92–95

Heitzler P, Simpson P (1991) The choice of cell fate in the epidermis of Drosophila. Cell 64(6):1083–1092

Hölldobler B, Wilson E (1990) The ants. Harvard University Press, Cambridge

Hölldobler B, Wilson E (2008) The superorganism: the beauty, elegance and strangeness of insect societies. W.W. Norton, New York

Holloway DM, Lopes FJ, da Fontoura Costa L et al (2011) Gene expression noise in spatial patterning: hunchback promoter structure affects noise amplitude and distribution in *Drosophila* segmentation. PLoS Comput Biol 7(2):e1001069

Huang MH (2010) Multi-phase defense by the big-headed ant, *Pheidole obtusospinosa*, against raiding army ants. J Insect Sci 10:1

Hudson A (1966) Proteins in the haemolymph and other tissues of the developing tomato hornworm, *Protoparce quinquemaculata* Haworth. Can J Zool 44(4):541–555

Jagers op Akkerhuis GAJM (2008) Analysing hierarchy in the organization of biological and physical systems. Biol Rev 83(1):1–12

Jenner RA, Wills MA (2007) The choice of model organisms in evo–devo. Nat Rev Genet 8(4):311–314

Jones FC, Grabherr MG, Chan YF et al (2012) The genomic basis of adaptive evolution in threespine sticklebacks. Nature 484(7392):55–61

Kilfoil ML, Lasko P, Abouheif E (2009) Stochastic variation: from single cells to superorganisms. HFSP J 3(6):379–385

Kohlsdorf T, Wagner GP (2006) Evidence for the reversibility of digit loss: a phylogenetic study of limb evolution in *Bachia* (Gymnophthalmidae: Squamata). Evolution 60(9):1896–1912

Kohlsdorf T, Lynch VJ, Rodrigues MT, Brandley MC, Wagner GP (2010) Data and data interpretation in the study of limb evolution: a reply to Galis et al. on the reevolution of digits in the lizard genus *Bachia*. Evolution 64(8):2477–2485

Landry CR (2009) Systems biology spins off a new model for the study of canalization. Trends Ecol Evol 24(2):63–66

Ledón-Rettig CC, Pfennig DW (2011) Emerging model systems in eco-evo-devo: the environmentally responsive spadefoot toad. Evol Dev 13(4):391–400

Ledón-Rettig CC, Pfennig DW, Crespi EJ (2010) Diet and hormonal manipulation reveal cryptic genetic variation: implications for the evolution of novel feeding strategies. Proc R Soc B Biol Sci 277(1700):3569–3578

Levinton JS (1986) Developmental constraints and evolutionary saltations: a discussion and critique. In: Gustafson JP, Stebbins GL, Ayala FJ (eds) Genetics, development and evolution. Plenum, New York, pp 253–288

Levy SF, Siegal ML (2008) Network hubs buffer environmental variation in *Saccharomyces cerevisiae*. PLoS Biol 6(11):e264

Lewontin R (1970) The units of selection. Annu Rev Ecol Evol Syst 1:1–18

Lewontin R (1974) Annotation: analysis of variance and analysis of causes. Am J Hum Genet 26(3): 400–411

Longo G, Miquel PA, Sonnenschein C, Soto AM (2012) Is information a proper observable for biological organization? Prog Biophys Mol Biol 109(3):108–114

Losick R, Desplan C (2008) Stochasticity and cell fate. Science 320(5872):65–68

Ma W, Lai L, Ouyang Q, Tang C (2006) Robustness and modular design of the *Drosophila* segment polarity network. Mol Syst Biol 2:70

MacMahon JA, Phillips DL, Robinson JV, Schimpf DJ (1978) Levels of biological organization: an organism-centered approach. BioScience 28(11):700–704

Masel J, Siegal M (2009) Robustness: mechanisms and consequences. Trends Genet 25(9):395–403

McAdams HH, Arkin A (1999) It's a noisy business! Genetic regulation at the nanomolar scale. Trends Genet 15(2):65–69

McGlashan JK, Spencer R-J, Old JM (2012) Embryonic communication in the nest: metabolic responses of reptilian embryos to developmental rates of siblings. Proc R Soc B 279:1709–1715

Moczek AP, Nijhout HF (2002) Developmental mechanisms of threshold evolution in a polyphenic beetle. Evol Dev 4(4):252–264

Nijhout HF (1998) Insect hormones. Princeton University Press, Princeton

Nijhout HF (1999) Control mechanisms of polyphenic development in insects. Bioscience 49(3):181–192

Nijhout HF, Emlen DJ (1998) Competition among body parts in the development and evolution of insect morphology. Proc Natl Acad Sci 95:3685–3689

Pal-Bhadra M, Leibovitch BA, Gandhi SG et al (2004) Heterochromatic silencing and HP1 localization in *Drosophila* are dependent on the RNAi machinery. Science 303(5658):669–672

Passera L (1977) Production des soldats dans les societes sortant d'hibernation chez la fourmi *Pheidole pallidulla* (NYL.) (Formicidae, Myrmicinae). Insect Soc 24(2):131–146

Passera L, Roncin E, Kaufmann B, Keller L (1996) Increased soldier production in ant colonies exposed to intraspecific competition. Nature 379:630–631

Peeters C, Liebig J, Hölldobler B (2000) Sexual reproduction by both queens and workers in the ponerine ant *Harpegnathos saltator*. Insect Soc 47:325–332

Pfennig DW, Murphy PJ (2000) Character displacement in polyphenic tadpoles. Evolution 54(5):1738–1749

Pie MR, Traniello JFA (2007) Morphological evolution in a hyperdiverse clade: the ant genus *Pheidole*. J Zool 271(1):99–109

Pigliucci M, Kaplan J (2006) Making sense of evolution: the conceptual foundations of evolutionary biology. University of Chicago Press, Chicago

Proulx SR, Phillips PC (2005) The opportunity for canalization and the evolution of genetic networks. Am Nat 165(2):147–162

Rajakumar R, San Mauro D, Dijkstra MB et al (2012) Ancestral developmental potential facilitates parallel evolution in ants. Science 335(6064):79–82

Renn SCP, Aubin-Horth N, Hoffman HA (2008) Fish and chips: functional genomics of social plasticity in an African cichlid fish. J Exp Biol 211:3041–3056

Rockman MV, Wray GA (2002) Abundant raw material for cis-regulatory evolution in humans. Mol Biol Evol 19(11):1991–2004

Roff DA (1996) The evolution of threshold traits in animals. Q Rev Biol 71:3–35

Roff DA, Fairbairn DJ (1991) Wing dimorphisms and the evolution of migratory polymorphisms among the Insecta. Am Zool 31(1):243–251

Rutherford SL, Lindquist S (1998) Hsp90 as a capacitor for morphological evolution. Nature 396(6709):336–342

Schlosser G, Wagner GP (2004) Modularity in development and evolution. University of Chicago Press, Chicago

Schluter D (2000) The ecology of adaptive radiation. Oxford University Press, Oxford

Schmalhausen I (1949) Factors of evolution : the theory of stabilizing selection. Blakiston Co., Philadelphia

Shbailat SJ, Abouheif E (2012) The wing-patterning network in the wingless castes of Myrmicine and Formicine Ant species is a mix of evolutionarily labile and non-labile genes. J Exp Zool B Mol Dev Evol 9999B:1–10

Shbailat SJ, Khila A, Abouheif E (2010) Correlations between spatiotemporal changes in gene expression and apoptosis underlie wing polyphenism in the ant *Pheidole morrisi*. Evol Dev 12(6):580–591

Siegal ML, Bergman A (2002) Waddington's canalization revisited: developmental stability and evolution. Proc Natl Acad Sci USA 99(16):10528–10532

Stiassny MLJ (2003) Atavism. In: Hall BK, Olsen WM (eds) Keywords and concepts in evolutionary developmental biology. Harvard University Press, Cambridge, pp 10–14

Suzuki Y, Nijhout HF (2006) Evolution of a polyphenism by genetic accomodation. Science 311:650–652

Tinbergen N (1971) Social behaviour in animals, with a special reference to vertebrates, Science paperbacks. Chapman and Hall, London

Tomić N, Meyer-Rochow VB (2011) Atavisms: medical, genetic, and evolutionary implications. Perspect Biol Med 54(3):332–353

Tomoyasu Y, Arakane Y, Kramer KJ, Denell RE (2009) Repeated co-options of exoskeleton formation during wing-to-elytron evolution in beetles. Curr Biol 19(24):2057–2065

Ungerer MC, Johnson LC, Herman MA (2008) Ecological genomics: understanding gene and genome function in the natural environment. Heredity 100:178–183

Valentine JW (2003) Architectures of biological complexity. Integr Comp Biol 43(1):99–103

von Dassow G, Meir E, Munro EM, Odell GM (2000) The segment polarity network is a robust developmental module. Nature 406(6792):188–192

Waddington CH (1942) Canalization of development and the inheritance of acquired characters. Nature 150:563–565

Waddington CH (1956) Genetic assimilation of the bithorax phenotype. Evolution 10(1):1–13

Waddington CH (1957) The genetic basis of 'assimilated bithorax' stock. J Genet 55(2):241–245

Waddington CH (1959) Evolutionary systems-animal and human. Nature 183(4676):1634–1638

Wagner A (2005) Distributed robustness versus redundancy as causes of mutational robustness. Bioessays 27(2):176–188

Wake DB, Wake MH, Specht CD (2011) Homoplasy: from detecting pattern to determining process and mechanism of evolution. Science 331(6020): 1032–1035

Weatherbee SD, Halder G, Kim J et al (1998) Ultrabithorax regulates genes at several levels of the wing-patterning hierarchy to shape the development of the *Drosophila* haltere. Genes Dev 12(10): 1474–1482

West-Eberhard MJ (2003) Developmental plasticity and evolution. Oxford University Press, New York

West-Eberhard MJ (2005) Developmental plasticity and the origin of species differences. Proc Natl Acad Sci USA 102(Suppl 1):6543–6549

Wheeler DE, Nijhout HF (1981) Soldier determination in ants: new role for juvenile hormone. Science 213:361–363

Wheeler DE, Nijhout HF (1983) Soldier determination in *Pheidole bicarinata*: effect of methoprene on caste and size within castes. J Insect Physiol 29(11): 847–854

Wheeler DE, Nijhout HF (1984) Soldier determination in *Pheidole bicarinata*: inhibition by adult soldiers. J Insect Physiol 30(2):127–135

Whitacre JM, Lin J, Harding A (2012) T cell adaptive immunity proceeds through environment-induced adaptation from the exposure of cryptic genetic variation. Front Genet 3:1–11

Whiting MF, Bradler S, Maxwell T (2003) Loss and recovery of wings in stick insects. Nature 421(6920):264–267

Wiens JJ (2011) Re-evolution of lost mandibular teeth in frogs after more than 200 million years, and re-evaluating Dollo's law. Evolution 65(5): 1283–1296

Wray GA (1999) Evolutionary dissociations between homologous genes and homologous structures. In: Bock GR, Cardew G (eds) Homology. Wiley, Chichester, pp 189–206

Wray GA, Abouheif E (1998) When is homology not homology? Curr Opin Genet Dev 8(6):675–680

Wray GA, Hahn MW, Abouheif E, Balhoff JP, Pizer M, Rockman MV, Romano LA (2003) The evolution of transcriptional regulation in eukaryotes. Mol Biol Evol 20(9):1377–1419

Wund MA, Baker JA, Clancy B et al (2008) A test of the "flexible stem" model of evolution: ancestral plasticity, genetic accommodation, and morphological divergence in the threespine stickleback radiation. Am Nat 172(4):449–462

Wund MA, Valena S, Wood S, Baker JA (2012) Ancestral plasticity and allometry in threespine stickleback reveal phenotypes associated with derived, freshwater ecotypes. Biol J Linn Soc 105(3):573–583

Yang AS, Martin CH, Nijhout HF (2004) Geographic variation in caste structure among ant populations. Curr Biol 14(6):514–519

Zera AJ, Tiebel KC (1988) Brachypterizing effect of group rearing, juvenile hormone III and methoprene in the wing-dimorphic cricket, *Gryllus rubens*. J Insect Physiol 34(6):489–498

Zera AJ, Harshman LG, Williams TD (2007) Evolutionary endocrinology: the developing synthesis between endocrinology and evolutionary genetics. Ann Rev Ecol Evol Syst 38:793–817

Zylstra U (1992) Living things as hierarchically organized structures. Synthese 91(1–2):111–133

Evolutionary and Ecological Genomics of Developmental Plasticity: Novel Approaches and First Insights From the Study of Horned Beetles

7

Armin P. Moczek, Teiya Kijimoto, Emilie Snell-Rood, Guilherme Rocha, Melissa Pespeni, and Karen Kafadar

Abstract

Phenotypic plasticity pervades organismal development and physiology where it facilitates an enormous range of adaptive responses to novel or stressful environments. Plasticity also impacts evolutionary processes, reducing the probability of population extinction in the face of environmental changes and sometimes increasing speciation rates in developmentally plastic lineages. Despite the adaptive significance of plasticity, organisms are not infinitely plastic; rather they are constrained in the kinds and ranges of environmental changes to which their body parts, organs, and tissues can respond. Understanding the nature, costs, and limits of developmental plasticity requires insight into (i) the developmental-genetic and genomic mechanisms underlying plastic responses as well as (ii) their interplay with ecological and social conditions. In this chapter we review and summarize recent progress in the development of horned beetles as a study system with which to explore the interactions between changing ecological conditions and plastic, genome-wide responses in gene expression and developmental function. In particular, we focus on plastic responses to nutritional variation, which in horned beetles differ widely as a function of body region, sex, and species. We begin by introducing the study system and summarize the developmental-genetic and genomic tool set currently available for horned beetles. We then present recently developed statistical approaches that can be used to guide

A.P. Moczek (✉) • T. Kijimoto • M. Pespeni
Department of Biology, Indiana University,
915 E. Third Street, Myers Hall 150, Bloomington, IN 47405-7107, USA
e-mail: armin@indiana.edu

E. Snell-Rood
Department of Evolution, Ecology and Behavior,
University of Minnesota, Minneapolis, MN, USA

G. Rocha • K. Kafadar
Department of Statistics, Indiana University,
Bloomington, IN, USA

C.R. Landry and N. Aubin-Horth (eds.), *Ecological Genomics: Ecology and the Evolution of Genes and Genomes*, Advances in Experimental Medicine and Biology 781,
DOI 10.1007/978-94-007-7347-9_7,

the design of multi-factorial genome-wide transcriptional comparisons when circumstances prohibit a fully balanced design. We present an example of such an approach in the horned beetle *Onthophagus taurus* and end by highlighting the growing opportunities for future ecological-genomic studies in horned beetles.

Keywords
Developmental plasticity • Nutrition • Microarray • Horned beetle • Sexual dimorphism • Onthophagus • Polyphenism • Doublesex

7.1 Introduction

Developmental plasticity is taxonomically widespread and believed to be of major ecological and evolutionary significance. Yet developmental plasticity and its emergent properties are difficult to study in traditional molecular-genetic model systems given their general lack of pronounced plastic responses and a solid ecological context. Hence, a deeper understanding of the causes, mechanisms, and consequences of plasticity requires the development of more appropriate model taxa and corresponding experimental tools and resources. In this chapter we review recent progress in the development of horned beetles as a study system with which to explore the interplay between ecological conditions and plastic, genome-wide responses in gene expression and developmental function. In the first part we provide a brief overview of the current critical questions in the field of developmental plasticity as they relate to long-standing frontiers in evolutionary ecology. We then introduce the biology of horned beetles, alongside key genetic, developmental, and genomic tools and resources that have been developed in recent years to investigate the multifactorial nature of horned beetle plasticity. In the second part we focus on an ecological variable of key relevance to most heterotrophic organisms – variation in nutrition. Here, we present recently developed statistical approaches that can be used to guide the design of multi-factorial genome-wide transcriptional comparisons of nutritional plasticity when circumstances prohibit fully balanced designs, and present the first results of such an effort in the horned beetle *Onthophagus taurus*. We end by discussing the growing opportunities for future ecological-genomic studies in horned beetles.

7.1.1 Plasticity's Significance in Development and Evolution

We define *developmental plasticity* as the ability of an individual or a genotype to adjust their development in response to the environment (see the Glossary for definitions of italicized terms). Developmental plasticity is taxonomically widespread, and virtually all organisms as well as developmental processes exhibit some degree of plasticity (Newman and Muller 2000; West-Eberhard 2003; Whitman and Ananthakrishnan 2009). Developmental plasticity ranges from simple responses to changes in ambient abiotic conditions such as temperature or pH, to highly choreographed adjustments of entire syndromes of traits, such as nutrition-dependent caste determination (Smith et al. 2008) or seasonal reproduction (Piersma and Drent 2003). Developmental responses may be gradual or discrete, reversible or not, and may or may not always be adaptive. In fact, *incorrect* developmental responses to the environment are at the heart of many human diseases, such as allergies, asthma, diabetes and obesity (Gilbert and Epel 2009; Gluckman et al. 2009).

In an extraordinarily wide range of circumstances, however, developmental plasticity

is adaptive, allowing organisms to maintain high performance in the face of environmental variability. As such, plasticity plays a key role in enabling individuals and populations to respond adaptively to environmental fluctuations, be they changes in climate, nutrient availability, social conditions, or predators (Charmantier et al. 2008; Sol 2009; Sih et al. 2011). In the process, developmental plasticity has significant consequences for evolutionary processes, on several levels (reviewed in West-Eberhard 2003; Pfennig et al. 2010; Moczek et al. 2011).

First, phenotypic plasticity promotes survival in novel and changing environments, which increases the chances that a population will eventually adapt to that environment (Price et al. 2003; Lande 2009). For instance, in vertebrates, variation in behavioral plasticity has been linked to differences in survival in new environments and subsequent diversification (Sol et al. 2005, 2008; Sol and Price 2008). Second, the developmental architecture underlying a plastic response can be utilized in the evolution of a fixed trait. For instance, the genes involved in a plastic aggressive response overlap with those that have diverged in expression between more or less aggressive honeybee subspecies (Alaux et al. 2009). Similarly, pathways involved in plasticity in skeletal development are similar to those that are responsible for skeletal divergence across species (Young and Badyaev 2007). Third, developmental plasticity can foster the accumulation of genetic variation because new mutations are more likely to be hidden from selection. Such cryptic genetic variation may be revealed in novel environments, leading to subsequent rapid adaptation (Snell-Rood et al. 2010; Draghi and Whitlock 2012). Fourth, the developmental machinery that enables plastic responses, once evolved, can be recruited to orchestrate plastic developmental responses in other, and possibly very different, contexts. A spectacular example of the repeated cooption of environment-sensitive development can be seen in holometabolous insects, where the same endocrine machinery plays a critical role in coordinating alternative reproductive decisions (e.g. whether to invest in growth/maintenance or reproduction), alternative developmental decisions (molting to larva, pupa, or adult) and polyphenic development (facultative diapause, host switch, caste and morph expression; reviewed in Stansbury and Moczek 2013). Fifth, plasticity enables diversification through *genetic assimilation*, a process whereby a trait that was initially induced by the environment becomes constitutive in expression through genetic changes in underlying developmental pathways (Pigliucci and Murren 2003; West-Eberhard 2003; Bateson and Gluckman 2011; Renn and Schumer 2013). Assimilation of induced traits is particularly likely if plasticity is costly (Lande 2009; Bateson and Gluckman 2011) as it is generally assumed (but see below). Lastly, plasticity can fuel diversification if the induction of alternate developmental pathways results in assortative mating or pre-zygotic isolation, for instance through changes in ornamentation, sensory systems or the timing or location of mating (Pfennig et al. 2010). In summary, while there are clearly some situations where plasticity can impede diversification through buffering of selection pressures (Huey et al. 2003; Pfennig et al. 2010), much emerging evidence suggests that in many cases, plasticity is capable of promoting adaptation, diversification, and innovation in response to novel and changing environments (West-Eberhard 2003; Pfennig et al. 2010; Moczek et al. 2011).

In order to understand how plasticity may impact diversification, we must also understand why plasticity varies within and across species. Despite the benefits of plasticity, organisms are not infinitely plastic; rather, plastic responses during development are limited in range and kind, and many environmental challenges result in no, or non-adaptive, responses. Biologists have long been interested in the costs and constraints that limit the evolution of plasticity (DeWitt 1998; Schlichting and Pigliucci 1998). Yet the costs of plasticity remain elusive (Pigliucci 2005; Van Buskirk and Steiner 2009) and the forces that constrain the evolution of plasticity remain

poorly understood. A major stumbling block to understanding the evolution of plasticity is an incomplete knowledge of the developmental mechanisms underlying plasticity, given that such mechanisms will determine the types of costs and constraints associated with plasticity (Snell-Rood et al. 2010; Snell-Rood 2012). For example, theoretical considerations predict that the evolution of alternate developmental pathways may be limited by *relaxed selection* relative to a specialized, less plastic genotype (Kawecki 1994; Whitlock 1996; Van Dyken and Wade 2010). In contrast, forms of plasticity that rely less on evolved switches and more on learning-like mechanisms (Frank 1996) come with substantial individual-level costs and life history tradeoffs, such as delayed reproduction and reduced fecundity (Mayr 1974; Johnston 1982; Snell-Rood 2012). However, the developmental genetic basis of many forms of plasticity remains unclear. Even for environment-induced differences in gene expression, we know that many plastic responses are a result of conserved, environmentally responsive pathways (like insulin signaling (Nijhout 2003; Shingleton et al. 2007)), but many other responses rely on stochastic processes (Eldar and Elowitz 2010; Feinberg and Irizarry 2010; Wang and Zhang 2011). Each of these mechanisms comes with distinct costs and evolutionary consequences.

A key hurdle in this process of determining the mechanisms underlying plasticity and their evolutionary consequences has been a general lack of model systems with pronounced plastic responses that also possess the relevant genomic and developmental tools. Horned beetles, most notably in the genus *Onthophagus*, have emerged as a valuable model system in this regard, combining rich diversity of plastic responses over a range of phylogenetic distances with an increasing array of developmental genetic and genomic tools and resources (reviewed in Kijimoto et al. 2012b). In the next section we provide a brief overview of the biology of horned beetles, and then summarize key techniques and resources available to investigate the multifactorial nature of horned beetle plasticity.

7.1.2 The Plastic Biology of Horned Beetles: A Primer

Horned beetles are not a monophyletic group; rather, species in at least seven, partly quite distantly related beetle families have evolved horns or horn-like structures (Fig. 7.1) (Snell-Rood and Moczek 2013). However, the majority of horned beetle species as well as diversity in horn structures are concentrated in two subfamilies within the Scarabaeidae: the Dynastinae (rhinoceros beetles) and Scarabaeinae (true dung beetles) (Arrow 1951). In both subfamilies, thousands of species develop horns, including many cases of extreme elaboration (Fig. 7.1). Where they exist, and no matter how diverse in shape and size, horns are used as weapons in male combat over access to females.

Horned beetles exhibit plasticity on a variety of levels of biological organization and developmental time scales (Fig. 7.1) (reviewed in Valena and Moczek 2012). Most obvious is the development of horns, which in most species is limited to males, and among conspecific males, is closely tied to the availability of food during larval development (Fig. 7.2). Nutrition-dependent development of horns can be isometric (i.e. large males are essentially proportionally enlarged versions of small males in all respects including horns), positively allometric (large males develop disproportionally larger horns) or discretely dimorphic: in this case males below and above a certain size threshold develop into alternative hornless and horned morphs, akin to the development of alternative worker and soldier castes in social insects. Male morphs also diverge plastically in other morphological traits, such as the relative sizes of wings, mouthparts and antennae (possibly due to resource allocation tradeoffs arising from investment into horns; Emlen 2001) as well as the relative sizes of testes, which play an especially important role in the behavioral ecology of individual, competing males. Whereas large, horned males rely on aggressive fighting behavior and the use of horns as weapons to secure mating opportunities, small, hornless males utilize non-aggressive sneaking behaviors to access females on the sly and rely greatly on

Fig. 7.1 Examples of the exuberance of horned beetle diversity. From *top* to *bottom*: *Phanaeus imperator* (South America), *Onthophagus watanabei* (Borneo), *Eupatorus gracilicornis* (Southeast Asia), *Trypoxylus* (*Allomyrina*) *dichotoma* (East Asia), *Golofa claviger* (South America)

enhanced sperm competition through enlarged ejaculate volumes (Simmons et al. 1999; Simmons and Emlen 2006). Alternative male morphs therefore reflect divergent syndromes of morphological, physiological, and behavioral traits, adapted to suit alternative, sexually selected competitive niches. Male dimorphisms are extremely common and can be so elaborate that alternative morphs have on occasion been described as belonging to separate species (Paulian 1935).

A second major level of variation in horned beetle development is found between sexes, which in many ways parallels the differences just described for male morphs (Fig. 7.2; Kijimoto et al. 2012b). Like the hornless, or *minor* male morph, females typically exhibit greatly reduced horn development or no horns altogether. Sexual dimorphisms are not due to plasticity in the strict sense and instead result from sex-specific development, most likely following XY sex-determination (Angus 2008). However, recent work has shown that male (morph)– and sex-specific horn development are underlain by the same developmental machinery, and that their co-evolution has greatly influenced the radiation of horned beetles (Kijimoto et al. 2012a). Lastly, enormous variation exists among horned beetle species in the precise location, number, and shape of horns, as well as degree of male– and sexual dimorphism, reflecting evolved modifications in the developmental mechanisms underlying plastic and *canalized* aspects of horn formation (Figs. 7.1 and 7.2; reviewed in Kijimoto et al. 2012b).

In recent years the horned beetle genus *Onthophagus* has emerged as a particularly accessible study system with which to examine the evolutionary and developmental genetics of plasticity, as well as the role of plasticity in diversification and innovation (e.g. Emlen et al. 2005; Moczek 2005; Kijimoto et al. 2012b). *Onthophagus* is home to over 2,000 extant, and highly diverse species, many of which are widely accessible and easy to maintain, observe, and rear. Moreover, a subset of species has been introduced to exotic locations either on purpose as part of bio control programs or by accident, providing rare opportunities to study contemporary evolution (including of plasticity) in action.

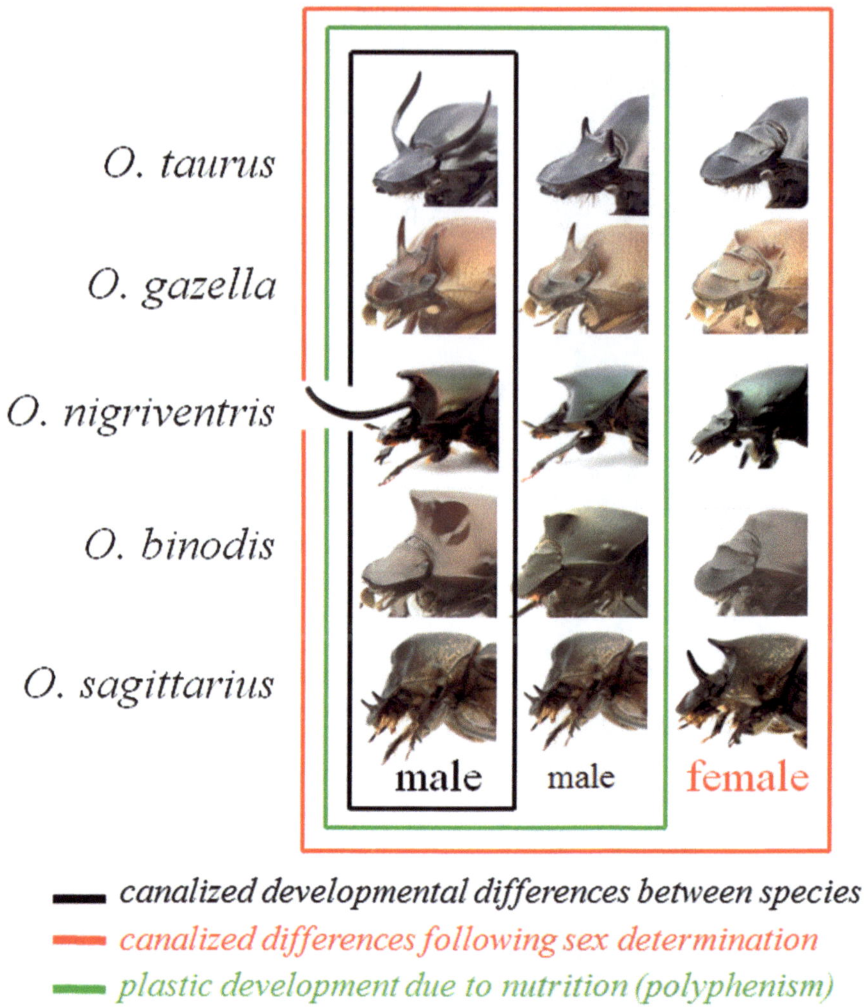

Fig. 7.2 Diversity in horned beetle morphologies and underlying causes, as illustrated by five *Onthophagus* species. Species differ in male and female morphologies due to evolved, canalized differences in developmental programs (*black box*, indicating species-specific differences in horn development in large males). Within each species, males and females exhibit more or less pronounced sexual dimorphisms due to canalized, sex-specific development following XX/XY sex-determination (*red box*). Lastly, males within each species exhibit more or less pronounced, facultative male dimorphisms, cued entirely by larval nutrition. All five species are widely available and easily maintained in captivity

To this end, a growing set of experimental tools and resources has become available over the past decade (reviewed in Kijimoto et al. 2012b). Next-generation transcriptomes of multiple species and populations and the use of custom microarrays or RNA sequencing now enable comprehensive, genome-wide comparisons of sequence and expression data, while RNA *interference* mediated transcript-depletion provides an effective and reliable means by which to examine the function of candidate pathways in a comparative, phylogenetic framework. In the next section we focus on recent efforts to utilize a subset of these resources to gain a better understanding of developmental-genetic underpinnings of sex– and body region-specific plasticity in one particular species: *Onthophagus taurus*.

7.2 Developmental Plasticity in Horned Beetles: Challenges and Approaches

Understanding the nature, costs and limits of plastic responses to environmental changes, and how and why plasticity evolves the way it does, requires a thorough understanding of the developmental genetic mechanisms that underlie the diversity of plastic responses seen in nature. In this section we focus on a study that aimed to compare and contrast the developmental-genetic mechanisms underlying diverse growth responses to nutritional variation in different body regions and sexes in *O. taurus*, a species with an extreme sexual and male dimorphism (Fig. 7.2). We begin by providing a brief background behind the rationale for this approach.

7.2.1 The Multifactorial Nature of Horned Beetle Plasticity

As introduced above, horned beetle plasticity occurs on different time scales and levels of biological organization. For example, male larvae initiate the development of future alternative hornless and horned morphologies in mid- to late larval development, with different tissues and body regions growing and differentiating at different rates. This process then gives rise to adult individuals which weeks later need to engage in facultative, morph-appropriate sneaking or fighting behaviors, etc.

Even at the same level of organization and time scale, such as growth responses to a shared nutritional gradient, plastic responses can vary dramatically. A case in point is the bull-headed dung beetle *Onthophagus taurus*, in which males and females differ substantially in body region-specific growth responses to nutritional variation. We chose to explore the regulation and diversification of plastic development by focusing on the diversity of growth responses seen among four such body regions, all of which derive from epidermal tissue: abdominal epidermis, legs, thoracic horns and head horns, for the following reasons.

In female *O. taurus*, all four body regions exhibit roughly proportional growth increases in response to increased nutrient availability during larval development. In males, in contrast, only abdominal epidermis shares the same growth response as seen in females, whereas male legs grow slightly – and male thoracic horns grow substantially – larger than their female counterparts when exposed to the same nutritional gradient. Finally, male head horn epidermis shows the most extreme growth response and exhibits explosive, non-linear growth once a certain nutrition threshold is exceeded. Combined, these sex- and body-region-specific responses to nutrient availability result in females expressing a continuous range of adult body sizes, such that large adult females represent proportionately enlarged versions of small females. Adult males exhibit the same range of nutritionally determined body sizes as do females, but instead metamorphose into two relatively discrete horned and hornless morphs.

Understanding the developmental mechanisms that enable plastic responses and their modification, be it during development as a function of sex and body region, or during evolution as a function of population- and species, requires that we realistically incorporate the complexities of plasticity into experimental designs. Focusing on *O. taurus*, we sought to execute a transcriptome-wide comparative study that would be able to robustly disentangle and analyze the transcriptional response associated with nutrition-dependent differential growth of different body regions in males and females. Before doing so, however, we had to overcome several experimental design limitations. In the next section we present a case study that hopes to demonstrate how careful experimental design can help overcome constraints imposed on transcriptional comparisons by limited resources or more generally, incomplete data, issues that will likely be common in ecological genomic studies in the future.

7.2.2 Robust Variance Estimation When Circumstances Preclude Balanced Designs

In this study we sought to characterize the nutritional responses of four different body regions in male and female *O. taurus*. Our experimental design therefore had to enable robust characterization of 16 conditions: two nutritional levels [large "L", small "S" x 2 sexes (male "M", female "F") x four body regions (abdominal epidermis "A", leg "L", thoracic horn "T" and head horn "H")]. We employed a microarray approach involving custom-made NimbleGen© arrays developed for *O. taurus* based on a comprehensive 454-transcriptome (Choi et al. 2010) to estimate the effects of nutrition, sex, and body region on gene expression. However, as explained in further detail below, executing this study using a traditional design was not possible due to cost limitations, which instead limited us to the use of only four arrays, or 48 subarrays. In the next section we first describe the general consideration that led to the final experimental design, which allowed us to greatly reduce the number of sample comparisons in our experiment without sacrificing statistical power for those contrasts we considered most relevant. We conclude this section by presenting measures that document the effectiveness of our approach.

7.2.2.1 Experimental Design: General Considerations

Completely balanced experimental designs were developed for precisely the kind of factorial experiment that is needed for studies of effects of different factors (e.g., nutrition, sex, body region) on responses of interest (e.g., gene expression levels; Box et al. 2005). Such designs aim primarily to estimate, with as little uncertainty as possible, main effects of these factors – e.g., the effect of nutrition (high vs. low) for all levels of sex and body region, the effect of sex for all levels of nutrition and body region, and the effect of body region (e.g. head horn vs. abdomen) for all levels of sex and nutrition. Secondary interest applies to two-way interactions among these main effects – e.g., the effect of nutrition for males and for females averaged over all body regions ($N \times G$), the effect of nutrition for different tissues averaged over both males and females ($N \times T$), and the effect of sex for different tissues averaged over both nutrition levels ($G \times T$). In the present circumstance, as well as in many other studies in ecological genomics, the focus was not on these main effects and two-way interactions; rather the primary interest lay in a 3-factor interaction: the effect of nutrition (N) on body-region and sex classes.

The technology of microarray experiments adds further complications: first, not all 16 treatment conditions can be examined in a single subarray. Instead, array technology limits comparisons to be executed only 2 at a time. The resulting design would need to take account of the fact that only 2 of the 16 conditions can be run in a single subarray. This situation arises in many other fields of experimentation as well, prompting the development of experimental designs with incomplete blocks (here, the "block" is the subarray). Second, array-based experiments pose the additional challenge of "dye bias" – i.e., the difference in the responses when conditions are tagged "red" vs. "green" may not be the same as when they are tagged "green" vs. "red". Taken together, these three features – interest in 3-way interaction, incomplete blocks of size 2, and dye bias – therefore challenge the classical experimental design paradigm to provide accurate, precise estimates of direct interest.

7.2.2.2 Possible Experimental Designs

To compare all 16 treatments, including dye-flips (also known as dye reversals), would require $16 \times 15 = 240$ subarrays, or 20 12-plex arrays, about five times the resources available to us at the start of this study. If we assume that the dye-bias effect does not interact with any of the other (main, 2-way, 3-way) effects of interest, then we could infer the effects of dye bias by a sensible labeling of treatments in the design (e.g., label a given treatment with Cy-3 on one subarray but with Cy-5 on another subarray), reducing the need for dye-flips among all 120 comparisons

among the 16 treatments. This assumption would enable us to reduce the number of 12-plex arrays to only ten, which is still 2.5 times the size of the study permitted with available resources. Thus, a balanced incomplete block design was considered infeasible.

Alternatively, we considered a specific type of partially balanced incomplete block design, known as a cyclic (or loop) design, in which each treatment occurs the same number of times and some pairs occur together zero, one, or two times. With a limit of only 48 subarrays, a balanced cyclic design would allow each of the 16 treatments to occur with six other treatments. Such a design initially represented a viable option. However, when compared with the design presented in the next section it became clear that a cyclic design would result in lower precision for the eight nutrition contrasts that were of primary interest in this study.

7.2.2.3 Alternative Experimental Design Options Through Application of the Square Combining Table

The *square combining table* (SCT) is a type of analysis of variance, specifically developed for data arising from pairwise comparisons where the relevant data can be computed as the difference between two states (Godfrey 1985). The SCT was originally developed as an analytical tool to enable robust variance decomposition in instances in which conventional analysis of variance approaches become unreliable, for instance when data are missing (Godfrey 1985). The effectiveness of the SCT in fitting tables with missing data values is a function of the difference data that do exist, and the degree to which they allow repeated, independent estimation of missing data. When applied correctly, the SCT enables a standard least-squares analysis of variance in the face of missing data and maintains formal orthogonality among row and column combinations, enabling simple comparisons among independent contrasts. Development, application, and limits of the SCT are discussed in detail in Godfrey (1985). While the SCT was developed originally as an analysis rather than as a design tool, it can be used to create experimental designs for comparing treatments within a block such as those that arise with microarray experiments and more generally any experimental approach where data are missing due to design constraints or partial experimental failure. Specifically, by arranging effect types *a priori* from most to least interesting from the viewpoint of the investigator, the SCT can be used to prioritize direct and indirect contrasts, as well as to determine the appropriate number of replicate observations for specific effects. Below we explain how we used the SCT to guide design and analysis of our experiment.

7.2.2.4 Prioritizing Direct Comparisons

Cost considerations limited our experimental design to four 12-plex *NimbleGen*© arrays, i.e. a total of 48 identical subarrays, for a maximum of 48 pairwise comparisons. At the same time, not all pairwise comparisons were considered equally biologically meaningful. Instead, the primary focus of this experiment was to estimate the effect of nutrition (i.e. comparing "L"(= "High nutrition") and "S"(= "low nutrition")) in each of eight sex x treatment conditions. Specifically, comparisons of the form LM-SM and LF-SF for all four body regions were of primary interest. Within this group, comparisons for head horns (LMH-SMH) and thoracic horns (LMT-SMT) were considered especially relevant, given the elevated and extreme levels of nutritional plasticity seen in developing thoracic and head horns, respectively. In contrast, characterization of the effect of sex in large (high nutrition) and small (low nutrition) individuals (i.e. LF-LM and SF-SM for all four body regions) were considered of secondary importance. Lastly, of least importance were comparisons among different sizes and sexes, i.e. comparisons of the form LM-SF and SM-LF for all four body regions.

With these considerations in mind we prioritized pairwise, direct comparisons (i.e. hybridizations onto the *same* subarray) such that (a) all possible body region comparisons would be executed for LM (i.e. high nutrition = large males), *including* dye flips, (b) all possible body region comparisons would be executed for SM

Table 7.1 Comparing variances of nutrition differences for two designs (SCT, Cyclic) by sex (F = female, M = male) and body region (A =abdomen, H = head horn, L = leg, T = thoracic horn d = nutrition difference; for example dFA = nutrition difference in female abdomen = LFA-SFA)

	dFA	dMA	dFH	dMH	dFL	dML	dFT	dMT
SCT-design	0.19365	0.17322	0.14034	0.13004	0.19409	0.19612	0.14036	0.13247
Cyclic design	0.15699	0.15699	0.15699	0.15699	0.15699	0.15699	0.15699	0.15699
Ratio	1.23350	1.10337	0.89391	0.82828	1.23631	1.24923	0.89403	0.84381

(low nutrition males) and LF (high nutrition females) *excluding* dye flips, and (c) all possible large-small comparisons (i.e. LM-SM, LF-SF, LM-LF, SM-SF) for each body region, including dye flips for LM-SM and LF-SF, but not LM-LF and SM-SF comparisons. However, this design did not permit direct tissue comparisons in small (=low nutrition) females, and subsequent analyses showed that corresponding estimates would not be obtainable from linear combinations of subsets of other observations in this design. Because the overall design offered more than adequate estimates of dye bias, we thus re-allocated six comparisons originally designated as replicate hybridizations (involving dye flip) to instead estimate body region differences in low-nutrition females. This adjusted design included a total of 33 red-green and 11 green-red hybridization among replicate hybridizations, which was deemed adequate for estimating dye bias while enabling at least some investigation into body region differences in low-nutrition females. Not surprisingly, the resulting final design was not perfectly balanced. Among 16 conditions, 12 are represented 5–6 times, while one each is represented four, seven, eight, and nine times respectively. This lack of balance had little effect on the precision of the estimates, and in fact enabled greater precision for some comparisons, as discussed below.

7.2.2.5 Effectiveness of SCT-Based Experimental Design

The effectiveness of our approach can be quantified by comparing the estimated variances in the effects of interest (the eight nutrition comparisons in the eight sex x body-region classes) that arise from both a conventional cyclic design (de Mendiburu 2013) and the SCT-based design. Both design matrices as well as details on variance computation are detailed in the Appendix. Using conventional least-squares estimates for the mean gene expression in each of the 16 conditions types, one can calculate not only the 16 means but also estimates of their precision, which then translate into estimates of precision for the contrasts of interest. In general, the higher the precision (i.e., the lower the variance of the estimated effect), the better the design. Table 7.1 compares the variances of the eight nutrition differences among the sex x body region categories, for both the SCT-based design and for a conventional cyclic design. The cyclic design results in similar variances for all pairs of differences, but many of which are of no interest for our study's purposes (e.g., LMA-SFA, LFH-SMT). In contrast, even though the SCT-based design gives slightly larger variances for the estimated nutrition differences among male and female abdomen and legs, the variance for the nutrition differences of greatest interest, i.e. head horns and thoracic horns for both males and females, are 11–18 % smaller compared to the cyclic design. Thus, although the cyclic design has better balance in the variances across all pairs, the SCT-based design has lower variances for the contrasts of greatest interest.

More generally, the SCT-based design allowed us to fit our experimental objectives within the constraints imposed by array technology as well as the limitations of our budget, while maximizing our ability to investigate the transcriptional response to nutritional variation across a diversity of traits. Our study illustrates the need for the design to take into consideration numerous sources of variability that can arise in any study, though clearly the exact issues to consider will depend on the measurement technology. Derivations for these calculations, as well as the equations for

estimating the contrasts and their estimates of precision, are detailed in the Appendix. Additional analyses that assess significance of the nutrition differences for all 42,010 contigs and their distribution across body regions and sexes are currently being conducted and will be reported elsewhere.

7.3 Genomics of Horned Beetle Plasticity: Recent Insights and Future Opportunities

In this last section we would like to summarize important recent studies and highlight several additional opportunities that exist in the study of horned beetles that would enable a further integration of developmental genetic and ecological genomic perspectives of the mechanisms and consequences of plasticity, while taking advantage of some of the statistical methodology discussed above.

7.3.1 Microevolution of Plasticity: Integrating Ecological Genomics and Behavioral Ecology

A major goal in ecological and evolutionary genomics is the identification of genes underlying ecologically important phenotypes. In the horned beetle system described above, both sexual and male dimorphisms involve marked differences in morphology, behavior, and reproductive tactics. Recall that while sexual dimorphisms are the product of canalized sex-specific differentiation, male dimorphisms are the product of nutritional variation experienced during larval development. In other words, genetically related males, depending on the nutritional conditions experienced during larval development, may develop into a large, horned (major) morph that fights for access to females or a small, hornless (minor) morph that sneaks copulations. The precise scaling relationship between body size and horn length is generally species specific and diagnostic, suggesting that this nutritional *polyphenism* is under tight genetic control (Emlen et al. 2005).

Importantly, species readily diverge in important components of this male nutritional polyphenism and the corresponding relationship between body size and horn length. For example, considerable differences exist among species in the morphological disparity between alternative morphs, manifest in species-specific differences in the average horn length of minor and major morphs, or the *amplitude* of the body size – horn length *allometry* (Valena and Moczek 2012). In Fig. 7.3a, for example, *O. nigriventris* males exhibit the most dramatic disparity between large and small morphs, followed by *O. taurus* and *O. gazella*. *O. sagittarius*, in contrast, has secondarily lost the male dimorphism; in this species male horns are overall rather small and scale linearly with body size. These differences among closely related species provide a rich context for comparative studies on the evolution of scaling relationships. Comparative genomics utilizing whole-transcriptome sequences of at least three *Onthophagus* species are currently being executed to explore the genetic underpinnings of horn polyphenisms and their diversification. Such genome-wide studies have the power to discover unexpected targets of selection as well as test predictions based on previous work.

For example, several developmental and physiological mechanisms have the potential to drive differences in scaling relationships, e.g. by altering the rate or duration of cell proliferation during growth (Emlen and Allen 2003) via changes in the expression and/or function of signaling molecules and transcription factors that coordinate outgrowth formation. A number of studies have already demonstrated that a diversity of patterning mechanisms and growth regulators normally involved in appendage formation have been co-opted to function in horn development (reviewed in Kijimoto et al. 2012b). Alternative, or additional mechanisms involve evolutionary changes in the function of endocrine regulators, such as juvenile hormone (Moczek and Nijhout 2002, see below) and in particular insulin signaling (Shingleton et al. 2005; Snell-Rood and Moczek 2012; Emlen et al. 2012). Two recent studies on species belonging to two distinct groups of horned beetles, which

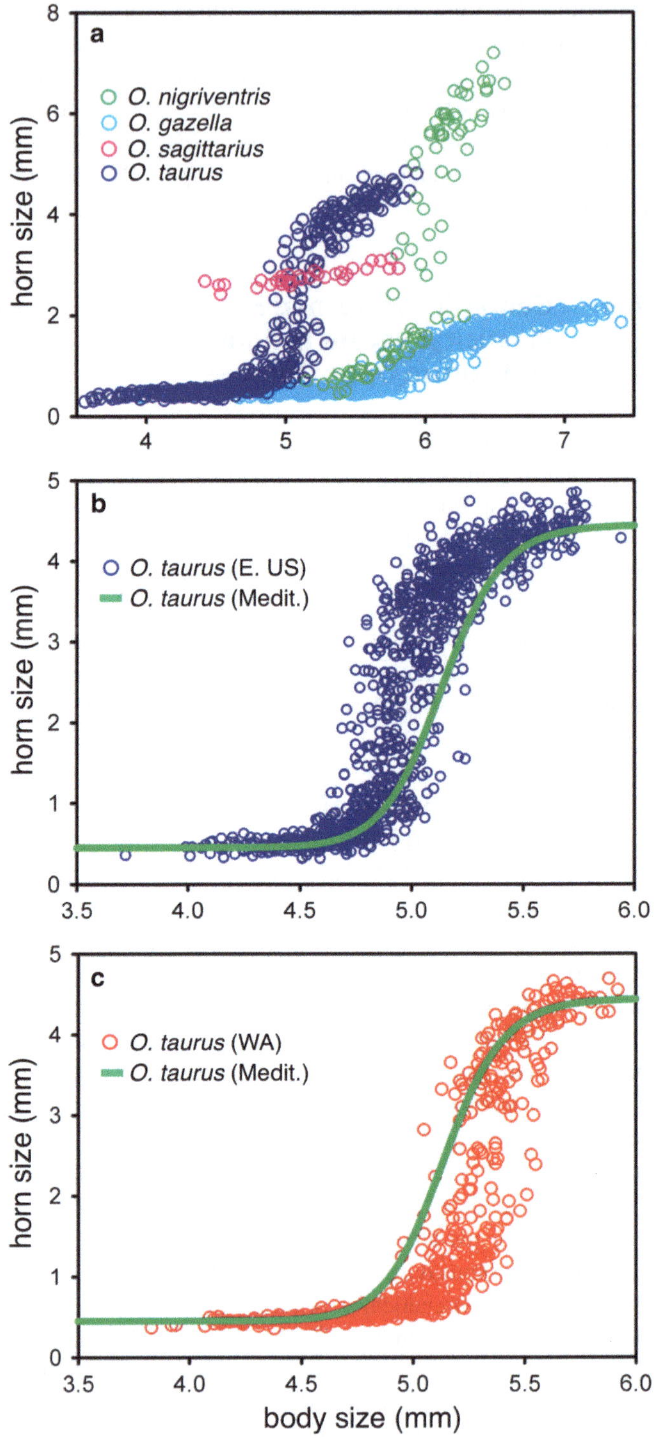

Fig. 7.3 Macro- and microevolutionary divergences in body size-horn length scaling relationships. (**a**) Body size – horn length scaling relationships highlight distinct degrees of male polyphenism among four *Onthophagus* species. (**b**, **c**) Diversity in body size thresholds separating hornless and horned male morphs in rapidly evolving introduced populations of *O. taurus* in the Eastern United States (**b**, *blue circles*) and Western Australia (**c**, *red circles*) relative to the average scaling relationship seen in ancestral Mediterranean populations (**b** and **c**, *green line*)

evolved horns independently (Dynastinae; *Trypoxylus* (*Allomyrina*) *dichotoma*; Emlen et al. 2012; Scarabaeinae; *Onthophagus nigriventris*; Snell-Rood and Moczek 2012), both implicate aspects of insulin signaling in the regulation of organ-specific sizes across nutritional gradients. For instance, results suggest that the insulin receptor may be particularly important in

disproportionate growth responses in plastic traits such as horns (Emlen et al. 2006, 2012), while the gene FOXO acts as a potential repressor of growth in traits such as genitalia (Snell-Rood and Moczek 2012). Together, these studies complement developmental work being performed in more traditional models systems such as *Drosophila* (Kopp 2011; Tang et al. 2011) because they find comparable results, but also show how these mechanisms may be co-opted to regulate differences between organs in their sensitivity to variation in nutritional conditions.

Similarly, horned beetles are beginning to provide important complementary insights into the epigenetic basis of developmental plasticity. Adaptive plastic responses to nutrition have been hypothesized to be regulated by environmentally-induced, heritable changes "above" the level of DNA, such as DNA methylation or histone acetylation (Junien et al. 2005; Burdge et al. 2007; Gilbert and Epel 2009). Recent work on horned beetles has shown that, like honey bees, (but unlike *Drosophila* and *Tribolium*), horned beetles possess the complete methylation machinery (Choi et al. 2010; reviewed in Valena and Moczek 2012), that methylation occurs, and that a fraction of it may underlay adaptive plastic responses to nutritional variation experienced during development (Snell-Rood et al. 2013).

Another divergence pattern common among species, as well as populations, involves the point of inflection of the body size-horn length allometry, or the threshold body size that separates small, minor (sneaking) males from large, major (fighting) males. A case in point are exotic populations of the beetle *O. taurus* introduced to Western Australia and the Eastern United States, which in less than 50 years have evolved highly divergent threshold body size (Fig. 7.3b, c). Threshold sizes have diverged in *opposite* directions relative to the ancestral, Mediterranean population, and to a degree that rivals divergences seen among closely related species (Fig. 7.3b, c) (Moczek and Nijhout 2003). Comparative ecological and behavioral studies suggest that threshold divergences have been driven by differences in the intensity of intra– and interspecific competition for breeding opportunities, which resulted in relatively low levels of male-male competition for females in the Eastern US but extremely high levels in Western Australia. In turn, these ecological differences may have resulted in selection for genotypes that express horns at relatively small body sizes to be favored in the US, but to be selected against in Western Australia.

Past as well as ongoing studies suggest that divergences among exotic *O. taurus* populations are not limited to male horn development, but also include larval physiology (Australian larvae require much longer to complete larval development and exhibit reduced sensitivity to hormonal manipulations; (Moczek and Nijhout 2002)) and female fertility and fecundity (Beckers and Moczek, unpublished). This differentiation among populations presents an excellent opportunity to study the early stages of polyphenism evolution, its developmental underpinnings, ecological causes, and its interactions with other diversifying traits. Integrating genome-wide data on genetic diversity (from next-gen sequencing efforts on W-Australian and Eastern US populations currently underway), gene expression (such as those described above for the multifactorial array experiment), and gene function (from a growing body of RNAi screens) aims to reveal mechanisms underlying such rapid morphological and behavioral evolution. Further, integrating data on patterns observed among populations with those observed among species can reveal processes that occur in parallel on micro- and macro-evolutionary scales. We predict similar functional targets related to cell growth and proliferation described above to also underlie trait diversification between rapidly evolving populations within a species. Such mechanisms may further interact with pheromones or other chemical cues that signal local densities of competing individuals, and thus act as a proxy of the degree of intra- and inter-specific competition (Butcher et al. 2007). Similarities in the genetic targets underlying trait diversification within and between species will reveal the genetic machinery that is repeatedly accessed at multiple evolutionary time-scales, while differences may indicate potentially novel evolutionary targets.

7.3.2 Macroevolution of Plasticity and the Diversification of Male and Sexual Dimorphisms

Male horn dimorphism is cued by variation in nutrition, causing well-fed male larvae to develop into horned males, whereas larvae subject to sub-optimal feeding conditions metamorphose into hornless males (Emlen 1994; Moczek 1998). In contrast, sex-specific development of horned males and hornless females is strictly tied to somatic sex determination following – most likely- a traditional XX/XY sex determination scheme (Angus 2008). However, on a more general level, both processes have much in common: in both cases the same genome (or nearly same genome if one includes the modest contribution of the Y chromosome) is used to allow developmental processes to generate very different phenotypic outputs depending on cues experienced either late in larval development (such as nutrition in the case of male dimorphisms) or very early during embryonic differentiation (as is the case for the somatic sex-determination cascade). Remarkably, recent work (Kijimoto et al. 2012a) has shown that these general similarities also extend to the molecular and developmental genetic level.

In insects, somatic sex determination involves the gene *doublesex* (*dsx*) as the terminal gene in the sex determination pathway that regulates the sex-limited expression of downstream target genes, which in turn enable sexually dimorphic development and behavior across diverse insects (Fig. 7.4a; Sanchez 2008). Even though the sex-determination pathway upstream of *dsx* is divergent across insect orders, the basic genetic architecture and function of *dsx* are highly conserved (Shukla and Nagaraju 2010). In particular, in all insects examined so far *dsx* structure and function involve the expression of male- and female-specific *Dsx* isoforms generated through alternative splicing (Fig. 7.4a).

Recent microarray-based transcriptional profiling of *Onthophagus* development suggested that, in line with previous studies, differential expression of male and female *dsx*-isoforms may underlie sex-specific differentiation in horned beetles. Unexpectedly, however, the same studies also raised the possibility that aspects of the same machinery have become co-opted to generate morph-specific development within males (Kijimoto et al. 2012a).

A subsequent analysis of the expression and function of alternate *dsx* isoforms yielded three major conclusions: first, alternative *dsx* transcripts indeed promote the presence of horns in males but inhibit their formation in females (Fig. 7.4b). As such, beetle horn development joins a growing list of secondary sexual traits whose sex-specific expression is regulated by *dsx*. Second, within males, the *level* of *dsx* expression appears to have evolved to function as a regulator of relative horn size, regulated in turn by larval nutrition. If the expression of the male *dsx* isoform is knocked down in *O. taurus*, nutrition-sensitive horn development is greatly reduced (Kijimoto et al. 2012a; Fig. 7.4b). Lastly, when these studies were replicated in a second species, *Onthophagus sagittarius*, it became clear that *dsx* represents a nexus in the evolution and diversification of both sex- and morph-specific development: *O. sagittarius* is an unusual, closely related and recently derived, species that exhibits a *reversed* sexual dimorphism: males have lost the ancestral male dimorphism and only develop small paired horns in front of their heads, whereas females have gained conspicuous medial head and thoracic horns. Sequencing experiments revealed that *O. sagittarius* expresses male- and female-specific *dsx* transcripts with splicing patterns and translated protein sequences highly similar to those in *O. taurus*, i.e. *dsx* expression appeared conserved across both species. However, comparative functional studies showed that *O. sagittarius dsx* functions have expanded beyond their conserved roles to include both modified as well as novel functions in the regulation of horn position, shape, and size (Kijimoto et al. 2012a).

Manipulations of *dsx* function are highly robust across species, have high penetrance, and yield long-lived adults, which offers interesting opportunities to further explore the developmental genetic mechanisms of plasticity, morphological integration, and plasticity evolution in

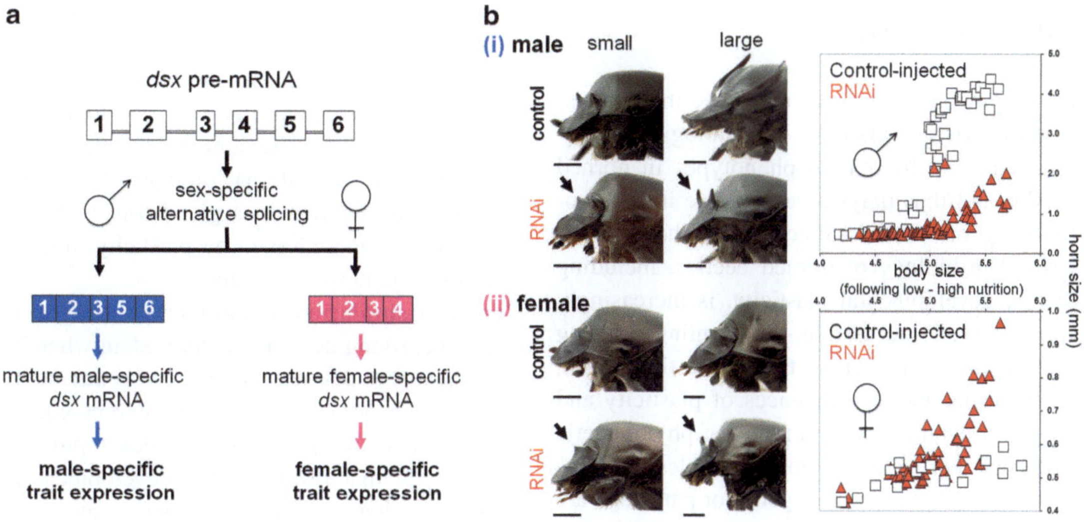

Fig. 7.4 Structure and function of the gene *doublesex* (*dsx*) and its role in the development of sexual and male dimorphisms. (**a**) In all insects examined so far *dsx* structure and function involve the expression of male- and female-specific Dsx isoforms generated through alternative splicing. Shown here is a schematic representation of *dsx* structure in *Drosophila melanogaster*. White-numbered boxes represent exons, whereas blue and pink-numbered boxes represent gene products in males and females, respectively. Sex-specific alternative splicing (presence or absence of exon 4) results in sex-specific trait expression. (**b**) Effects of *dsx* double-stranded (ds) RNA injection (RNAi) on horn development in adult *O. taurus* (i) males and (ii) females. Left: Representative animals obtained after mock (control) injections (*top row*) and *dsx* dsRNA injections, respectively (*bottom row*). Small individuals are shown on the *left* and large individuals on the *right*. *Filled arrows* indicate locations of head horn development in RNAi individuals. Right: Bivariate plots of body size (x-axis) and head horn length (y-axis) for (i) male and (ii) female *O. taurus*. Control and RNAi individuals are plotted as white squares and red triangles, respectively. *dsx* dsRNA injections substantially reduced nutrition-responsive horn development in males but induced it in females (Modified after Kijimoto et al. 2012a)

nature. For example, efforts are under way to investigate the degree to which morphological plasticity (horned vs. hornless morphs), behavioral plasticity (sneaking vs. fighting) and sex-specific behaviors (e.g. courting) are co-regulated by *dsx* via a detailed behavioral analysis of *dsx*-deficient males and females. Similarly, the conservation of *dsx* expression across species on one side, and the diversification of *dsx* function on the other, invite a comparative analysis of *dsx*'s target repertoire. Specifically, experiments are being conducted utilizing next-gen sequencing approaches to identify which genes change expression following *dsx*-knockdown as a function of body region, sex, nutritional conditions, and ultimately, species. Similar to the analysis of nutrition-dependent gene expression across diverse conditions described above, this effort will make use of the same statistical toolbox to generate robust results in the face of incomplete data.

Lastly, the same approaches may allow us to investigate the possible involvement of *dsx* and its target repertoire in the context of *threshold evolution*, for instance as detailed for exotic *O. taurus* populations in the preceding section. Recall that exotic populations have diverged heritably with respect to the body size (= larval nutrition) threshold that separates hornless from horned developmental fates, a developmental decision we now know is at least in part regulated via the differential expression of male-specific *dsx*-isoforms. Collectively, these efforts will help inform our understanding of the similarities and differences in the mechanisms by which diversity is generated within and across sexes, and the evolutionary lability or conservation of these mechanisms.

7.4 Conclusion

Developmental plasticity mediates the expression of a rich diversity of morphological, physiological, and behavioral phenotypes in horned beetles and thus plays a central role in the evolutionary and behavioral ecology of these organisms. The biology of horned beetles, including that of developmental plasticity, is increasingly experimentally accessible, representing growing opportunities with which to explore the causes, mechanisms and consequences of plasticity and plasticity evolution over a range of phylogenetic distances. For example, horned beetles allow us to address whether certain genes or pathways are biased or specific in their expression to different nutritional conditions, whether they are subject to relaxed selection, whether these patterns are shared across species and/or types of plastic responses, and whether the developmental-genetic underpinnings of plastic responses have enabled phenotypic diversification or innovation in other developmental contexts. At the same time, many horned beetle species are easily and inexpensively maintained in captivity, and several of the most interesting species are broadly distributed geographically. Research conducted thus far has only begun to scratch the surface of what horned beetles can teach us about the interplay between development, environment, phenotypic variation, and evolution, and we hope that this chapter will encourage future research efforts into the biology of horned beetles and beetle horns.

Acknowledgments We thank the editors for the opportunity to contribute this chapter and two anonymous reviewers for helpful comments on an earlier draft. Research presented here was made possible in large part through funding from the National Science Foundation.

Glossary: Important Concepts at the Interface of Ecological Genomics and Developmental Plasticity

Phenotypic plasticity The capacity of a genotype or individual to express different phenotypes across a range of environments. Adaptive plasticity results in maintenance of high performance across these conditions.

Canalization The capacity of a developmental system to produce the same phenotype despite variation in the external or internal environment. Canalization in a given trait (e.g., fitness) may be underlain by plasticity in other traits (e.g., horn expression).

Genetic assimilation Evolutionary changes to an underlying developmental system whereby a phenotype that was initially environmentally induced becomes constitutively expressed.

Polyphenism A specific form of developmental plasticity that results in the development of discrete alternate phenotypes in response to an environmental cue.

Relaxed selection Any form of lower selection intensity relative to a population, trait or ancestral state subject to higher selection intensity. Relaxed selection encompasses both reduced purifying selection and positive selection, leading to a lower likelihood of loss of deleterious alleles and fixation of beneficial alleles, respectively. Also referred to as relaxed selective constraint.

Allometry The relationship of traits to body size. Changes in scaling or allometric relationships can be a method of describing a plastic response of a trait in response due to variation in nutritional effects on body size.

Square combining table A statistical method for pairwise comparisons that allows researchers to focus on specific comparisons of interest and cope with an unbalanced design.

Appendix: Effectiveness of SCT-Based Experimental Design: Computational Analysis

As indicated in Sect. 7.2.2.5, the effectiveness of our approach can be quantified by comparing the estimated variances in the effects of interest (the eight nutrition comparisons in the eight sex x body-region classes) that arise from both a conventional cyclic design (de Mendiburu 2013)

and the SCT-based design (developed with reference to the Square Combining Table, described in Godfrey 1985). Both design matrices, shown at the end of this appendix, can be expressed with 48 rows (corresponding to the 48 subarrays) and 16 columns (for the 16 body region x sex x nutrition types), whose elements x_{ik}, $i = 1, \ldots, 48$ and $k = 1, \ldots, 16$, are 0 (type not on subarray), 1 (type is labeled "red"), or –1 (type is labeled "green"). The log2-response on the first contig from the first array and first subarray reflects $\log_2(LFL/LFS) = \log_2(LFL) - \log_2(LFS)$ or $\log_2(AML/AMS) = \log_2(AML) - \log_2(AMS)$, depending on whether X is the SCT-based design, X_{SCT}, or the cyclic design, X_{cyc}, respectively. With multiple observations on LFL, LFS, AML, AMS (etc.), we need to estimate a reliable summary of the observations for each of the 16 conditions, taking into account that LFL may be paired with LFS for one subarray but with LML on another subarray (and likewise for all beetle types). If we denote the 48 gene responses (on the 48 subarrays) for a single contig (gene) by the vector y_j (here, $j = 1, \ldots, 42{,}010$), the design matrix by X (same design matrix for all contigs), and the vector of the 16 means of the responses from the 16 types (two nutrition levels × 2 sexes × 4 body regions) for contig j by M_j, then we can express the observed responses y_j as a linear model as follows:

$$y_j = X_{SCT} M_j + (error) \ or \ y_j = X_{cyc} M_j + (error)$$

depending upon whether the design was the SCT-based design or the cyclic design. Whichever X design is used, the least-squares estimate of the vector of the 16 means, denoted as m_j, is calculated in the usual fashion:

$$m_j = \left(X'X\right)^{-1} X' y_j$$

In fact, both design matrices have a redundancy in them, in that the last column can be obtained from a linear combination of the first 15 columns, so the matrix inverse $(X'X)^{-1}$ is actually calculated as $\left(\tilde{X}'\tilde{X}\right)^{-1}$ where $\tilde{X}$ is the design matrix X without the last column. (This problem formulation is equivalent to setting the last mean, SMT, equal to zero. Because we are interested in differences between two type means at a time, the actual value for SMT will be irrelevant.) Conventional least squares analysis (i.e., m_j is the vector of 16 means that minimizes the sum of the squared error terms from this linear model) allows us to calculate the variances of the estimated means m_j as the diagonal of the matrix $\left(\tilde{X}'\tilde{X}\right)^{-1} s_j^2$, where s_j^2 provides a measure of the variability in the error terms from the linear model and is estimated from the mean of the squared error terms:

$$s_j^2 = \sum\nolimits_{k=1}^{48} \left(y_{jk} - \sum\nolimits_{i=1}^{15} x_{i=1} m_{jk} \right)^2 / 33$$

(the denominator "33" arises from 48 differences less 15 means being estimated; recall that one mean, SMT, is set to zero). In fact, we are not interested in the variances of the 15 means (SFA, ..., LMT), but rather in the variances of the eight nutrition differences LFA-SFA, ..., LMT-SMT, which can be expressed in matrix terms as Cm_j, where the (8 rows × 16 columns) matrix C is:

	LFA	SFA	LMA	SMA	LFH	SFH	LMH	SMH	LFL	SFL	LML	SML	LFT	SFT	LMT	SMT
LFA-SFA:	1	−1	0	0	0	0	0	0	0	0	0	0	0	0	0	0
LMA-SMA:	0	0	1	−1	0	0	0	0	0	0	0	0	0	0	0	0
LFH-SFH:	0	0	0	0	1	1	0	0	0	0	0	0	0	0	0	0
LMH-SMH:	0	0	0	0	0	0	1	−1	0	0	0	0	0	0	0	0
LFL-SFL:	0	0	0	0	0	0	0	0	1	−1	0	0	0	0	0	0
LML-SML:	0	0	0	0	0	0	0	0	0	0	1	−1	0	0	0	0
LFT-SFT:	0	0	0	0	0	0	0	0	0	0	0	0	1	−1	0	0
LMT-SMT:	0	0	0	0	0	0	0	0	0	0	0	0	0	0	1	−1

Table 7.2 Cyclic design

	LFA	SFA	LMA	SMA	LFH	SFH	LMH	SMH	LFL	SFL	LML	SML	LFT	SFT	LMT	SMT
1	0	0	1	−1	0	0	0	0	1	0	0	0	0	0	0	0
2	0	0	0	0	0	0	0	0	1	−1	0	0	0	0	0	0
3	−1	0	0	0	0	0	0	0	0	0	0	0	0	0	0	1
4	0	0	0	0	0	0	0	0	0	0	0	0	0	1	−1	0
5	0	0	0	0	0	0	0	0	0	0	1	−1	0	0	0	0
6	0	0	0	0	0	0	0	0	0	0	0	0	1	−1	0	0
7	0	0	0	0	0	0	0	1	−1	0	0	0	0	0	0	0
8	1	−1	0	0	0	0	0	0	0	0	0	0	0	0	0	0
9	0	0	0	1	−1	0	0	0	0	0	0	0	0	0	0	0
10	0	1	−1	0	0	0	0	0	0	0	0	0	0	0	0	0
11	0	0	0	0	0	0	0	0	0	1	−1	0	0	0	0	0
12	0	0	0	0	0	0	0	0	0	0	0	1	−1	0	0	0
13	0	0	0	0	1	−1	0	0	0	0	0	0	0	0	0	0
14	0	0	0	0	0	1	−1	0	0	0	0	0	0	0	0	0
15	0	0	0	0	0	0	1	−1	0	0	0	0	0	0	0	0
16	0	0	0	0	0	0	0	0	0	0	0	0	0	0	1	−1
17	0	1	0	0	0	−1	0	0	0	0	0	0	0	0	0	0
18	−1	0	0	0	0	0	0	0	0	0	0	0	1	0	0	0
19	0	0	0	1	0	0	0	−1	0	0	0	0	0	0	0	0
20	0	0	0	0	1	0	0	0	−1	0	0	0	0	0	0	0
21	0	0	1	0	0	0	−1	0	0	0	0	0	0	0	0	0
22	0	0	0	0	0	1	0	0	0	−1	0	0	0	0	0	0
23	1	0	0	0	−1	0	0	0	0	0	0	0	0	0	0	0
24	0	0	0	0	0	0	0	0	0	0	0	1	0	0	0	−1
25	0	0	0	0	0	0	1	0	0	0	−1	0	0	0	0	0
26	0	0	0	0	0	0	0	0	0	1	0	0	0	−1	0	0
27	0	0	0	0	0	0	0	0	0	0	1	0	0	0	−1	0
28	0	−1	0	0	0	0	0	0	0	0	0	0	0	1	0	0
29	0	0	0	0	0	0	0	1	0	0	0	−1	0	0	0	0
30	0	0	0	−1	0	0	0	0	0	0	0	0	0	0	0	1
31	0	0	0	0	0	0	0	0	1	0	0	0	−1	0	0	0
32	0	0	−1	0	0	0	0	0	0	0	0	0	0	0	1	0
33	0	−1	0	0	0	0	0	0	0	0	1	0	0	0	0	0
34	0	0	0	1	0	0	0	0	0	0	−1	0	0	0	0	0
35	0	0	−1	0	0	0	0	0	0	0	0	1	0	0	0	0
36	0	0	0	0	1	0	0	0	0	0	0	−1	0	0	0	0
37	0	0	0	0	0	0	0	0	1	0	0	0	0	0	0	−1
38	0	0	1	0	0	0	0	0	0	−1	0	0	0	0	0	0
39	0	0	0	0	0	0	1	0	0	0	0	0	0	−1	0	0
40	0	1	0	0	0	0	0	0	−1	0	0	0	0	0	0	0
41	0	0	0	0	−1	0	0	0	0	0	0	0	0	1	0	0
42	0	0	0	0	0	−1	0	0	0	0	0	0	0	0	1	0
43	0	0	0	0	0	1	0	0	0	0	0	0	−1	0	0	0
44	0	0	0	0	0	0	0	1	0	0	0	0	0	0	−1	0
45	−1	0	0	0	0	0	0	0	0	1	0	0	0	0	0	0
46	1	0	0	0	0	0	0	−1	0	0	0	0	0	0	0	0
47	0	0	0	−1	0	0	0	0	0	0	0	0	1	0	0	0
48	0	0	0	0	0	0	−1	0	0	0	0	0	0	0	0	

Table 7.3 Design based on square combining table

	LFA	SFA	LMA	SMA	LFH	SFH	LMH	SMH	LFL	SFL	LML	SML	LFT	SFT	LMT	SMT
1	0	0	0	0	0	0	0	0	1	−1	0	0	0	0	0	0
2	0	−1	0	0	0	1	0	0	0	0	0	0	0	0	0	0
3	0	0	0	0	−1	0	1	0	0	0	0	0	0	0	0	0
4	0	0	0	0	0	0	0	0	0	0	0	0	1	−1	0	0
5	0	−1	0	0	0	0	0	0	0	0	0	0	0	1	0	0
6	0	0	0	0	0	0	0	0	0	0	1	−1	0	0	0	0
7	0	0	0	0	0	0	0	0	0	0	0	0	0	0	1	−1
8	−1	0	1	0	0	0	0	0	0	0	0	0	0	0	0	0
9	0	0	0	0	1	−1	0	0	0	0	0	0	0	0	0	0
10	0	0	0	0	0	0	0	0	−1	0	1	0	0	0	0	0
11	0	0	0	0	0	0	1	−1	0	0	0	0	0	0	0	0
12	0	0	0	0	0	0	0	0	0	0	0	0	−1	0	1	0
13	0	0	0	0	0	0	0	0	0	0	−1	0	0	0	1	0
14	1	0	0	0	0	0	0	0	−1	0	0	0	0	0	0	0
15	0	0	0	0	0	0	1	0	0	0	−1	0	0	0	0	0
16	−1	0	0	0	1	0	0	0	0	0	0	0	0	0	0	0
17	0	0	−1	0	0	0	1	0	0	0	0	0	0	0	0	0
18	0	0	0	0	1	0	0	0	−1	0	0	0	0	0	0	0
19	0	0	1	0	0	0	0	0	0	0	−1	0	0	0	0	0
20	0	0	0	0	0	0	1	0	0	0	0	0	0	0	−1	0
21	0	0	0	0	0	1	0	0	0	0	0	0	0	−1	0	0
22	1	0	0	0	0	0	0	0	0	0	0	0	−1	0	0	0
23	0	0	0	0	0	0	0	0	−1	0	0	0	1	0	0	0
24	0	0	0	0	1	0	0	0	0	0	0	0	−1	0	0	0
25	0	0	0	0	0	−1	0	1	0	0	0	0	0	0	0	0
26	0	−1	0	1	0	0	0	0	0	0	0	0	0	0	0	0
27	0	0	0	0	0	0	0	0	0	0	0	0	0	0	−1	1
28	−1	1	0	0	0	0	0	0	0	0	0	0	0	0	0	0
29	0	0	0	0	0	0	0	0	0	1	0	0	0	−1	0	0
30	0	0	0	1	0	0	0	0	0	−1	0	0	0	0	0	0
31	0	0	0	0	0	0	0	0	0	0	0	0	0	−1	0	1
32	0	0	0	0	0	0	0	0	0	0	0	0	−1	1	0	0
33	0	0	−1	1	0	0	0	0	0	0	0	0	0	0	0	0
34	0	0	0	0	−1	1	0	0	0	0	0	0	0	0	0	0
35	0	0	0	0	0	0	−1	1	0	0	0	0	0	0	0	0
36	0	1	0	0	0	0	0	0	0	−1	0	0	0	0	0	0
37	0	0	0	0	0	0	0	1	0	0	0	0	0	0	0	−1
38	0	0	0	0	0	0	−1	0	0	0	0	0	0	0	1	0
39	0	0	0	0	0	0	0	1	0	0	0	−1	0	0	0	0
40	0	0	0	−1	0	0	0	0	0	0	0	0	0	0	0	1
41	0	0	−1	0	0	0	0	0	0	0	0	0	0	0	1	0
42	0	0	0	0	0	−1	0	0	0	1	0	0	0	0	0	0
43	0	0	0	0	0	0	−1	0	0	0	1	0	0	0	0	0
44	0	0	0	1	0	0	0	0	0	0	0	−1	0	0	0	0
45	0	0	1	0	0	0	−1	0	0	0	0	0	0	0	0	0
46	0	0	0	0	0	0	0	0	0	0	0	1	0	0	0	−1
47	0	0	0	0	0	0	0	0	0	0	1	0	0	0	−1	0
48	0	0	0	−1	0	0	0	1	0	0	0	0	0	0	0	0

(recall that SMT is set to zero so again we let $\tilde{C}$ denote the above matrix without the last column). The variances of these eight contrasts are on the diagonal of the matrix,

$$\tilde{C}\left[\left(\tilde{X}'\tilde{X}\right)^{-1}\right]\tilde{C}'s_j^2$$

Thus, we can compare the eight values on the diagonal of this matrix where $\tilde{X}$ is either $\tilde{X}_{SCT}$ or $\tilde{X}_{cyc}$, as shown in Tables 7.2 and 7.3.

References

Alaux C, Sinha S, Hasadsri L, Hunt GJ, Guzman-Novoa E, DeGrandi-Hoffman G, Uribe-Rubio JL, Southey BR, Rodriguez-Zas S, Robinson GE (2009) Honey bee aggression supports a link between gene regulation and behavioral evolution. Proc Natl Acad Sci USA 106:15400–15405

Angus RB (2008) A chromosomal analysis of the Onthophagus similis-opacicollis-fracticornis species group (Coleoptera: Scarabaeidae). Tijdschrift voor Entomologie 151(2):235–244

Arrow GJ (1951) Horned beetles : a study of the fantastic in nature. W. Junk, The Hague

Bateson PPG, Gluckman PD (2011) Plasticity, robustness, development and evolution. Cambridge University Press, Cambridge

Box GEP, Hunter WG, Hunter JS (2005) Statistics for experimenters: design, innovation, and discovery, 2nd edn. Wiley, New York

Burdge GC, Hanson MA, Slater-Jefferies JL, Lillycrop KA (2007) Epigenetic regulation of transcription: a mechanism for inducing variations in phenotype (fetal programming) by differences in nutrition during early life? Br J Nutr 97:1036–1046

Butcher RA, Fujita M, Schroeder FC, Clardy J (2007) Small-molecule pheromones that control dauer development in *Caenorhabditis elegans*. Nat Chem Biol 3(7):420–422

Charmantier A, McCleery RH, Cole LR, Perrins C, Kruuk LE, Sheldon BC (2008) Adaptive phenotypic plasticity in response to climate change in a wild bird population. Science 320(5877):800–803

Choi JH, Kijimoto T, Snell-Rood E, Tae H, Yang Y, Moczek AP, Andrews J (2010) Gene discovery in the horned beetle *Onthophagus taurus*. BMC Genomics 11:703

de Mendiburu F (2013) Statistical procedures for agricultural research: package 'agricolae' in R. http://cran.r-project.org/web/packages/agricolae/agricolae.pdf, Version 1.1–4, April 2013

DeWitt TJ (1998) Costs and limits of phenotypic plasticity: tests with predator-induced morphology and life history in a freshwater snail. J Evol Biol 11(4):465–480

Draghi JA, Whitlock MC (2012) Phenotypic plasticity facilitates mutational variance, genetic variance, and evolvability along the major axis of environmental variation. Evolution 66(9):2891–2902

Eldar A, Elowitz MB (2010) Functional roles for noise in genetic circuits. Nature 467(7312):167–173

Emlen DJ (1994) Environmental control of horn length dimorphism in the beetle *Onthophagus acuminatus* (Coleoptera: Scarabaeidae). Proc R Soc Lond B Biol Sci 256:131–136

Emlen DJ (2001) Costs and the diversification of exaggerated animal structures. Science 291(5508):1534–1536

Emlen DJ, Allen CE (2003) Genotype to phenotype: physiological control of trait size and scaling in insects. Integr Comp Biol 43(5):617–634

Emlen DJ, Hunt J, Simmons LW (2005) Evolution of sexual dimorphism and male dimorphism in the expression of beetle horns: phylogenetic evidence for modularity, evolutionary lability, and constraint. Am Nat 166(Suppl 4):S42–S68

Emlen DJ, Szafran Q, Corley LS, Dworkin I (2006) Insulin signaling and limb-patterning: candidate pathways for the origin and evolutionary diversification of beetle 'horns'. Heredity 97:179–191

Emlen DJ, Warren IA, Johns A, Dworkin I, Lavine LC (2012) A mechanism of extreme growth and reliable signaling in sexually selected ornaments and weapons. Science 337(6096):860–864

Feinberg AP, Irizarry RA (2010) Evolution in health and medicine Sackler colloquium: stochastic epigenetic variation as a driving force of development, evolutionary adaptation, and disease. Proc Natl Acad Sci USA 107(Suppl 1):1757–1764

Frank S (1996) The design of natural and artificial adaptive systems. In: Rose M, Lauder G (eds) Adaptation. Academic, New York, pp 451–505

Gilbert SF, Epel D (2009) Ecological developmental biology: integrating epigenetics, medicine, and evolution. Sinauer Associates, Sunderland

Gluckman PD, Hanson MA, Buklijas T, Low FM, Beedle AS (2009) Epigenetic mechanisms that underpin metabolic and cardiovascular diseases. Nat Rev Endocrinol 5(7):401–408

Godfrey K (1985) Fitting by organized comparisons: the square combining table, Chap 2. In: Hoaglin DC, Mosteller F, Tukey JW (eds) Exploring data tables, trends, and shapes. Wiley, New York, pp 37–66

Huey RB, Hertz PE, Sinervo B (2003) Behavioral drive versus behavioral inertia in evolution: a null model approach. Am Nat 161:357–366

Johnston TD (1982) Selective costs and benefits in the evolution of learning. Adv Study Behav 12:65–106

Junien C, Gallou-Kabani C, Vige A, Gross MS (2005) Nutritional epigenomics: consequences of unbalanced diets on epigenetics processes of programming during lifespan and between generations. Ann D Endocrinol 66:S19–S28

Kawecki TJ (1994) Accumulation of deleterious mutations and the evolutionary cost of being a generalist. Am Nat 144:833–838

Kijimoto T, Moczek AP, Andrews J (2012a) Diversification of *double sex* function underlies morph-, sex-, and species-specific development of beetle horns. Proc Natl Acad Sci USA 109(50):20526–20531

Kijimoto T, Pespeni M, Beckers O, Moczek AP (2012b) Beetle horns and horned beetles: emerging models in developmental evolution and ecology. Wiley Interdisciplinary Reviews: Developmental. Biology: n/a-n/a

Kopp A (2011) *Drosophila* sex combs as a model of evolutionary innovations. Evol Dev 13:504–522

Lande R (2009) Adaptation to an extraordinary environment by evolution of phenotypic plasticity and genetic assimilation. J Evol Biol 22(7):1435–1446

Mayr E (1974) Behavior programs and evolutionary strategies. Am Sci 62:650–659

Moczek AP (1998) Horn polyphenism in the beetle *Onthophagus taurus*: diet quality and plasticity in parental investment determine adult body size and male horn morphology. Behav Ecol 9: 636–641

Moczek AP (2005) The evolution and development of novel traits, or how beetles got their horns. BioScience 11:935–951

Moczek AP, Nijhout HF (2002) Developmental mechanisms of threshold evolution in a polyphenic beetle. Evol Dev 4(4):252–264

Moczek AP, Nijhout HF (2003) Rapid evolution of a polyphenic threshold. Evol Dev 5:259–268

Moczek AP, Sultan S, Foster S, Ledon-Rettig C, Dworkin I, Nijhout HF, Abouheif E, Pfennig DW (2011) The role of developmental plasticity in evolutionary innovation. Proc R Soc B 278(1719):2705–2713

Newman SA, Muller GB (2000) Epigenetic mechanisms of character origination. J Exp Zool 288(4):304–317

Nijhout HF (2003) The control of body size in insects. Dev Biol 261(1):1–9

Paulian R (1935) Le polymorphisme des males de coléopteres. Hermann et Cie, Paris

Pfennig DW, Wund MA, Snell-Rood EC, Cruickshank T, Schlichting CD, Moczek AP (2010) Phenotypic plasticity's impacts on diversification and speciation. Trends Ecol Evol 25(8):459–467

Piersma T, Drent J (2003) Phenotypic flexibility and the evolution of organismal design. Trends Ecol Evol 18(5):228–233

Pigliucci M (2005) Evolution of phenotypic plasticity: where are we going now? Trends Ecol Evol 20:481–486

Pigliucci M, Murren CJ (2003) Genetic assimilation and a possible evolutionary paradox: can macroevolution sometimes be so fast as to pass us by? Evolution 57:1455–1464

Price TD, Qvarnstrom A, Irwin DE (2003) The role of phenotypic plasticity in driving genetic evolution. Proc R Soc B 270(1523):1433–1440

Renn S, Schumer M (2013) Genetic accommodation and behavioural evolution: insights from genomic studies. Anim Behav 85:1012–1022

Sanchez L (2008) Sex-determining mechanisms in insects. Int J Dev Biol 52(7):837–856

Schlichting CD, Pigliucci M (1998) Phenotypic evolution: a reaction norm perspective. Sinauer Associates, Sunderland

Shingleton AW, Das J, Vinicius L, Stern DL (2005) The temporal requirements for insulin signaling during development in *Drosophila*. PLoS Biol 3(9):e289

Shingleton AW, Frankino WA, Flatt T, Nijhout HF, Emlen DJ (2007) Size and shape: the developmental regulation of static allometry in insects. Bioessays 29(6):536–548

Shukla JN, Nagaraju J (2010) Double sex: a conserved downstream gene controlled by diverse upstream regulators. J Genet 89(3):341–356

Sih A, Ferrari MCO, Harris DJ (2011) Evolution and behavioural responses to human-induced rapid environmental change. Evol Appl 4(2):367–387

Simmons LW, Emlen DJ (2006) Evolutionary trade-off between weapons and testes. Proc Natl Acad Sci USA 103(44):16346–16351

Simmons LW, Tomkins JL, Hunt J (1999) Sperm competition games played by dimorphic male beetles. Proc R Soc B 266(1415):145–150

Smith CR, Toth AL, Suarez AV, Robinson GE (2008) Genetic and genomic analyses of the division of labour in insect societies. Nat Rev Genet 9(10):735–748

Snell-Rood EC (2012) Selective processes in development: implications for the costs and benefits of phenotypic plasticity. Integr Comp Biol 52(1):31–42

Snell-Rood EC, Moczek AP (2012) Insulin signaling as a mechanism underlying developmental plasticity: the role of FOXO in a nutritional polyphenism. PLoS One 7(4):e34857

Snell-Rood EC, Moczek AP (2013) Horns and the role of development in the evolution of beetle contests. In: Hardy ICW, Briffa M (eds) Animal Contests. Cambridge University Press, UK

Snell-Rood EC, Van Dyken JD, Cruickshank T, Wade MJ, Moczek AP (2010) Toward a population genetic framework of developmental evolution: the costs, limits, and consequences of phenotypic plasticity. Bioessays 32(1):71–81

Snell-Rood EC, Troth A, Moczek AP (2013) DNA methylation as a mechanism of nutritional plasticity: limited support from horned beetles. J Exp Zool Part B-Mol Dev Evol 320B:22–34

Sol D (2009) Revisiting the cognitive buffer hypothesis for the evolution of large brains. Biol Lett 5(1):130–133

Sol D, Price TD (2008) Brain size and the diversification of body size in birds. Am Nat 172(2):170–177

Sol D, Bacher S, Reader SM, Lefebvre L (2008) Brain size predicts the success of mammal species introduced into novel environments. Am Nat 172:S63–S71

Sol D, Duncan RP, Blackburn TM, Cassey P, Lefebvre L (2005) Big brains, enhanced cognition, and response of birds to novel environments. Proc Natl Acad Sci USA 102(15):5460–5465

Stansbury M, Moczek AP (2013) The evolvability of arthropods. In: Minelli A, Boxshall G, Fusco G (eds) Arthropod biology and evolution: molecules, development, morphology. Springer Verlag, Berlin, pp 479–494

Tang HY, Smith-Caldas MSB, Driscoll MV, Salhadar S, Shingleton AW (2011) FOXO regulates organ-specific phenotypic plasticity in *Drosophila*. PLoS Genet 7:12

Valena S, Moczek AP (2012) Epigenetic mechanisms underlying developmental plasticity in horned beetles. Genet Res Int 2012:576303

Van Buskirk J, Steiner UK (2009) The fitness costs of developmental canalization and plasticity. J Evol Biol 22(4):852–860

Van Dyken JD, Wade MJ (2010) The genetic signature of conditional expression. Genetics 184: 557–570

Wang Z, Zhang J (2011) Impact of gene expression noise on organismal fitness and the efficacy of natural selection. Proc Natl Acad Sci USA 108(16):E67–E76

West-Eberhard MJ (2003) Developmental plasticity and evolution. Oxford University Press, Oxford/New York

Whitlock MC (1996) The red queen beats the jack-of-all-trades: the limitations on the evolution of phenotypic plasticity and niche breadth. Am Nat 148: S65–S77

Whitman D, Ananthakrishnan TN (2009) Phenotypic plasticity of insects: mechanisms and consequences. Science Publishers, Enfield

Young RL, Badyaev AV (2007) Evolution of ontogeny: linking epigenetic remodeling and genetic adaptation in skeletal structures. Integr Comp Biol 47(2): 234–244

Neurogenomics of Behavioral Plasticity

8

Rayna M. Harris and Hans A. Hofmann

Abstract

Across animals, there is remarkable diversity in behavior. Modern genomic approaches have made it possible to identify the molecular underpinnings of varied behavioral phenotypes. By examining species with plastic phenotypes we have begun to understand the dynamic and flexible nature of neural transcriptomes and identified gene modules associated with variation in social and reproductive behaviors in diverse species. Importantly, it is becoming increasingly clear that some candidate genes and *gene networks* are involved in complex social behaviors across even divergent species, yet few comparative transcriptomics studies have been conducted that examine a specific behavior across species. We discuss the implications of a range of important and insightful studies that have increased our understanding of the neurogenomics of behavioral plasticity. Despite its successes, behavioral genomics has been criticized for its lack of hypotheses and causative insights. We propose here a novel avenue to overcome some of these short-comings by complementing "forward genomics" studies (i.e., from phenotype to behaviorally relevant gene modules) with a "reverse genomics" approach (i.e., manipulating novel gene modules to examine effects on behavior, hormones, and the genome itself) to examine the functional causes and consequences of differential gene expression patterns. We discuss how several established approaches (such as pharmacological manipulations of a novel candidate pathway, fine scale mapping of novel candidate gene expression in the brain, or identifying direct targets of a novel transcription factor of interest)

R.M. Harris • H.A. Hofmann (✉)
Department of Integrative Biology, The University of Texas at Austin, Institute for Cellular and Molecular Biology, 2415 Speedway, C0990, Austin, TX 78712, USA
e-mail: hans@utexas.edu

C.R. Landry and N. Aubin-Horth (eds.), *Ecological Genomics: Ecology and the Evolution of Genes and Genomes*, Advances in Experimental Medicine and Biology 781, DOI 10.1007/978-94-007-7347-9_8, © Springer Science+Business Media Dordrecht 2014

can be used in combination with the analysis of the accompanying neurogenomic responses to reveal unexpected biological processes. The integration of forward and reverse genomics will move the field beyond statistical associations and yield great insights into the neural and molecular control of social behavior and its evolution.

Keywords
Transcriptomics • Reverse genomics • Neuroethology • Social behavior • Dispersal • Mate choice • Evolution

8.1 Introduction

Across the animal kingdom, there is remarkable diversity in naturally occurring behavioral phenotypes. Many animals live in complex social environments, and they make decisions based on the context of their interactions with other individuals. How do they make these decisions, and why do they behave the way they do are questions that have long fascinated biologists (Tinbergen 1963). A recent review by O'Connell and Hofmann (2011a) outlines a variety of ways in which these questions can be addressed by combining genomic and evolutionary approaches with studies examining brain and behavior. Modern genomic techniques such as *microarrays* (see the Glossary for definitions of italicized terms) and, more recently, *next-generation sequencing* have made it possible to examine the molecular underpinnings of plasticity in animal behavior and decision-making as well as their evolution (Hitzemann et al. 2013). By examining neural *transcriptomes* of polymorphic species we have begun to understand the dynamic and flexible nature of genome activity in the brain and identified *gene modules* (set of co-regulated genes or proteins (Segal et al. 2004)) that are associated with variation in social and reproductive behaviors in diverse species (O'Connell and Hofmann 2011b). While it is increasingly clear that some candidate genes and gene networks are involved in complex social behaviors across even divergent species (O'Connell and Hofmann 2011b; Toth and Robinson 2007), few comparative transcriptomics studies have been conducted to test this notion of conserved molecular pathways on a genomic scale.

Behavioral genomics has clearly transformed our understanding of social plasticity, yet the field has also been criticized for its apparent lack of concrete hypotheses and the uninformative gene lists that often result from these studies. While it is indeed relatively easy to obtain a wealth of transcriptional information, identifying the genes or gene networks that are causal in the behavioral context under study is much more challenging. In the same manner that geneticists advance the field by using reverse genetics (Alonso and Ecker 2006), it is thus becoming increasingly important that these "forward genomic" studies are followed up with "reverse genomic" studies to examine the functional causes and consequences of differential gene expression patterns. In other words, once novel candidate genes or pathways have been identified, we must use experimental tests on a genomic scale to further dissect the contribution of each gene to the behavioral phenotype.

Here, we discuss forward and reverse genomic studies that have shed light on various aspects of social behavior and its underpinnings and suggest promising avenues for future research into the evolution of neuroethological systems. There are many examples of forward genomic experiments and a dearth of reverse genomic experiments, which we argue are necessary for examining causality and function. We highlight several studies that have applied a "reverse genomics" approach successfully in diverse model systems, complementing approaches such as pharmacological manipulations of a novel candidate pathway, distribution mapping of novel candidate gene expression in the brain, or identification of direct targets of a novel transcription factor of interest with transcriptomics (Fig. 8.1). The

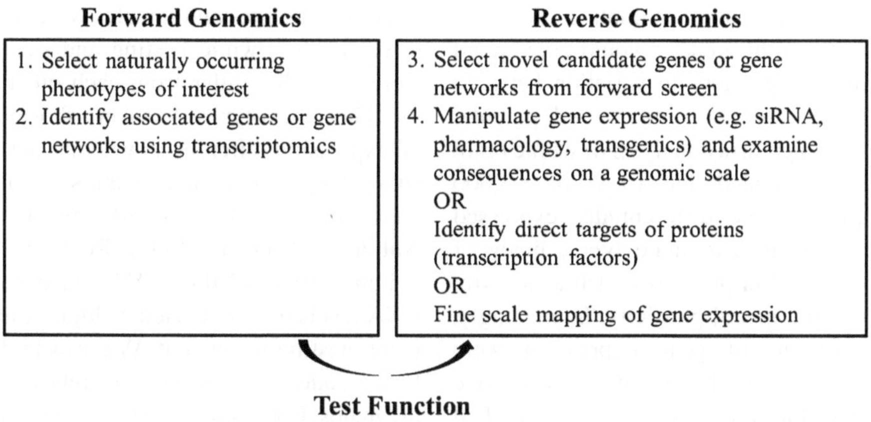

Fig. 8.1 Forward and reverse genomics of behavioral plasticity. *1.* Forward genomic approaches begin with the selection of two or more phenotypes for comparison. *2.* Then gene expression differences are compared on a genomic scale to identify genes and gene networks that are associated with the naturally occurring behavioral phenotype. *3.* Reverse genomics begins with the selection of novel candidate genes, gene networks, or pathways based on the gene expression analysis. *4.* To better understand the function of the observed gene expression patterns, one can manipulate gene expression, identify DNA/protein interactions, or examine brain region specific differences

need to examine the neurogenomic responses that result from these perturbations is increasingly becoming clear. The combination of forward with reverse genomics will move the field beyond statistical associations and yield great insights into the neural and molecular control of social behavior and its evolution.

8.2 Functional Genomics of Neural and Behavioral Plasticity

Plasticity in the nervous system comprises the functional and structural changes in information processing after the initial formation of neuronal contacts. When approached from an integrative perspective, the analysis of these mechanisms usually begins by describing and analyzing the neural, endocrine, and behavioral traits that can potentially be realized by an organism or in a population depending on environmental conditions. We can distinguish several (often overlapping) time scales on which plasticity can occur in response to social or environmental stimuli (Hofmann 2003): Changes that occur in real time (e.g., modulation, learning/memory formation) via variation in neural and/or hormonal activity; slower changes that involve regulation of gene expression as well as possible structural and physiologic changes; and changes that can alter developmental trajectories and shift neural functioning throughout life history, even in adult animals (e.g., seasonal and use-dependent changes). A number of studies have integrated concepts from neurobiology, ethology, and evolutionary biology with powerful genomic technologies in order to gain a more comprehensive understanding of the roles that genetic and environmental factors play in neural and behavioral plasticity.

8.2.1 Alternative Reproductive Tactics

Organisms that share the same genotype can develop into divergent phenotypes, depending on environmental conditions (Brockmann 2001; Ross 1990). Atlantic salmon (*Salmo salar*) exhibits extreme alternative life histories and *reproductive tactics* based on their growth rate and duration as juveniles. Young males of the same age can be found either as mature sneakers or immature males that will be anadromous the next year. Aubin-Horth and colleagues (2005) hypothesized that brain gene expression patterns would vary considerably between age-matched mature males (sneakers), immature males (future

anadromous males) and immature females. Specifically, these differences would correspond to organism-level phenotypic variation between divergent life history and developmental trajectories. A microarray analysis of whole brain transcriptomes revealed that 15 % of ~3,000 genes examined were differentially expressed in the brains of the two male types, many of which are involved in processes such as growth, reproduction, and neural plasticity. Interestingly, consistent patterns of gene expression were found for individuals of the same reproductive tactic despite the potentially high individual variation that is often associated with genomic studies on wild caught animals. Notably, gene expression patterns in immature males were quite different both from immature females and mature sneakers; this pattern indicates that delayed maturation and sea migration, the 'default' life cycle, may actually result from an active inhibition of development into a sneaker (Aubin-Horth et al. 2005, 2009).

Like the Atlantic salmon, the ocellated wrasse, *Symphodus ocellatus*, is another fish species in which males display plasticity in life history trajectory and reproductive tactic. *S. ocellatus* males engage in one of three alternative tactics during a reproductive season: nesting, satellite, and sneaker. While males utilize a single tactic per reproductive season, reproductive tactic is plastic because males can transition to other tactics between seasons and thus have multiple potential life history trajectories depending on early growth prior to their first winter or first reproductive year. Satellites and sneakers spawn parasitically in nesting males' nests, but only nesting males provide parental care. Nesting and satellite males show transient cooperative defense of nests against sneakers. To better understand the neuroendocrine and genomic mechanisms that give rise to these dramatic differences in phenotype, Stiver and colleagues (in prep) analyzed neural gene expression profiles and circulating sex steroid hormone levels in these three male phenotypes and in females. Multivariate analyses of the genes that were differentially expressed between any two phenotypes revealed striking similarities and differences in expression profiles between phenotypes. Specifically, brain transcriptomes of satellites and females were most similar to each other, while nesting and sneaker males were most dissimilar from each other and from the other phenotypes. Sneakers showed more total expression differences, whereas nesting males showed higher magnitude expression differences. Based on work by Aubin-Horth et al. (2007), Aubin-Horth et al. (2005), Renn et al. (2008), Schumer et al. (2011), AVT and parvalbuminm RNA levels were expected be highest in the dominant, nesting males, but AVT was highest in the female, and parvalbumin was highest in the satellite males. Ribosomal-, histone-, and proteasome-related genes, which were expected to correlate with future growth (Alonzo et al. 2000; Renn et al. 2008) were indeed up-regulated in sneakers and satellite males.

With respect to circulating sex steroid hormones, 11-ketotestosterone (but not testosterone) was highest in nesting males, while estradiol was highest in females. Overall, these genomic and endocrine findings reveal the surprising extent to which neural gene expression patterns vary across reproductive tactics, providing important insights into the molecular mechanisms underlying variation in cooperative and reproductive behavior (Stiver et al. in prep).

8.2.2 From Nurse to Forager

Some animals undergo fascinating changes in brain and behavior across their lifetime. The non-reproductive workers of honeybee (*Apis mellifera*) societies provide a compelling example as they transition through a series of distinct behavioral tasks as they age (polyethism). Worker bees begin their adult lives tending to within-hive chores such as nursery and queen care and then transition to the role of a forager. This age-related transition to foraging is associated with dramatic changes in brain morphology and brain gene expression. For example, the mushroom bodies, a region in the insect brain associated with complex social behavior and memory (Haehnel and Menzel 2012), increase in size (Fahrbach 2006). There are also substantial changes in gene expression (>85 % of approximately 5,500 genes showed

differences) associated with the transition from nurse to forager that are largely independent of age-related changes. Principal component analysis revealed discrete influences of age, behavior, genotype, environment, and experience (Whitfield et al. 2006). Interestingly, the hive bee to forager transition is accompanied by changes in genes related to energy metabolism and genes driven by the actions of juvenile hormone, highlighting the importance of hormones in driving neural plasticity (Ament et al. 2010). Inspired by findings in *Drosophila* (Osborne et al. 1997), Ben-Shahar et al. (2002) showed that the age-related transition from hive worker to forager is associated with increased expression levels of the foraging gene (for). Furthermore, treatment with a guanosine 3′, 5′-monophosphate (cGMP) -dependent protein kinase (PKG) that is encoded by *for* caused foraging behavior (Ben-Shahar et al. 2002).

8.2.3 Social Hierarchies

It is well known that behavior and physiology are regulated by both environment and social context, and an important study by Renn and colleagues demonstrated that neural gene expression is regulated by social environment (Renn et al. 2008). The authors used the African cichlid fish *Astatotilapia burtoni,* a model system for the study of how social interactions regulate neural and behavioral plasticity (Hofmann 2003; Robinson et al. 2008). *A. burtoni* males are either socially dominant, territorial, reproductively active, and brightly colored or subordinate, non-territorial, reproductively suppressed, and cryptically colored. Amazingly, these phenotypic differences are reversible, and males ascend and descend many times during their life. Renn et al. examined whole brain gene expression in dominant and subordinate males as well as in brooding females, and integrated the genomic data with quantitative behavioral measures. Using this integrative approach, the authors identified co-regulated gene sets (gene modules) that are significantly associated with either *dominance* or reproductive state. While the regulation of neuroendocrine genes was predicted from previous research, the results also revealed unexpected and novel roles for two classic neurotransmitter systems (GABA and glutamate/kainate) in mediating behavioral plasticity. Also, the application of the Gene Ontology framework (Ashburner et al. 2000) underscored the importance of hormonal regulation and highlighted the hitherto under-appreciated roles of cytoskeletal components and neuronal remodeling activity in addition to neurochemical pathways. Importantly, the authors found a high degree of individual variation in expression levels of genes that are differentially regulated between these phenotypes even though the dominant and subordinate phenotypes are robustly defined. These results demonstrated the molecular complexity in the brain associated with different social phenotypes, including gene modules that underlie reproduction and submissive behavior (Renn et al. 2008). Taken together, this genome-scale analysis of molecular systems in the brain identified complex patterns of gene expression that are associated with a socially regulated switch in behavioral phenotype.

As a follow up study, Huffman and colleagues (2013) designed an experiment to analyze the role of aromatase, the enzyme that converts testosterone into estradiol, in mediating aggression and reproductive behavior in male *A. burtoni.* Using quantitative radioactive *in situ* hybridization, the authors found that subordinate males have higher aromatase expression than dominant males in the magnocellular and gigantocellular regions of the preoptic area. Then, they pharmacologically inhibited aromatase activity by giving intraperitoneal injections of fadrozole (FAD) to dominant males and found that FAD treatment decreases aggressive, but not reproductive, behaviors compared to saline controls. Furthermore, they found that circulating estradiol levels decreased while testosterone levels increased in response to FAD treatment. Moreover, FAD-treated males had increased aromatase expression in the gigantocellular portion of the preoptic area (POA), possibly a compensatory response. Together, these results suggest that aromatase promotes aggression in *A. burtoni* males through actions in the preoptic area (Huffman et al. 2013). While this study

did not examine the genomic response to FAD treatment, it did test for function associated with the significant correlations found between dominance behavior and aromatase gene expression identified by Renn et al. 2008.

In an elegant study on the molecular basis of social dominance, Aubin-Horth et al. (2007) used the cooperatively breeding African cichlid *Neolamprologus pulcher* to identify brain gene expression profiles associated with aggression and dominance behavior independent of sex. In this species, dominant individuals (males and females) display similar behaviors, have high testosterone levels and have high brain arginine vasotocin expression when compared to subordinate helpers, but dominant females have lower levels of 11-ketotestosterone than males. Furthermore, brain gene expression profiles of dominant females are most similar to those of the males (independent of social rank), indicating that dominant breeder females are masculinized at the molecular and hormonal level while being at the same time reproductively competent. By investigating different levels of biological organization, from behavior to hormones and gene expression, this study provided new insights into the mechanisms underlying vertebrate social dominance, and the molecular and endocrine masculinization of the female brain depending on social status is likely not limited to fishes. This finding underscores the need for a comparative approach in a wide range of vertebrates with diverse patterns of social organization to determine where similar molecular and endocrine substrates regulate social life and where they have evolved independently (Aubin-Horth et al. 2007).

8.2.4 Social Defeat

To characterize the neural circuitry and cellular process by which social experience alters the activity of the mesolimbic dopamine pathway, Nestler and colleagues (Berton et al. 2006) used a chronic social defeat paradigm. In this paradigm, a mouse that is repeatedly exposed to a more aggressive individual will display increased anxiety and decreased exploratory behaviors. These depression-like phenotypes are associated with differences in BDNF (brain-derived neurotrophic factor) concentrations in the *nucleus accumbens* (NAcc), a brain region central to processing the salience and rewarding properties of a stimulus. This research has provided good evidence for socially induced remodeling of the physiological, molecular, and cellular mechanism within this mesolimbic dopamine pathway that affects stimulus processing. These modifications included changes in activity of transcription factors, histone modification and DNA methylation, giving rise to short and longer term changes in gene expression (Nestler 2012a). A microarray study from the same group (Krishnan et al. 2007) revealed that resilient mice (i.e., individuals who maintain normal physiological function despite defeat experience) showed selective up-regulation of multiple voltage-gated K^+ channel subunits in the ventral tegmental area (VTA; the source of dopamine affecting the NAcc) after chronic social defeat, but maintained low BDNF release from the VTA as in controls. This inspired them to examine the electrochemical properties of the VTA neurons. The increase K+ channel correlated with decreased firing of VTA neurons. Studies like this show the power of integrating electrophysiology with functional genomics and protein assays to better understand behavioral, cellular, and molecular responses to social challenges.

8.3 Molecular Mechanisms of Decision-Making

Animals are confronted daily with social challenges and opportunities where they must make adaptive decisions to ultimately increase their fitness. The brain integrates external social or environmental information with internal physiology by changes in neural gene expression and organization. Variation in neural gene expression patterns can have profound influences on how an individual responds to a stimulus and explains why we see so much diversity in animal behavior between individuals of the same species, across an individual's lifetime, and over

generations. Such molecular changes allow animals to integrate social information into an appropriate behavioral response, orchestrate neural changes that promote reproduction, and respond to social and other cues in ways that ultimately may serve to maximize fitness. In this section we review several studies that examine the rapid changes in neural activity and gene expression that are associated with behavioral decision-making.

8.3.1 Neuroeconomics

We begin this section with a discussion of neuroeconomics, an interdisciplinary field that combines cognitive neuroscience tools and economic theory to study the processes that govern behavioral decision making in the human brain (Fehr and Camerer 2007). Experimental games are often used in this research to measure how the salience of a reward (often monetary) influences a player's behavior. There are many types of games that can be used to study decision-making processes. These games create paradigms on how social status, age, and sex influence social decision making. The prisoner's dilemma is an excellent game theory example that demonstrates why two individuals might cooperate even when it is not in the individual's best interest. Thus, decision making is complex because individuals are motivated not only by personal gains but also by some reward derived from cooperating in certain social situations (Brede 2013). Humans frequently sacrifice material and personal gains to endorse or to oppose societal causes. The neural basis of charitable donation behavior has been the subject of experimental neurogenomic economics studies using a modified prisoner's dilemma paradigm and functional Magnetic Resonance Imaging (fMRI). The players were subjected to fMRI while choosing to donate or not to donate to real charitable organizations. Surprisingly, the mesolimbic reward system was engaged when the player donated to a charity and when the player received a monetary reward, suggesting that that the act of being charitable is itself rewarding. While social neuroeconomic studies have provided the evidence for neural circuits involved in decision making (Moll et al. 2006), they provide little insight into the genetic and genomic underpinnings, and we therefore return to animal model systems.

8.3.2 To Sing or Not to Sing?

A classical method of measuring neuronal responses is through electrophysiological recordings. Such studies often focus on presenting an animal with a behaviorally relevant sensory stimulus and measuring neuronal activity in various brain regions. Songbirds provide a powerful model system in this regard, where songs produced by males vary based on the social context. In the zebra finch (*Taeniopygia guttata*), neuronal activity is markedly different in brain regions involved in song learning when the male sings a song directed at a conspecific compared to undirected song (Hessler and Doupe 1999). However, recording neural activity simultaneously in several nodes of the birdsong circuit of an awake and behaving animal in a naturalistic environment is not feasible in most cases. To determine what brain regions or neuronal populations may respond to a particular social stimulus or which brain areas are active during singing, many researchers therefore use detection of *immediate early genes* (IEGs) as markers of neuronal activity (Jarvis and Nottebohm 1997; Mello et al. 1992). IEGs (e.g., *c-fos*, *jun*, *egr-1*, *arc*; Loebrich and Nedivi 2009) are typically transcription factors that are thought to quickly respond to internal and external stimuli and thus coordinate neuronal plasticity. Dong et al. (2009) expanded this experimental framework using a microarray approach. They showed that exposure to novel song induces rapid expression changes in thousands of genes, many of which are involved in transcription and RNA processing as well as cellular homeostasis. These authors concluded that natural stimuli such as birdsong can result in major changes in the metabolic state of the brain (Dong et al. 2009).

Gene expression studies have also revealed that the transcription factor *FoxP2* is critical for singing in songbirds. Within the song-specialized striato-pallidal Area X, *FoxP2* levels decrease after 2 h. of undirected singing (Teramitsu et al. 2010; Teramitsu and White 2006), and the magnitude of down-regulation is correlated to how much the birds sang (Teramitsu et al. 2010). Hilliard et al. 2012 used this finding as a starting point for examining the genomic differences between singing and non-singing males. The songs of singing males were undirected, presumably to remove any confound caused by the presence of a social stimulus. RNA was extracted from Area X for microarray analysis (Hilliard et al. 2012). In order to look for broad patterns in the dataset, the authors employed weighted gene coexpression network analysis (WGCNA; Zhang and Horvath 2005). First, sets of co-regulated genes were clustered into modules. Then, singing duration and number of motifs sung were correlated with the gene modules. These modules may consist of genes that are regulated by the same transcription factor (s), genes that regulate the phenotype directly, or genes that are consequences of the phenotype but otherwise unrelated in function to each other. By examining such covariance patterns, the researchers were able to identify two large gene modules that were positively associated with singing and one that was negatively associated. As in previous studies, *FoxP2* mRNA levels were negatively correlated with singing duration and the singing-associated modules. Finally, using a network approach, the authors were able to identify a network of genes that was correlated with *FoxP2* activity. Taken together, this study has provided many novel insights into how the down-regulation of *FoxP2* via singing can give rise to a whole suite of changes in gene expression in a particular brain region (Hilliard et al. 2012).

8.3.3 To Stay or to Disperse?

In landscapes where older populations may go extinct and new populations become established, do dispersal and colonization select upon existing genetic variation? Wheat et al. 2011 used an unusually integrative approach to study dispersal-related life history variation in a meta-population of the Glanville fritillary butterfly (*Melitaea cinxia*). Using microarray analysis, *quantitative PCR*, and physiological measurements in a common garden design, the authors identified metabolic and endocrine factors that may contribute to the disperser and non-disperser phenotype of new and old populations, respectively. Specifically, females from new populations (dispersers) had higher expression of genes involved in egg provisioning in thorax tissue and higher expression of genes involved in maintenance of flight muscle proteins in the thorax than females from established populations (non-dispersers). These findings were complemented with physiological measures, which showed that females from new populations had accelerated egg maturation, higher juvenile hormone titers, and enhanced flight metabolism. By identifying molecular candidate mechanisms of fitness variation maintained by dispersal dynamics in a heterogeneous environment, this study uncovered fascinating and intricate connections between physiology, genomics, ecology and evolution (Wheat et al. 2011).

In addition to genetic variation, other studies have found that neural and genomic plasticity can result in phenotypic variation across generations of butterflies. The spectacular fall and spring migratory patterns of the monarch butterfly (*Danaus plexippus*) provide a compelling example. These migrations span three to four generations because the journey takes longer than the life span of each migrant (Brower 1995). How is it then that they can so accurately navigate the path taken by their ancestors without a single veteran migrant? As migrating butterflies are always on their maiden voyage, a genetic program that integrates two mechanisms in the brain (a molecular clock and a sun compass) provides the basis for the annual migration from Canada to Mexico and back (Reppert et al. 2010). Fall migrant butterflies are reproductively inactive whereas summer monarchs are reproductively active, a switch triggered by juvenile hormone and a cascade of hormon-

ally regulated genes involved in immunity and metabolism. Moreover, microarray analyses have revealed 40 genes that are differentially expressed between summer and fall migrants in relation to migratory behavior (independent of juvenile hormone).

8.3.4 Territorial Defense

Transcriptome studies suggest that the brain can rapidly respond to social stimuli by modulating transcriptional regulatory networks. This type of response requires the interaction between transcription factors and the cis-regulatory sequences of DNA, including promoter and enhancer regions. Bell and colleagues (Sanogo et al. 2012) used a bioinformatics approach to scan the promoters of differentially expressed genes identified in a microarray study that examined the genomic response to territorial intrusion in stickleback (*Gasterosteus aculeatus*). It is important to note that this study did not examine gene expression of the whole brain; rather it examined the transcriptomes of the telencephalon, diencephalon, cerebellum, and brain stem. The researchers found significant correlations between male behavioral response and spatially explicit gene expression patterns in that a large number of differentially expressed genes showed opposite patterns across brain regions. For instance, pro-opiomelanocortin (*pomc*) mRNA was up-regulated in the diencephalon but down-regulated in the telencephalon in response to the intruder. To further explore the mechanisms that could give rise to coordinated change in transcriptional regulatory networks, the authors identified cis-regulatory motifs that were located within 5,000 bp upstream of the differentially expressed genes. This analysis resulted in a list of candidate transcription factors that may be involved in the aggressive response to a behavioral challenge, which can now be used to generate novel hypotheses for future studies into the neurogenomic response to a territorial intrusion (Sanogo et al. 2012). For example, cis-regulatory analysis identified two potential regulators of *pomc* (POU domain, class 3 transcription factor 2 (POU3F2) and the estrogen receptor (ER)), which have previously been shown to regulate *pomc* expression (De Souza et al. 2005). Future studies could employ pharmacological manipulations to determine the functional relevance of ER regulation of *pomc* in the context of territorial defense. Alternatively, one could conduct *chromatin immunoprecipitation sequencing (ChIP-seq)* analysis using an antibody for POUF32 and/or ER to determine on a genomic scale to which extent *pomc* and other genes within the same module are directly regulated by POU3F2 and/or ER.

Songbirds provide another powerful model system to understand the genomics of territorial behavior. For example, male song sparrows of the species *Melospiza melodia* are territorial year-round, yet the neuroendocrine responses to a territorial intruder vary between breeding and non-breeding season (Wingfield and Hahn 1994). Exposure to an intruder in the breeding but not the non-breeding season leads to increases in luteinizing hormone and testosterone. This suggests that the mechanisms that control neuroendocrine responses to social stimuli differ between seasons. In fact, a microarray study by Mukai et al. (2009) demonstrated that an intruder challenge drives differential genomic responses in the hypothalamus depending on season. In autumn and spring, 173 and 67 genes, respectively, were differentially expressed in the control versus territorial intrusion. Because a larger number of genes were differentially expressed between seasons (262), the authors suggested that the underlying seasonal effects on neural gene expression are major contributors to the difference in neuroendocrine responses to social stimuli (Mukai et al. 2009). Overall, these studies show that remarkable genomic plasticity is associated with territorial defense across a broad range of species.

8.3.5 Mating Preferences

Across taxa, variation in the way females choose mates can drive evolutionary change both within and between species. For decades, the research focus has been to identify the male

traits that arouse sexual interest in females (reviewed in Andersson 1994). More recently, however, researchers have begun to identify the physiological and neural processes underlying female choice. The swordtail *Xiphophorus nigrensis*, a poeciliid fish from Mexico, has become one of the most powerful model systems for this kind of research (Houde 1988). In this species, females prefer large males with elaborate sexual traits and courtship behaviors over smaller, more cryptic males that use forced copulation. To investigate the neural and molecular underpinnings that give rise to this preference, Cummings and colleagues (2008) conducted whole brain transcriptome analysis on females given a dichotomous choice between large and small males. What they found was a surprising down-regulation of gene expression when exposed to large males. It is possible that this was the result of a release of transcriptional silencing in response to courtship advances by the males that prepare the female for mating (Wong and Hofmann 2010). Validation experiments using quantitative PCR showed a correlation between individual variation in female preference behavior and the expression levels of several genes, including *neuroserpin*, an extracellular serine protease inhibitor implicated in modulating synaptogenesis and synaptic plasticity (Miranda and Lomas 2006) and exploratory behavior in mice (Madani et al. 2003). However, this study did not examine where in the brain these genes were expressed or how they might differ between closely related species with different *mating systems* (Cummings et al. 2008). To further investigate these findings using in situ hybridization, Wong et al. mapped *neuroserpin* gene expression in female brains, focusing on brain regions of the social behavior network (section 8.5.1, Newman 1999). Quantitative differences in *neuroserpin* gene expression in the *preoptic area* and the medial and lateral zones of the dorsal telencephalon were significantly correlated with female preference behavior (Wong et al. 2012).

In another follow up study, Lynch et al. (2012) compared mate preference behavior between the choosy swordtail females with the Western mosquitofish (*Gambusia affinis*), a poeciliid fish that uses coercive mating tactics. These contrasting behavioral phenotypes provide an excellent comparative model to further investigate the role of *neuroserpin* in mate preference. Using quantitative PCR on whole brain samples, they found that *neuroserpin* levels were positively associated with mate preference behavior in female swordtails but were down-regulated in mosquitofish females expressing male biases. These results suggest that the presence of males in mosquitofish species may inhibit *neuroserpin* expression. Because both gene expression and female behavioral responses to males exhibit opposing patterns between these species, this genetic pathway may potentially act as a substrate for the evolution of mate preference behavior (Lynch et al. 2012). It would be interesting to compare brain region-specific transcriptomes of these females to further investigate the genomic contribution to *neuroserpin*-mediated mate preferences.

8.4 Comparative Approaches

Are there conserved gene modules that are involved in complex social behaviors across distantly related species? Comparative studies that examine closely and distantly related species can provide great insight into the conservation of genome function (O'Connell and Hofmann 2012a). Although striking similarities in neurochemistry and plasticity are seen across wide evolutionary distances, differentiating between conserved and independently evolved traits depends on a well resolved phylogeny with the underlying behavioral mechanisms known for many branches. However, it has been suggested that in cases of behavioral transitions that have occurred independently multiple times (e.g., monogamy), even across large evolutionary distances, similar gene networks have been recruited repeatedly (Toth and Robinson 2007; O'Connell and Hofmann 2011a). Ancestral signaling molecules such as peptide or steroid hormones and biogenic amines likely acted within an ancient neural framework in response to social stimuli (O'Connell and Hofmann 2012a). Over the course of animal evolution,

this simple behavioral framework may have been modified in various ways in order to adapt to new environmental challenges or opportunities that represented rewarding or aversive salience (Barron et al. 2010). In the following section, we will discuss two studies that have compared brain transcriptomes across species in order to gain insight into evolutionary conserved and novel gene expression patterns that are associated with behavioral phenotypes. While there are clearly several obstacles associated with comparative transcriptomics (e.g., increased cost and reliably identifying orthologous genes), this approach promises exciting new insights.

8.4.1 Mating System Evolution

Analysis of gene expression through heterologous hybridization in particular has enabled genome-scale studies in many ecologically and evolutionarily interesting species. Using a cichlid fish microarray platform, Machado et al. (2009) examined neural gene expression levels between individual males and females from a pair of sister species of the Ectodini tribe of Lake Tanganyika cichlids: the polygynous *Enantiopus melanogenys* and the monogamous *Xenotilapia flavipinnis*. Their results indicated that the gene expression profiles are species-specific to a large extent, as relatively few genes show conserved expression patterns associated with either sex. This finding that sex-specific gene expression was highly variable across species indicates that social organization, such as mating system, may play an important role in sculpting transcription profiles in the brain. However, it could also mean that there are core sets of genes whose expression is coordinated across species. Future studies comparing more species will provide us with a better understanding of how these gene sets relate to social phenotypes (Machado et al. 2009).

8.4.2 Evolution of Eusocial Behavior

Comparative genomic analyses can provide great insights into the evolution of mechanisms that regulate social behavior. Toth et al. examined brain gene expression profiles of *Polistes metricus*, a primitive eusocial wasp. Then, the authors compared the results to the database of brain gene expression data for *Apis mellifera*, the advanced eusocial honeybee. To examine genomic variation associated with foraging/provisioning behavior and reproductive status, the authors studied four female wasp groups (foundress, gyne, queen, and worker) using a custom-made *P. metricus* microarray. They found striking differences in the expression across the four groups, many of which showed significant associations with foraging/provisioning status and a handful associated with reproductive status. Next, the authors compared these two differentially expressed gene lists with genes previously shown to be differentially expressed in association with honeybee division of labor and found a striking and significant overlap of genes associated with foraging/provisioning across the two species. Their results suggest that there is indeed common molecular code or a conserved 'genetic toolkit' for division of labor in two independently evolved social insect species (Toth et al. 2010). Future forward and reverse genomic studies that compare distantly related species in a similar behavioral context could provide detailed insights into the mechanisms regulating plastic social behaviors.

8.4.3 Meta-Analyses

While the use of microarray technology may be in decline, this should not stop anyone from analyzing the data collected in these experiments. Meta-analyses of transcriptomic datasets collected within and across institutions can provide a rich source of biological insight when statistical tests are used to rigorously evaluate a single overarching hypothesis.

In Sect. 8.2.1 we already introduced the astonishing life history transitions exhibited by Atlantic salmon (*Salmo salar*), which in their second year of life all females and most males migrate to the sea, where they grow considerably in size before returning to their native stream for reproduction. As discussed above, a subset of

males will remain in freshwater and mature into a small sneaker phenotype (Aubin-Horth et al. 2005). Similarly, some of the migrating fish do not enter the seas directly (early migrants) but instead wait a year before entering the sea (late migrants; Garcia de Leaniz et al. 2007). Immature and sneakers males as well as females differ considerably in brain genes expression profiles (Aubin-Horth et al. 2005). In one of the first meta-analyses of behaviorally relevant transcriptome data, Aubin-Horth et al. (2009) compared the brain expression profiles of all mature phenotypes with that of immature phenotypes and discovered a molecular correspondence between the transition to the sneaker life history in year 1 and the early vs. late migrant transition in year 2. Specifically, these authors discovered a set of 20 genes that are regulated in a concordant fashion in both life history transitions (Aubin-Horth et al. 2009), suggesting that there might be a 'life history transition module' that becomes engaged every time an animal undergoes a major transitions, whether it is in the context of reproduction or migration.

A much more sophisticated meta-analysis was conducted by Ament and colleagues (2012), who developed and applied informatics techniques for discovering meta-associations across transcriptomic experiments collected from many years of research. Deploying these techniques for brain transcriptome profiles from about 400 individual of the relatively docile European honeybee (*Apis mellifera mellifera*) and the more aggressive Africanized honeybee of different ages and worker classes, the authors show that both behavioral/developmental and evolutionary plasticity is regulated by complex interactions between a few common transcription factors, such that distinct combinations of cis-regulatory motifs can give rise to different maturation processes. These findings indicate that phenotypic traits (such as aggression) utilize a common toolkit of regulatory genes, and that variation in the regulatory network can give rise to phenotypic diversity (Ament et al. 2012).

Another fine example of the utility of meta-analysis of large transcriptome datasets comes from the Songbird Neuro-Genomics Initiative (Replogle et al. 2008) where Drnevich et al. (2012) investigated neural gene expression profiles of six different songbird species by analyzing a comprehensive dataset collected by 11 laboratories under a variety of experimental conditions. For example, using the WGCNA approach discussed in Sect. 8.3.2, the authors identified transcriptions factors with high connectivity that may be responsible for coordinating other genes within gene expression modules in area X, a brain region known to be important for song learning. This analysis also found that brain region strongly influenced gene expression patterns, more so than did species (Drnevich et al. 2012). These individual and combined datasets provide a wealth of insights into the relationships between neural anatomy, social behavior in response to environmental cues, and gene expression.

8.5 Reverse Genomics

Like genomic approaches across biology, the field of behavioral genomics has been criticized for its exploratory nature and lack of causality. Given recent advances in next-generation sequencing that allow large amounts of expression and other genome-scale data to be collected at a reasonable expense, it is now high time for researchers in this area to move beyond gene lists and Venn diagrams. The meta-analyses discussed in the previous section provide one promising avenue. But what other approaches could help us to test for function associated with the significant correlations between genomic state and behavior or decision making? Reverse genomic approaches provide a novel and powerful avenue to complement the forward genomic studies discussed above. In order to examine the function of these novel candidates, researchers may choose from a variety of approaches (Fig. 8.1). One option is to manipulate gene expression (e.g., using pharmacology, transgenic techniques, or *siRNA*) and examine the behavioral and genomic consequences of perturbed gene expression. If one is interested in determining the cause

of changes in gene expression, researchers can examine transcription factor binding profiles through ChIP-seq analysis or examination of methylation profiles using *bisulfite sequencing.* These approaches provide insight into whether the observed gene expression changes are due to loss or gain of a *transcription factor binding site* or a change in promoter methylation or histone acetylation, respectively.

Many of the studies discussed in Sects. 8.2–8.4 of this chapter identified interesting and significant correlations between gene networks and a behavioral motif, but only a few have followed up with studies examining the causal or functional relationship. It is worth noting, however, that while we are able to generate lists of sometimes thousands of differentially expressed genes, we can usually experimentally manipulate only a handful of genes, a limitation that requires prioritization of the genes to be manipulated and thus a compelling rationale for selecting such candidate genes in an unbiased manner. In the next section, we will discuss a few studies that have already utilized reverse genomics approaches to better understand the correlations of behavior with one to a few genes identified using forward genomic approaches.

8.5.1 Examining Brain Region-Specific Transcriptomes

It is clear that regions of the brain, having specific biological functions, express a unique suite of genes to perform these functions (Nadler et al. 2006), and in many of the whole- or grossly dissected brain studies described above, lack of spatial resolution was often cited as a reason for not recovering a candidate gene previously associated with the observed behavior or phenotype. This could be because expression of a gene in one brain region can mask its expression in the other regions of the brain. In order to link gene expression to activity within a neural circuit we must look at a higher resolution. Oldham et al. (2006) were among the first to conduct such a spatially explicit analysis. In search for factors that drive evolutionary changes and conservation of gene expression they used a WGCNA approach to compare the gene networks of multiple brain regions (white matter, cerebellum, caudate nucleus, anterior cingulate cortex, and the cortex) in humans and in chimpanzees. The authors noted that genes with high intramodular connectivity were conserved in the human and chimpanzee brain, a finding that supports the idea of conserved molecular mechanisms that govern primate brain organization. Likewise, dramatic differences in gene coexpression networks between the two species are strikingly consistent with the rapid expansion of the cerebral cortex in the lineage leading to humans. By using a comparative approach to examining gene co-expression networks across brain regions, the authors gain valuable insight into how differential network activity in discrete brain regions can be a driver of evolutionary change (Oldham et al. 2006).

Beyond primates, the *dopaminergic reward system* functions to evaluate the salience of a stimulus in the mesolimbic dopamine system, with a key role for dopaminergic projections from the midbrain ventral tegmental area to the regions of the forebrain (Lammel et al. 2011). The social behavior network controls male mating behavior, female sexual behavior, parental behavior, and various forms of aggression. Its involvement in regulating animals' social responses can be understood as a series of hormonally regulated behaviors that are shaped by development, experience and environmental signals (Newman 1999). Together these circuits make up a larger social decision-making network that is highly conserved across vertebrates (O'Connell and Hofmann 2011a, 2012a). Furthermore, this social-decision making network overlaps with what Hoke and Pitts (2012) refer to as the sensory-motor relay, which is important for integrating auditory signals and generating a behavioral output (Hoke and Pitts 2012). While many studies have used immediate-early gene induction to measure neural activity in different social contexts, few have investigated genomic differences across brain regions (Nadler et al. 2006). As methods for whole transcriptome analysis of gene expression from single neurons

or small tissue samples become more reliable (e.g., Morris et al. 2011; Whitaker et al. 2011), we expect to see more studies examining transcriptomic variation within specific neural networks.

8.5.2 Perturbing Molecular Pathways

Many studies have investigated differences between animals displaying varying amounts of aggression (Aubin-Horth et al. 2007; Greenberg et al. 2012; Renn et al. 2008; Sanogo et al. 2012; Toth et al. 2010). These and other studies have implicated a strong role for androgenic and estrogenic regulation of aggressive behavior. O'Connell and Hofmann (2012b) investigated how sex steroids modulate social behaviors, circulating steroids, and the preoptic area transcriptome in dominant and subordinate *A. burtoni* males. They found that social status predicts how sex steroid receptors regulate complex behaviors; androgens and progestins modulated courtship behavior in dominant but not subordinate males, while estrogens modulated aggressive behavior in both dominant and subordinate males. Because of the similar effect of estrogens on aggressive behavior in both phenotypes, the authors then examined the preoptic area transcriptome of estrogen receptor antagonist treated and control treated males. In dominant males, 8.25 % of all genes examined were differentially regulated by treatment while only 0.56 % was differentially expressed in subordinate males. Moreover, the preoptic area transcriptome responses to estrogen receptor perturbation showed very little overlap between dominant and subordinate males. The estrogen receptor was down-regulated in subordinate males, which may have contributed to the lack of gene expression changes associated with the pharmacological manipulation. It seems that inhibition of the estrogen receptor (in combination with other physiological characteristics of subordinate males such as low circulating testosterone levels and the absence of brain activation by the androgen receptor) leads to a remarkable genome-wide suppression of both transcriptional activity and variation in the POA. These results showed for the first time that individuals of the same species can exhibit different behavioral, hormonal, and transcriptomic responses to a perturbation (O'Connell and Hofmann 2012b).

The development of transgenic techniques for the study of behavior in adult animals has and will continue to greatly facilitate our understanding of brain region specific regulation of genes and behavior. Larry Young and colleagues have developed techniques for over-expression of genes in the monogamous prairie vole, *Microtus ochrogaster*, a model system for the study of affiliative behavior (McGraw and Young 2010). Previous studies from the vole community found that the oxytocin receptor expression in the NAcc promoted alloparental behavior and partner preference formation in female prairie voles. Using a viral vector for gene delivery, the researchers found that over-expressing the oxytocin receptor in the NAcc of adult female prairie voles facilitated pair bond formation but had no effect on alloparental behavior. This result demonstrated that oxytocin receptor expression elicited acute activational effects on affiliative behaviors. To examine whether or not it also elicited organizational effects, they used viral vector gene transfer to increase oxytocin receptor density in the NAcc of prepubertal female prairie voles. As adults, these females exhibited both increased alloparental behavior and partner preference. These results are consistent with the hypothesis that oxytocin can have both long-term organizational effects as well as acute activational effects on affiliative behaviors and parental behaviors (Keebaugh and Young 2011). A promising next step would be to compare the transcriptomes of the females.

8.5.3 Functional Genomics Beyond Nucleic Acids

Some of the studies described above identified gene networks that were highly correlated with specific transcription factors. ChIP-seq is an excellent technique for identifying direct targets

of transcription factors to better understand the relationship of these gene networks and their associated behavioral implications (Landt et al. 2012). Work from Eric Nestler's lab and others has found evidence for the role played by several prominent transcription factors, including a Fos family protein (ΔFosB), cAMP response element binding protein (CREB), and nuclear factor kappa B (NFκB), among several others, in the brain reward circuitry (Nestler 2012b). By integrating data from behavioral assays and DNA expression arrays with detailed analysis of chromatin remodeling and histone modification at drug-regulated gene promoters, these researchers were able to identify genes that are regulated by drugs of abuse via the induction of ΔFosB. These findings established that chromatin remodeling can play an important regulatory role underlying drug-induced behavioral plasticity and provided novel insight into the mechanisms by which ΔFosB regulated expression of specific target genes in reward pathways and contributes to addiction (Nestler 2008). Likewise, the study by Ament and colleagues discussed above found associations between behavior and the transcription factors *Creb*, *br*, *dl*, *Xbp1*, and others, suggesting that these genes are particularly promising candidates for functional characterization in future experiments (Ament et al. 2012). While these approaches have become feasible even in non-traditional model systems, few studies use ChIP-seq in behaviorally relevant contexts. It is clear, however, that future experiments should further investigate the interactions between transcription factors and DNA.

8.6 Into the Future

Research into the functional neurogenomics of social behavior has given us great insights into the evolutionarily conserved and plastic mechanisms that modulate neural and molecular responses to changes in an animal's social environment. We want to review and briefly summarize the major insights we have gained over the past decade and then discuss where we think the field might be heading.

8.6.1 Emerging Themes of Behavioral Genomics

What are some of the general insights that have emerged from the more than a decade of research behavioral genomics? First, we now know that the genome can change much more rapidly and dramatically in response to environmental stimuli than anyone thought possible (e.g., ca. 10 % of protein coding genes in only 30 min; Cummings et al. 2008). These dynamic properties likely reflect the real-time adjustments in the activity of gene networks in response to – and in preparation for – changes in the activity of both neural circuits and neuroendocrine systems (Hofmann 2010). Furthermore, a large fraction of the genome is involved in these responses, not merely a few genes (Renn et al. 2008; Whitfield et al. 2003). Particular functional groups or gene families appear to be involved in different kinds of plastic phenotypes as suggested by Aubin-Horth et al. (2009) and Sanogo et al. (2012). It also appears that a small set of transcription factors governs global changes in response to different environmental or social stimuli, giving rise to co-regulated gene sets or modules (Ament et al. 2012). Importantly, gene expression profiles can vary considerably across brain regions (Oldham et al. 2006), underscoring the importance of examining individual brain nuclei or even single neurons in future studies. Finally, there is increasing evidence that conserved or deeply homologous gene modules can be associated with behavioral phenotypes that have evolved independently (O'Connell and Hofmann 2012a; Toth and Robinson 2007). No one could have predicted any of these surprising and fundamental insights during the early days of behavioral genomics, but we believe that the best is yet to come.

8.6.2 New Horizons

With the rapid advances in sequencing technology, RNA-seq, ChIP-seq and related technologies are poised to replace microarray-based approaches for functional analyses of the dynamic genome. For example, a recent

review by Hitzemann et al. (2013) illustrates why RNA-seq is a superior strategy. While microarray analysis of gene expression is a mature technology, is relatively inexpensive, and has well developed analysis pipelines, it is limited by the need for primary sequence information and poor detection of rare transcripts, allelic variation, and splice variants. RNA-seq on the other hand requires no prior knowledge of expected transcripts, has wide dynamic range of detection, and provides information on individual sequence variation. However, it is still more expensive, requires bioinformatic expertise and high performance computing infrastructure (Hitzemann et al. 2013).

Both microarray and RNA-seq technology face similar difficulties when it comes to comparative studies across distantly related species. Hofmann and colleagues demonstrated the feasibility of heterologous hybridization for comparative analysis of gene expression. In the experiments, only genes with minimal sequence divergence could be compared (Renn et al. 2004). The same will probably be true for RNA-seq since part of the pipeline requires that orthologs be called, but this technology will have as added benefit information on sequence variation. Along those same lines, the choice of reference genome will be an important decision to make for comparative studies. One can use a well annotated genome from a more or less distantly related species or one can assemble reference transcriptomes *de novo* from the data collected. In any case, researcher need to keep in mind that the method chosen can have profound impacts on the outcome of the analysis (Grabherr et al. 2011).

Many of the studies described above obtained correlational results that suggested that, for example, a given transcription factor or set of transcription factors might be responsible for regulating dramatic genomic changes in response to stimuli (e.g., Ament et al. 2012; Nestler 2008; Sanogo et al. 2012). A number of techniques are available for testing the functional implications of such inferences. One option is to manipulate gene expression (e.g., using pharmacology, transgenic techniques, or siRNA) and examine the consequences of perturbed gene expression at the level of both behavior and transcriptome. If antibodies of the candidate transcription factor are available, one could use ChIP followed by PCR or *deep sequencing* to identify its direct targets. Also, several techniques for characterizing the response on a more spatially refined level are available, which allows the analysis of gene expression changes within and across the nodes of a neural circuit implicated in behavioral regulation. This list of reverse genomic approaches is by no means exhaustive, rather it is meant to raise awareness that methods well established in other fields (such as genetics, neuroscience, or microbiology) can be applied to of the integrative study of behavior and evolutionary and ecological genomics in general.

As more and more transcriptional datasets are made publically available, we are confident that these big data sets will be harnessed for biological discovery and that new approaches will be developed that will facilitate the comparison of data collected on different platforms (both existing and those yet to be invented). In conclusion, we urge researchers in the area of ecological and evolutionary functional genomics to combine forward genomics approaches (i.e., from phenotype to behaviorally relevant gene modules) with reverse genomic approaches (i.e., manipulating of novel gene modules to examine effects on behavior, hormones, and the genome itself). With such an integrative approach we will gain fundamentally new insights into the relationship between gene expression and behavior and their evolution. We can gain a lot of novel and fundamental insights into behavioral plasticity by examining genome activity across brain regions and discovering whether variation in gene expression profiles is due to differential regulation of chromatin structure and/or transcription factors. The future of this endeavor is sure to yield many great discoveries.

Acknowledgments Research in our lab has been supported by NSF, NIH-NIGMS, and the Alfred P. Sloan Foundation. We want to thank the editors for giving us the opportunity to review the state of our field; Jeff Gross, Suzy Renn, and members of the Matz Lab and the Hofmann Lab for discussion; and Ryan Wong and two anonymous reviewers for helpful comments on earlier versions of this chapter.

Glossary

Bisulfite sequencing The use of a bisulfite treatment of DNA followed by deep sequencing to determine the methylation pattern.

Chromatin immunoprecipitation sequencing (ChIP-seq) The use of high-throughput sequencing technologies to sequence the regions of the genome that interact with a given protein of interest, often a transcription factor.

Deep sequencing The process of obtaining both the sequence and frequency of RNA or DNA molecules in a given tissue at a given time through any number of next-generation sequencing technologies.

Dopaminergic reward processing The role that dopamine plays in the integration of environmental and physiological cues and the encoding of the rewarding properties of a stimulus to generate an adaptive behavioral response.

Gene network A statistical representation of correlated gene expression data for identifying sets of co-regulated genes or gene modules.

Gene module A set of co-regulated genes.

Immediate early genes (IEGs) Genes, usually encoding transcription factors, that are rapidly and transiently activated in response to a wide variety of cellular and extracellular stimuli.

Mating system A classification of the time, place, and number of partners an individual has during reproduction.

Microarray An array of thousands of RNA, cDNA, or DNA probes, usually printed on a glass slide with which the activity of thousands of genes can be assayed simultaneously.

Next-generation (NextGen) Sequencing (also referred to as high-throughput sequencing) Any of a number of technologies that yield millions of sequences concurrently by parallelizing the sequencing process, thereby significantly lowering the cost of sequencing while increasing the amount of data.

Nucleus accumbens (NAcc) A mesolimbic brain region that receives massive dopaminergic input from the VTA and is intimately involved in evaluating stimulus salience and reward processing.

Preoptic area (POA) A region of the forebrain that is important for regulating many social behaviors in males and females as well as other basic physiological functions such as energy homeostasis and thermoregulation.

Quantitative PCR (qPCR) A molecular technique used to amplify and simultaneously quantify a targeted DNA or RNA molecule.

Reproductive tactic Behavioral strategy used by individuals to increase their reproductive success.

RNA sequencing (RNA-seq) The use of high-throughput sequencing for quantitative analysis of short cDNA reads.

Small interfering RNA (siRNA) A class of double stranded RNA molecules, usually 20–25 base pairs, that interferes with the expression of genes with complementary sequence.

Social dominance High status or hierarchical rank in a social group.

Striato-pallidal Area X A region of the songbird brain that has been linked to singing. It is part of the basal ganglia, a set of nuclei that have been widely implicated in motor control and learning.

Transcription factor binding site Short stretches of DNA where other molecules, specifically transcription factors that regulate gene activity, can bind.

Transcriptome The set of all the expressed RNA molecules (or a subset, e.g., mRNA) in a given tissue or cell.

Ventral tegmental area (VTA) A region of the brain that is major source of dopamine in the brain. It plays an important role in evaluating the salience of environmental stimuli and signaling motivational events.

References

Alonso JM, Ecker JR (2006) Moving forward in reverse: genetic technologies to enable genome-wide phenomic screens in Arabidopsis. Nat Rev Genet 7: 524–536

Alonzo SH, Taborsky M, Wirtz P (2000) Male alternative reproductive behaviours in a Mediterranean wrasse, Symphodus ocellatus: Evidence from otoliths for multiple life-history pathways. Evol Ecol Res 2:997–1007

Ament SA, Wang Y, Robinson GE (2010) Nutritional regulation of division of labor in honey bees: toward a systems biology perspective. Wiley Interdiscip Rev Syst Biol Med 2:566–576

Ament SA, Blatti CA, Alaux C et al (2012) New meta-analysis tools reveal common transcriptional regulatory basis for multiple determinants of behavior. Proc Natl Acad Sci USA 109:E1801–E1810

Andersson M (1994) Sexual selection. Princeton University Press, Princeton

Ashburner M, Ball CA, Blake JA et al (2000) Gene ontology: tool for the unification of biology. The gene ontology consortium. Nat Genet 25:25–29

Aubin-Horth N, Landry CR, Letcher BH, Hofmann HA (2005) Alternative life histories shape brain gene expression profiles in males of the same population. Proc R Soc B 272:1655–1662

Aubin-Horth N, Desjardins JK, Martei YM et al (2007) Masculinized dominant females in a cooperatively breeding species. Mol Ecol 16:1349–1358

Aubin-Horth N, Letcher BH, Hofmann HA (2009) Gene-expression signatures of Atlantic salmon's plastic life cycle. Gen Comp Endocrinol 163:278–284

Barron AB, Sovik E, Cornish JL (2010) The roles of dopamine and related compounds in reward-seeking behavior across animal phyla. Front Behav Neurosci 4:163

Ben-Shahar Y, Robichon A, Sokolowski MB, Robinson GE (2002) Influence of gene action across different time scales on behavior. Science 296:741–744

Berton O, McClung CA, Dileone RJ et al (2006) Essential role of BDNF in the mesolimbic dopamine pathway in social defeat stress. Science 311:864–868

Brede M (2013) Short versus long term benefits and the evolution of cooperation in the prisoner's dilemma game. PLoS One 8:e56016

Brower L (1996) Monarch butterfly orientation: missing pieces of a magnificent puzzle. J Exp Biol 199:93–103

Brockmann HJ (2001) The evolution of alternative strategies and tactics. Adv Study Behav 30:1–51

Cummings ME, Larkins-Ford J, Reilly CRL et al (2008) Sexual and social stimuli elicit rapid and contrasting genomic responses. Proc R Soc B 275:393–402

De Souza FS, Santangelo AM, Bumaschny V et al (2005) Identification of neuronal enhancers of the proopiomelanocortin gene by transgenic mouse analysis and phylogenetic footprinting. Mol Cell Biol 25:3076–3086

Dong S, Replogle KL, Hasadsri L et al (2009) Discrete molecular states in the brain accompany changing responses to a vocal signal. Proc Natl Acad Sci USA 106:11364–11369

Drnevich J, Replogle KL, Lovell P et al (2012) Impact of experience-dependent and -independent factors on gene expression in songbird brain. Proc Natl Acad Sci USA 109(Suppl):17245–17252

Fahrbach SE (2006) Structure of the mushroom bodies of the insect brain. Annu Rev Entomol 51:209–232

Fehr E, Camerer CF (2007) Social neuroeconomics: the neural circuitry of social preferences. Trends Cogn Sci 11:419–427

Garcia de Leaniz C, Fleming I, Einum S et al (2007) A critical review of adaptive genetic variation in Atlantic salmon: implications for conservation. Biol Rev Camb Philos Soc 82:173–211

Grabherr MG, Haas BJ, Yassour M et al (2011) Full-length transcriptome assembly from RNA-Seq data without a reference genome. Nat Biotechnol 29:644–652

Greenberg JK, Xia J, Zhou X et al (2012) Behavioral plasticity in honey bees is associated with differences in brain microRNA transcriptome. Genes Brain Behav 11:660–670

Haehnel M, Menzel R (2012) Long-term memory and response generalization in mushroom body extrinsic neurons in the honeybee *Apis mellifera*. J Exp Biol 215:559–565

Hessler NA, Doupe AJ (1999) Social context modulates singing-related neural activity in the songbird forebrain. Nat Neurosci 2:209–211

Hilliard AT, Miller JE, Fraley ER et al (2012) Molecular microcircuitry underlies functional specification in a basal ganglia circuit dedicated to vocal learning. Neuron 73:537–552

Hitzemann R, Bottomly D, Darakjian P et al (2013) Genes, behavior and next-generation RNA sequencing. Genes Brain Behav 12:1–12

Hofmann HA (2003) Functional genomics of neural and behavioral plasticity. J Neurobiol 54:272–282

Hofmann HA (2010) The neuroendocrine action potential. Winner of the 2008 Frank Beach Award in Behavioral Neuroendocrinology. Horm Behav 58:555–562

Hoke KL, Pitts NL (2012) Modulation of sensory-motor integration as a general mechanism for context dependence of behavior. Gen Comp Endocrinol 176:465–471

Houde AE (1988) The effects of female choice and male-male competition on the mating success of male guppies. Anim Behav 36:888–896

Huffman LS, O'Connell LA, Hofmann HA (2013) Aromatase regulates aggression in the African cichlid fish *Astatotilapia burtoni*. Physiol Behav 112-113C: 77–83

Jarvis ED, Nottebohm F (1997) Motor-driven gene expression. Proc Natl Acad Sci USA 94:4097–4102

Keebaugh AC, Young LJ (2011) Increasing oxytocin receptor expression in the nucleus accumbens of pre-pubertal female prairie voles enhances alloparental responsiveness and partner preference formation as adults. Horm Behav 60:498–504

Krishnan V, Han M-H, Graham DL et al (2007) Molecular adaptations underlying susceptibility and resistance to social defeat in brain reward regions. Cell 131:391–404

Lammel S, Ion DI, Roeper J, Malenka RC (2011) Projection-specific modulation of dopamine neuron

synapses by aversive and rewarding stimuli. Neuron 70:855–862

Landt SG, Marinov GK, Kundaje A et al (2012) ChIP-seq guidelines and practices of the ENCODE and modENCODE consortia. Genome Res 22:1813–1831

Loebrich S, Nedivi E (2009) The function of activity-regulated genes in the nervous system. Physiol Rev 89:1079–1103

Lynch KS, Ramsey ME, Cummings ME (2012) The mate choice brain: comparing gene profiles between female choice and male coercive poeciliids. Genes Brain Behav 11:222–229

Machado HE, Pollen AA, Hofmann HA, Renn SC (2009) Interspecific profiling of gene expression informed by comparative genomic hybridization: a review and a novel approach in African cichlid fishes. Int Comp Biol 49:644–659

Madani R, Kozlov S, Akhmedov A et al (2003) Impaired explorative behavior and neophobia in genetically modified mice lacking or overexpressing the extracellular serine protease inhibitor neuroserpin. Mol Cell Neurosci 23:473–494

McGraw LA, Young LJ (2010) The prairie vole: an emerging model organism for understanding the social brain. Trends Neurosci 33:103–109

Mello CV, Vicario DS, Clayton DF (1992) Song presentation induces gene expression in the songbird forebrain. Proc Natl Acad Sci USA 89:6818–6822

Miranda E, Lomas DA (2006) Neuroserpin: a serpin to think about. Cell Mol Life Sci CMLS 63:709–722

Moll J, Krueger F, Zahn R et al (2006) Human fronto-mesolimbic networks guide decisions about charitable donation. Proc Natl Acad Sci USA 103:15623–15628

Morris J, Singh JM, Eberwine JH (2011) Transcriptome analysis of single cells. J Vis Exp 50:2634

Mukai M, Replogle K, Drnevich J et al (2009) Seasonal differences of gene expression profiles in song sparrow (*Melospiza melodia*) hypothalamus in relation to territorial aggression. PLoS One 4:14

Nadler JJ, Zou F, Huang H et al (2006) Large-scale gene expression differences across brain regions and inbred strains correlate with a behavioral phenotype. Genetics 174:1229–1236

Nestler EJ (2008) Transcriptional mechanisms of addiction: role of ΔFosB. Philos Trans R Soc B Biol Sci 363:3245–3255

Nestler (2012a) Stress makes its molecular mark. Nature 490:171

Nestler EJ (2012b) Transcriptional mechanisms of drug addiction. Clin Psychopharmacol Neurosci 10: 136–143

Newman SW (1999) The medial extended amygdala in male reproductive behavior. A node in the mammalian social behavior network. Ann NY Acad Sci 877: 242–257

O'Connell LA, Hofmann HA (2011a) The vertebrate mesolimbic reward system and social behavior network: a comparative synthesis. J Comp Neurol 519:3599–3639

O'Connell LA, Hofmann HA (2011b) Genes, hormones, and circuits: an integrative approach to study the evolution of social behavior. Front Neuroendocrinol 32:320–335

O'Connell LA, Hofmann HA (2012a) Evolution of a vertebrate social decision-making network. Science 336:1154–1157

O'Connell LA, Hofmann HA (2012b) Social status predicts how sex steroid receptors regulate complex behavior across levels of biological organization. Endocrinology 153:1341–1351

Oldham MC, Horvath S, Geschwind DH (2006) Conservation and evolution of gene coexpression networks in human and chimpanzee brains. Proc Natl Acad Sci USA 103:17973–17978

Osborne KA, Robichon A, Burgess E et al (1997) Natural behavior polymorphism due to a cGMP-dependent protein kinase of *Drosophila.* Science 277:834–836

Renn SC, Aubin-Horth N, Hofmann HA (2004) Biologically meaningful expression profiling across species using heterologous hybridization to a cDNA microarray. BMC Genomics 5:42

Renn SC, Aubin-Horth N, Hofmann HA (2008) Fish and chips: functional genomics of social plasticity in an African cichlid fish. J Exp Biol 211:3041–3056

Replogle K, Arnold AP, Ball GF et al (2008) The Songbird Neurogenomics (SoNG) Initiative: community-based tools and strategies for study of brain gene function and evolution. BMC Genomics 9:131

Reppert SM, Gegear RJ, Merlin C (2010) Navigational mechanisms of migrating monarch butterflies. Trends Neurosci 33:399–406

Robinson GE, Fernald RD, Clayton DF (2008) Genes and social behavior. Science 322:896–900

Ross RM (1990) The evolution of sex-change mechanisms in fishes. Environ Biol Fish 29:81–93

Sanogo YO, Band M, Blatti C et al (2012) Transcriptional regulation of brain gene expression in response to a territorial intrusion. Proc R Soc B 279:4929–4938

Schumer M, Krishnakant K, Renn SCP (2011) Comparative gene expression profiles for highly similar aggressive phenotypes in male and female cichlid fishes (*Julidochromis*). J Exp Biol 214:3269–3278

Segal E, Friedman N, Koller D, Regev A (2004) A module map showing conditional activity of expression modules in cancer. Nat Genet 36:1090–1098

Teramitsu I, White SA (2006) FoxP2 regulation during undirected singing in adult songbirds. J Neurosci 26:7390–7394

Teramitsu I, Poopatanapong A, Torrisi S, White SA (2010) Striatal FoxP2 is actively regulated during songbird sensorimotor learning. PLoS One 5:e8548

Tinbergen N (1963) On aims and methods of ethology. Z Tierpsychol 20:410–433

Toth AL, Robinson GE (2007) Evo-devo and the evolution of social behavior. Trends Genet 23:334–341

Toth AL, Varala K, Henshaw MT et al (2010) Brain transcriptomic analysis in paper wasps identifies genes associated with behaviour across social insect lineages. Proc R Soc B 277:2139–2148

Wheat CW, Fescemyer HW, Kvist J et al (2011) Functional genomics of life history variation in a butterfly metapopulation. Mol Ecol 20:1813–1828

Whitaker KW, Neumeister H, Huffman LS et al (2011) Serotonergic modulation of startle-escape plasticity in an African cichlid fish: a single-cell molecular and physiological analysis of a vital neural circuit. J Neurophysiol 106:127–137

Whitfield CW, Cziko A-M, Robinson GE (2003) Gene expression profiles in the brain predict behavior in individual honey bees. Science 302:296–299

Whitfield CW, Ben-Shahar Y, Brillet C et al (2006) Genomic dissection of behavioral maturation in the honey bee. Proc Natl Acad Sci USA 103:16068–16075

Wingfield JC, Hahn TP (1994) Testosterone and territorial behaviour in sedentary and migratory sparrows. Anim Behav 47:77–89

Wong RY, Hofmann HA (2010) Behavioural genomics: an organismic perspective. In: Encyclopedia of life sciences. Wiley, Chichester

Wong RY, Ramsey ME, Cummings ME (2012) Localizing brain regions associated with female mate preference behavior in a swordtail. PLoS One 7: e50355

Zhang B, Horvath S (2005) A general framework for weighted gene co-expression network analysis. Stat Appl Genet Mol Biol 4: Article 17

Ecological Genomics of Host Behavior Manipulation by Parasites

9

François Olivier Hébert and Nadia Aubin-Horth

Abstract

Among the vast array of niche exploitation strategies exhibited by millions of different species on Earth, parasitic lifestyles are characterized by extremely successful evolutionary outcomes. Some parasites even seem to have the ability to 'control' their host's behavior to fulfill their own vital needs. Research efforts in the past decades have focused on surveying the phylogenetic diversity and ecological nature of these host-parasite interactions, and trying to understand their evolutionary significance. However, to understand the proximal and ultimate causes of these behavioral alterations triggered by parasitic infections, the underlying molecular mechanisms governing them must be uncovered. Studies using ecological genomics approaches have identified key candidate molecules involved in host-parasite molecular cross-talk, but also molecules not expected to alter behavior. These studies have shown the importance of following up with functional analyses, using a comparative approach and including a time-series analysis. High-throughput methods surveying different levels of biological information, such as the transcriptome and the epigenome, suggest that specific biologically-relevant processes are affected by infection, that sex-specific effects at the level of behavior are recapitulated at the level of transcription, and that epigenetic control represents a key factor in managing life cycle stages of the parasite through temporal regulation of gene expression. Post-translational processes, such as protein-protein interactions (interactome) and post translational modifications (e.g. protein phosphorylation, phosphorylome), and processes modifying gene expression and translation, such as interactions with microRNAs

F.O. Hébert • N. Aubin-Horth (✉)
Département de Biologie, Institut de Biologie Intégrative et des Systèmes, Université Laval 1030, avenue de la Médecine, Québec, QC G1V 0A6 Canada
e-mail: francois-olivier.gagnon-hebert.1@ulaval.ca; Nadia.Aubin-Horth@bio.ulaval.ca

C.R. Landry and N. Aubin-Horth (eds.), *Ecological Genomics: Ecology and the Evolution of Genes and Genomes*, Advances in Experimental Medicine and Biology 781, DOI 10.1007/978-94-007-7347-9_9,

(microRNAome), are examples of promising avenues to explore to obtain crucial insights into the proximal and ultimate causes of these fascinating and complex inter-specific interactions.

Keywords
Host-parasite interactions • Behavioral manipulation • Parasitology • Transcriptome • Proteome • Epigenome • Phosphorylome • Interactome

9.1 Introduction

Humans are mammals characterized by a strong capability of self-awareness, and the notion of having one's personality abducted or controlled by another entity is terrifying and inconceivable. Petrifying ideas of mind usurpation have fueled the cinematographic industry for decades; for instance, renowned Canadian filmmaker David Cronenberg projected onscreen in 1975 the apocalyptic end of humanity through massive endoparasite infections in a movie entitled *Shivers* (http://imdb.com). When the movie came out, the original poster suggested that the concept of a genetically modified parasite turning innocent suburban residents into mindless fiends was "beyond the power of priest or science to exorcise". However, the concept of a parasitic infection altering the host's physical appearance or radically changing its behavior has been around much longer than modern horror and science fiction. Parasites from numerous distinct phyla have thrived for the past 500,000,000 years by defrauding and evading elaborate defense mechanisms deployed by their hosts, giving birth to intricate coevolutionary systems (Schmidt-Hempel 2011).

Recent studies suggest that parasites can be 20 times more numerous than predators in an ecosystem and that they have an influence on interactions between trophic levels (Kuris et al. 2008; Marcogliese and Cone 1997). A parasite can be defined as an organism living in close proximity with another organism from which it gains benefits, usually leading to decreased physical condition of the host. The numerous effects that parasites have on their hosts are multidimensional (Thomas et al. 2010): infected hosts may show slower growth (Wright et al. 2007), reduced reproductive output (Candolin and Voigt 2001; Heins et al. 2010), and even changes in morphology (Dingemanse et al. 2009, see Cézilly et al. 2013 for a complete review). Parasites often have complex life cycles that require several hosts to complete the dispersal and reproductive stages of their development. Some parasites seem to successfully pass through these stages by manipulating different aspects of the biology of their host, including behavior (Lefèvre et al. 2008; Poulin 2010). Hindsbo (1972) and Holmes and Bethel (1972) were the first to empirically describe how infections by an acanthocephalan parasite results in the alteration of host behavior and appearance, thereby increasing their transmission to their definitive host. The major phenotypic consequences associated with these types of infections can be referred to as "behavioral alterations", defined as any changes in host behavior that result from parasitic infection.

Significant progress has been accomplished towards describing and understanding peculiar associations between vastly different species (Adamo 2013). Some of these inter-specific interactions involve stunning behavioral alterations that only seem possible in science fiction stories and Hollywood scenarios. One exceptional example is the parasitic wasp *Cotesia congregata* that lays its eggs in the body of the caterpillar of *Manduca sexta*, so that larvae can feed on the host's haemolymph during their development, without damaging internal organs. Once larval development is completed, the wasps use their two mouthparts to perforate the caterpillar's body wall in order to exit their host (reviewed in Beckage and Templeton 1986). The wasps subsequently build a cocoon

that remains attached to the host's body, from which adult wasps emerge after 4–5 days of development. The sudden appearance of multiple wounds in the host's body stimulates the release of cytokines, among which paralytic peptides believed to be used to help heal the wounds by increasing haemocyte stickiness and reducing locomotive activity, thus enhancing recovery (Skinner et al. 1991). Wasps can also control other arthropods. Larvae of the ichneumonid wasp *Hymenoepimecis sp*. induce their spider host (*Plesiometa argyra*) to build a unique cocoon-like web, instead of the normal web that it usually produces, to act as a larval cocoon that will protect them during their development (Eberhard 2000). Fungal parasites like *Ophiocordyceps unilateralis* that infects arboreal *Camponotus leonardi* ants can also cause complex behavioral alteration. In this case, the infected ant climbs down its normal tree habitat and bites vigorously into the principal vein of a leaf at about 25 cm above the ground, where it quickly dies, allowing the fungus to wrap it into a hyphal growth, securing it to the leaf and providing a safe and stable growth environment from which it can efficiently propagate (Andersen et al. 2009). Another impressive case is the nematode (*Myrmeconema neotropicum*), which turns the abdomen of a tropical arboreal ant (*Cephalotes atratus*) bright red after infection, causing it to stand on a leaf among patches of red berries, with its abdomen elevated at a precise angle, so that frugivorous birds – the final hosts – can ingest their red bellies full of the parasite's eggs (Yanoviak et al. 2008).

The aforementioned examples of behavioral alterations have mainly focused on invertebrate hosts, but parasites also alter the behavior of vertebrate hosts. The protozoan parasite *Toxoplasma gondii* infects rats as intermediate hosts and can convert the rat's natural disdain for the odor of cat urine into a positive signal associated with desirable and attractive smells, thus increasing the parasite's chances of transmission to the final host, a cat, through predation (House et al. 2011; Vyas et al. 2007). A trematode-infected killifish's swimming behavior is altered in such a way that it is 30 times more likely to be predated by a bird, the larval trematode's final host (Lafferty and Morris 1996; Shaw et al. 2009). Clearly, complex parasitic infections triggering behavioral changes in the host can target many different behaviors and involve vastly different parasitic and host species, from small and simple protozoan organisms to primates such as humans (Table 9.1). This striking phylogenetic diversity suggests that behavioral alterations in parasitized individuals evolved on several independent occasions.

Three different explanations have been proposed to explain behavioral alterations (reviewed in Poulin 2010). First, the most parsimonious explanation suggests that the change in behavior is simply a side effect, a by-product of the pathology experienced by the infected animal. Indeed, most sick animals do behave differently than their healthy conspecifics. Second, behavioral alteration can be an adaptive response of the host to infection – compensatory response, aimed at fighting the parasite and its negative physiological consequences. Finally, it is possible that the change in behavior is an adaptation developed by the parasite to increase its trophic transmission rate, which can be referred to as "adaptive behavioral manipulation". In this case, the parasite specifically acts on the host to alter its behavior in a way that augments the parasite's fitness.

Many research studies in the past have focused their attention on describing ecological interactions between the host and the parasite and trying to understand their adaptive or evolutionary significance (Hughes et al. 2012; Moore 2002; Poulin 2007). However, to decipher the proximate (mechanistic) and ultimate (evolutionary) causes of such a fascinating phenomenon, including determining if it represents an adaptation, it is essential to uncover the mechanisms directly or indirectly involved in these host-parasite interactions. The goal of this chapter is thus to integrate past and present knowledge to reflect on how to use large-scale ecological genomics tools to understand the fundamental molecular mechanisms – covering as many biologically relevant levels as possible (Fig. 9.1) – implicated in behavioral alterations of parasite-infected hosts.

Table 9.1 Diversity of parasites, hosts and behaviors reported to change following parasitic infection

Parasite		Host			
Phylum (class)	Species	Phylum (class)	Species	Behavior	References
Arthropoda (Insecta)	*Cotesia congregata*	*Arthropoda (Insecta)*	*Manduca sexta*	Stops moving and serves as a shelter for wasp larvae	Beckage and Templeton (1986)
Ascomycota (Sordari-omycetes)	*Ophiocordyceps unilateralis*	*Arthropoda (Insecta)*	*Camponotus leonardi*	Climbs down the tree, bites vigorously into a leaf and awaits its death in an optimal growing and disseminating fungal environment	Andersen et al. (2009)
Nematoda (Adenophorea)	*Myrmeconema neotropicum*	*Arthropoda (Insecta)*	*Cephalotes atratus*	Turns bright red, stops moving and imitates a berry by exposing its gaster, ending up eaten by definitive host	Yanoviak et al. (2008)
Apicomplexa (Conoidasida)	*Toxoplasma gondii*	*Chordata (Mammalia)*	*Rattusnorvegicus*	Becomes attracted to cat urine	Vyas et al. (2007)
Plathelminthes (Cestoda)	*Echinococcus granulosus*	*Chordata (Mammalia)*	*Alcesalces* (moose)	More likely to be killed by humans	Rau and Caron (1979)
Plathelminthes (Cestoda)	*Schistocephalus solidus*	*Chordata (Actinopterygii)*	*Gasterosteus aculeatus*	Swims near the surface where it is more vulnerable to predators	Barber et al. (2004)
Microsporidia (Microsporea)	*Glugea anomala*	*Chordata (Actinopterygii)*	*Gasterosteus aculeatus*	Enhanced predator avoidance response	Milinski (1985)
Apicomplexa (Conoidasida)	*Eimeriavermiformis*	*Chordata (Mammalia)*	*Mus musculus*	Attracted to predator odours	Kavaliers and Cowell (1995)
Arthropoda *(Insecta)*	*Hymenoepimecisargyraphaga*	*Arthropoda (Arachnida)*	*Plesiometa argyra*	Builds a special cocoon for the development of parasitoid larvae just a few hours before they kill it and eat it after which they begin their development within the cocoon	Eberhard (2000)
Virus	Barley yellow dwarf virus (BYDV)	*Arthropoda (Insecta)*	*Rhopalosiphum padi*	Virus infected aphids prefer non-infected wheat plants and non-infected aphids prefer infected wheat plants	Ingwell et al. (2012)
Nematoda (Chromadorea)	*Trichostrongylus tenuis*	*Chordata (Aves)*	*Lagopus lagopus scotica*	Infected individuals are less aggressive	Fox and Hudson (2001)
Apicomplexa (Conoidasida)	*Toxoplasma gondii*	*Chordata (Mammalia)*	*Homo sapiens sapiens*	More inclined in taking risks while driving	Flegr et al. (2002)

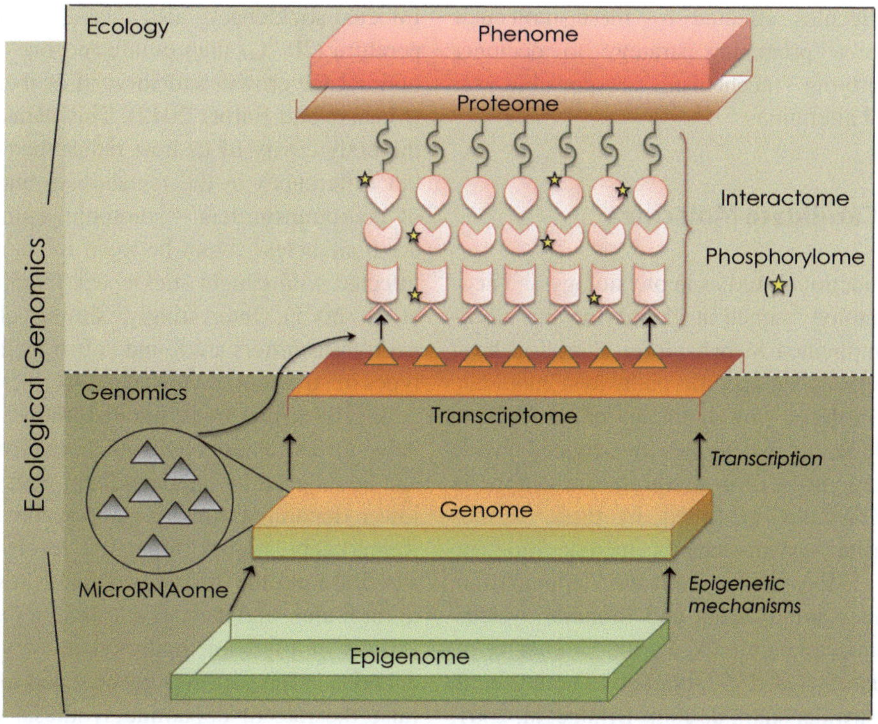

Fig. 9.1 Concept of ecological genomics. The genome – *orange box* – encodes fundamental information dictating the nature of the molecules that can be produced by a given organism. Genome-wide patterns of epigenetic marks – epigenome, *green envelope* – then influence which regions of the genome are accessible, while large-scale gene transcription (transcriptome, *orange lid*) determines the abundance and/or the timing of protein production within the organism (proteome, *orange brackets*). The genome also produces a vast array of microRNAs, the microRNAome (*gray triangles*), that regulate translation – symbolized by *gray triangles* blocking the link between the transcriptome and the interactome/proteome – and accelerate mRNA degradation. Following various post-translational modifications – phosphorylome, *yellow stars* –, all of the proteins transcribed and translated will interact together – interactome, *red* geometrical forms – to produce a wide range of phenotypic traits – phenome, *red box*. These phenotypes are the ultimate targets on which selection will act. All levels of biological organization are thus intricately inter-related to each other and in order to fully understand ecological questions, it is essential to integrate these different levels in a global perspective, which constitutes the basis of ecological genomics

9.2 Studying the Mechanisms Involved in Host Behavior Alteration

When parasitic infections occur in nature, a molecular dialog between the host and its parasite begins, often taking the form of an arms race (Biron and Loxdale 2013). The underlying mechanisms responsible for behavioral changes triggered by various types of infections have yet to be fully understood. To address this issue, it is crucial to bear in mind that the central nervous system (CNS) integrates and processes essential information, such as perceiving and filtering environmental stimuli, controlling locomotor activities and managing physiological pathways according to the input of sensory receptors. It can thus be hypothesized that many changes in host behavior will, to some extent, be correlated with molecular changes in the CNS (Biron and Loxdale 2013). Furthermore, changes in behavior likely have consequences for several systems within the organism, including the CNS. Trying to identify signaling molecules that are involved in these transfers of information and

other molecules affected by these molecular signals is a promising strategy to decipher the underlying mechanisms responsible for behavioral alterations.

9.2.1 Candidate Molecules

Pioneer functional studies in physiology and neuroendocrinology aimed at identifying the mechanisms implicated in behavioral alteration have focused mainly on specific pathways, candidate genes or molecules of significant effect (Adamo 2013). These targeted studies investigated a wide phylogenetic array of invertebrate and vertebrate hosts. Molecules identified in these various host-parasite systems chiefly include biogenic amines – dopamine, serotonin, epinephrine, octopamine, amino acids and immune system-associated cytokines (silk worm: Skinner et al. 1991; Adamo et al. 1997, crickets: Thomas et al. 2003, freshwater snail: de Jonk-Brink et al. 2001, killifish: Shaw et al. 2009; Shaw and Overli 2012, mouse: Skallova et al. 2006). For instance, the impact of a trematode infection (*Microphallus papillorobustus*) on serotonin levels in a gammarid host (*Gammarus insensibilis*) has been shown, and is believed to be responsible for the change in the host's responses to mechanical and photic stimuli (Helluy and Thomas 2003). A similar candidate approach applied to crickets (*Nemobius sylvestris*) committing "suicide" by jumping in a water body after being infected by the nematomorph hairworm (*Paragordius tricuspidatus*) revealed the implication of multiple changes in brain levels of amino acids – some acting as neurotransmitters in insects or potential precursors of biogenic amines – during this peculiar behavioral alteration (Thomas et al. 2003).

Threespine stickleback fish (*Gasterosteus aculeatus*) are the intermediate host of the flatworm *Schistocephalus solidus*. Sticklebacks infected by this flatworm lose their anti-predator response (Barber et al. 2004; Godin and Sproul 1988) and wild infected individuals are observed close to the water surface during the day at a higher frequency than non-infected conspecifics and have a higher chance of being captured by a bird predator, the final host of their parasite (Quinn et al. 2012). Infected sticklebacks also choose a warmer temperature (20 °C) than uninfected ones, which is optimal for growth and survival of their parasite (Macnab and Barber 2012). This parasite infects the body cavity of its host rather than the brain, but differences in the metabolism and quantity of neurotransmitters – serotonin, epinephrine – are nonetheless found between infected and uninfected wild-caught stickleback females (Overli et al. 2001). Interestingly, similar changes in neurotransmitters are found in fish facing various types of stressors (reviewed in Overli et al. 2001).

In the same line, suicidal feline attraction behavior in rats infected with *Toxoplasma gondii* can be reduced by injecting them with drugs that lower dopamine activity (Webster et al. 2006). Remarkably, higher dopamine levels are measured around the *Toxoplasma* parasites encysted in rat brains, and these levels are the result of the activity of a tyrosine hydroxylase enzyme that is encoded in the parasites genome and functions to raise the rate of dopamine synthesis within the host's brain (Prandovsky et al. 2011). In this case, it is known that the parasite itself is producing a molecule that will modulate the neuroendocrinological system of the host and result in a change in behavior. Whether such direct manipulation is causing the differences in the other host-parasite systems presented above remains to be tested (see Sect. 9.2.3). Some of the aforementioned studies suggest a link between the parasite affecting the host's immune system and the subsequent modulation of neutrally active molecules capable of altering behavior directly or as a by-product of their function (Adamo 2013). This may particularly be the case for the serotonergic system (Helluy 2013).

Candidate gene studies mostly use high performance liquid chromatography (HPLC) (Shaw et al. 2009; Skallova et al. 2006, see Mant et al. 2007 for a general review on the method), immunohisto/cytochemistry (Helluy and Thomas 2003, see Buchwalow and Bocker 2010 for a general overview on the method) or pharmacological manipulations coupled with behavioral assessments (Helluy and Holmes 1990) to identify and localize potentially interesting molecules. Several studies using these approaches have identified candidate molecules

putatively involved in proximate mechanisms responsible for parasite-induced behavioral changes (Adamo 2013; Biron and Loxdale 2013). However, as exemplified in this review, post-translational processes, such as protein-protein interactions and post-translational modifications (e.g. protein phosphorylation), and processes modifying gene expression and translation, such as epigenetic marks and interactions with micro RNAs, are examples of promising avenues to explore in order to deepen our understanding of behavioral alteration. Recent bio-technological developments have further promoted the rapid acquisition of massive high-throughput datasets in various fields of research. These large-scale, assumption-free approaches not only provide further evidence suggesting that closely related chemical messengers are involved in interactions in parasites-hosts pairs belonging to different taxonomic groups (see Ponton et al. 2006 for innovative work and Poulin 2011 for a complete review), but also expose new unsuspected candidates and potentially very important interactions among molecules.

9.2.2 A Large-Scale Perspective on Parasite-Mediated Behavior Alteration

Behaviors have a complex underlying molecular architecture (e.g. in mice: Weber et al. 2013, reviewed in Bendesky and Bargmann 2011) and as a consequence, numerous molecular targets could be involved in shaping behavioral alterations in hosts by parasites and interactions are probably a fundamental part of these traits (Albert et al. 2009; Aubin-Horth et al. 2005; Filby et al. 2010; Mukai et al. 2009; Zwarts et al. 2011). Information is thus gained by studying the correlated variation of several molecules with behavior, such as variations in mRNA levels, and to characterize these covarying molecules as *modules* (examples in the study of behavior: deJong et al. 2010; Renn et al. 2008, see Aubin-Horth and Renn 2009 for a synthesis within an ecological and evolutionary perspective and Zhang and Horvath 2005 for the use of Weighted Correlation Network Analysis to define modules). This approach, of course, necessitates monitoring several of these molecules simultaneously. It is also important to uncover which molecular networks are involved and which functions are affected, although this may only become apparent with a higher order functional analysis – such as the identification of enrichment for some biological functions in the molecules that differ significantly in their abundance or activity in infected individuals (The Gene Ontology Consortium, 2008). The multi-genic and modular nature of the altered behavioral traits thus calls for a global, unbiased and quantitative approach to understand the mechanisms underlying these alterations, which has become more feasible with the development of high-throughput technologies in the last two decades (Hawkins et al. 2010). These large-scale approaches include surveying changes in gene expression (transcriptome), in levels of microRNAs (miRNAome), in epigenetic marks such as DNA methylation (epigenome), in protein levels and localization (proteome), in post-transcriptional modification of these proteins such as activation or inactivation by phosphorylation (phosphorylome) and in the interactions of these proteins with each other (interactome) (Fig. 9.1). Because of the global survey approach it provides, these methods can include the study of candidate molecules while also allowing the discovery of new ones that have not been implicated in these phenotypes before, therefore combining a hypothesis-driven approach with the possibility of uncovering new molecular pathways associated with behavior alteration in parasitized hosts.

In many instances of behavioral alteration/manipulation, data suggests that parasites influence several phenotypic dimensions in their host (Cézilly et al. 2013), which means that to maximize our understanding of their proximal and ultimate causes, these multidimensional consequences of infection by a parasite and their molecular causes should be studied jointly. For example, the previously described infection of an ant by the nematode *M. neotropicum*, results in the ant altering its locomotion pattern and mimicking a red berry by raising its red abdomen (reviewed in Yanoviak et al. 2008).

In this example – as in many other instances of behavioral alteration – the success of the changes in host behavior for the parasite depends on the synergetic effects of numerous altered traits in the host in addition to behavior. Such concurrent changes in behavioral, physiological and morphological traits to form an elaborate phenotype is referred to as "phenotypic integration" (Ketterson et al. 2009; Pigliucci and Preston 2004). These phenotypic correlations are thought to play an important role in many species, as they maintain stasis and influence the evolutionary trajectory of numerous traits (Ketterson et al. 2009). It is thus crucial to uncover the underlying causes of these correlated changes at different levels to understand how the parasite can simultaneously affects it's host behavior and morphology by breaking apart the correlations observed in uninfected individuals or by creating new correlations between traits.

A specific case of phenotypic integration applied to the study of behavior is the concept of behavioral syndromes. Consistent differences in behavioral traits among individuals define their personality – e.g. exploratory, aggressive, bold (Sih et al. 2004, see Box 9.1). These personality traits can also be correlated across individuals in a population, such that the most aggressive individuals are also the boldest and vice-versa, which is referred to as "behavioral syndromes" (see Barber and Dingemanse 2010 and Poulin 2013 for reviews of the concept applied to host-parasite interactions). One could ask if a parasite infection can alter these integrated phenotypes by breaking apart the behavioral syndrome or if on the contrary they take advantage of their existence by "high jacking" the molecular cause of these correlations.

Because of all these reasons, research efforts aiming at a complete mechanistic understanding of host-parasite interaction should thus be focused on large-scale, assumption-free and genome-wide analyses – i.e. from a large genetic and phenotypic perspective – that take into account the multiple levels of the host traits that are influenced by their parasites. Also, for molecules or molecular processes that have been shown to vary between infected and uninfected individuals, or between stages of infection, the next essential step is to go beyond correlative proof by directly determining if the observed change is a cause or a consequence of the behavioral alteration in the host. To do so, it is crucial to determine if manipulating the molecule/process directly affects behavior and the associated traits forming the integrated phenotype – physiology, morphology, and any other relevant trait such as reproduction (Pavey et al. 2012) (see Sect. 9.5).

On a molecular level, correlations between traits as discussed above might be explained by three non-mutually exclusive scenarios: they are part of the same gene network controlled by either (i) pleiotropic effects – a gene (or hormone) has an effect on several traits, (ii) linkage disequilibrium – a non-random association between alleles, or (iii) genetic interactions involving several genes. Empirical evidence on host-parasite interactions showed that modifying the expression of only one neurotransmitter or hormonal compound can result in cascading effects on multiple related behavioral traits (Adamo 2002; Shaw et al. 2009), supporting a pleiotropic effect. To tease apart these three possibilities, functional analyses would be essential (Sect. 9.5).

Box 9.1 Potential Impacts of Parasites on Host Behavioral Syndromes

Behavioral syndrome (Fig. 9.2) corresponds to correlations between behaviors across individuals within a given population (reviewed in Réale et al. 2007). Parasite infections can alter correlated traits that are part of a population behavioral syndrome in different ways, as discussed in Poulin (2010, 2013). First, the infection can simply increase the global phenotypic value of one of the behaviors, without affecting the correlation between the traits (Fig. 9.2a). Second, it can strengthen the correlation by reducing the variance around the trend line (Fig. 9.2b). Third, it can

(continued)

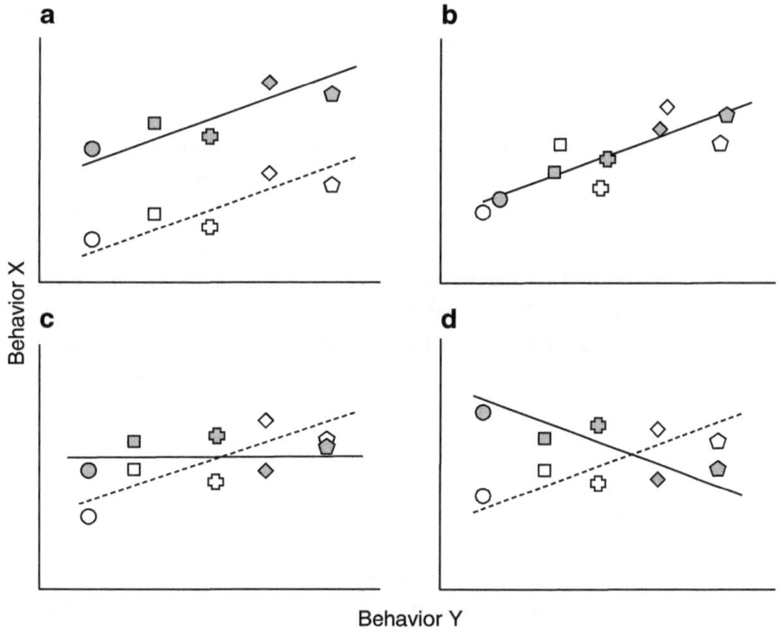

Fig. 9.2 Impact of parasite infections on behavioral correlations. (a) Values of only one behavior increase, (b) Stronger correlation between both behaviors, (c) Association between behaviors is uncoupled, (d) Inversion of the correlation. White symbols: before infection. Gray symbols: after infection. (Modified from Poulin (2010))

Box 9.1 (continued)

uncouple the association between traits. Their values are thus no longer significantly correlated (Fig. 9.2c). Fourth, it can reverse the direction in which the traits correlate (Fig. 9.2d). These theoretical modifications in behavioral syndromes do not represent an exhaustive list of all the exiting possibilities. They provide examples of how to address the question of host behavioral alteration on a large-scale, i.e. by measuring changes in how multiple behavioral traits correlate as a result of parasite infection, instead of comparing mean trait values between parasitized and non-parasitized individuals.

9.2.3 Groundbreaking Proteomics Studies

The field of proteomics, defined as the large-scale study of protein structure and functions – for an overview of methods, see Diz et al. 2012, is a vital component of the study of host-parasite interactions, given the wide implications of proteins in physiological pathways of the cells. Characterization of the parasite and host proteomes expressed by their genome within specific environmental conditions is called 'parasito-proteomics' (Biron et al. 2005a, c). Two pioneer studies relying on a parasito-proteomics approach aimed at deciphering the molecular mechanisms responsible for the water-seeking behavior induced in orthopteran species – e.g. crickets, grasshoppers – by hairworm parasites (Biron et al. 2005b, 2006). Using 2-D gel electrophoresis (2-DE) combined with mass spectrometry, they showed that molecules secreted by the parasite – analogous to Wnt family protein, involved in signal transduction, could directly influence some of the host's CNS functions. Two parasitic species that each infects two distinct orthopteran species use opposite tactics for similar behavioral alterations – i.e. induction of suicidal tendencies: inhibition of apoptosis and induction of apoptosis. Studies focusing on these approaches highlighted the importance of proteins associated with neurogenesis and neurotransmitters, while also acknowledging the role of various modifications in the Wnt signalling pathway that have crucial

functions in the central nervous system of the host (Biron et al. 2006). These empirical results were among the first pieces of evidence strongly suggesting that host behavioral changes are modulated via direct and indirect biochemical alterations.

9.2.4 Studying Molecular Convergence of Behavior Manipulation with Proteomics

The aforementioned example describing how orthopteran suicide behavior is modulated by a parasite shows that the same parasite (the hairworm) can manipulate two different hosts (cricket and grasshopper), in part, by using the same molecular functions. However, the opposite can also be true, when phylogenetically distant parasites show convergence in host behavior-manipulation. A trematode (*Microphallus papillorobustus*) and an acanthocephalan (*Polymorphus minutus*) have been shown to infect closely related gammarid species (*Gammarus insensibilis and G. pulex*) and alter their behavior in a similar manner: the infected hosts swim to the surface where they are eaten by the avian final host of the parasite (Ponton et al. 2006). Both infections result in negative geotaxis, aberrant locomotion and a higher predation probability for their host, and alter behavior only when they are at a stage in their life cycle at which they are ready to infect their final host. With this repeated and independent evolution of behavior alteration, one can ask if the same molecules are affected in the closely related hosts, if particular functional characteristics can be uncovered and if a specific part of the molecular network is the weakest link, or the corner stone of the host's behavior manipulation.

Ponton et al. (2006) studied the brain proteome of uninfected and infected individuals. Data on large-scale changes measured during distinct stages of behavioral alteration can be used to characterize the molecular *signature* that is specific to this modification. This approach of looking at – sometimes small – changes in a large array of molecules has been used in cancer research to show that cancers that had similar symptoms, but different survival outcomes and response to treatment, had in fact completely independent molecular causes and signature (Lamb et al. 2006). Using 2-D protein electrophoresis and mass spectrometry, Ponton et al. (2006) determined the identity of the differentially expressed peptides found in infected individuals by mass similarity with previously sequenced peptides in insects and other groups – a technique called peptide mass fingerprinting. Interestingly, their results show partially similar brain proteome changes – proteins involved in CNS development, activation of immune-related proteins – in the two host species. Their results show that while the specific molecules are distinct between the two systems, a higher-order analysis focusing on the functions of these molecules allows detection of a partial overlap in the molecular signature of infection. They detected convergence in the pathways potentially 'manipulated' by the parasite's presence, for example, the pathway of nitric oxide synthesis, which has recently been shown to act as a neurotransmitter, but is also involved in the immune system (Ponton et al. 2006). Convergence between different parasite lineages in terms of behavioral changes induced and in terms of pathways or fundamental mechanisms used to achieve greater transmission rate can support the hypothesis that behavior alteration is an adaptation, although it should be coupled with other types of evidence (Losos 2011).

Ponton et al. (2006) also found a distinct proteome-alteration signature – for example of vision-related proteins, which is in accordance with differences in the two gammarid species. The host species showing these vision-related protein changes (*G. insensibilis*) also show a reversal to positive phototaxis after infection by the acantocephalan. Remarkably, one of the proteins involved in vision and found to be altered in the infected gammarid was also found to be differentially expressed in a suicidal cricket (*N. sylvestris*) infected by a nematomorph worm (*P. tricuspidatus*), whose vision and phototaxis known to be altered – discussed above, see Biron et al. 2006 for a detailed comparative synthesis.

This study can be used to illustrate a very important aspect of studying the molecular changes

associated with behavior alteration: controls. The strength of this study indeed lies in its comparative approach, but also on the addition of uninfected individuals constrained to live at the surface of the water body to mimic the surface-dwelling behavior of infected gammarids. This control aims to remove effects of water depth and its correlates – food, temperature, light – that could greatly affect the brain proteome. Furthermore, they also focus only on males and on single infections. Remarkably, Ponton et al. (2006) find that infected hosts and controls living at the surface were more similar in their proteome than they were to uninfected individuals living deeper in the water column, showing the importance of having well thought-out controls. Comparing the signature associated with parasite infection against the molecular signature associated with the response of the host to well-known stressors such as starvation, temperature, pH change or social stress, should allow differentiating the signatures that are unique to the behavioral manipulation from the ones that are similar among stressors.

In the systems formed of a gammarid host and either a trematode or an acanthocephalan parasite, host behavior is not altered at the initiation of infection, but rather at a specific time during the parasite's development. Therefore, a time series following the effects on the host's brain proteome is an essential and natural next step. As development is a dynamic process that is initiated anterior in time to the moment that the phenotype is expressed, it is also crucial to determine which molecular changes are associated to the early stage when behavior is altered. This allows clear identification of molecules involved in *implementing* phenotype changes, which are possibly expressed only at particular stages of development, separately from genes implicated in *maintaining* the phenotypes. For example, web making-behavior alterations is seen over time in a wasp-parasitized spider and the capacity for recovery of the host if the interaction is stopped is also time-dependent (Eberhard 2010). This is an ideal system to study the molecules affected in the spider and the dose change over time. A time-series experiment facilitates an understanding of the sequential recruitment of molecules and molecular processes underlying the phenotype and especially how their interactions are modified, from the initiation of parasitic infection to the final production of the phenotype, and provides insight into the diverse molecular biological processes underlying behavioral alteration. A time-series study of the molecular effects of the infection of a cricket by an hairworm – described above, including the effects of the circadian cycle, illustrates the power of such an approach (Biron et al. 2006). Furthermore, in order to be considered an adaptation, changes in host behavior must be precisely triggered by the parasite when it reaches the exact developmental phase at which it is beneficial to undergo these changes – e.g. when the parasite is ready to move to its next life stage (Poulin 2010). Therefore, in addition to testing fitness effects for the parasite (see Sect. 9.6), a time series is also important when one is aiming to test the hypothesis that the observed behavior alteration is an adaptation.

9.2.5 Transcriptomics

In order to observe changes in protein levels within a cell, a DNA sequence must first be transcribed into messenger RNA, which will then be translated into a protein. This gene expression process will result in specific amounts of mRNA. Simultaneously quantifying the expression level of thousands of genes to describe the "transcriptome" is referred to as "transcriptomics", which can be obtained through various methods (Aubin-Horth and Renn 2009; Vijay et al. 2013). Studying the plastic changes in the transcriptome in response to environmental factors is one of the most currently accessible ways to uncover the molecular bases of phenotypic change in non-model systems (Aubin-Horth and Renn 2009). For example, large-scale gene expression levels in the brain have been shown to vary significantly between individuals differing in behavior, such as between the nurse life stage and the forager task in honeybees (Whitfield et al. 2003) or between subordinate and dominant individuals in a group-living fish species (Aubin-Horth et al. 2007).

A transcriptomic approach has been used to study the molecular effects of a *Toxoplasma gondii* infection in the brain of its mouse host (Xiao et al. 2012). Focusing on the frontal cortex, this study uncovered different transcription signatures of infection by *Toxoplasma* in males and females, showing the importance of including sex as a factor in the analysis. Interestingly, behavioral alterations following the *Toxoplasma* infection was also sex-specific, with only females showing a reverse in their aversion to cat odor. This large-scale approach showed that the expression level of less than 1 % of the 24,000 genes studied changed after the infection, exposing a quite focused effect of the parasite, 1 month after the infection. Furthermore, only five genes were found to be affected in both male and female mice. When doing an enrichment analysis to find statistically significant over representation of certain biological processes, Xiao et al. (2012) nonetheless found that the "neural system development" category was common to both sexes. This higher order analysis is a good example of the additional biologically important information that is gained from studying the covariation of several genes compared to focusing only on individual genes. In line with behavior alterations, enrichment of some biological process categories were specific to a given sex, with female transcription profiles being associated with upregulation of olfaction-associated genes and changes in frontal cortex maturation, neurogenesis and sensory and motor coordination pathways. In males, changes in gene expression were related to down regulation of olfactory receptors and the dopamine pathway, which has been previously shown to be impacted by *Toxoplasma* infections in rodents. While several experiments and hypotheses remain to be done, this experiment shows that sex-specific effects of *Toxoplasma* infection on behavior are also found at the gene expression level, in a coherent manner with behavior alterations – aversion to predator odour, social transmission of food preference, activity level – observed in live animals (Xiao et al. 2012).

9.2.6 Combining Transcriptomics and Epigenomics

Large-scale epigenetics and transcriptomics approaches have also been used to investigate the molecular basis of the *Trichinella spiralis* life cycle, a widely distributed parasitic nematode (Gao et al. 2012). The field of epigenetics could be defined as "the study of mitotically and/or meiotically heritable changes in gene function that cannot be explained by changes in DNA sequence" (Russo et al. 1996). Epigenetic analyses focus on the identification of specific 'epigenetic marks' provided by DNA methylation, histone modifications – acetylation, phosphorylation, methylation, ubiquitination, crotonylation – and histone variants (Allis et al. 2007; Bannister and Kouzarides 2011; Huang et al. 2012; Tan et al. 2011), collectively called the "epigenome". The term 'methylome' in turn designates epigenetic marks on a genome-wide scale. Newly developed methods to survey these global epigenetic changes in the context of behavioral alterations or disorders are gaining more attention as constant progress is being made in the field (Stolzenberg et al. 2011). The brain epigenome has been shown to vary in relation to behavior, for example functionally-relevant methylome differences have been reported between honey bee workers and foragers (Lockett et al. 2012).

By characterizing the methylome of *T. spiralis* sampled at different stages, it was found that different methylation patterns correlate with different gene expression levels throughout the genome, depending on the nature of the loci and the developmental stages of the parasite (Gao et al. 2012). Going large scale promoted the discovery that *T. spiralis* was the only species, among 11 nematode species, to possess and express a gene (dnmt3) encoding de novo methylation machinery. Concurrently, a high-resolution analysis of methylation on a genome-wide scale suggested that epigenetic control in this species, given the presence of dnmt3, might be a key factor in managing the development of its different life-cycle stages by temporally regulating

gene expression (Gao et al. 2012). It has been shown that 'behavior-altering stages' are closely related to a parasite's life cycle stages (Biron and Loxdale 2013). As a consequence, understanding the large-scale changes occurring within the parasite's genome – or epigenome –and affecting its physiology is extremely relevant when looking at the proximate causes of behavioral changes in the host. Interestingly, early data on *T. spiralis* showed that the offspring of a female mouse infected by the parasitic nematode exhibit greater exploratory behavior when reared by an infected foster mother (Rau 1985). Such peculiar results suggest that infection of the mother triggers both pre- and post-natal behavioral alteration on the offspring. Knowing that post-natal interactions have significant impacts on the epigenomic state of genes of the offspring and, consequently, on behavior persisting in adulthood (Weaver et al. 2004), it was suggested that infection of a mother could lead to different DNA methylation profiles responsible for gene expression variations (Poulin and Thomas 2008). These transcriptomic differences could therefore be responsible for behavior alterations.

Results from such large-scale studies show that parasitic infections have the potential to trigger epigenomic changes – even transgenerational effects – affecting methylation patterns, which can significantly alter gene expression. Monitoring such epigenomic patterns across multiple individuals – host and parasite – offers the possibility of finding small changes, spanning the entire epigenome, which can be the starting point of cascading physiological events leading to the expression of behavioral phenotypes.

9.3 Investigating Behavior Alteration in an Ecological Genomics Context: A Preview

The large-scale analyses of gene expression, methylation and proteins we presented above have shown the importance of studying these different levels to understand the mechanisms underlying behavior alterations associated with parasitic infections. However, these molecular changes are only three of the numerous biological levels of organization in which significant changes affecting the organism phenotype can happen (Fig. 9.1). Here we present other promising and mostly untapped molecular processes that could be a major role in behavior alterations.

9.3.1 MicroRNAome

MicroRNAs are a class of non-coding single-stranded small RNAs. Ever since it was shown that they are numerous and diverse in animals (Lee and Ambros 2001), their various functions have been under scrutiny. MicroRNAs found in animals mostly act post-transcriptionally by inhibiting translation and accelerating mRNA degradation. MicroRNAs have been implicated in neural processes, including development and synaptic plasticity (Giraldez et al. 2005), and in human neural disorders (methods reviewed in Kan et al. 2012; Qureshi and Mehler 2011; Schratt 2009). The brain microRNAome can be modulated by the environment and is associated with changes in behavior, as shown by plastic changes uncovered between life stages in honey bees (nurses and foragers: Greenberg et al. 2012, dancers and foragers: Li et al. 2012), between rats that react to a repeated acute stress with learned helplessness versus those who react by avoidance of the stressor (Smalheiser et al. 2011), and in the brain of a songbird after song exposure (Gunaratne et al. 2011). The potential importance of miRNAs to the study of behavioral alterations is further underscored by the fact that *Toxoplasma gondii* is known to modify the miRNA levels of its host during infection (Zeiner et al. 2010) and that parasites may manipulate the miRNA repertoire of their host to their advantage (Hakimi and Cannella 2011). Furthermore, it can be hypothesized that parasites may not only affect the expression of the host's own repertoire of miRNAs, but may also be able to transfer their own miRNAs to act on the translation of the hosts's transcriptome. Prokaryotes and

eukaryotes evolved small RNAs independently (Shabalina and Koonin 2008). Therefore, even prokaryote parasites could potentially have an analogous silencing tool that could function in eukaryotic cells. It has recently been shown that exosomes are vesicles that can transport molecules extra-cellularily, including miRNAs. These exosomes are produced by numerous organisms, including protozoans (Hu et al. 2012). One could hypothesize that this transport machinery could potentially be used to transfer miRNAs – or other small RNAs – from the parasite to the host, resulting in cross-species communication and a potent manipulation mechanism (Liang et al. 2013). Whether all these puzzle pieces indeed fit together to result in an efficient system that facilitates behavioral alteration in the parasitized host is still unknown, but the potential implications of such a communication system are wide-ranging and fascinating.

9.3.2 Phosphorylome

Protein activities are regulated by many types of posttranslational modifications, which include for instance protein phosphorylation. Protein kinases modulate this biochemical process by transferring phosphate groups to other proteins, thus influencing their level of activity, interaction with other proteins and cell location. This molecular mechanism is used to regulate signaling networks and cascades of cellular processes, which can ultimately influence various phenotypic traits (Cohen 1992, 2000). The rapid and reversible nature of phosphorylation allows a fine-tuned control of protein activity, localization, stability, conformation and interaction with other proteins (Jensen 2006). By controlling the active and inactive states of the proteins, these post-translational modifications are known to be involved in the development and maintenance of plastic traits. For example, different phosphorylation levels of candidate proteins have been shown to covary with behavior, such as for tyrosine hydroxylase – a rate-limiting enzyme for the production of dopamine – in singing female starlings (Ellis and Riters 2013). Several detection techniques have been developed over the years in order to assess the roles of kinases and identify their targets using large scale approaches – e.g. in vitro assays, in vitro oriented peptide arrays, in-gel kinase assays, in vivo labeling, whole cell mass spectrometry, and detection by phospho-specific antibodies (Sopko and Andrews 2008). Using both a gel-electrophoresis approach and a gel-free digestion method, Lasonder et al. (2012) were able to characterize the 'phosphorylome' or 'phosphoproteome' – a description of the state of phosphorylation of all the proteins in a cell of a given tissue – of *Plasmodium falciparum* in order to understand the signaling pathways that regulate the development of this parasite responsible for malaria. They were thus able to confirm the widespread role of protein phosphorylation in regulating the parasite development via the involvement of cAMP and phosphatidylinositol signaling pathways (Lasonder et al. 2012). Specific changes in the proteome and phosphoproteome of the honeybee, *Apis melifera ligustica*, highlighted through the use of 2-DE, multiplex fluorescent staining, mass spectrometry and qRT-PCR, also revealed the importance of post-translational modifications in triggering the transition of embryo to larvae (Gala et al. 2013). Based on these types of findings, such novel approaches could be used to investigate the possible implications of phosphorylation in the regulation of key life-cycle transitions of 'manipulative' parasites. We could also gain useful insight into specific post-translational modifications imposed by the parasite on its host during various 'manipulative' stages.

9.3.3 Interactome

Protein-protein interactions are central to life as they govern most cell functions, and, as a case in point, many diseases originate from the disruption of protein-protein interactions (Diss et al. 2013). The compendium of these interactions is called the 'interactome' (Mathivanan et al. 2006; Ramirez et al. 2007). New methods have been developed to quantitatively measure this

interactome in cells (e.g. Bisson et al. 2011; Diss et al. 2013). For example, such a method has been used to uncover how the interactions between one protein known to be involved in Huntington's disease and all other proteins change in the brain of an infected mouse model of this human disease when compared to healthy controls (Shirasaki et al. 2012). Although still in its infancy and largely developed in model species, the field of 'interactomics' (Collura and Boissy 2007; Maarten Altelaar et al. 2012) offers a great potential for finding protein networks involved in host-parasite molecular cross-talk.

Interactions between proteins could also happen between species, effectively creating another level of molecular cross-talk between the parasite and its host, called "host-parasite interactome" (Biron et al. 2006; Diss et al. 2013). Such a 2-species interactome has been uncovered in a plant facing it's pathogen (Mukai et al. 2009), showing the feasibility and the information gained from such an approach. A parallel analysis of the kinetic of host-parasite interactome, which would consist in monitoring the dynamic of host-parasite protein-protein interactions at various stages of infection, would be a crucial molecular signature to characterize in order to understand how species communicate and interact at the protein-protein level during the development and maintenance of behavioral alterations. Information pertaining to the nature of the interactions between various types of host and parasite proteins at different stages during the alteration process will undoubtedly help identify key signaling pathways or physiological cascades exploited by parasites to successfully complete their life cycle.

9.4 What About Parasites?

In most studies, whether at the candidate or large-scale level, the focus is on the molecular changes in the host. However, what about changes in the parasite? What is the nature and range of molecular changes happening in a parasite during its development within a host? Uncovering the molecular changes occurring at a genome-wide scale in the parasite when it reaches the stage where behavioral alterations – potentially – augment its transmission would narrow the choice of candidate molecules implicated in the cross-talk with its host. The production of a tyrosine hydroxylase enzyme by *Toxoplasma* within its host brain, which affects the production of dopamine, a neuropeptide with wide-ranging effects on behavior (Prandovsky et al. 2011), is a perfect example of the importance of also focusing on the parasite. Studying the proteome of the parasite reveals important biological functions that are implicated in the infection and potentially directly related to behavioral alteration, as shown by the innovative study of the proteome of different nematomorph parasites and their host (synthesized in Biron et al. 2006). Studying the circulatory system in species where the parasite is not located in the brain will also bring new information about the molecular interactions taking place between the host and its parasite. While studying proteins circulating in the host has recently been used in human parasitology (Hood et al. 2012), very little information exists for other systems (Poulin 2010).

9.5 Functional Analyses

Once molecules are known to covary with a parasite infection (in the host of the parasite), it is essential to determine if that molecule plays a causal role in the observed phenotypes, which is the focus of functional analyses. For instance, a goal may be to replicate the behavioral alterations experimentally by directly manipulating the level of the identified candidate molecule, or to prevent behavior change by blocking the molecule's effects. In many non-model systems, which are often of interest in the study of host-parasite interaction, the only available technique to perform a functional analysis is pharmacological manipulation. Using such an approach is similar in thinking to using gene mutants, transgenic animals or RNAi techniques – which in turn could be used in several interesting model systems of host-parasite interactions. Indeed, a pharmacological manipulation allows the functional characterization of infection-associated molecular changes

and therefore extends beyond correlative proof by directly determining if the observed change is a cause or a consequence of the behavioral alteration in the host. It also allows to determine which aspect of the phenotype is being affected. For example, in the acanthocephalus -*Gammarus pulex* system presented above, an injection of serotonin to an uninfected individual, but not of octopamin, resulted in a behavior alteration mimicking the one observed in infected individuals (Tain et al. 2006). In this species, individuals are photophobic, but infected and pharmacologically manipulated individuals become photophilic. In contrast, these modifications have no effect on geotaxism. One of the interesting aspects of pharmacological manipulation is the possibility to use a dose–response approach, by varying the amount of drug used to provide further insight. Furthermore, manipulating more than one molecule at a time facilitates the differentiation of behavioral effects that are unique to each product, are similar among molecules or processes, or unique to combined manipulation because of interactions, a well-known effect in pharmacological research (Pritchard et al. 2013). Indeed, it may be the ratio of two molecules that are important rather than the absolute level (Williams 2008). Such an experimental manipulation will contribute to determining if different molecules involved in the behavioral alteration act in the same or in parallel pathways, additively or synergistically, or in an antagonistic fashion.

9.6 Adaptive Behavioral Manipulation

Research on altered host behavior has been dominated by the idea that infected hosts are literally 'manipulated' by their parasite in such a way that it facilitates the parasite's transmission or dispersal, allowing it to successfully complete its life cycle. Specific loci within the parasite genome would therefore modulate – directly or indirectly – precise phenotypic traits in the host 'on purpose', which, according to Dawkins (Dawkins 1982; Ramirez et al. 2007), is solid evidence of the extended phenotype concept. In order to rigorously state that some host behavioral changes are adaptive for certain parasites, basic criteria must be met. First, there must be variation in the 'manipulation' capacities within a given parasite population on which natural selection can act. It is possible that a very beneficial trait has become fixed in a population, which would therefore leave apparent genetic signatures – e.g. loci showing signs of strong positive selection that can be detected by surveying allele frequencies, fixation index and haplotype variations (see Burke et al. 2010 for examples of methods in *Drosophila*). Second, the phenotypic traits associated with manipulation capacities should be heritable – partially or completely (Butler et al. 2009; Cosseau et al. 2010; Ebert 2008; Gomez-Diaz et al. 2012; Richards et al. 2011). Third and most important, manipulative capacities should confer fitness benefits to the parasites that possess them. As a consequence, parasites that are able to efficiently manipulate their host behavior should show a fitness gain compared to conspecifics that cannot – whether it is measured as higher survival rate, greater number of descendants, higher transmission rate, etc. To address the question of evolutionary consequences of host-parasite manipulation, it is crucial to consider the population patterns of host-parasite molecular cross-talk in an attempt to link the environmental conditions to the population variability of the host and the parasite response. A complete evolutionary analysis should therefore include investigation on how the environment affects the intensity of the parasite-induced behavioral changes on an individual and a population scales, using various "-omics" tools as discussed above.

The same pharmacological manipulations used for functional analyses can be harnessed to perform "phenotypic engineering". In this approach, one creates variance in a trait among individuals, which allows testing its fitness consequences (Ketterson and Nolan 1999; Schmitt et al. 1999; Sinervo et al. 1992). Engineering is an excellent tool to test the central hypothesis that behavioral alterations enhance the parasite's fitness, and the application of engineering techniques can lead to unexpected results. For example, it has long been known

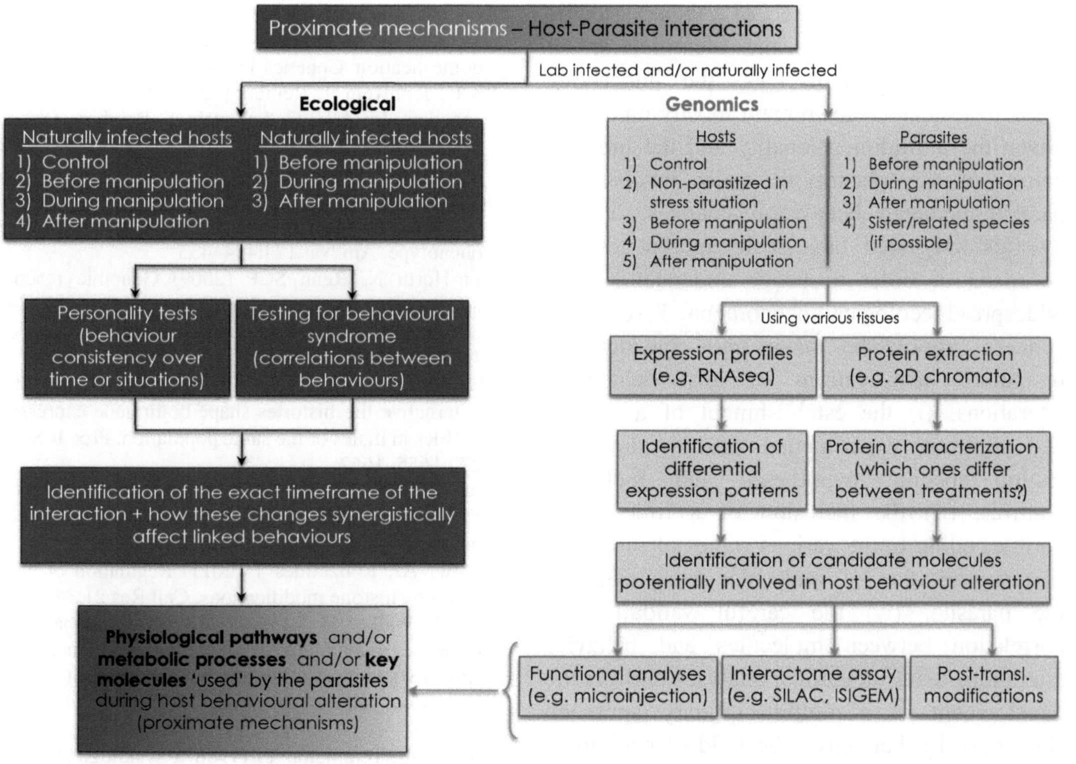

Fig. 9.3 Pipeline analysis of an integrated approach combining different complementary fields of research. A central question that needs to be addressed is which tissues are best suited for any "–omics" analysis that one wants to conduct in a given host-parasite system? As emphasized by the diagram, the CNS – Central Nervous System – remains a key compartment to investigate given its integrative role in terms of analysis and response to sensory information

that manipulating the serotonergic axis of gammarids affects their behavior in a way that mimics the effects of a parasitic infection by an acanthocephalan parasite (Helluy and Holmes 1990). Perrot-Minot et al. (2012) manipulated the serotonergic axis of a *Gammarus pulex* host, resulting in the same inversion of photophobia measured in individuals parasitized by the acanthocephalan *Pomphorhynchus tereticollis*. Predation rate by fish, the parasite's final host, was then tested in a semi-natural setting using these engineered gammarids. Surprisingly, they were not more predated than uninfected *Gammarus*, although they exhibited an altered behavior that has been proposed to enhance the parasite's fitness by increasing transmission rate. As emphasized by Poulin (2010) and this experiment, the most important criterion to test is whether the behavioral change in a host confers a fitness benefit to the parasite. Unfortunately, too few examples of such a direct test of the adaptive significance of host behavioral alterations exist.

9.7 Conclusion

Parasites impose significant physiological, morphological and behavioral constraints on their hosts. Accumulating evidence suggests that molecules such as biogenic amines – dopamine, serotonin, epinephrine, octopamine –, amino acids and immune system-associated cytokines, involved in major central nervous system functions and physiological pathways are altered or used by parasites, potentially leading to aberrant host behavior changes. Although precious knowledge has been gained in the past years concerning the ecology and the

evolutionary significance of these trophic interactions and despite the identification of several candidate molecules, the underlying molecular mechanisms remain mostly unknown. Given the multidimensionality and the intricate nature of the phenotypic traits involved in parasite-mediated behavioral changes, a shift towards integrated large-scale approaches can be made in order to better understand these widespread ecological phenomena. Five major concerns should be taken into consideration in future investigations of host behavioral alterations: (i) the establishment of a careful and rigorous experimental design aimed at testing specific hypotheses and including controls, (ii) the inclusion of a time series corresponding to the various stages of infection, (iii) the addition of molecular knowledge on the parasite, (iv) the careful validation of correlation between molecules and behavior alterations using functional analyses and (v) the development of a multidisciplinary approach (Fig. 9.3). In that sense, the field of ecological genomics (Fig. 9.1) allows research programs to take into account various levels of biological organization in an attempt to reconstruct intricate biological links that combine to produce complex organisms interacting with their biotic and abiotic environment.

Acknowledgments We thank David Biron, Caitlin Friesen and Iain Barber for reviewing a previous version of the manuscript and Christian R. Landry for excellent editorial comments and for inviting us to write this chapter. FOH is supported by a FRQ-NT Doctoral research Scholarship and Discovery Grants of the Natural Sciences and Engineering Research Council of Canada to N. Aubin-Horth and C.R. Landry.

References

Adamo SA (2002) Modulating the modulators: parasites, neuromodulators and host behavioral change. Brain Behav Evol 60:370–377

Adamo SA (2013) Parasites: evolution's neurobiologists. J Exp Biol 216:3–10

Adamo SA, Linn CE, Beckage NE (1997) Correlation between changes in host behaviour and octopamine levels in the tobacco hornworm Manduca sexta parasitized by gregarious braconid parasitoid wasp *Cotesia congregata*. J Exp Biol 200:117–127

Albert FW, Carlborg O, Plyusnia I et al (2009) Genetic architecture of tameness in a rat model of animal domestication. Genetics 182:541–554

Allis C, Jenuwein T, Reinberg D (2007) Overview and concepts. In: Allis C, Jenuwein T, Reinberg D (eds) Epigenetics. Cold Spring Harbor Laboratory, New York, pp 23–61

Andersen SB, Gerritsma S, Yusah KM et al (2009) The life of a dead ant: the expression of an adaptive extended phenotype. Am Nat 174:424–433

Aubin-Horth N, Renn SCP (2009) Genomic reaction norms: using integrative biology to understand molecular mechanisms of phenotypic plasticity. Mol Ecol 18:3763–3780

Aubin-Horth N, Landry CR, Letcher BH et al (2005) Alternative life histories shape brain gene expression profiles in males of the same population. Proc R Soc B 272:1655–1662

Aubin-Horth N, Desjardins JK, Martei YM et al (2007) Masculinized dominant females in a cooperatively breeding species. Mol Ecol 16:1349–1358

Bannister AJ, Kouzarides T (2011) Regulation of chromatin by histone modifications. Cell Res 21:381–395

Barber I, Walker P, Svensson PA (2004) Behavioural responses to simulated avian predation in female three spined sticklebacks: the effect of experimental *schistocephalus solidus* infections. Behaviour 141:1425–1440

Beckage NE, Templeton TJ (1986) Physiological effects of parasitism by *Apanteles congregatus* in terminalstage tobacco hornworm larvae. J Insect Physiol 32:299–314

Bendesky A, Bargmann CI (2011) Genetic contributions to behavioural diversity at the gene-environment interface. Nat Rev Genet 12:809–820

Biron DG, Loxdale HD (2013) Host-parasite molecular cross-talk during the manipulative process of a host by its parasite. J Exp Biol 216:148–160

Biron DG, Joly C, Galéotti N et al (2005a) The proteomics: a new prospect for studying parasitic manipulation. Behav Process 68:249–263

Biron DG, Marché L, Ponton F et al (2005b) Behavioral manipulation in a grasshopper harbouring hairworms: a proteomics approach. Proc R Soc B 272: 2117–2126

Biron DG, Moura H, Marché L et al (2005c) Towards a new conceptual approach to "parasitoproteomics". Trends Parasitol 21:162–168

Biron DG, Ponton F, Marché L et al (2006) 'Suicide' of crickets harbouring hairworms: a proteomics investigation. Insect Mol Biol 15:731–742

Bisson N, James DA, Ivosev G et al (2011) Selected reaction monitoring mass spectrometry reveals the dynamics of signaling through the GRB2 adaptor. Nat Biotechnol 29:653–658

Buchwalow I, Bocker W (2010) Immunohistochemistry: basics and methods. Springer, New York

Burke MK, Dunham JP, Shaharestani P et al (2010) Genome-wide analysis of a long-term evolution experiment with *Drosophila*. Nature 467:587–592

Butler G, Rasmussen MD, Lin MF et al (2009) Evolution of pathogenicity and sexual reproduction in eight *Candida* genomes. Nature 459:657–662

Candolin U, Voigt H-R (2001) No effect of a parasite on reproduction in stickleback males: a laboratory artefact? Parasitology 122:457–464

Cézilly F, Favrat A, Perrot-Minnot MJ (2013) Multidimensionality in parasite-induced phenotypic alterations: ultimate versus proximate aspects. J Exp Biol 216:27–35

Cohen P (1992) Signal integration at the level of protein kinases, protein phosphatases and their substrates. Trends Biochem Sci 17:408–413

Cohen P (2000) The regulation of protein function by multisite phosphorylation – a 25 years update. Trends Biochem Sci 25:596–601

Collura V, Boissy G (2007) From protein-protein complexes to interactomics. Springer, Dordrecht

Consortium TGO (2008) The gene ontology project in 2008. Nucleic Acids Res 36:D440–D444

Cosseau C, Azzi A, Rognon A et al (2010) Epigenetic and phenotypic variability in populations of *Schistosoma mansoni* – a possible kick-off for adaptive host/parasite evolution. Oikos 119:669–678

Dawkins R (1982) The extended phenotype. Oxford University Press, New York

De Jonk-Brink M, Bergamin-Sassen M, Solis Soto M (2001) Multiple strategies of schistosomes to meet their requirements in the intermediate snail host. Parasitology 123:S129–S141

Dejong S, Fuller TV, Janson E et al (2010) Gene expression profiling in C57BL/6J and A/J mouse inbred strains reveals gene networks specific for brain regions independent of genetic background. BMC Genomics 2010:1–14

Dingemanse NJ, Oosterhof C, Van Der Plas F et al (2009) Variation in stickleback head morphology associated with parasite infection. Biol J Linn Soc 96:759–768

Diss G, Filteau M, Freschi L et al (2013) Integrative avenues for exploring the dynamics and evolution of protein interaction networks. Curr Opin Biotechnol 24(4):775–783

Diz AP, Martinez-Fernadez M, Rolan-Alvarez E (2012) Proteomics in evolutionary ecology: linking the genotype with the phenotype. Mol Ecol 21:1060–1080

Eberhard WG (2000) Spider manipulation by a wasp larva. Nature 406:255–256

Eberhard W (2010) Recovery of spiders from the effects of parasitic wasps: implications for fine-tuned mechanisms of manipulation. Anim Behav 79:375–383

Ebert D (2008) Host–parasite coevolution: insights from the Daphnia–parasite model system. Curr Opin Microbiol 11:290–301

Ellis JMS, Riters LV (2013) Patterns of phosphorylated tyrosine hydroxylase vary with song production in female starlings. Brain Res 1498:41–49

Filby A, Paull GC, Hickmore TFA et al (2010) Unravelling the neurophysiological basis of aggression in a fish model. BMC Genomics 11:1–17

Flegr J et al (2002) Increased risk of traffic accidents in subjects with latent toxo plasmosis: a retrospective case-control study. BMC Infect Dis 2:11

Fox A, Hudson PJ (2001) Parasites reduce territorial behaviour in red grouse (Lagopus lagopus scoticus). Ecol Lett 4:139–143

Gala A, Fang Y, Woltedji D et al (2013) Changes of proteome and phosphoproteome trigger embryo–larva transition of honeybee worker (*Apis mellifera ligustica*). J Proteomics 78:428–446

Gao F, Liu X, Wu X-P et al (2012) Differential DNA methylation in discrete developmental stages of the parasitic nematode *Trichinella spiralis*. Genome Biol 13:R100

Giraldez AJ, Cinalli RM, Glasner ME et al (2005) MicroRNAs regulate brain morphogenesis in zebrafish. Science 308:833–838

Godin J-GJ, Sproul CD (1988) Risk taking in parasitized sticklebacks under threat of predation: effects of energetic need and food availability. Can J Fish Aquat Sci 66:2360–2367

Gomez-Diaz E, Jorda M, Peinado M et al (2012) Epigenetics of host-pathogen interactions: the road ahead and the road behind. PLoS Pathog 8:e1003007

Greenberg JK, Xia J, Zhou X et al (2012) Behavioral plasticity in honey bees is associated with differences in brain microRNA transcriptome. Genes Brain Behav 11:660–670

Gunaratne PH, Lin YC, Benham AL et al (2011) Song exposure regulates known and novel microRNAs in the zebra finch auditory forebrain. BMC Genomics 12:277

Hakimi MA, Cannella D (2011) Apicomplexan parasites and subversion of the host cell microRNA pathway. Trends Parasitol 27:481–486

Hawkins RD, Hon GC, Ren B (2010) Next-generation genomics: an integrative approach. Nat Rev Genet 11:476–486

Heins DC, Baker JA, Toups MA et al (2010) Evolutionary significance of fecundity reduction in threespine stickleback infected by the diphyllobothriidean cestode *Schistocephalus solidus*. Biol J Linn Soc 100:835–846

Helluy S (2013) Parasite-induced alterations of sensorimotor pathways in gammarids: collateral damage of neuroinflammation? J Exp Biol 216:67–77

Helluy S, Holmes J (1990) Serotonin, octopamine, and the clinging behavior induced by the parasite *Polymorphus paradoxus* (Acanthocephala) in *Gammarus lacustris* (Crustacea). Can J Fish Aquat Sci 68:1214–1220

Helluy S, Thomas F (2003) Effects of *Microphallus papillorobustus* (Platyhelminthes: Trematoda) on serotonergic immunoreactivity and neuronal architecture in the brain of *Gammarus insensibilis* (Crustacea: Amphipoda). Proc R Soc B 270:563–568

Hindsbo O (1972) Effects of *Polymorphus* (Acanthocephala) on colour and behaviour of *Gammarus lacustris*. Nature 238:333

Holmes J, Bethel W (1972) Modification of intermediate host behaviour by parasites. In: Canning E, Wright C (eds) Behavioural aspects of parasite transmission. Academic, London, pp 123–149

Hood LE, Omenn GS, Moritz RL et al (2012) New and improved proteomics technologies for understanding complex biological systems: addressing a grand challenge in the life sciences. Proteomics 12:2773–2783

House PK, Vyas A, Sapolsky R (2011) Predator cat odors activate sexual arousal pathways in brains of *Toxoplasma gondii* infected rats. PLoS ONE 6:e23277

Hu G, Drescher K, Chen X (2012) Exosomal miRNAs: biological properties and therapeutic potential. Front Genet 3:1–9

Huang C, Xu M, Zhu B (2012) Epigenetic inheritance mediated by histone lysine methylation: maintaining transcriptional states without the precise restoration of marks? Philos Trans R Soc B 368:1–6

Hughes D, Brodeur J, Thomas F (2012) Host manipulation by parasites. Oxford University Press, Oxford

Ingwell LL et al (2012) Sc Rep 2(518):1–6

Jensen ON (2006) Interpreting the protein language using proteomics. Nat Rev Mol Cell Biol 7:391–403

Kan AA, Van Erp S, Derijck AA et al (2012) Genome-wide microRNA profiling of human temporal lobe epilepsy identifies modulators of the immune response. Cell Mol Life Sci 69:3127–3145

Kavaliers M, Cowell DD (1995) Parasitology 111:257–263

Ketterson E, Nolan V (1999) Adaptation, exaptation, and constraint: a hormonal perspective. Am Nat 154:S1–S22

Ketterson E, Atwell J, Mcglothlin J (2009) Phenotypic integration and independence: hormones, performance, and response to environmental change. Integr Comp Biol 49:365–379

Kuris AM, Hechinger RF, Shaw JC et al (2008) Ecosystem energetic implications of parasite and free-living biomass in three estuaries. Nature 454:515–518

Lafferty KD, Morris AK (1996) Altered behavior of parasitized killifish increases susceptibility to predation by bird final hosts. Ecology 77:1390–1397

Lamb J, Crawford ED, Peck D et al (2006) The connectivity map: using gene-expression signatures to connect small molecules, genes, and disease. Science 313:1929–1935

Lasonder E, Green JL, Camarda G et al (2012) The *Plasmodium falciparum* schizont phosphoproteome reveals extensive phosphatidylinositol and cAMP-protein kinase A signaling. J Proteome Res 11:5323–5337

Lee RC, Ambros V (2001) An extensive class of small RNAs in *Caenorhabditis elegans*. Science 294:862–863

Lefèvre T, Roche B, Poulin R et al (2008) Exploiting host compensatory responses: the 'must' of manipulation? Trends Parasitol 24:435–439

Li L, Liu F, Li W et al (2012) Differences in microRNAs and their expressions between foraging and dancing honey bees, *Apis mellifera* L. J Insect Phys 58:1438–1443

Liang H, Zen K, Zhang J et al (2013) New roles for microRNAs in cross-species communication. RNA Biol 10:1–4

Lockett GA, Kicharski R, Malszka R (2012) DNA methylation changes elicited by social stimuli in the brains of worker honey bees. Genes Brain Behav 11:235–242

Losos J (2011) Convergence, adaptation, and constraint. Evolution 65:1827–1840

Maarten Altelaar AF, Munoz J, Heck AJR (2012) Next-generation proteomics: towards an integrative view of proteome dynamics. Nat Rev Genet 14:35–48

Macnab V, Barber I (2012) Some (worms) like it hot: fish parasites grow faster in warmer water, and alter host thermal preferences. Global Change Biol 18:1540–1548

Mant CT, Chen Y, Yan Z et al (2007) HPLC analysis and purification of peptides. Meth Mol Ecol 386:3–55

Marcogliese DJ, Cone DK (1997) Food webs: a plea for parasites. Trends Ecol Evol 12:320–325

Mathivanan S, Periaswamy B, Gandhi TK et al (2006) An evaluation of human protein-protein interaction data in the public domain. BMC Bioinform 7(Suppl 5):S19

Milinski M (1985) Risk of predation of parasitized sticklebacks (*Gasterosteus aculeatus L.*) under competition for food. Behaviour 93:203–215

Moore J (2002) Parasites and the behavior of animals. Oxford University Press, Oxford

Mukai M, Replogle K, Drnevich J et al (2009) Seasonal differences of gene expression profiles in song sparrow (*Melospiza melodia*) hypothalamus in relation to territorial aggression. PLoS ONE 4:e8182

Overli O, Pall M, Borg B et al (2001) Effects of *Schistocephalus solidus* infection on brain monoaminergic activity in female three-spined sticklebacks *Gasterosteus aculeatus*. Proc R Soc B 268:1411–1415

Pavey SA, Bernatchez L, Aubin-Horth N et al (2012) What is needed for next-generation ecological and evolutionary genomics? Trends Ecol Evol 27: 673–678

Perrot-Minnot MJ, Maddaleno M, Balourdet A et al (2012) Host manipulation revisited: no evidence for a causal link between altered photophobia and increased trophic transmission of amphipods infected with acanthocephalans. Funct Ecol 26:1007–1014

Pigliucci M, Preston K (2004) Phenotypic integration: studying the ecology and evolution of complex phenotypes. Oxford University Press, New York

Ponton F, Lefèvre T, Lebarbenchon C et al (2006) Do distantly related parasites rely on the same proximate factors to alter the behaviour of their hosts? Proc R Soc B 273:2869–2877

Poulin R (2007) Evolutionary ecology of parasites. Princeton University Press, Princeton

Poulin R (2010) Parasite manipulation of host behavior: an update and frequently asked questions. In: Brockmann J (ed) Advances in the study of behavior. Elsevier, Burlington, pp 151–186

Poulin R (2011) The many roads to parasitism: a tale of convergence. In: Rollinson D, Hay SI (eds) Advances in parasitology. Academic, Burlington, pp 1–49

Poulin R (2013) Parasite manipulation of host personality and behavioural syndromes. J Exp Biol 216:18–26

Poulin R, Thomas F (2008) Epigenetic effects of infection on the phenotype of host offspring: parasites reaching across host generations. Oikos 117:331–335

Prandovsky E, Gaskell E, Martin H et al (2011) The neurotropic parasite *Toxoplasma gondii* increases dopamine metabolism. PLoS ONE 6:e23866

Pritchard JR, Bruno PM, Gilbert LA et al (2013) Defining principles of combination drug mechanisms of action. Proc Natl Acad Sci USA 110:E170–E179

Quinn TP, Kendall NW, Rich HB Jr et al (2012) Diel vertical movements, and effects of infection by the cestode *Schistocephalus solidus* on daytime proximity of three-spined sticklebacks *Gasterosteus aculeatus* to the surface of a large Alaskan lake. Oecologia 168:43–51

Qureshi IA, Mehler MF (2011) Non-coding RNA networks underlying cognitive disorders across the lifespan. Trends Mol Med 17:337–346

Ramirez F, Schlicker A, Assenov Y et al (2007) Computational analysis of human protein interaction networks. Proteomics 7:2541–2552

Rau ME (1985) The effects of *Trichinella spiralis* infection of pregnant mice on the future behavior of their offspring. J Parasitol 71:774–778

Rau ME, Caron FR (1979) Can J Zool 57:2466–2468

Réale D, Reader SM, Sol D et al (2007) Integrating animal temperament within ecology and evolution. Biol Res 82:291–318

Renn SCP, Aubin-Horth N, Hofmann HA (2008) Fish and chips: functional genomics of social plasticity in an African cichlid fish. J Exp Biol 211: 3041–3056

Richards TA, Soanes DM, Jones MDM et al (2011) Horizontal gene transfer facilitated the evolution of plant parasitic mechanisms in the oomycetes. Proc Natl Acad Sci USA 108:15258–15263

Russo V, Martienssen R, Rigg A (1996) Introduction. In: Russo V, Martienssen R, Rigg A (eds) Epigenetic mechanisms of gene regulation. Cold Spring Harbor Laboratory, New York, pp 1–4

Schmidt-Hempel P (2011) Evolutionary parasitology: the integrated study of infections, immunology, ecology, and genetics. Oxford University Press, New York

Schmitt J, Dudley S, Pigliucci M (1999) Manipulative approaches to testing adaptive plasticity: phytochrome-mediated shade-avoidance responses in plants. Am Nat 154:S43–S54

Schratt G (2009) microRNAs at the synapse. Nature 10:842–849

Shabalina SA, Koonin E (2008) Origins and evolution of eukaryotic RNA interference. Trends Ecol Evol 23:578–587

Shaw JC, Overli O (2012) Brain-encysting trematodes and altered monoamine activity in naturally infected killifish *Fundulus parvipinnis*. J Fish Bio:81: 2213–2222

Shaw JC, Korzan WJ, Carpenter RE et al (2009) Parasite manipulation of brain monoamines in California killifish (*Fundulus parvipinnis*) by the trematode *Euhaplorchis californiensis*. Proc R Soc B 276:1137–1146

Shirasaki DI, Greiner ER, Al-Ramahi I et al (2012) Network organization of the huntingtin proteomic interactome in mammalian brain. Neuron 75:41–57

Sih A, Bell AM, Johnson JC et al (2004) Behavioral syndromes: an integrative overview. Q Rev Biol 79:241–277

Sinervo B, Doughty P, Huey RB et al (1992) Allometric engineering: a causal analysis of natural selection on offspring size. Science 258:1927–1931

Skallova A, Kodym P, Frynta D et al (2006) The role of dopamine in *Toxoplasma*-induced behavioural alterations in mice: an ethological and ethopharmacological study. Parasitology 133:525–535

Skinner WS, Dennis PA, Li JP et al (1991) Isolation and identification of paralytic peptides from hemolymph of the lepidopteran insects *Manduca sexta, Spodoptera exigua*, and *Heliothis virescens*. J Biol Chem 266:12873–12877

Smalheiser NR, Lugli G, Rizavi HS et al (2011) MicroRNA expression in rat brain exposed to repeated inescapable shock: differential alterations in learned helplessness vs. non-learned helplessness. Int J Neuropsychop 14:1315–1325

Sopko R, Andrews BJ (2008) Linking the kinome and phosphorylome – a comprehensive review of approaches to find kinase targets. Mol BioSyst 4:920–933

Stolzenberg DS, Grant PA, Bekiranov S (2011) Epigenetic methodologies for behavioral scientists. Horm Behav 59:407–416

Tain L, Perrot-Minnot MJ, Cezilly F (2006) Altered host behaviour and brain serotonergic activity caused by acanthocephalans: evidence for specificity. Proc R Soc B 273:3039–3045

Tan M, Luo H, Lee S et al (2011) Identification of 67 histone marks and histone lysine crotonylation as a new type of histone modification. Cell 146:1016–1028

Thomas F, Ulitsky P, Augier R et al (2003) Biochemical and histological changes in the brain of the cricket *Nemobius sylvestris* infected by the manipulative parasite *Paragordius tricuspidatus* (Nematomorpha). Int J Parasitol 33:435–443

Thomas F, Poulin R, Brodeur J (2010) Host manipulation by parasites: a multidimensional phenomenon. Oikos 119:1217–1223

Vijay N, Poelstra JW, Kunstner A et al (2013) Challenges and strategies in transcriptome assembly and differential gene expression quantification. A comprehensive in silico assessment of RNA-seq experiments. Mol Ecol 22:620–634

Vyas A, Kim SK, Giacomini N et al (2007) Behavioral changes induced by *Toxoplasma* infection of rodents are highly specific to aversion of cat odors. Proc Natl Acad Sci USA 104:6442–6447

Weaver ICG, Cervoni N, Champagne FA et al (2004) Epigenetic programming by maternal behavior. Nat Neurosci 7:847–854

Weber JN, Peterson PK, Hoekstra HE (2013) Discrete genetic modules are responsible for complex burrow evolution in *Peromyscus* mice. Nature 493: 402–406

Webster JP, Lamberton PHL, Donnelly CA et al (2006) Parasites as causative agents of human affective disorders? The impact of anti-psychotic, mood-stabilizer and anti-parasite medication on *Toxoplasma gondii's* ability to alter host behaviour. Proc R Soc B 273: 1023–1030

Whitfield CW, Cziko A-M, Robinson GE (2003) Gene expression profiles in the brain predict behavior in individual honey bees. Science 302:296–299

Williams TD (2008) Individual variation in endocrine systems: moving beyond the 'tyranny of the Golden Mean'. Phil Trans R Soc B 363:1687–1698

Wright HA, Wootton RJ, Barber I (2007) Compensatory growth in threespine sticklebacks (*Gasterosteus aculeatus*) inhibited by experimental Schistocephalus infections. Can J Fish Aquat Sci 64: 819–826

Xiao J, Kannan G, Jones-Brando L et al (2012) Sex-specific changes in gene expression and behavior induced by chronic *Toxoplasma* infection in mice. Neuroscience 206:39–48

Yanoviak SP, Kaspari M, Dudley R et al (2008) Parasite-induced fruit mimicry in a tropical canopy ant. Am Nat 171:536–544

Zeiner GM, Norman KL, Thomson JM et al (2010) *Toxoplasma gondii* infection specifically increases the levels of key host microRNAs. PLoS ONE 5:e8742

Zhang B, Horvath S (2005) A general framework for weighted gene co-expression network analysis. Stat Appl Genet Molec Biol 4: Article17 1–43

Zwarts L, Magwire MM, Carbone MA et al (2011) Complex genetic architecture of *Drosophila* aggressive behavior. Proc Natl Acad Sci USA 11: 17070–17075

Ecological Epigenetics

10

Holly J. Kilvitis, Mariano Alvarez, Christy M. Foust, Aaron W. Schrey, Marta Robertson, and Christina L. Richards

Abstract

Biologists have assumed that heritable variation due to DNA sequence differences (i.e., genetic variation) allows populations of organisms to be both robust and adaptable to extreme environmental conditions. Natural selection acts on the variation among different genotypes and ultimately changes the genetic composition of the population. While there is compelling evidence about the importance of genetic polymorphisms, evidence is accumulating that epigenetic mechanisms (e.g., chromatin modifications, DNA methylation) can affect ecologically important traits, even in the absence of genetic variation. In this chapter, we review this evidence and discuss the consequences of epigenetic variation in natural populations. We begin by defining the term epigenetics, providing a brief overview of various epigenetic mechanisms, and noting the potential importance of epigenetics in the study of ecology. We continue with a review of the ecological epigenetics literature to demonstrate what is currently known about the amount and distribution of epigenetic variation in natural populations. Then, we consider the various ecological contexts in which epigenetics has proven particularly insightful and discuss the potential evolutionary consequences of epigenetic variation. Finally, we conclude with suggestions for future directions of ecological epigenetics research.

Keywords

Epigenetics • Ecology • Phenotypic variation

10.1 Introduction

Understanding ecology requires the examination of complex questions that motivate studies at every biological level, from molecules through ecosystems. At the molecular level, controlled laboratory and sequencing studies have revealed

H.J. Kilvitis (✉) • M. Alvarez • C.M. Foust • M. Robertson • C.L. Richards
Department of Integrative Biology, University of South Florida, 4202 E Fowler Ave, Tampa, FL 33620, USA
e-mail: hkilviti@mail.usf.edu

A.W. Schrey
Department of Biology, Armstrong Atlantic State University, Savannah, GA, USA

C.R. Landry and N. Aubin-Horth (eds.), *Ecological Genomics: Ecology and the Evolution of Genes and Genomes*, Advances in Experimental Medicine and Biology 781,
DOI 10.1007/978-94-007-7347-9_10, © Springer Science+Business Media Dordrecht 2014

that the genome is a dynamic system of interacting elements. Although we know that environmental factors have tremendous power to shape the genome (McClintock 1984; Karrenberg et al. 2007; Boyko and Kovalchuk 2011), our understanding of how complex phenotypes arise from a given genotype, and how important environmental factors are in this process, remains limited (Richards et al. 2009, 2012a; Pigliucci 2010; Martin et al. 2011).

Biologists have assumed that heritable variation due to DNA sequence differences (i.e., genetic variation) allows populations of organisms to be both robust and adaptable to extreme environmental conditions. Natural selection acts on the variation among different genotypes and ultimately changes the genetic composition of the population. While there is compelling evidence about the importance of genetic polymorphisms, evidence is accumulating that epigenetic mechanisms (e.g., chromatin modifications, DNA methylation) can affect ecologically important traits, even in the absence of genetic variation. In this chapter, we review this evidence and discuss the consequences of epigenetic variation in natural populations. We begin by defining the term epigenetics, providing a brief overview of various epigenetic mechanisms, and noting the potential importance of epigenetics in the study of ecology. We continue with a review of the ecological epigenetics literature to demonstrate what is currently known about the amount and distribution of epigenetic variation in natural populations. Then, we consider the various ecological contexts in which epigenetics has proven particularly insightful and discuss the potential evolutionary consequences of epigenetic variation. Finally, we conclude with suggestions for future directions of ecological epigenetics research.

10.1.1 History of Epigenetics

The term 'epigenetics' was originally coined in the early 1940s by Conrad Waddington, who is widely recognized for investigating the concepts of canalization and genetic assimilation in the context of developmental biology (Waddington 1942, 1953; Jablonka and Lamb 2002). Waddington's use of the term 'epigenetics' was very broad, referring to all developmental events which lead to the manifestation of an organism's phenotype (Holliday 2006). Since Waddington, interpretations of the term epigenetics have evolved, especially in light of discoveries of molecular processes that regulate gene activity and the inheritance of cellular phenotypes (Jablonka and Lamb 2002). These findings revealed that there is an alternative form of inheritance not easily explained by traditional Mendelian genetics (Holliday 2006). Molecular biologist Robin Holliday, was among the first to recognize this concept of non-genetic inheritance by defining epigenetics as the study of alterations in gene expression and the mitotic inheritance of gene expression patterns (Holliday 1994; Jablonka and Lamb 2002).

In recent years, epigenetics has been commonly defined as the study of heritable changes in gene expression not explained by changes in the DNA sequence (Holliday 1994; Richards 2006; Bird 2007; Bossdorf et al. 2008). However, the definition of epigenetics continues to be a topic of debate among biologists, primarily because of the inclusion of the term "heritable" in most modern definitions (Bird 2007). Depending on the field of study (e.g., molecular/cellular biology, developmental biology, or ecological/evolutionary biology), some believe that the study of epigenetics encompasses all processes aside from DNA sequence that produce the phenotype in organisms (Hallgrímsson and Hall 2011), heritable or not. Others argue that because epigenetics is often associated with "soft inheritance," heritability is a necessary component of the definition (Ho and Burggren 2010; Kovalchuk 2012). Due to these opposing views, we differentiate the term "epigenetics" from "epigenetic inheritance" (sensu Jablonka and Raz 2009). Jablonka and Raz (2009) argue that epigenetics is the study of both cellular-level and organismal-level processes underlying developmental plasticity and canalization that lead to enduring developmental effects. Alternatively,

Fig. 10.1 The classic example of phenotypic effects of natural epigenetic variation and epigenetic inheritance. Cubas et al. (1999) showed that the change from normal bilateral (*right*) to radial symmetry (*left*) of *Linaria vulgaris* was associated with methylation and silencing of the *Lcyc* gene (Photos Reprinted from Palevitz 1999)

epigenetic inheritance is an extension of epigenetics that occurs when variations in phenotype, not caused by DNA sequence variation, are mitotically and/or meiotically inherited by future generations (Jablonka and Raz 2009). For the purpose of this chapter, we define epigenetics as the study of molecular-level mechanisms that affect gene expression without altering the underlying DNA sequence, which may lead to potentially heritable changes in phenotype (Bossdorf et al. 2008; Richards et al. 2010a; Richards 2011; Ledón-Rettig et al. 2013).

10.1.2 Epigenetic Mechanisms

There are several epigenetic mechanisms that can alter gene expression (e.g., chromatin remodeling, histone modifications, small interfering RNAs), yet DNA methylation of cytosines is by far the most-common mechanism studied in ecological epigenetics (Schrey et al. 2013). In many eukaryotes, DNA methylation occurs at the fifth carbon of a cytosine – as 5-methylcytosine – located within one of the following DNA sequence motifs: CpG (in animals) or CpCpG, CpHpH, CpNpG (in plants) (Rapp and Wendel 2005; Zilberman and Henikoff 2007; Zhang et al. 2008; Laird 2010; Bock 2012). Extensive research shows that DNA methylation plays an integral role in numerous biological processes including organismal development (Monk et al. 1987), genomic imprinting (Li et al. 1993), mammalian X-chromosome inactivation (Kaslow and Migeon 1987), and polyploidy/hybridization in plants (Salmon et al. 2005) (reviewed in Rapp and Wendel 2005; Zilberman and Henikoff 2007; Bock 2012; Richards et al. 2012a). A complex relationship between DNA methylation and gene expression patterns has been demonstrated (Nätt et al. 2012) and the interaction between DNA methylation and transcription machinery can directly influence an organism's phenotype (e.g., floral symmetry in *Linaria vulgaris* – Fig. 10.1 – (Cubas et al. 1999); reviewed in Jaenisch and Bird 2003; Bossdorf et al. 2008; Bock 2012). The stability of DNA methylation varies across the genome, but some loci can be directly influenced by the environment, remain stable throughout an individual's lifetime, and be inherited by future generations (Johannes et al. 2009; Angers et al. 2010; Verhoeven et al. 2010). Molecular tools have been developed to screen both genome-wide and gene-specific patterns of methylation that can be applied to study many biological taxa (reviewed in Laird 2003; Rapp and Wendel 2005; Bossdorf et al. 2008; Schrey et al. 2013). These characteristics make DNA methylation particularly useful for studying epigenetics in an ecological context.

Although the current ecological epigenetics literature is primarily focused on DNA

methylation, other epigenetic modifications can alter gene expression. Histone modifications alter the way DNA is packaged and change the accessibility of the packaged DNA for transcription. These modifications can also interact with DNA methylation (Richards and Elgin 2002; Rapp and Wendel 2005). The activity of transposable elements, regions of DNA that have the ability to move within the genome and integrate into new sites, are regulated primarily by small interfering RNAs or by DNA methylation (Kazazian 2004; Kejnovsky et al. 2012; Richards et al. 2012a; Slotkin et al. 2012). Transposable elements have the potential to alter gene expression and function when inserted within coding regions, so regulation of these areas of the genome is highly important (Kazazian 2004; Feschotte 2008). Small interfering RNAs are active in DNA methylation pathways and histone methylation pathways. Similarities between these pathways in animals and plants suggest evolutionary conservation in these epigenetic processes (Saze et al. 2012).

10.1.3 Epigenetics in the Study of Ecology

The field of ecological genomics has provided valuable insight into the genetic basis of ecologically and evolutionarily relevant phenotypic variation within and among natural populations (Ungerer et al. 2008). However, there are ecologically relevant phenomena that cannot be entirely explained by genetic variation (Bossdorf et al. 2008). Ecological epigenetics aims to address how epigenetic processes may also be important mediators of phenotypic variation in populations (Bossdorf et al. 2008). Just as genomics is a sub-discipline in genetics, epigenomics is a discipline within the broader field of epigenetics, which focuses specifically on characterizing epigenetic processes on a genome-wide scale. Thus for the purpose of this review, we use the more encompassing term ecological epigenetics which include genome-wide assessments of epigenetic variation (e.g., MS-AFLP studies), as well as more gene-specific approaches used to determine the effects of epigenetic processes on phenotypic variation. One of the critical differences between ecological genomics generally and ecological epigenetics is that epigenetic variation is typically more labile and responsive to external environmental factors (e.g., via alteration of DNA methylation), often within ecological time scales (Fig. 10.2) (Bossdorf et al. 2008; Angers et al. 2010). Research on the epigenetic basis of ecologically relevant traits has posited that epigenetically-mediated response to environment can be heritable across generations and may have major implications for our understanding of evolutionary processes (Richards 2006; Bossdorf et al. 2008; Jablonka and Raz 2009).

Until recently, the vast majority of our knowledge about epigenetic mechanisms has stemmed from laboratory studies of model organisms, such as mice (Morgan et al. 1999) and *Arabidopsis thaliana* (Lippman et al. 2004). However, with the birth of ecological epigenetics, researchers are now attempting to decipher the role and significance of epigenetic processes in the context of ecology and evolution (Richards 2006; Bossdorf et al. 2008; Jablonka and Raz 2009; Richards et al. 2012a, b).

10.2 The Extent and Structure of Epigenetic Variation Within and Among Natural Populations

Among the most fundamental objectives underlying ecological epigenetics is understanding the importance of epigenetic variation in natural environments (Bossdorf et al. 2008; Richards et al. 2012a). To achieve this objective, various molecular techniques have been used to assess DNA methylation patterns at the individual and population level (reviewed in Dahl and Guldberg 2003; Laird 2003; Liu and Maekawa 2003; Zilberman and Henikoff 2007; Schrey et al. 2013). Thus far, methylation sensitive-AFLP (MS-AFLP; Reyna-Lopez et al. 1997) has been the most commonly used method in ecological epigenetics (reviewed by Schrey et al. 2013). This technique has been

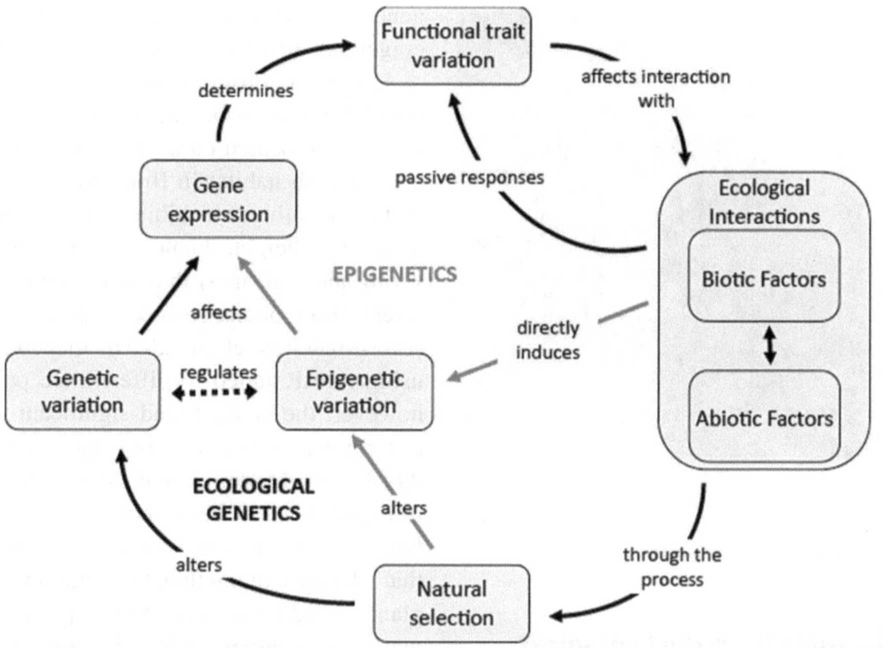

Fig. 10.2 Genetic processes (in *black*) interact with epigenetic processes (in *grey*) and the environment. Here, focus on functional trait variation emphasizes the need for data that links epigenetic loci and specific ecologically relevant phenotypes. We also highlight that ecological interactions include biotic and abiotic factors which can both play roles in creating and maintaining epigenetic variation (Modified with permission from Bossdorf et al. 2008)

used in a variety of studies to determine the extent and structure of epigenetic variation in populations. MS-AFLP is a modification of the standard AFLP protocol (Vos et al. 1995) in that it uses methylation-sensitive isoschizomeric enzymes (e.g., substituting *MspI* and *HpaII* for *MseI*) to detect genome-wide variation in DNA methylation (Cervera et al. 2002; Salmon et al. 2005). The enzymes *MspI* and *HpaII* have different sensitivities to cytosine methylation of the CCGG recognition sequence (McClelland et al. 1994; Roberts et al. 2007). Comparing the banding pattern from independent reactions with *MspI* and *HpaII* allows for the identification of the methylation state at a particular restriction site (Salmon et al. 2008). MS-AFLP screens variation in DNA methylation at many restriction sites, generating a multi-locus epigenotype for each individual.

There are several benefits associated with using MS-AFLP to study epigenetic variation in natural populations (Schrey et al. 2013). First, this technique enables researchers to assess epigenetic variation in non-model organisms that lack a sequenced genome. The technique is also economical, which is important for ecological studies with large sample sizes. MS-AFLP also requires the same equipment and technical skills as traditional AFLP, which makes it easier for labs to couple epigenetic questions to their genetic (i.e., AFLP) work. MS-AFLP is especially useful for population epigenetic studies because it provides a genome-wide scan and allows for many individuals to be screened at multiple loci concurrently. Lastly, MS-AFLP data obtained from appropriately designed studies can demonstrate that heritable epigenetic variation – specifically DNA methylation – may provide an additional source of variation important in the process of natural selection. An extensive review of the benefits and weaknesses of MS-AFLP can be found in Schrey et al. (2013).

Fig. 10.3 Spanish violets (*Viola cazorlensis*) studied by Herrera and Bazaga (2010, 2011) (Image courtesy of © Carlos M. Herrera. All Rights Reserved)

10.2.1 Insights from the Ecological Epigenetics Literature

Understanding the mechanisms of local adaptation to different habitats is an enduring quest in ecology. Since epigenetic mechanisms can respond to the environment and cause heritable variation in traits, epigenetics may contribute to the process of adaptation. This will be reflected in an association of epigenetic structure by habitat, which has been supported in several recent studies (Herrera and Bazaga 2010; Lira-Medeiros et al. 2010; Massicotte et al. 2011; Richards et al. 2012b). For example, Herrera and Bazaga (2010) examined the distribution of genetic and epigenetic variation within and among wild populations of Spanish violet (*Viola cazorlensis* – Fig. 10.3) using AFLP and MS-AFLP. They detected population differentiation at both the genetic and epigenetic levels, but also that epigenetic variation exceeded genetic variation. Identifying more variation at epigenetic markers suggests that epigenetic mechanisms could contribute a significant amount of variation in this species. They also found an association between the patterns of variation observed at epigenetic markers and the genetic loci that were implicated in adaptive differentiation in floral traits between the populations (Herrera and Bazaga 2010). This study was one of the first to find that adaptive genetic divergence may be associated with epigenetic differentiation between populations.

Epigenetic differentiation also exists between populations of white mangroves (*Laguncularia racemosa*) located either in a river basin or near a salt marsh habitat in Brazil (Fig. 10.4). Plants in the river basin exhibited several phenotypic traits (e.g., height, diameter at breast height, leaf width and leaf area) that were significantly different from plants located near the salt marsh (Lira-Medeiros et al. 2010). Genetic analysis, using AFLP, failed to differentiate populations; however, the study found significant epigenetic differentiation between populations using MS-AFLP (Lira-Medeiros et al. 2010). These findings illustrate an association between epigenetic differences and environmental factors, and suggest that changes in methylation among salt marsh plants could be important in response to the salt marsh environment (Lira-Medeiros et al. 2010, see also Salmon et al. 2005).

In one of the few studies of wild animal populations, Massicotte et al. (2011) found a higher rate of epigenetic variation than genetic variation in the clonal fish, *Chrosomus eos-neogaeus*, from multiple lakes in Canada using MS-AFLP. Going beyond MS-AFLP, the authors also incorporated bisulfite sequencing, which clearly shows the benefits gained by using multiple techniques. They first excised and sequenced 15 randomly chosen MS-AFLP fragments. These fragments were then compared to the zebrafish genome and 11 were found to have putative similarity to zebrafish sequences. One locus that showed homology with DIRS1, a transposable element in zebrafish was then bisulfite sequenced and the researchers were able to identify five epigenetic variants at this locus (Massicotte et al. 2011). Thus, the combination of techniques was able to add context to the epigenetic differences observed by MS-AFLP and indicated that at least some of the differences detected by MS-AFLP may target important genetic elements.

One of the fundamental differences between genetic and epigenetic variation is that the latter is more environmentally labile and potentially reversible (Richards et al. 2010a, b). Therefore, patterns of epigenetic differentiation among field populations measured in different

Fig. 10.4 Adult white mangroves (*Laguncularia racemosa*) from the salt marsh (*left panel*) and from the riverine habitat (*right panel*; Lira-Medeiros et al. 2010) (Images courtesy of © Catarina F. Lira-Medeiros. All Rights Reserved)

environments – like the ones observed in the majority of these studies – include both a reversible component due to phenotypic plasticity and a non-reversible or relatively stable component due to heritable epigenetic differentiation. In this respect, analyses of epigenetic variation are similar to analyses of phenotypic variation, and common garden experiments are necessary to separate plastic and heritable components of variation (Richards et al. 2010a, b). For this reason, future studies should account for environmentally induced epigenetic effects by growing organisms in common environments and performing MS-AFLP analyses under these conditions. For example, Richards et al. (2012b), collected Japanese knotweed rhizomes from 16 different populations across three different habitat types, but grew them in a common glasshouse environment before sampling for MS-AFLP analyses. This design confirms that methylation patterns were in fact persistent and not just induced by habitat.

10.3 Ecological Consequences of Epigenetic Variation

Despite the recent progress that has been made in understanding the magnitude of epigenetic variation within and among populations, the extent to which epigenetic processes are associated with ecologically-relevant traits is surprisingly understudied. There have been few ecological studies that have directly linked epigenotype to phenotype or that have assessed the effects of epigenetically-mediated phenotypic differences on organismal fitness. The classic example of how epigenetics may affect ecologically important traits is the epimutation occurring in the plant *Linaria vulgaris* (Fig. 10.1) which was first observed over 250 years ago by Linneaus (Cubas et al. 1999). The epimutation results in silencing of the *Lcyc* gene, which regulates flower symmetry. Plants with a hypomethylated epimutation have flowers with radial symmetry

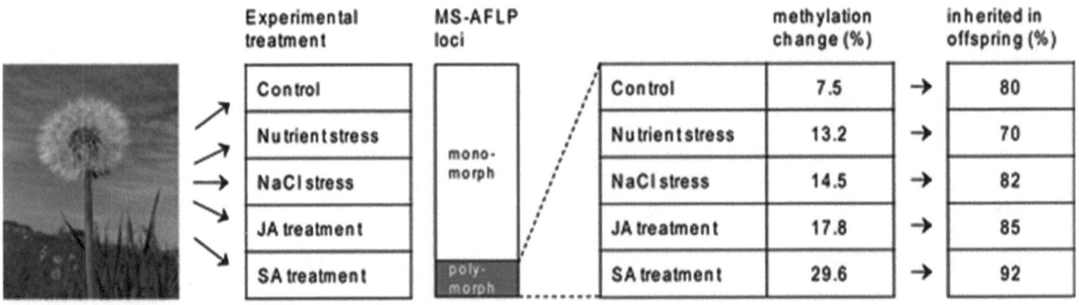

Fig. 10.5 Induction of DNA methylation changes by ecological stresses and their heritability in asexual dandelions (*Taraxacum officinale*) from Verhoeven et al. (2010). A single apomictic dandelion genotype was exposed to different experimental environments. Although these 'susceptible' loci showed some background level of methylation change also in the control group, the rate of methylation change that was observed within the subset of susceptible loci increased significantly due to stress treatment, particularly due to treatment with jasmonic (JA) or salicylic (SA) acid. Most of the induced methylation changes were inherited in apomictic offspring that were not exposed to stress but raised in a common control environment (Reprinted with permission from Richards et al. 2012a)

whereas those without the epimutation exhibit dorsoventral symmetry (Cubas et al. 1999). This epigenetic change in flower morphology is inherited by future generations, which has important implications for flower pollination and evolutionary questions pertaining to *L. vulgaris*. However, this study did not directly attempt to characterize how these phenotypic differences influence fitness, which should be a consideration in future ecological epigenetics research.

While we know of no study that has made a direct causal link between epigenotype, phenotype and fitness in an ecological setting, several exemplary studies indicate that making these connections could enhance our understanding of ecological processes. We review below how epigenetic mechanisms may play a role in several important ecological phenomena: response to environmental stimuli, trophic interactions, niche breadth, invasive species, behavioral variation, disease susceptibility, and speciation events.

10.3.1 Response to Environmental Stimuli

A unique characteristic that differentiates epigenetic from genetic variation is that epigenetic processes (i.e., DNA methylation) are more responsive to the environment (Bossdorf et al. 2008; Richards et al. 2010b). Verhoeven et al. (2010) were some of the first researchers to assess how interaction with abiotic and biotic environmental factors influences DNA methylation. Verhoeven et al. (2010) used genetically identical lines of dandelion (*Taraxacum officinale*) to assess the impact of five different biotic and abiotic conditions on epigenetic variation: low nutrients, salt stress, jasmonic acid (to mimic herbivore damage), salicylic acid (to mimic pathogen damage), and control treatment. Using MS-AFLP, the authors found significantly more methylation changes genome-wide in treated plants than in controls (Fig. 10.5). Moreover, data collected on offspring from each of the treated plants raised in a common garden environment showed that the majority of the changes were also inherited (Verhoeven et al. 2010).

In a similar study, Dowen et al. (2012) exposed *Arabidopsis thaliana* wild type and mutants defective in methylation maintenance machinery to bacterial pathogens, avirulent bacteria, and salicylic acid to determine the effects on the epigenome. Each treatment resulted in different methylation patterns, suggesting that environmental stimuli not only affect global methylation, but that the epigenome responds uniquely to each stimulus and regulates gene expression dynamically (Dowen et al. 2012). In addition, salicylic acid elicited a response in transposable elements and/or their proximal genes through differential methylation, lending support to the

idea that dynamic methylation of TEs may be an important component of response to this stress.

10.3.2 Trophic Interactions

In addition to response to environmental stimuli, epigenetic mechanisms may play a role in response to trophic interactions, such as herbivory. Herrera and Bazaga (2011) continued their work on epigenetic and genetic variation in Spanish populations of *V. cazorlensis* by investigating the response to herbivory. DNA methylation in populations of *V. cazorlensis* exposed to long-term ungulate herbivory varied considerably, which was partially explained by browsing damage (Herrera and Bazaga 2011). The methylation state at some variable loci was also associated with specific AFLP markers that were associated with herbivory levels (Herrera and Bazaga 2011). This study was one of the first to compare both genetic and epigenetic variation contemporaneously in response to variation in a natural environmental stressor and emphasizes the importance of disentangling these two components of an organism's response. They used structural equation modeling (SEM) to show that genotype contributed directly to herbivory damage and epigenotype, but could not discriminate the relationship between epigenotype and herbivory damage. One of the SEM models predicted a consequential role between epigenetic variation and herbivory suggesting that the epigenetic patterns are induced by herbivory. Another, equally likely SEM model predicted a causal role between epigenetic variation and herbivory such that the likelihood of herbivory depended on epigenotype (Herrera and Bazaga 2011). Another possibility is that random epigenetic mutations arise and build up rapidly within isolated populations, potentially resulting in (neutral) epigenetic differences between populations that correlate with genetic differentiation of the populations (Richards et al. 2012a). Overall, the study suggests that a complex relationship exists among genotype, epigenotype, and herbivory damage requiring controlled studies to tease apart the relationships.

A persistent problem we face in understanding the importance of epigenetics in ecology is that there is a complex relationship between genetic and epigenetic effects. We must therefore separate the distinct contributions of each source of variation in order to understand if epigenetic effects provide something not attributable to genetic effects. Richards (2006) summarized the problem another way: some epigenetic effects are entirely determined by genotype, while others may be "facilitated" by specific genotypes, or may be completely independent from genotype. Some portion of variation in DNA methylation is likely attributable to underlying DNA sequence variation, for example differences in the genetic sequence of methyltransferase enzymes (Bird 2002). Therefore, disentangling these different possibilities is complicated not only because we know little about genetic-epigenetic interactions, but also because the genetic basis of most complex traits is still not well understood. To date we know of no studies that have characterized the effects of natural variation in the epigenetic machinery. However, studies in non-model systems can explore variation among genotypes for response at epigenetic markers to different environmental factors. Using a classic phenotypic plasticity design with clonal replicates of *Spartina alterniflora*, Richards and colleagues have shown that genotypes vary in the magnitude of response to community make-up and that there is a correlation between phenotypic variation and epigenetic variation among genotypes (Richards et al. Unpublished). Further study is needed to determine the relative contribution of genotype to epigenotype and the extent to which this interaction governs phenotypic variation in natural populations.

10.3.3 Niche Breadth

Another ongoing endeavor in ecology is to determine the mechanisms that underlie the ability of some organisms to occupy a broad niche within a community. A recent study of a nectar-living yeast (*Metschnikowia reukaufii*)

Fig. 10.6 Six focal species from which flower nectar sugar environments were studied: from *left* to *right*, *top row*, *Digitalis obscura*, *Gladiolus illyricus*, *Aquilegia vulgaris*, *bottom row*, *Helleborus foetidus*, *Atropa baetica*, and *Primula vulgaris* (Herrera et al. 2012) (Images courtesy of © Carlos M. Herrera. All Rights Reserved)

showed that methylation changes are a critical component of its ability to use resources from a wide range of host environments, particularly harsh environments (Fig. 10.6) (Herrera et al. 2012). Herrera et al. (2012) grew yeast lines in multiple media of varying concentrations of sucrose, glucose, and fructose, and applied the demethylating agent 5-Azacytidine (5-AzaC). 5-AzaC had an inhibitory effect on growth, which was more pronounced in the challenging environment of high sugar concentrations. These data suggest that DNA methylation in *M. reukaufii* responds to different nectar conditions, and the epigenetic response allows the yeast to occupy a wide range of nectars and flowers (Fig. 10.7).

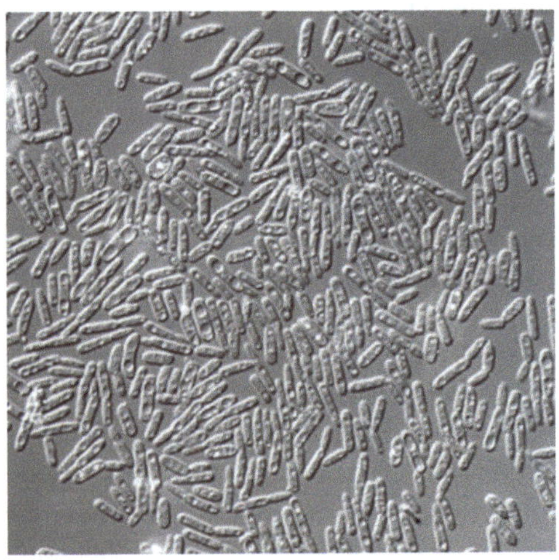

Fig. 10.7 The nectar-living yeast *Metschnikowia reukaufii* (Herrera et al. 2012) (Image courtesy of © Carlos M. Herrera. All Rights Reserved)

10.3.4 Invasive Species

Introduced species should be at a major disadvantage throughout the invasion process because they may not be well-adapted to their new habitat and often experience reduced genetic variation and inbreeding due to the small size of the founding population (Pérez et al. 2006; Schrey et al. 2012). Despite these challenges, invasive species are surprisingly successful in colonizing new environments. In some cases, invasive species display

extensive phenotypic variation, which results in a genetic paradox (Pérez et al. 2006; Richards et al. 2008, 2012b; Schrey et al. 2012; Liebl et al. 2013). Several studies suggest that epigenetic variation may compensate for reduced genetic diversity and potentially mediate phenotypic plasticity in traits associated with "invasiveness" (e.g., rapid growth or reproductive output, increased competitive ability, etc.) (Pérez et al. 2006; Richards et al. 2012b; Schrey et al. 2012).

Fig. 10.8 Japanese knotweed (*Fallopia japonica*) populations across beach, salt marsh and roadside habitats on Long Island, NY (Image courtesy of Christina L. Richards)

Japanese knotweed (*Fallopia japonica*) is a highly invasive plant species in Europe and has recently colonized the northeastern United States, where it occupies roadside, marsh, and beach habitats (Fig. 10.8). Richards et al. (2008, 2012b) sampled populations from these diverse habitats in Long Island, NY and grew them in common garden. Plants from the different populations displayed almost no genetic diversity, but maintained a high degree of epigenetic and phenotypic variation, and phenotypic plasticity in response to controlled salt treatments (Richards et al. 2008, 2012b). These findings suggest that epigenetic variation, rather than genetic variation, may be facilitating the rapid colonization of knotweed across a range of environments (Figs. 10.9 and 10.10).

Similarly, the house sparrow (*Passer domesticus*) is one of the most globally distributed bird species, having been successfully introduced on every continent except Antarctica (Anderson 2006; Schrey et al. 2012). Extensive phenotypic variability has been observed across native and introduced ranges, indicating the species has overcome many limitations associated with population bottlenecks (Johnston and Selander 1973; Martin 2005; Martin et al. 2005). Schrey et al. (2012) found that Nairobi (introduction 50 years ago) and Tampa (introduction 150 years ago) populations shared similar levels of epigenetic variation, while Nairobi had less genetic diversity than Tampa. Within Kenya, epigenetic diversity was negatively correlated with genetic diversity and positively correlated with inbreeding across the range expansion (Liebl et al. 2013). These results suggest that epigenetic variation could be a factor underlying the phenotypic diversification often observed in these recently introduced populations. However, more research is needed to determine how methylation at specific restriction sites is functionally linked to phenotypic variation (Schrey et al. 2012; Liebl et al. 2013).

Fig. 10.9 AFLP markers indicating no polymorphism across ramets collected from a roadside (*left*) and marsh (*right*) populations of *Fallopia japonica* (Richards et al. 2008) (Modified from Richards et al. 2012a)

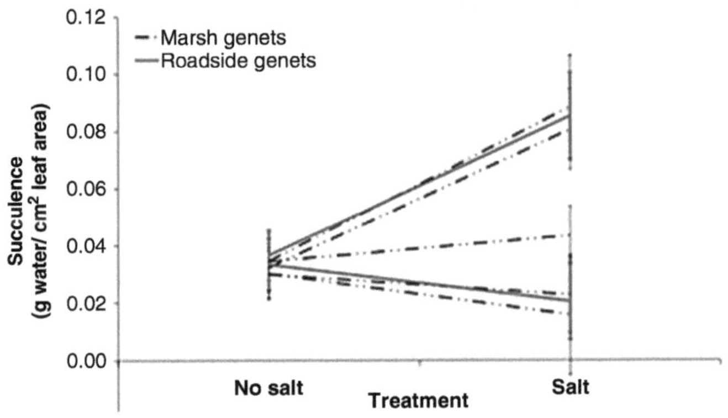

Fig. 10.10 Response of replicates of rhizomes collected at a marsh (*-dashed*) and nearby roadside (*solid*) sites. AFLP showed that all individuals from these two populations were the same genotype (Richards et al. 2008) (Modified from Richards et al. 2012a)

10.3.5 Behavioral Variation

Behavior is often considered one of the most flexible and environmentally-sensitive phenotypic traits (West-Eberhard 2003). Unlike changing other aspects of an organism's phenotype (e.g., morphology), the ability to alter behavior allows for a more rapid and less costly response to environmental cues (West-Eberhard 2003). Extensive research has demonstrated that behavioral variation is fairly common within and among populations and is often associated with different selection pressures exerted by the environment (Ledón-Rettig et al. 2013). Laboratory studies have shown that epigenetic mechanisms can affect behavioral variation in association with environmental conditions. For example, differences in larval diet can influence behavioral variation associated with the caste system in honeybees via epigenetic mechanisms (Kucharski et al. 2008; Miklos and Maleszka 2011). Despite being genetically identical, larvae that are fed royal jelly develop into reproductive queens which are behaviorally more aggressive, whereas those that consume lower quality diets become non reproductive workers that spend much of their lives displaying behaviors associated with foraging (Kucharski et al. 2008; Miklos and Maleszka 2011). These differences in behavioral phenotype were correlated with differences in DNA methylation in the brain, such that queens had reduced DNA methylation of certain genes when compared to workers (Kucharski et al. 2008; Miklos and Maleszka 2011). Interestingly, experimental injection of larvae with a small interfering-RNA caused downregulation of the DNA methyltransferase system, which led to the production of more queens than a control group (Kucharski et al. 2008; Miklos and Maleszka 2011).

Another well-known study demonstrates how variation in maternal care behaviors within the first week of life of neonatal rats can lead to individual phenotypic differences in behavior and

stress responsiveness, which persist into adulthood (Weaver et al. 2004). Offspring of mothers who displayed high licking and grooming behavior (high-LG), grew up to be less fearful and had more attenuated stress responses when compared to offspring of mothers who displayed low licking and grooming (low-LG) behavior (Weaver et al. 2004). These major phenotypic differences were associated with a difference in the methylation status of the glucocorticoid receptor (GR) promoter in the hippocampus (Weaver et al. 2004, 2005, 2006). A cross-fostering study revealed that offspring phenotype was determined by the behavioral (high or low-LG) phenotype of the foster mother – rather than the biological mother – providing evidence that epigenetic, rather than genetic processes, are responsible for these phenotypic differences (Weaver et al. 2004).

10.3.6 Disease Susceptibility

Environmental perturbations experienced within critical periods of development have been implicated in the risk of disease development (e.g., cancer, heart disease, diabetes and schizophrenia) (Jirtle and Skinner 2007; Skinner et al. 2010). The genetic and environmental basis of certain diseases is well-documented in epidemiological studies (Skinner et al. 2010; Tost 2010), yet the molecular mechanisms by which environmental factors contribute to disease etiology have only recently been explored (Skinner et al. 2010). There is mounting evidence that environmentally-sensitive epigenetic processes play an important role in regulating disease susceptibility (Jirtle and Skinner 2007; Skinner et al. 2010; Tost 2010). One of the best characterized examples comes from the study of the metastable *Avy* allele of the agouti gene in mice (Morgan et al. 1999; Jirtle and Skinner 2007; Skinner et al. 2010). In animals with the *Avy* allele, expression is mediated by variable DNA methylation of a transposable element located upstream of the agouti gene. Low levels of methylation of the *Avy* allele result in yellow coat color whereas increasing methylation of the transposable element causes a shift toward the wild-type pseudo-agouti (brown) coat color (Fig. 10.11) (Morgan et al. 1999; Jirtle and Skinner 2007; Skinner et al. 2010). Furthermore, the unmethylated state is also associated with obesity and increased susceptibility to diabetes and tumor formation (Jirtle and Skinner 2007; Skinner et al. 2010). Maternal nutritional supplementation with methyl-donors and the phytoestrogen, genistein, increases DNA methylation, which leads to a shift in offspring coat color from yellow to brown and significantly reduces the incidence of obesity, diabetes and cancer in pseudo-agouti offspring (Waterland and Jirtle 2003, 2004; Dolinoy et al. 2006, 2007). While many of the details of the epigenetic regulation of pseudo-agouti coat color have been worked out in laboratory experiments, the implications for natural populations of mammals have not been explored at all (Ledón-Rettig et al. 2013). Intuitively, the link between the mother's diet and disease susceptibility in offspring should have important ecological implications for wild populations. Thus, understanding the molecular mechanisms underlying the dramatic response to diet and its influence on disease trajectories will impact our understanding of disease dynamics.

Another series of studies have shown that exposure to certain chemicals in early life also influences disease susceptibility through direct effects on the epigenome (Jirtle and Skinner 2007; Skinner et al. 2010; Tost 2010). Environmental toxins with endocrine disruptor activity (e.g., fungicides, pesticides, plastic by-products and pharmacological substances) have been found to impact disease phenotype and fitness in adulthood (Crews et al. 2007; Skinner et al. 2010). For example, developmental exposure to environmentally relevant amounts of bisphenol A (BPA), a residue found in many plastic materials, produced changes in DNA methylation associated with increased susceptibility to cancer in rats (Ho et al. 2006; Skinner et al. 2010). Transient embryonic exposure to fungicides and pesticides also led to reduced spermatogenic capacity

Fig. 10.11 Phenotypes of isogenic *Avy* littermates range from *pure yellow* and obese (*left*) through *mottled yellow*/agouti to lean fully agouti (called pseudoagouti, *right*) (Reprinted with permission from © (Cropley et al. 2010). All Rights Reserved)

and male infertility in rats, and the decreased male fertility was transmitted transgenerationally via alterations in DNA methylation in the male germ-line (Anway et al. 2005; Skinner et al. 2010). Although data from the biomedical literature have improved our understanding about the role of epigenetic mechanisms in the environmental basis of disease, there have been no studies to investigate the degree to which these processes influence disease susceptibility or the ecology of infectious diseases in natural populations. Studying these processes in an ecological context may reveal how increasing anthropogenic disturbances are impacting the health of populations and may be useful for ongoing conservation efforts.

10.3.7 Speciation Events

Epigenetic mechanisms are often involved in polyploidy and hybridization events in plants (Rapp and Wendel 2005; Richards et al. 2012a). Epigenetic mechanisms may be involved with dosage regulation of replicate genes which could allow for the separate genomes to persist or merge without gene interaction problems (Liu 2003). Ainouche and colleagues have demonstrated the importance of epigenetic mechanisms in the genus *Spartina*, which has evolved through multiple allopolyploid and hybridization speciation events (Fortune et al. 2007). In particular, *Spartina alterniflora* and *S. maritima* have formed two distinct hybrids (*S. x townsendii* and *S. x neyrautii*) in the past century, and *S. anglica* has since formed as an allopolyploid from *S. x townsendii* (Ainouche et al. 2004). *Spartina anglica* has increased physiological tolerance over its progenitors to multiple stresses of the intertidal zone, and has become extremely invasive around the world (Ainouche et al. 2009). Using MS-AFLP, transposon display, and the Agilent rice microarray, Ainouche's group showed that changes in DNA sequence in the hybrid species were more or less additive compared to the parental species, but genome methylation and gene expression were not (Salmon et al. 2005; Parisod et al. 2010; Chelaifa et al. 2010a, b). These studies suggest that changes in DNA methylation may help explain the dramatic differences in phenotype that allow members of this genus to successfully occupy novel habitats (Salmon et al. 2005). However, these ecologically-oriented questions have not yet been addressed.

Paun et al. (2010) provide another example of how epigenetic mechanisms may be important in polyploid speciation. The orchids *Dactylorhiza majalis*, *D. traunsteineri s.l.*, and *D. ebudensis* (Fig. 10.12) all arose from independent hybridization events of the diploids *D. fuchsii* and *D. incarnata* followed by allopolyploidization. *Dactylorhiza majalis* has a wide range while *D. ebudensis* is a narrow endemic living in a single coastal dune slack habitat. *Dactylorhiza traunsteineri s.l.* has an intermediate range, but narrow tolerances of both soil moisture and pH, and

Fig. 10.12 The allotetraploid *Dactylorhiza traunsteineri* at a natural site in Yorkshire, England (Image courtesy of © (Paun et al. 2010). All Rights Reserved)

grows in calcareous fens and marshes (Paun et al. 2010). Genome-wide methylation patterns, obtained using MS-AFLP, showed differentiation of the three species (Paun et al. 2010). Because the different species have arisen from independent hybridization events of the same parents, the authors suggest that epigenetic mechanisms could be important to the process of differentiation and contrasting environmental tolerance of these species. However, the extent that the epigenetic differences were the cause or the consequence of the lineages inhabiting different environments remain to be elucidated.

10.4 Evolutionary Consequences of Epigenetic Variation

Current data on epigenetic variation and its influence on phenotype have provocative implications for evolution. In at least some cases, induced epigenetic changes have been shown to be heritable through meiosis without reset for both plants and animals (Crews et al. 2007; Feng et al. 2010; Verhoeven et al. 2010; Grossniklaus et al. 2013). A number of theoretical models have been proposed to describe the evolutionary value of epigenetic variation in natural populations (Jablonka and Lamb 1989; Jablonka et al. 1992, 1995; Lachmann and Jablonka 1996; Pal and Miklos 1999; Day and Bonduriansky 2011; Geoghegen and Spencer 2012). Recent models have been limited, primarily due to a paucity of information on the behavior of epigenetic marks, but they have demonstrated that because the epigenetic code can be more dynamic and reversible than the DNA code, it can add adaptive flexibility (Jablonka and Lamb 1989; Jablonka et al. 1995; Lachmann and Jablonka 1996; Pal and Miklos 1999). Variation in epigenetic mechanisms can contribute to phenotypic variation, which is not necessarily adaptive (Pal 1998; Rapp and Wendel 2005; Richards et al. 2010b). However, epigenetic memory could be adaptive in changing environments, where epigenetic variation creates a buffering system against high rates of environmental change (Jablonka et al. 1995; Lachmann and Jablonka 1996). Epigenetic modifications could 'hold' a potentially advantageous phenotype for multiple generations, allowing time for more durable genetic processes to stabilize the phenotype (i.e., canalization or genetic assimilation; Waddington 1942, 1953; West-Eberhard 2005; Richards et al. 2012a). The ability to generate heritable epigenetic variation can speed up the process of reaching a fitness peak in the adaptive landscape, facilitate peak shifts, or facilitate the transition from one fit genotypic state to another (Pal and Miklos 1999), and create the potential for novel evolutionary outcomes in the absence of genetic variation (Tal et al. 2010; Geoghegen and Spencer

2012). Still, most models assume epigenetic motifs all have the same likelihood of reset, and that they can be easily reset even though the rate at which epigenetic maintenance and erasure occurs has been shown to vary across different sites within the genome (Rakyan et al. 2001; Feng et al. 2010; Zhou et al. 2011; Grossniklaus et al. 2013).

10.5 Conclusions and Future Directions

It is becoming clearer that our knowledge about important ecological processes will be informed by understanding how the epigenome functions in natural environments. Significant progress has been made in understanding the extent and distribution of epigenetic variation in natural populations as well as the potential ecological and evolutionary consequences of such variation. However, due to the relative infancy of the field of ecological epigenetics, there are still many questions that remain unanswered.

Like ecological genomics, the future of ecological epigenetics will require carefully designed studies that can account for genotype and environment effects. With experimental studies on genotypic replicates exposed to different environments, future studies can investigate the behavior of epialleles and interactions with annotated components of the genome including functional genes, regulatory elements and non-coding regions such as transposable elements. Creative new approaches to modeling the importance of both genetic and non-genetic inheritance will lend important insight for understanding the dynamic nature of genome function (*sensu* Day and Bonduriansky 2011).

Although models for epigenetically controlled traits have found that epigenetic effects may enhance the adaptive possibilities of a variety of taxa, particularly in response to novel environments (Jablonka and Lamb 1989; Geoghegen and Spencer 2012), these models are limited by a lack of data on epigenetic response to environmental factors. Ultimately, models will better inform our understanding of evolution if we can characterize behavior of epigenetic marks at specific genomic elements (e.g., the promoters of ecologically important genes or activity of transposable elements). To date, this type of information has been available only for model organisms such as *Arabidopsis thaliana* (Lippman et al. 2004; Slotkin and Martienssen 2007; Vaughn et al. 2007) or mice (Morgan et al. 1999; Weaver et al. 2004). The use of next generation sequencing technology, like restriction-site-associated DNA sequencing (RAD-seq), has expanded the possibilities for non-model systems that have no reference genome. RAD-seq reduces the complexity of the genome sampled and increases the power to identify repeat or duplicate sequences (Etter et al. 2011). RAD-seq can also incorporate paired-end sequencing (RAD-PE), allowing for the assembly of larger continuous sequences from short Illumina sequences (Etter et al. 2011). RAD-PE has not been used to study methylation patterns yet, but methylation sensitive enzymes have been used to target low copy number, gene rich regions to exclude highly repetitive DNA regions that are highly methylated (Chutimanitsakun et al. 2011). This indicates that the same methodology could be used in an experimental context to compare replicates of the same genotype exposed to different conditions to allow for a genome wide probing of changes in methylation.

While most models assume that all epigenetic marks behave similarly, a recent model proposed by Day and Bonduriansky (2011) posits that some genetic elements are more likely to acquire methyl groups than others. Future studies can test this model with data from MS-AFLP or novel next generation sequencing approaches on organisms from natural populations and over clonal generations of experimental transplants and greenhouse experiments. This will allow for a genome-wide characterization of the stability and behavior of different methylation marks, how they affect phenotype and how this varies by genotype. Considering that the research community has made little progress in understanding how the genome actually functions to create complex traits and adapt to complex environments (Richards et al. 2009, 2012a; Pigliucci 2010; Martin et al. 2011), characterizing the role of epi-

genetic effects in natural systems could transform our understanding of ecological and evolutionary processes.

Acknowledgements The authors would like to thank Nadia Aubin-Horth and Christian R. Landry for the invitation to write this chapter. We also thank LB Martin, M Boruta, AJ Brace and CAC Coon for feedback on previous versions of the manuscript. This work was supported by the National Science Foundation (GRFP-1144244 to HJK).

References

Ainouche ML, Baumel A, Salmon A, Yannic G (2004) Hybridization, polyploidy and speciation in *Spartina* (Poaceae). New Phytol 161(1):165–172

Ainouche ML, Fortune PM, Salmon A, Parisod C, Grandbastien MA, Fukunaga K, Ricou M, Misset MT (2009) Hybridization, polyploidy and invasion: lessons from *Spartina* (Poaceae). Biol Invasions 11:1159–1173

Anderson TR (2006) Biology of the ubiquitous house sparrow: from genes to populations. Oxford University Press, New York

Angers B, Castonguay E, Massicotte R (2010) Environmentally induced phenotypes and DNA methylation: how to deal with unpredictable conditions until the next generation and after. Mol Ecol 19(7):1283–1295

Anway MD, Cupp AS, Uzumcu M, Skinner MK (2005) Epigenetic transgenerational actions of endocrine disruptors and mate fertility. Science 308:1466–1469

Bird A (2002) DNA methylation patterns and epigenetic memory. Genes Dev 16(1):6–21

Bird A (2007) Perceptions of epigenetics. Nature 447(7143):396–398

Bock C (2012) Analysing and interpreting DNA methylation data. Nat Rev Genet 13(10):705–719

Bossdorf O, Richards CL, Pigliucci M (2008) Epigenetics for ecologists. Ecol Lett 11:106–115

Boyko A, Kovalchuk I (2011) Genome instability and epigenetic modification: heritable responses to environmental stress? Curr Opin Plant Biol 14(3):260–266

Cervera MT, Ruiz-Garcia L, Martinez-Zapater JM (2002) Analysis of DNA methylation in *Arabidopsis thaliana* based on methylation-sensitive AFLP markers. Mol Gen Genomics 268:543–552

Chelaifa H, Monnier A, Ainouche ML (2010a) Transcriptomic changes following recent natural hybridization and allopolyploidy in the salt marsh species *Spartina x townsendii* and *Spartina anglica* (Poaceae). New Phytol 186:161–174

Chelaifa H, Mahe F, Ainouche ML (2010b) Transcriptome divergence between the hexaploid salt-marsh sister species *Spartina maritima* and *Spartina alterniflora* (Poaceae). Mol Ecol 19:2050–2063

Chutimanitsakun Y, Nipper RW, Cuesta-Marcos A, Cistué L, Corey A, Filichkina T, Johnson EA, Hayes PM (2011) Construction and application for QTL analysis of a restriction site associated DNA (RAD) linkage map in barley. BMC Genomics 12:4

Crews D, Gore AC, Hsu TS, Dangleben NL, Spinetta M, Schallert T, Anway MD, Skinner MK (2007) Transgenerational epigenetic imprints on mate preference. Proc Natl Acad Sci USA 104(14):5942–5946

Cropley JE, Suter CM, Beckman KB, Martin DI (2010) CpG methylation of a silent controlling element in the murine Avy allele is incomplete and unresponsive to methyl donor supplementation. PLoS ONE 5(2):e9055

Cubas P, Vincent C, Coen E (1999) An epigenetic mutation responsible for natural variation in floral symmetry. Nature 401:157–161

Dahl C, Guldberg P (2003) DNA methylation analysis techniques. Biogerontology 4:233–250

Day T, Bonduriansky R (2011) A unified approach to the evolutionary consequences of genetic and nongenetic inheritance. Am Nat 178:E18–E36

Dolinoy DC, Weidman JR, Waterland RA, Jirtle RL (2006) Maternal genistein alters coat color and protects Avy mouse offspring from obesity by modifying the fetal epigenome. Environ Heal Perspect 144(4):567–572

Dolinoy DC, Weidman JR, Jirtle RL (2007) Epigenetic gene regulation: linking early developmental environment to adult disease. Reprod Toxicol 23(3):297–307

Dowen RH, Pelizzola M, Schmitz RJ, Lister R, Dowen JM, Nery JR, Dixon JE, Ecker JR (2012) Widespread dynamic DNA methylation in response to biotic stress. Proc Natl Acad Sci USA 109(32):E2183–E2191

Etter PD, Preston JL, Bassham S, Cresko WA, Johnson EA (2011) Local de novo assembly of RAD paired-end contigs using short sequencing reads. PLoS ONE 6(4):e18561

Feng S, Cokus SJ, Zhang X, Chen PY, Bostick M, Goll MG, Hetzel J, Jain J, Strauss SH, Halpern ME et al (2010) Conservation and divergence of methylation patterning in plants and animals. Proc Natl Acad Sci USA 107(19):8689–8694

Feschotte C (2008) Transposable elements and the evolution of regulatory networks. Nat Rev Genet 9:397–405

Fortune PM, Schierenbeck KA, Ainouche AK, Jacquemin J, Wendel JF, Ainouche ML (2007) Evolutionary dynamics of waxy and the origin of hexaploid Spartina species (Poaceae). Mol Phylogenet Evol 43(3):1040–1055

Geoghegen JL, Spencer HG (2012) Population-epigenetic models of selection. Theor Popul Biol 81:232–242

Grossniklaus U, Kelly W, Ferguson-Smith AC, Pembrey M, Lindquist S (2013) Transgenerational epigenetic inheritance: how important is it? Nat Rev Genet 14:228–235

Hallgrímsson B, Hall BK (2011) Epigenetics: linking genotype and phenotype in development and evolution. University of California Press, Berkeley

Herrera CM, Bazaga P (2010) Epigenetic differentiation and relationship to adaptive genetic divergence in discrete populations of the violet *Viola cazorlensis*. New Phytol 187:867–876

Herrera CM, Bazaga P (2011) Untangling individual variation in natural populations: ecological, genetic and epigenetic correlates of long-term inequality in herbivory. Mol Ecol 20:1675–1688

Herrera CM, Pozo MI, Bazaga P (2012) Jack of all nectars, master of most: DNA methylation and the epigenetic basis of niche width in a flower living yeast. Mol Ecol 21(11):2602–2616

Ho DH, Burggren WW (2010) Epigenetics and transgenerational transfer: a physiological perspective. J Exp Biol 213(1):3–16

Ho SM, Tang WY, Belmonte de Frausto J, Prins GS (2006) Developmental exposure to estradiol and bisphenol A increases susceptibility to prostate carcinogenesis and epigenetically regulates phosphodiesterase type 4 variant 4. Cancer Res 66(11): 5624–5632

Holliday R (1994) Epigenetics: an overview. Dev Genet 15(6):453–457

Holliday R (2006) Epigenetics: a historical overview. Epigenetics 1(2):76–80

Jablonka E, Lamb MJ (1989) The inheritance of acquired epigenetic variations. J Theor Biol 139:69–83

Jablonka E, Lamb MJ (2002) The changing concept of epigenetics. Ann N Y Acad Sci 981:82–96

Jablonka E, Raz G (2009) Transgenerational epigenetic inheritance: prevalence, mechanisms, and implications for the study of heredity and evolution. Q Rev Biol 84:131–176

Jablonka E, Lachmann M, Lamb MJ (1992) Evidence, mechanisms and models for the inheritance of acquired characters. J Theor Biol 158(2): 245–268

Jablonka E, Oborny B, Molnar I, Kisdi E, Hofbauer J, Czaran T (1995) The adaptive advantage of phenotypic memory in changing environments. Philos Trans R Soc Lond B Biol Sci 350(1332):133–141

Jaenisch R, Bird A (2003) Epigenetic regulation of gene expression: how the genome integrates intrinsic and environmental signals. Nat Genet 33:245–254

Jirtle RL, Skinner MK (2007) Environmental epigenomics and disease susceptibility. Nat Rev Genet 8(4):253–262

Johannes F, Porcher E, Teixeira FK, Saliba-Colombani V, Simon M, Agier N, Bulski A, Albuisson J, Heredia F, Audigier P, Bouchez D, Dillmann C, Guerche P, Hospital F, Colot V (2009) Assessing the impact of transgenerational epigenetic variation on complex traits. PLoS Genet 5:e1000530

Johnston RF, Selander RK (1973) Evolution in the house sparrow. III. Variation in size and sexual dimorphism in Europe and North and South America. Am Nat 107(955):373–390

Karrenberg S, Lexer C, Rieseberg LH (2007) Reconstructing the history of selection during homoploid hybrid speciation. Am Nat 169(6):725–737

Kaslow DC, Migeon BR (1987) DNA methylation stabilizes X chromosome inactivation in eutherians but not in marsupials: evidence for multistep maintenance of mammalian X dosage compensation. Proc Natl Acad Sci 84(17):6210–6214

Kazazian HH (2004) Mobile elements: drivers of genome evolution. Science 303(5664):1626–1632

Kejnovsky E, Hawkins JS, Feschotte C (2012) Plant transposable elements: biology and evolution. In: Wendel JF (ed) Molecular biology and evolution of the plant genome. Springer Verlag, Vienna/New York, pp 17–34

Kovalchuk I (2012) Transgenerational epigenetic inheritance in animals. Front Genet 3:76

Kucharski R, Maleszka J, Foret S, Maleszka R (2008) Nutritional control of reproductive status in honeybees via DNA methylation. Science 319(5871):1827–1830

Lachmann M, Jablonka E (1996) The inheritance of phenotypes: an adaptation to fluctuating environments. J Theor Biol 181:1–9

Laird PW (2003) The power and the promise of DNA methylation markers. Nat Rev Cancer 3(4):253–266

Laird PW (2010) Principles and challenges of genomewide DNA methylation analysis. Nat Rev Genet 11(3):191–203

Ledón-Rettig CC, Richards CL, Martin LB (2013) Epigenetics for behavioral ecologists. Behav Ecol 24(1): 1–14

Li E, Beard C, Jaenisch R (1993) Role for DNA methylation in genomic imprinting. Nature 366(6453):362–365

Liebl AL, Schrey AW, Richards CL, Martin LB (2013) Patterns of DNA methylation throughout a range expansion of an introduced songbird. Integr Comp Biol 53(2):351–358

Lippman Z, Gendrel AV, Black M, Vaughn MW, Dedhia N, McCombie WR, Lavine K, Mittal V, May B, Kasschau KD, Carrington JC, Doerge RW, Colot V, Martienssen R (2004) Role of transposable elements in heterochromatin and epigenetic control. Nature 430:471–476

Lira-Medeiros CF, Parisod C, Fernandes RA, Mata CS, Cardoso MA, Ferreira PCG (2010) Epigenetic variation in mangrove plants occurring in contrasting natural environment. PLoS ONE 5:e10326

Liu B (2003) Epigenetic phenomena and the evolution of plant allopolyploids. Mol Phylogenet Evol 29(3):365–379

Liu Z, Maekawa M (2003) Polymerase chain reaction-based methods of DNA methylation analysis. Anal Biochem 317(2):259–265

Martin LB (2005) A taste for novelty in invading house sparrows, *Passer domesticus*. Behav Ecol 16(4):702–707

Martin LB, Gilliam J, Han P, Lee K, Wikelski M (2005) Corticosterone suppresses cutaneous immune function in temperate but not tropical House Sparrows, *Passer domesticus*. Gen Comp Endocrinol 140(2):126–135

Martin LB, Liebl AL, Trotter JH, Richards CL, McCoy K, McCoy MW (2011) Integrators: physiological determinants of phenotypic plasticity. Integr Comp Biol 51:514–527

Massicotte R, Whitelaw E, Angers B (2011) DNA methylation: a source of random variation in natural populations. Epigenetics 6(4):421–427

McClelland M, Nelson M, Raschke E (1994) Effect of site-specific modification on restriction endonucleases and DNA modification methyltransferases. Nucleic Acids Res 22(17):3640–3659

McClintock B (1984) The significance of responses of the genome to challenge. Science 226:792–801

Miklos GL, Maleszka R (2011) Epigenomic communication systems in humans and honey bees: from molecules to behavior. Horm Behav 59(3):399–406

Monk M, Boubelik M, Lehnert S (1987) Temporal and regional changes in DNA methylation in the embryonic, extraembryonic and germ cell lineages during mouse embryo development. Development 99(3):371–382

Morgan HD, Sutherland HE, Martin DIK, Whitelaw E (1999) Epigenetic inheritance at the agouti locus in the mouse. Nat Genet 23(3):314–318

Nätt D, Rubin CJ, Wright D, Johnsson M, Beltéky J, Andersson L, Jensen P (2012) Heritable genome-wide variation of gene expression and promoter methylation between wild and domesticated chickens. BMC Genomics 13:59

Pal C (1998) Plasticity, memory and the adaptive landscape of the genotype. Proc R Soc B Biol Sci 265(1403):1319–1323

Pal C, Miklos I (1999) Epigenetic inheritance, genetic assimilation and speciation. J Theor Biol 200:19–37

Palevitz BA (1999) Helical science. Scientist 13(19):31

Parisod C, Alix K, Just J, Petit M, Sarilar V, Mhiri C, Ainouche M, Chalhoub B, Grandbastien M (2010) Impact of transposable elements on the organization and function of allopolyploid genomes. New Phytol 186(1):37–45

Paun O, Bateman RM, Fay MF, Hedrén M, Civeyrel L, Chase MW (2010) Stable epigenetic effects impact adaptation in allopolyploid orchids (Dactylorhiza: Orchidaceae). Mol Biol Evol 27:2465–2473

Pérez JE, Nirchio M, Alfonsi C, Muñoz C (2006) The biology of invasions: the genetic adaptation paradox. Biol Invasions 8(5):1115–1121

Pigliucci M (2010) Genotype-phenotype mapping and the end of the 'genes as blueprint' metaphor. Philos Trans R Soc B 365:557–566

Rakyan VK, Preis J, Morgan HD, Whitelaw E (2001) The marks, mechanisms and memory of epigenetic states in mammals. Biochem J 356:1–10

Rapp RA, Wendel JF (2005) Epigenetics and plant evolution. New Phytol 168:81–91

Reyna-Lopez GE, Simpson J, Ruiz-Herrera J (1997) Differences in DNA methylation patterns are detectable during the dimorphic transition of fungi by amplification of restriction polymorphisms. Mol Gen Genet 253(6):703–710

Richards EJ (2006) Inherited epigenetic variation – revisiting soft inheritance. Nat Rev Genet 7:395–401

Richards EJ (2011) Natural epigenetic variation in plant species: a view from the field. Curr Opin Plant Biol 14:204–209

Richards EJ, Elgin SCR (2002) Epigenetic codes for heterochromatin formation and silencing: rounding up the usual suspects. Cell 108(4):489–500

Richards CL, Walls R, Bailey JP, Parameswaran R, George T, Pigliucci M (2008) Plasticity in salt tolerance traits allows for invasion of salt marshes by Japanese knotweed s.l. (*Fallopia japonica* and *F.* × *bohemica*, Polygonaceae). Am J Bot 95:931–942

Richards CL, Hanzawa Y, Ehrenreich IM, Katari M, Engelmann KE, Purugganan MD (2009) Perspectives on ecological and evolutionary systems biology. In: Gutierrez RA, Coruzzi GM (eds) Plant systems biology, vol 35, Annual plant reviews. Blackwell, Oxford, pp 331–351

Richards CL, Bossdorf O, Verhoeven KJF (2010a) Understanding natural epigenetic variation. New Phytol 187:562–564

Richards CL, Bossdorf O, Pigliucci M (2010b) What role does heritable epigenetic variation play in phenotypic evolution? Bioscience 60:232–237

Richards CL, Verhoeven KJF, Bossdorf O (2012a) Evolutionary significance of epigenetic variation. In: Wendel JF (ed) Molecular biology and evolution of the plant genome. Springer Verlag, Vienna/New York, pp 257–274

Richards CL, Schrey A, Pigliucci M (2012b) Invasion of diverse habitats by few Japanese knotweed genotypes is correlated with epigenetic differentiation. Ecol Lett 15:1016–1025

Roberts RJ, Vincze T, Posfai J, Macelis Roberts RJ, Vincze T, Posfai J, Macelis D (2007) REBASE-enzymes and genes for DNA restriction and modification. Nucleic Acids Res 35(Database Issue): D269–D270

Salmon A, Ainouche ML, Wendel JF (2005) Genetic and epigenetic consequences of recent hybridization and polyploidy in *Spartina* (Poaceae). Mol Ecol 14: 1163–1175

Salmon A, Clotault J, Jenczewski E, Chable V, Manzanares-Dauleux M (2008) *Brassica oleracea* displays a high level of DNA methylation polymorphism. Plant Sci 174(1):61–70

Saze H, Tsugane K, Kanno T, Nishimura T (2012) DNA methylation in plants: relationship to small RNAs and histone modifications, and functions in transposon inactivation. Plant Cell Physiol 53(5):766–784

Schrey AW, Coon C, Grispo M, Awad M, McCoy E, Mushinsky H, Richards CL, Martin LB (2012) Epigenetic variation may compensate for decreased genetic variation with introductions: a case study using house sparrows (*Passer domesticus*) on two continents. Genet Res Intl Vol 2012, Article ID 979751, 7 pages

Schrey AW, Alvarez M, Foust C, Kilvitis HJ, Lee JD, Liebl AL, Martin LB, Richards CL, Robertson M (2013) Ecological epigenetics: beyond MS-AFLP. Integr Comp Biol 53(2):340–350

Skinner MK, Manikkam M, Guerrero-Bosagna C (2010) Epigenetic transgenerational actions of environmental factors in disease etiology. Trends Endocrinol Metab 21(4):214–222

Slotkin RK, Martienssen R (2007) Transposable elements and the epigenetic regulation of the genome. Nat Rev Genet 8:272–285

Slotkin RK, Nuthikattu S, Jiang N (2012) The evolutionary impact of transposable elements on gene and genome regulation. In: Wendel JF (ed) Molecular biology and evolution of the plant genome. Springer Verlag, Vienna/New York

Tal O, Kisdi E, Jablonka E (2010) Epigenetic contribution to covariance between relatives. Genetics 184(4):1037–1050

Tost J (2010) DNA methylation: an introduction to the biology and the disease-associated changes of a promising biomarker. Mol Biotechnol 44(1):71–81

Ungerer MC, Johnson LC, Herman MA (2008) Ecological genomics: understanding gene and genome function in the natural environment. Heredity 100(2): 178–183

Vaughn MW, Tanurdzic M, Lippman Z, Jiang H, Carrasquillo R, Rabinowicz PD, Dedhia N, McCombie WR, Agier N, Bulski A, Colot V, Doerge RW, Martienssen RA (2007) Epigenetic natural variation in *Arabidopsis thaliana*. PLoS Biol 5:1617–1629

Verhoeven KJF, Jansen JJ, van Dijk PJ, Biere A (2010) Stress-induced DNA methylation changes and their heritability in asexual dandelions. New Phytol 185:1108–1118

Vos P, Hogers R, Bleeker M, Reijans M, Vandelee T, Hornes M, Frijters A, Pot J, Peleman J, Kuiper M et al (1995) AFLP-a new technique for DNA-fingerprinting. Nucleic Acids Res 23(21):4407–4414

Waddington CH (1942) Canalization of development and the inheritance of acquired characteristics. Nature 150:563–565

Waddington CH (1953) Genetic assimilation of an acquired character. Evolution 7:118–126

Waterland RA, Jirtle RL (2003) Transposable elements: targets for early nutritional effects on epigenetic gene regulation. Mol Cell Biol 23(15):5293–5300

Waterland RA, Jirtle RL (2004) Early nutrition, epigenetic changes at transposons and imprinted genes, and enhanced susceptibility to adult chronic diseases. Nutrition 20:63–68

Weaver ICG, Cervoni N, Champagne FA, D'Alessio AC, Sharma S, Seckl JR, Dymov S, Szyf M, Meaney MJ (2004) Epigenetic programming by maternal behavior. Nat Neurosci 7(8):847–854

Weaver IC, Champagne FA, Brown SE, Dymov S, Sharma S, Meaney MJ, Szyf M (2005) Reversal of maternal programming of stress responses in adult offspring through methyl supplementation: altering epigenetic marking later in life. J Neurosci 25(47):11045–11054

Weaver IC, Meaney MJ, Szyf M (2006) Maternal care effects on the hippocampal transcriptome and anxiety-mediated behaviors in the offspring that are reversible in adulthood. Proc Natl Acad Sci USA 103(9): 3480–3485

West-Eberhard MJ (2003) Developmental plasticity and evolution. Oxford University Press, Oxford

West-Eberhard MJ (2005) Developmental plasticity and the origin of species differences. Proc Natl Acad Sci USA 102:6543–6549

Zhang D, Yang Q, Ding Y, Cao X, Xue Y, Cheng Z (2008) Cytological characterization of the tandem repetitive sequences and their methylation status in the *Antirrhinum majus* genome. Genomics 92(2):107–114

Zhou VW, Goren A, Bernstein BE (2011) Charting histone modifications and the functional organization of mammalian genomes. Nat Rev Genet 12(1):7–18

Zilberman D, Henikoff S (2007) Genome-wide analysis of DNA methylation patterns. Development 134:3959–3965

11 The Reproducibility of Adaptation in the Light of Experimental Evolution with Whole Genome Sequencing

Guillaume Achaz, Alejandra Rodriguez-Verdugo, Brandon S. Gaut, and Olivier Tenaillon

Abstract

A key question in evolutionary biology is the reproducibility of adaptation. This question can now be quantitatively analyzed using experimental evolution coupled to whole genome sequencing (WGS). With complete sequence data, one can assess convergence among replicate populations. In turn, convergence reflects the action of natural selection and also the breadth of the field of possible adaptive solutions. That is, it provides insight into how many genetic solutions or adaptive paths may lead to adaptation in a given environment. Convergence is both a property of an adaptive landscape and, reciprocally, a tool to study that landscape. In this chapter we present the links between convergence and the properties of adaptive landscapes with respect to two types of microbial experimental evolution. The first tries to reconstruct a full adaptive landscape using a handful of carefully identified mutations (the reductionist approach), while the second uses WGS of replicate experiments to infer properties of the adaptive landscape. Reductionist approaches have highlighted the importance of epistasis in shaping the adaptive landscape, but have also uncovered a wide diversity of landscape architectures. The WGS approach has uncovered a very high diversity of beneficial mutations that affect a limited set of genes or functions and also suggests

G. Achaz
UMR7148 and Atelier de Bioinformatique, UPMC, Paris, France CIRB, Collège de France, Paris, France

A. Rodriguez-Verdugo
INSERM, UMR-S 722, Paris 75018, France

University Paris Diderot, Sorbonne Paris Cité, UMR-S 722, Site Xavier Bichat, Paris 75018, France

Department of Ecology and Evolutionary Biology, UC Irvine, Irvine, CA 92697, USA

O. Tenaillon (✉)
INSERM, UMR-S 722, Paris 75018, France

University Paris Diderot, Sorbonne Paris Cité, UMR-S 722, Site Xavier Bichat, Paris 75018, France
e-mail: olivier.tenaillon@inserm.fr

B.S. Gaut
Department of Ecology and Evolutionary Biology, UC Irvine, Irvine, CA 92697, USA

C.R. Landry and N. Aubin-Horth (eds.), *Ecological Genomics: Ecology and the Evolution of Genes and Genomes*, Advances in Experimental Medicine and Biology 781,
DOI 10.1007/978-94-007-7347-9_11, © Springer Science+Business Media Dordrecht 2014

some shortcomings of the reductionist approach. We conclude that convergence may be better defined at an integrated level, such as the genic level or even at a phenotypic level, and that integrated mechanistic models derived from systems biology may offer an interesting perspective for the analysis of convergence at all levels.

Keywords
Genomics • Evolution • Convergence • Adaptive landscape

11.1 Introduction

A key question in evolutionary biology is the reproducibility of adaptation. The importance of this question initially applied to its existentialist dimension, as it is directly linked to the likelihood of our own existence and also to the plausibility of the biological world. Stephen J. Gould wrote about assessing the reproducibility of adaptation: "...call this experiment 'replaying life's tape'. You press the rewind button and, making sure you thoroughly erase everything that actually happened, go back to any time and place in the past – say, to the seas of the Burgess Shale. Then let the tape run again and see if the repetition looks at all like the original. If each replay strongly resembles life's actual pathway, then we must conclude that what really happened pretty much had to happen. But suppose that the experimental versions all yield sensible results strikingly different from the actual history of life? What could we then say about the predictability of self-conscious intelligence? Or of mammals? Or of life on land? Or simply of multicellular persistence for 600 million difficult years?" (Gould 1989). This proposition by Gould is not just a *Gedankenexperiment*; it can be quantitatively addressed in the laboratory using experimental evolution coupled to whole genome sequencing (WGS). However, in the context of experimental evolution the question is no longer about the emergence of self-conscious organisms but more about the shape of an adaptive landscape and the stochastic nature of adaptation on this landscape.

11.2 Convergence and Parallelism

Convergence and parallelism are terms that characterize the reproducibility of adaptation (Haas and Simpson 1946). Convergence usually refers to the independent appearance of similarities in distant lineages that have faced a similar environment (Arendt and Reznick 2008). In other words, convergence occurs when two species – despite initial differences between them – converge towards a shared phenotype. A classical example of convergence is the appearance of fins in fish, reptiles and mammals, all of which have evolved in aquatic environments. Parallelism refers generally to similarities in adaptation of independent populations from the same species in similar environments (Arendt and Reznick 2008). In this case, these initially identical populations diverge by the process of mutation, adaptation and genetic drift but then accumulate parallel genetic changes when subjected to the same environment. At the genetic level, the difference between convergence and parallelism initially suggested that convergence relies on different genetic mechanisms, because the species were highly diverged; in contrast, parallelism would rely on similar genetic targets due to the genetic similarities between populations. Yet, the evolution of color pattern in mammals has suggested that the difference between convergence and parallelism is not that marked at the genetic level: a single species may use different genetic paths to generate similar phenotypic changes (Steiner et al. 2007) and very distant species such as mice and bears may use

the same gene to change color (Ritland et al. 2001; Hoekstra et al. 2006).

Given that the two terms represent a somewhat false dichotomy, we subsequently use the term convergence to refer to the reproducibility of adaptation.

11.2.1 Multiple Facets of Convergence

Convergence is a term that is widely used but the associated concept is indeed quite complex. When genetic convergence is observed, it is classically considered to be strong evidence of adaptation. As it is usually very difficult to classify mutations as beneficial, neutral or deleterious, multiple independent occurrences of a mutation suggest the filtering action of natural selection (Pelosi et al. 2006), thus constituting a strong signal of adaptation. This remains the dominant usage of convergence in evolution. Yet, the existence of genetic convergence does not only reflect the action of natural selection – it also implies the lack of possible alternative beneficial mutations. That is, if many different solutions are possible, then convergence is not to be expected. Hence, the existence of convergence informs about the number of possible adaptations to an environment. These two facets of convergence – i.e., as a marker of adaptation and as an indicator of adaptive limitations – make it a powerful tool to study the structure of the adaptive landscape and the molecular bases of adaptation.

The interpretation of genetic convergence can also be complicated by the fact that it can be defined at several levels – e.g., at a specific nucleotide site, within a particular gene or even among genes that contribute to an integrated function or network (Tenaillon et al. 2012). While this appears to be just a matter of semantics, the level at which convergence is considered has consequences for inferring adaptive processes. In its strictest sense, convergence is the presence of the exact same mutation at the same nucleotide site; in this case, adaptation can be treated as a discrete genotypic space (Fig. 11.1a). However, when mutations are pooled within genes (that may harbor several alleles), or even within functional units, the study of convergence implicitly moves toward a continuous multidimensional trait space for the adaptive process (Fig. 11.1b).

The meaning of convergence can also vary according to the time scale used to analyze it. Convergence in state occurs when the same changes are present in independent populations at the same time. In contrast, convergence in path refers to the succession of changes that is similar among populations. Hereafter we will refer to 'state' and 'path' convergence and will clarify their differences below.

11.3 Convergence and Simple Fitness Landscapes

Convergence provides insight into the shape and slope of adaptive landscapes. Adaptive landscapes were first introduced by Sewall Wright (1932) to illustrate the impact of mutations on evolutionary fitness. While they were initially defined at the genotypic level, landscapes can be extended to phenotypes. In the genetic case, with a set of n biallelic loci, the landscape is defined as a discrete hypercube (Wright 1932) (Fig. 11.1a); in the phenotypic case, the landscape can be envisioned as a continuous multidimensional space (Fisher 1930; Orr 2005) (Fig. 11.1b). Once an adaptive landscape is characterized, the dynamics of adaptation – and the path followed by populations on this landscape - can be computed.

Interactions among mutations, or epistasis, play a key role in shaping an adaptive landscape (Kauffman 1993; Gros et al. 2009). Epistasis is the influence of alleles at other sites on the effect of a mutation at a given site (Fig. 11.2). In population genetics, for two biallelic loci, epistasis can be mathematically defined as:

$$e = \log [w_{11}/(w_{10}w_{01})]$$

where w_{11} is the fitness of the genotype with a mutant allele at two different sites, w_{10} and w_{01} are the fitnesses with alternative alleles at the two

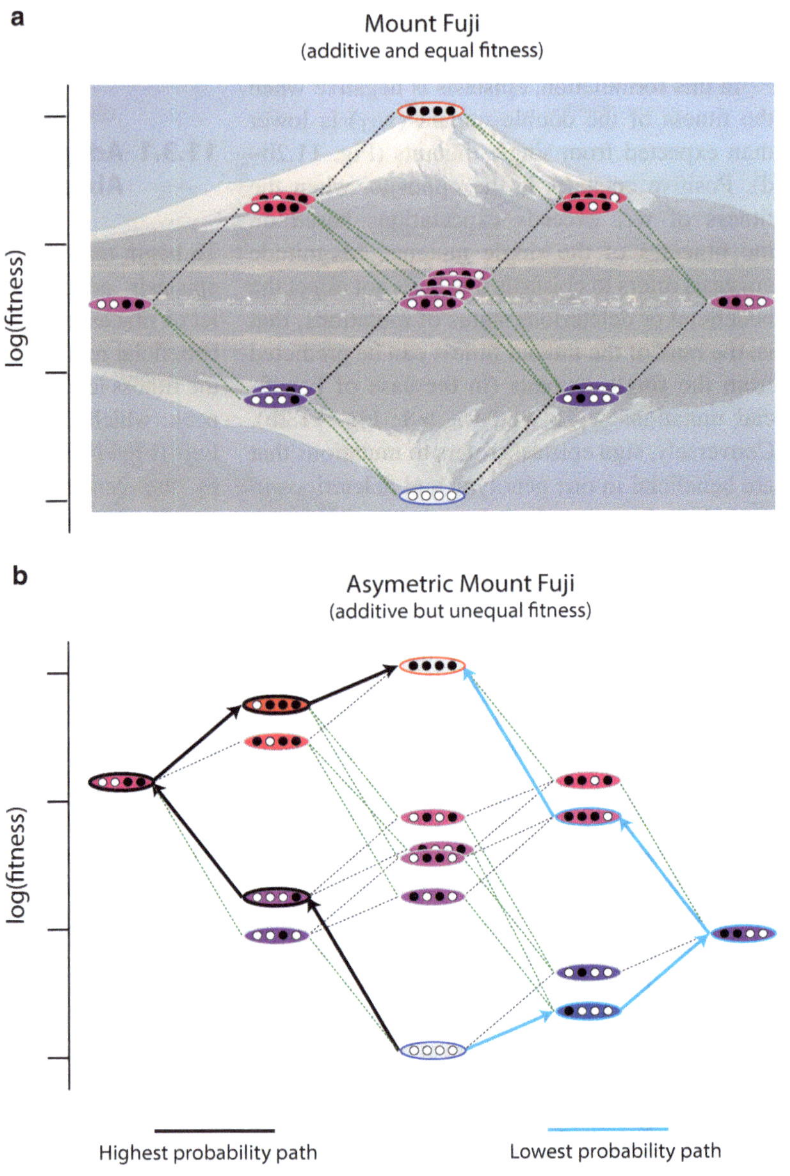

Fig. 11.3 Path Probability under Mount Fuji Landscapes. When mutations have additive fitness, the resulting landscape harbors a unique peak and a unique sink. (**a**) If all mutations have equal fitness effects, the highly symmetric structure of the landscape is reminiscent of Mount Fuji. All paths that lead from one genotype to another are equally likely. (**b**) When mutations are additive but vary in their fitness effects, the landscape is no more symmetric each adaptive paths has a specific probability. Typically, if mutations are not limiting, the path that has highest fitness improvement at each step (the *black path* on the *left*) is associated with the highest probability and reciprocally, the one that has the lowest fitness improvement at each step (the *blue path* on the *right*) is associated with the lowest probability

first and the small-effect mutations later. If populations are small the impact of the variance in fitness effects among beneficial mutations may not be important, because mutations are limiting such that as soon as one survives drift it will be fixed. As a result, the probability of surviving genetic drift predominates over selection in small populations (Weinreich 2005). However, as the population size gets bigger, more and more beneficial mutations will be simultaneously present in the population, and may compete in what is called 'clonal interference' (Gerrish and Lenski 1998; Desai and Fisher 2007). As the fittest mutations outcompete other mutations, both forms of convergence will be higher because mutations will be fixed approximately according to the ranks of their fitness effects. If populations get even larger, double mutants will appear, making path convergence difficult to observe. Thus, both state and path convergence are most likely to be observed

in populations that are large but not so large as to commonly house double-mutants (Paixao et al. manuscript under preparation).

11.3.2 Epistasis Affects the Likelihood of Observing Convergence

Both forms of convergence may strongly be affected by epistasis. The constraints imposed by epistasis on convergence are often referred to as 'historical contingency'. In other words, the appearance of a mutation by chance has a strong impact on consequent steps of adaptation. Thus, epistatic interactions may strongly affect the likelihood of convergence. Moreover, epistasis can affect the adaptive landscape in several ways. It can generate a rugged fitness landscape (Poelwijk et al. 2011) or it may affect the shape and slope of a single-peaked landscape.

Epistasis impacts convergence on a single-peak fitness landscape by modifying the effect and number of beneficial mutations that are accessible to evolution. For example, in the presence of sign epistasis some mutations will be beneficial in only a subset of genetic backgrounds and therefore all the paths in which they are not beneficial will not be adaptive paths. Hence, the presence of sign epistasis in an adaptive landscape should increase both state and path convergence to some extent. Multiple adaptive peaks may result from reciprocal sign epistasis (Poelwijk et al. 2011), which is a necessary, but not sufficient, condition for the existence of multiple peaks (Fig. 11.2). Reciprocal sign epistasis is also likely to affect convergence by driving populations to diverge as they evolve on separate peaks. Hence, the reproducibility of adaptation is intimately linked to the adaptive landscape.

11.4 Experimental Evolution as a Tool to Study Convergence

Given the complexity of epistatic interactions and adaptive landscapes, how does one study convergence? One approach is comparative studies (Besnard et al. 2009), which may identify similar changes in distant species. Yet, when one examines the molecular details of adaptive events, it is often difficult to be sure that they resulted from the same selective pressures; that is, did the different organisms truly experience and adapt to the same environmental challenge? As a consequence, if different adaptive responses are observed among species or among populations, it is difficult to know if they result from slight differences in selective pressures or from the stochastic nature of adaptation. Alternatively, if similarities are observed, it is difficult to know if they result from the same selective pressure or appeared as collateral effects of different selective pressures. Therefore, to truly understand convergence, adaptive landscapes and the stochastic nature of the adaptive process, highly controlled environments are necessary.

Experimental evolution in the laboratory can provide such controlled conditions. Although some interesting studies have been performed with higher eukaryotes (Rose et al. 2005), we will focus on experimental evolution with microbes, mostly bacteria and yeast, with an occasional mention of viruses. The benefit of using microbes in experimental evolution comes from their short generation time, their small size, their large populations, and the ability to stop their evolution by storing them at -80°C and reviving them on demand. These properties allow population sizes of millions to billions of cells to be evolved for hundreds to thousands of generations and compared to their ancestor. Finally, as microbes have relatively small genome sizes, it has become possible to sequence complete genomes with next-generation sequencing (Herring et al. 2006; Barrick et al. 2009; Tenaillon et al. 2012).

Another advantage of experimental evolution is the ability to perform experiments in replicate. For example, the longest evolutionary experiment in the lab is the one initiated by Richard Lenski in 1988 (Lenski et al. 1991; Barrick 2009 #2), in which 12 replicate populations have been evolved for more than 50,000 generations to low nutrient conditions. We will later on refer to this experiment as the Long Term Experimental Evolution, or LTEE.

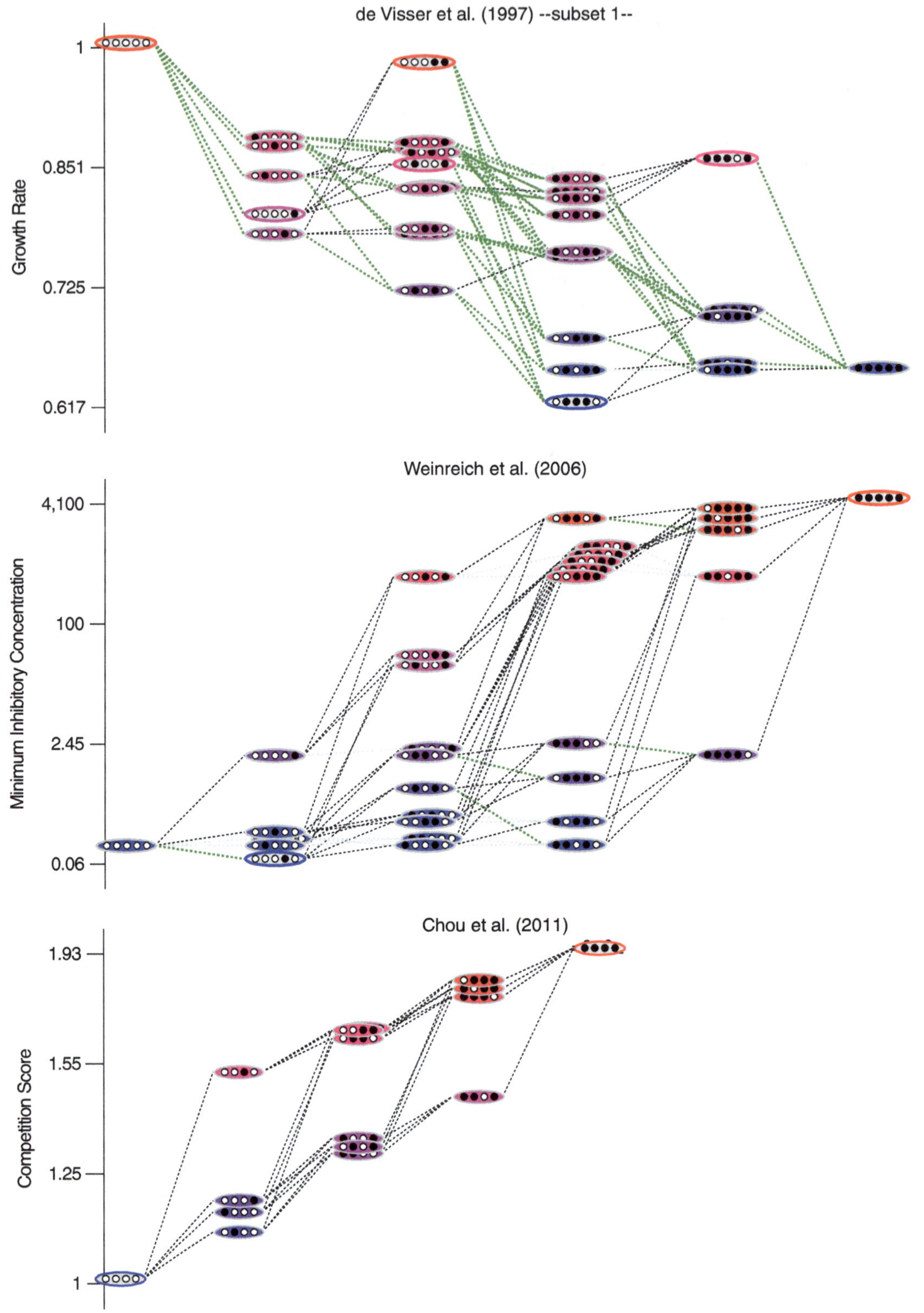
de Visser et al. (1997) --subset 1--
Growth Rate
1
0.851
0.725
0.617
Weinreich et al. (2006)
Minimum Inhibitory Concentration
4,100
100
2.45
0.06
Chou et al. (2011)
Competition Score
1.93
1.55
1.25
1

As soon as replicated experiments have been sequenced, information on the reproducibility of adaptation can be gathered at a genetic level. The pioneering work of Wichman et al. (1999) nicely illustrates the study of convergence with experimental evolution. Wichman and colleagues (1999) evolved two independent populations (one in Texas and one in Idaho) of bacteriophage phiX174, a single stranded DNA genome of 5,386 bp, and sequenced their genomes after 11 days of adaptation in a chemostat in which the phage adapted to a new host at a high temperature. They first observed that 50 % of the 14–15 nucleotides changes present in the sequenced clones from each population were identical. Hence, end-point sequencing revealed a very high level of convergence. To go further, they then followed the time frame of the appearance of each of the common mutations in each of the populations and demonstrated that the similarity of the adaptive paths was poor, because the mutations appeared in different orders. Hence, their experiment yielded different messages about convergence, depending on whether they focused on convergence in state or convergence in path.

11.4.1 Experimental Adaptive Landscapes

The study of Wichman et al. (1999) demonstrates the feasibility of studying convergence using the approach of experimental evolution. Similar approaches have been useful in inferring properties of an adaptive landscape. Generally, these studies are of one of two types. In the first, the adaptive landscape is thoroughly described for a handful of mutations, which are then used to infer the number of possible paths on the landscape (Poelwijk et al. 2007; Kogenaru et al. 2009). In the second, empirical observations of convergence are used to infer some of the landscape's properties.

11.4.1.1 Reconstructing Adaptive Landscapes

Using site-directed mutagenesis, it is possible to reconstruct an adaptive landscape. With N observed mutations, a landscape is an N-dimensional hypercube, whose full characterization requires the construction and fitness determination of 2^N mutants (assuming two states per mutated site). To date, the experiments that have constructed adaptive landscapes have focused on four to nine mutations that were found together in an adapted genotype. Pioneering work by Lee (Lee et al. 1997), Lunzer (Lunzer et al. 2005), de Visser (de Visser et al. 1997, 2009) and Weinreich (Weinreich et al. 2006) have used this approach. However, de Visser, rather than using beneficial mutations that co-evolved together, generated a landscape from unrelated deleterious mutations with selectable phenotypes (Fig. 11.4a). Several similar studies have been published more recently (Lozovsky et al. 2009; O'Maille et al. 2008; da Silva et al. 2010; Tan et al. 2011; Khan et al. 2011; Chou et al. 2011). Many of the recent studies have focused on mutations that provide antibiotic resistance or adaptation to new culture conditions.

Interestingly, this experimental approach has revealed that the pattern of epistasis seems to vary as to whether mutations are found within a single gene or across a series of genes. Mutations within a single gene or enzyme frequently result in sign epistasis (Lee et al. 1997; Lunzer

Fig. 11.4 Three experimentally resolved adaptive landscapes. All proxies for 'fitness' values are represented here in log-scale. The landscape from de Visser et al. (1997) is a combination of deleterious mutation held in different genes of *Aspergillus niger*. Although there is a general tendency to have smaller fitness with more mutations, the landscape harbors several peaks and sinks; it is rugged. The landscape of Weinreich et al. (2006) explores the effect of mutations that have coevolved together in *E. coli* toward a high resistance to cefotaxime. The landscape has one peak and one sink but shows pervasive sign epistasis. From the sink to the peak, only 18 direct paths are possible (out of 120). The landscape of Chou et al. (2011) reports how 4 mutations that have coevolved to allow *Methylobacterium extorquens* to grow on methanol interact together. It shows no sign epistasis and is close to a model with no epistasis

Finally, integration of convergence at the gene or functional level based on the presence of mutations could also be misleading not only about their side effects but also in their primary phenotypic action. For example, the same genes have been target for mutations in different environmental conditions. In several experiments, mutations in the RNA polymerase have been found. Two of these studies have enough replicates to perform some comparative statistical analysis. Conrad et al. (2010), evolved 45 lineages of *E. coli* to glycerol minimal media and found that 37 had a mutation in the *rpoBC* complex. In our thermal adaptation experiment (Tenaillon et al. 2012), 91 linages out of 114 had mutations in one of these two genes. One could easily conclude there is functional convergence between these two experiments. However, the patterns of mutations were quite distinct. Out of the 37 mutants found in *rpoBC* in the glycerol adapted lineages, two were in *rpoB* and 35 in *rpoC*. All of the *rpoC* mutants were deletions, 31 of them being an identical 9 bp deletion. In the thermal adaptation experiment, 75 lines had a mutation in *rpoB*, 21 lineages had a mutation in *rpoC* and five lineages had a mutation in both. Of these, one of the 21 *rpoC* mutations was a 3 bp deletion in a distinct region from the deletions observed by Conrad et al. (2010). Hence, the ratio of mutations among the two genes (Fisher test (FT) $p < 10^{-13}$), the type of mutations within *rpoC* (FT, $p < 10^{-13}$) and the location of mutations are so different between experiments that it strongly suggests that the mutations must have different phenotypic effects despite being in the same genes.

11.4.1.6 A Continuous Approach

Are there any fruitful alternatives to these discrete, small-scale attempts to reconstruct an adaptive landscape? Over the last decade, a phenomenological continuous model, the Fisher Geometric model of adaptation, has retained the attention of the scientific community. In his book "The Genetical Theory of Natural Selection", R.A. Fisher (1930) outlined a model in which an organism is characterized by a set of independent phenotypes, each corresponding to an axis in an n-dimensional vector space. An individual, with its specific phenotypic values, is represented by a point in this space and a mutation as a vector (Fig. 11.6). All phenotypes are supposed to be under stabilizing selection, i.e., there is an optimal value for each phenotype. As a consequence, an optimal combination of phenotypes exists and is defined to be the center of the space. The distance to that optimum determines the fitness of individuals. Several functions can be used to describe how fitness declines with the distance to the optimum. For the sake of similarity with multilocus models, this function has been selected to be of the form.

$$W(d) = e^{-d^Q}$$

where d is the distance to the optimum and Q is an epistasis parameter (Tenaillon et al. 2007; Gros et al. 2009; Gros and Tenaillon 2009).

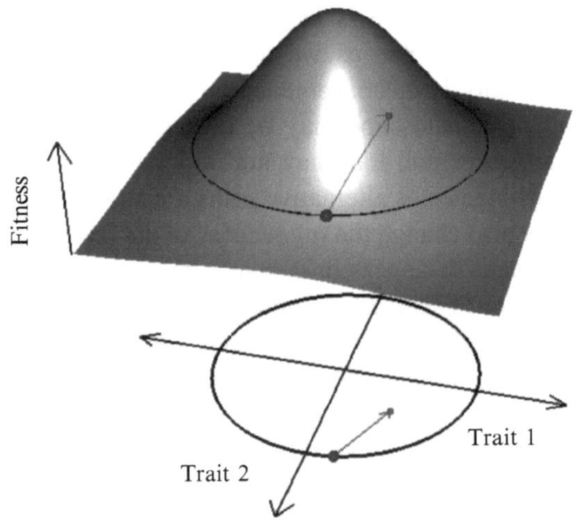

Fig. 11.6 **Fisher's Geometric model of adaptation.** Fisher's geometric model is an idealized model of phenotypic landscape. In this model, an organism is defined by a number of independent phenotypes (here 2). The number of phenotypes is what is called phenotypic complexity. The model assumes the existence of an optimal combination of phenotype having maximal fitness and therefore the existence of a single peak. The more organisms are distant from that optimal combination, the lower are their fitnesses. Here an organism presented in *blue*, mutated to a new combination of phenotypes presented in *red*. Under the 3D presentation, a 2D vision of the phenotypic change is presented

of the landscape is beyond reach with current technology. Yet, as we have noted, restricting the analysis of convergence to a fraction of the adaptive landscape may be misleading. Adding more sites and loci may not only decrease path convergence as alternative paths are added, but it may also change the ruggedness of the landscape as peaks disappear or appear with an increasing number of loci. As we mentioned in the case of *spoT* alleles, adding a site may add new peaks, but it could also reduce the number of peaks. A landscape made of two mutations with reciprocal sign epistasis suggests a double-peaked fitness landscape. Yet adding a third site may result in a single peak (Fig. 11.5a).

11.4.1.5 Reducing the Complexity of Discrete Adaptive Landscape

As we mentioned previously, a fitness landscape is a discrete hypercube with a set of n biallelic loci. As evolution experiments increase their replication and the ability to detect genetic variants, this hypercube quickly becomes intractable. To reduce the dimensionality of the hypercube to be analyzed it may be tempting to aggregate mutations according to their function. For instance the different alleles of *spoT* in the LTEE could be all considered similar. Going even further all the loci affecting the same functional units could be aggregated into a single locus with two alleles. In the case of thermal adaptation, for instance, the various alleles affecting the six genes that contribute to bacterial shape are never found in combination; thus they can be considered as single mutation at a bi-allelic locus, and the mutation rate toward the mutated allele can be defined as the rate of production of any of the underlying mutations. Analyzing convergence on the resulting landscape clearly means a shift in the way we study convergence. Rather than genetic convergence, we move towards a vision of convergence at a phenotypic level. This is not shocking, as it is through their phenotypic effects that genotypes are selected. Reverting to the Gould quote, the original question was not whether self-conscious organism with DNA identical to ours would emerge with a replay of the tape of life but rather whether a phenotype – self-consciousness – would inevitably emerge.

Yet, to be able to aggregate mutations affecting a gene or a functional unit, mutations need to have similar fitness and phenotypic effects. This is often not the case. First, mutations in the same gene may have different selective effects (Rodriguez-Verdugo et al. 2013). Even in the case of a gene inactivation, an early or late stop codon, a frameshift or a full gene deletion may have different fitness effects. Moreover, the phenotypes of the different mutations may differ. In the LTEE, it was shown that alternative *SpoT* selected alleles had different side effects such as different fitness effects in other environments (Ostrowski et al. 2005, 2008). In the thermal adaptation experiment, most lineages adapted through changes in the RNA polymerase. However, among these changes 10 % provided resistance to the antibiotic rifampicin (Rodriguez-Verdugo et al. 2013). This means that depending on the allele chosen, populations may fully survive an antibiotic treatment or be eradicated. This illustrates how drastic the collateral effects of different alleles may be. This observation is not restricted to bacteria, as parallel adaptation of yeast to fungicide nystatin also leads to diversification among lineages on alternative environments (Gerstein et al. 2012).

Rather than considering all mutations in a gene or functional unit to be a single allele, it may be more realistic to consider them as alternative alleles at single locus. This would reduces the dimensionality of the landscape to $n = \prod_L n_i$, where **n** is the number of genotypes, **L** the number of loci and $\mathbf{n_i}$ the number of allele per loci. This formalism allows the specificity of alleles to be taken into account, and reduces the number of genotypes. However, this simplification may be limiting as sometimes multiple mutations may arise within a gene. This was the case in RNA polymerase in the thermal adaptation experiment and in the rod shape genes in the LTEE (Philippe et al. 2009). In addition, the number of genotypes to be characterized remains extremely large and suggests that this formalism provides only a modest improvement over the hypercube approach.

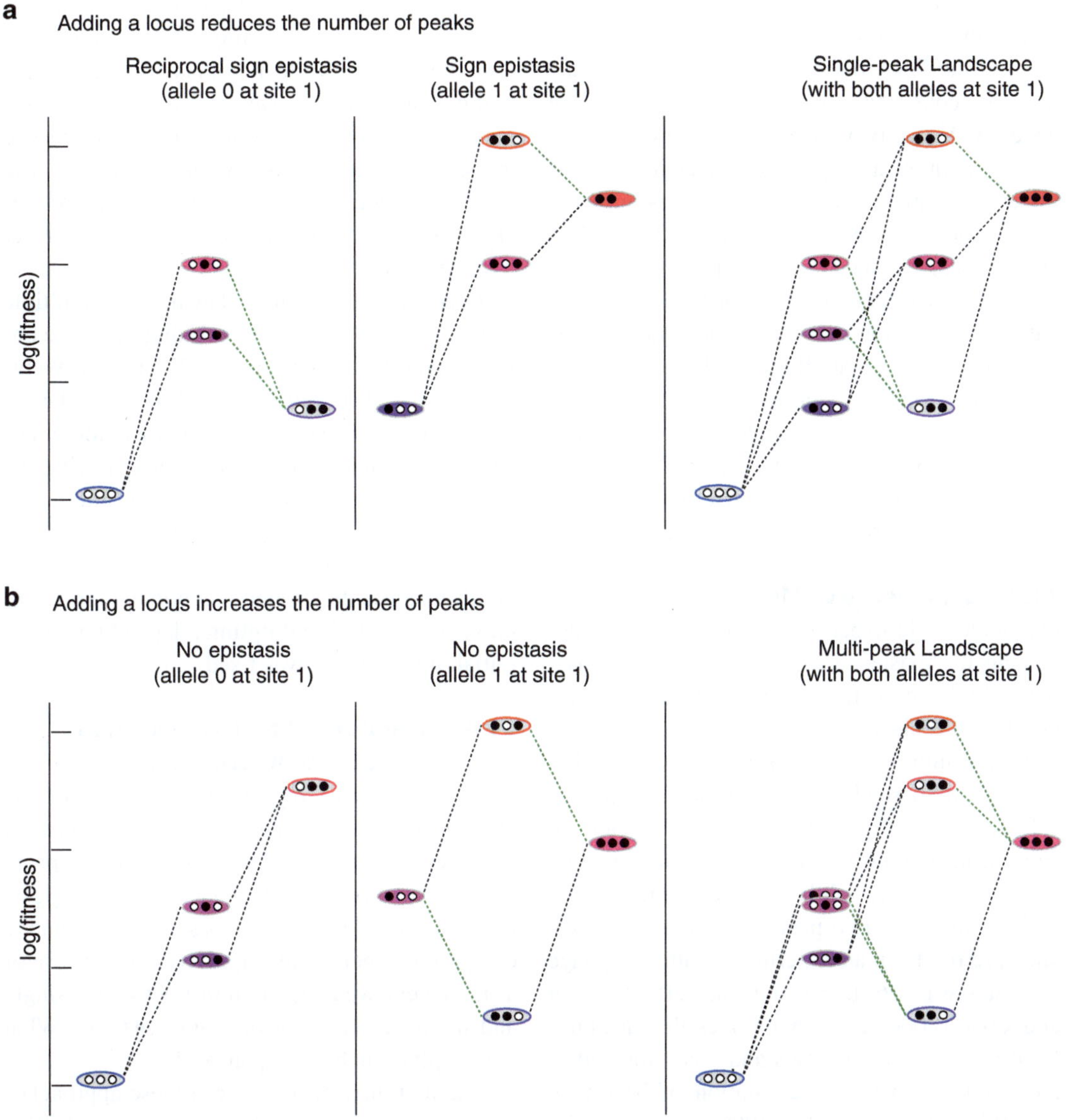

Fig. 11.5 How landscapes change with more loci. The addition/discovery of a new polymorphic locus can change the number of peaks in the landscape. (**a**) Although loci 2 and 3 show (reciprocal) sign epistasis, taking into account the two alleles in locus 1 leads to a single peak landscape. (**b**) When the allele at locus 1 is fixed, loci 2 and 3 show no epistasis (but their fitness rank depends on the allele at locus 1). When the locus 1 is included in the landscape, the landscape harbors two peaks, a strong reciprocal sign epistasis

included mutations on a genome scale that were not isolated in combination discovered several peaks (de Visser et al. 1997).

For technical reasons it is difficult to generate a full landscape for many sites. The number of mutants to be constructed and characterized increase by 2^N. The biggest such landscape reconstructed thus far has been with 9 mutations, and it could not be fully reconstructed because only 418 of 512 combinations were achievable (O'Maille et al. 2008). Yet, adaptation at the genome level or even at the gene level reveals that tens to thousands of mutations may be recruited in the adaptive process. Thus, complete reconstruction

two alternatives adaptive trajectories and fitness peaks (in a broad sense). One consisted of mutations in the RNA polymerase β−β′ subunit, and the other consisted of mutations in the *rho* factor gene, with the two trajectories being partially exclusive. Interestingly a systemic analysis of the impact of a mutation in *rho* on the fitness effect of transposon-based gene inactivation (Freddolino et al. 2012) revealed that a single mutation in that gene was enough to drastically change the pattern of selection over the whole genome. This confirms the idea that affecting global regulators such as *rho*, which is a termination factor, or RNA polymerase can have a large epistatic impact by making some previously beneficial mutations deleterious and by, reciprocally, making some previously deleterious mutations beneficial.

11.4.1.3 Evidence of Multiple Peaks

Our large-scale experiment is not alone in suggesting the existence of multiple fitness peaks. Several other evolution experiments with a lower number of replicates have also suggested the existence of multiple peaks (Burch and Chao 2000; Kvitek and Sherlock 2011). In these experiments, a fewer number of replicates were sufficient to suggest adaptation by competing pathways that were subsequently validated experimentally.

For example, a two-peaked adaptive landscape was inferred from an experiment with two phage populations of ϕ6 (Burch and Chao 2000). Burch and Chao (2000) used two clones that accumulated deleterious mutations and were allowed to recover through the accumulation of beneficial mutations (Burch and Chao 1999). These founder clones (clone A and clone B) were used to find five new replicate populations and were evolved for 100 generations. Derived populations from clone A evolved to higher fitness, while derived populations from clone B evolved to lower fitness, suggesting that clones A and B resided on different fitness peaks.

In a study of yeast adaptation to glucose limitation, Kvitek and Sherlock (2011) also found evidence for two competing adaptive genetic paths. They found that glucose transport into the cell could be improved by two means – either inactivation of a negative regulator of the glucose sensing signal transduction pathway or by an amplification of genes encoding a high affinity glucose transporter. While individual mutations were highly beneficial, they were deleterious in combination. Based on these observations, the glucose transporter experiments appear to reveal a case of negative epistasis within a functional unit because both mutations increase hexose transporter transcription.

Interestingly one can find evidence of multiple peaks even at the gene level. Using the same enzyme and selective pressures as Weinreich (Weinreich et al. 2006), Salverda et al. (2011) adapted β−lactamase TEM1 to cefotaxim by random mutagenesis. While many lines adapted with the mutations studied by Weinreich, many alternative mutations also appeared to be beneficial. Moreover, while Weinreich suggested a single peak, in this analysis they found that populations could move toward different optima, depending on the initial mutation that was fixed.

11.4.1.4 Reduced Landscapes Versus Replicate Adaptation

There are some discrepancies between the different approaches. Thorough analysis of a handful of beneficial mutations suggests a single peak, while adaptation experiments commonly suggest multiple genetic solutions based on alternative competing genetic pathways. Even in the case of a single gene we have a contrast between a single and multiple peaks for the same enzyme. What may explain such discrepancies?

The first difference between these approaches may be due to sampling biases in adaptive landscape reconstruction. Typically the mutations chosen to study were found in combination in a single adapted population. This may drastically limit the chance to find multiple adaptive peaks. For instance in the case of the LTEE, the landscape had a single peak (Khan et al. 2011). Yet that landscape included a *spoT* mutation and it has been shown that among the replicates many alternatives alleles of *spoT* were recovered (Ostrowski et al. 2008). Had these alleles been included in the landscape reconstruction, a different picture may have emerged (Fig. 11.5b). Indeed the only landscape reconstruction that

with daily 100-fold dilution to fresh medium. After 2,000 generations of adaptation at 42.2°C, we sequenced one clone of each population.

Why choose such a system? The high temperature stress was chosen because under these conditions we had no a priori ideas about the mutational targets that natural selection would affect; temperature affects not only the folding of every protein but also the kinetics of all reactions within the cell. Choosing a stress for which the targets of selection are known (like rifampicin resistance, for example) may bias the study of convergence. We chose an extreme environment and a reasonably long adaptive period, because we wanted both a fast response to selection and several mutations to be present in each of the sequenced clones. The low level of initial adaptation and the low resource conditions had the benefit of limiting the potential emergence of ecological interactions within each population. In rich media such interactions may occur faster than in limited resource conditions (Le Gac et al. 2008). For example, under LTEE conditions, several thousands of generations of adaptation were needed to observe ecological interactions (Le Gac et al. 2012), while only several hundred generations were necessary in richer media (Le Gac et al. 2008). Moreover strong maladaptation to the initial conditions also limits the emergence of ecological divergence, because the first steps of the adaptive process are more likely to be linked to general adaptation. While ecological interactions are interesting, they complicate analyses of adaptation and convergence and should be avoided, if possible. For example, if two morphotypes emerge and are maintained through a frequency dependent interaction (Rainey and Travisano 1998), sampling a single clone randomly from each population may generate a pattern of divergence among populations, when in fact all the replicate populations are completely convergent with respect to the emergence of two morphs.

Using the full sequences of the 115 clones, we were able to uncover several interesting facets about convergence. First, because we identified a large number of mutations, convergence could be studied at the level of individual mutations, single genes, an operon or even on a functional level – i.e., groups of genes that contribute to the same function. As expected, the more integrated the definition of similarity, the higher was the convergence among lineages. However a substantial gain was observed when comparing convergence at the level of individual mutations (around 2 % convergence) to convergence at the gene level (>20 %) or at the functional level (>30 %). A direct consequence of this discrepancy was that we could identify about 30 functional units that were affected repeatedly by adaptive mutations, but that the number of mutations affecting those functional units was extremely large. In our large study, we observed several hundred different beneficial sites, but simple extrapolations suggested that several thousands of potentially beneficial sites are present in the genome. It is important to note that the large diversity of mutations affecting a gene or function was not only restricted to the inactivation of genes but also extended to essential genes such as the RNA polymerase or the genes involved in the bacterial rod shape. Within these genes we could estimate that tens to hundreds of different mutations may have adaptive potential.

Second, through statistical analysis of the association of these mutations among lineages, we were able to show that epistasis was pervasive. Alternative mutations within a gene or a functional unit appeared to be strongly negative, because combinations of mutations within a gene were rarely found. This observation has an intuitive explanation within a functional unit: if two different mutations have the same phenotypic effect, once one is selected for, the second is likely to be useless, if not harmful.

However, evidence of epistasis was not limited to within gene interactions; epistasis among mutations across genes and functional units was also strong enough to leave a statistical signal. For example, the presence of mutations in *rpo*D, the capsule operons, or the genes involved in cell shape were conditional on the presence of mutations in the RNA polymerase β or β' subunits. This suggests some positive epistasis among these genes and their functions. Altogether, the pattern of these interactions suggested

Once epistatic interactions are inferred experimentally, the number of paths can be studied *in silico*. To study convergence computationally, most simulations use the "low mutation, high selection" regime. In this regime, the population is mostly monomorphic; following the appearance of a mutant, the populations may instantly shift to the mutant state depending on its fitness. The *in silico* studies have shown, as mentioned previously, that the structure of the landscape will affect drastically the chance of observing convergence (Lobkovsky and Koonin 2012; Lobkovsky et al. 2011). This is also obvious from empirical studies, because one striking contrast among studies comes from the number of possible paths to the optimal genotypes. In Chou et al. (2011) all possible forward paths were possible (24 out of 24) because each represented an increase in fitness. However, among the 120 possible forward paths in the Kahn et al. (2011) landscape, only 90 were possible. Finally, only 18 (out of 120) paths led to increasing fitness in the case of β–lactamase TEM1 (Weinreich et al. 2006, Fig. 11.4).

One fundamental question is: Why are the landscapes so different? A tempting answer is that, considering a handful of beneficial mutations, there ought to be more interactions if they lie within the same gene than if they are spread throughout the genome. In a broader sense, adaptation at the genome level is likely to be possible at many different functional levels, with the different improvements less likely to interact than those within a single gene. Yet this answer is not fully satisfactory because apparently unrelated mutations in different genes can also show pervasive sign epistasis (de Visser et al. 1997). Only new and ideally larger adaptive landscapes will help to quantify the differences.

11.4.1.2 Convergence in Experimental Systems: Adaptation in Replicates

An alternative to constructing a landscape from a handful of experimentally combined mutations is replicated evolutionary experiments. Experiments of parallel evolution have been performed and interpreted in terms of convergence and the properties of adaptive landscapes. In most studies, just like in the case of phiX174 mentioned above (Wichman et al. 1999), some shared mutations are recovered among replicates while others are specific to a single lineage (Woods et al. 2006; Herring et al. 2006; Conrad et al. 2009; Charusanti et al. 2010; Anderson et al. 2010; Toprak et al. 2012). It can be difficult to make conclusions from mutations in single lineages unless there are a large number of replicates or in the absence of a further functional characterization, because mutations in single lineages may have simply hitchhiked with beneficial mutations and therefore may not be adaptive. Yet, even these straightforward empirical comparisons shed light on the complexity of convergence that is missing from landscape reconstruction experiments. Most of these studies reveal that lineages infrequently share the same mutation but may mutate the same gene (Woods et al. 2006; Toprak et al. 2012), suggesting that the gene may be the appropriate unit to measure convergence. Taken together, these studies indicate that the unit of convergence needs to be clearly defined, because the unit confers implicit assumptions about convergence and the epistatic properties of the adaptive landscapes.

Most inferences about adaptive landscapes from replicated evolutionary experiments come either from studies with a large number of replicates or, alternatively, from studies in which a clear signal of convergence emerged with few replicates and that signal could be confirmed experimentally or with more targeted replicates. An example of the former (a large number of replicates) is represented by our recent work (Tenaillon et al. 2012), in which we adapted 115 independent replicates of *E. coli* to a low resource media with an elevated temperature. The ancestral clone to all replicates was preadapted for 2,000 generations to the same low resource media that was used in the LTEE. A temperature of 42.2°C was chosen for heat stress, because it was the maximal temperature that allowed a 100-fold increase in population size over a daily growth cycle. In other words, at 42.3°C the populations could continue to grow, but they would have slowly decayed to extinction

et al. 2005; Weinreich et al. 2006; O'Maille et al. 2008; Lozovsky et al. 2009; da Silva et al. 2010; Tan et al. 2011), such that mutations become beneficial conditionally on the presence of other mutations. In other words some mutations compensate the fitness deficits produced by previous mutations. For example, Weinreich (2005) observed that the β–lactamase TEM1 could improve its ability to hydrolyze the antibiotic cefotaxim by 40,000 times with a total of 5 different mutations (Fig. 11.4b). Yet, in the wild type TEM1 background only two of the individual mutations were beneficial. In this case, sign epistasis could be understood functionally. For example, the stabilizing mutation M182T had no effect in the ancestral background, but was highly beneficial once a destabilizing mutation was fixed, because the destabilizing mutations modified the active site to improve the efficiency of M182T on cefotaxim. Similarly, a mutation in the promoter was only beneficial once the enzyme started to have activity on the new substrate (Brown et al. 2009). Overall, out of the 120 different forward paths, only 18 different paths were found to be accessible due to the presence of sign epistasis.

Another interesting case of sign epistasis was observed for the 5S ribosomal RNA of the marine bacterium *Vibrio* (Lee et al. 1997). Lee et al. used the sequences of 5SrRNAs found in *V. proteolyticus* and *V. alginolyticus* that differed at four nucleotides, and constructed the 16 possible allelic combinations ($2^4 = 16$) between these two sequences. They found that only 5 out of 24 possible mutational pathways were functional, again suggesting sign epistasis between mutations. Similarly, using phylogenetic information, nine mutations in Sesquiterpene synthetase were identified as having lead to a catalytic shift in the evolution of that enzyme. Biochemical analysis of 418 combinations of the nine mutations, out of the 512 possible combinations, also revealed a rugged landscape with multiple peaks (O'Maille et al. 2008).

In the case of TEM1, the landscape had a single peak despite the high prevalence of sign epistasis. In other single-gene landscapes, multiple peaks have been found, confirming a high frequency of sign epistasis and even the presence of reciprocal sign epistasis, which is a necessary condition for the existence of multiple peaks (Poelwijk et al. 2011). For example, in the case of DHFR gene evolution in the malaria parasite, two peaks were found when selection was due to the drug pyrimethamine and three peaks when it was to resistance to the drug antifolate chlorcycloguanil (Costanzo et al. 2011).

In contrast to single-gene studies, sign epistasis seems to be less prevalent among mutations from different genes. This conclusion is primarily based on two landmark studies. In one, Khan et al. (2011) produced all possible mutants from the first five mutations involved in the adaptation of *Escherichia coli* to the LTEE conditions. Similarly, Chou et al. (2011) built the 2^4 genotypes of the four primary mutations that permits *Methylobacterium extorquens* to grow on methanol (Fig. 11.4c). It is worth noting that in these cases, producing the combination of mutants is a much harder task than in the case of a single gene cloned on a plasmid. Moreover, while single gene studies often use a biochemical activity such as IC50 (concentration allowing 50 % of survival) or MIC as proxy for fitness, true fitness assays must be performed in the case of genome analysis.

Nonetheless, epistasis was pervasive in both of these studies. Both landscapes had a single peak, and epistasis was consistently negative in magnitude. Most mutations appeared to be beneficial in all background in Chou et al. (2011), but the magnitude of their fitness effects decreased as mutations accumulated. As a result, the authors concluded that there are 'diminishing returns' with respect to the addition of each additional new advantageous mutation. However, we suggest that this conclusion be considered with caution, as the relationship between the fitness defined in mathematical models and the measured experimental values is not necessarily trivial (Wagner 2010). In population genetics, epistasis can only be meaningful when measured as a departure of additivity on log fitness. However for historical reasons, fitness in experimental evolution is often measured as ratios of log fitness.

As soon as replicated experiments have been sequenced, information on the reproducibility of adaptation can be gathered at a genetic level. The pioneering work of Wichman et al. (1999) nicely illustrates the study of convergence with experimental evolution. Wichman and colleagues (1999) evolved two independent populations (one in Texas and one in Idaho) of bacteriophage phiX174, a single stranded DNA genome of 5,386 bp, and sequenced their genomes after 11 days of adaptation in a chemostat in which the phage adapted to a new host at a high temperature. They first observed that 50 % of the 14–15 nucleotides changes present in the sequenced clones from each population were identical. Hence, end-point sequencing revealed a very high level of convergence. To go further, they then followed the time frame of the appearance of each of the common mutations in each of the populations and demonstrated that the similarity of the adaptive paths was poor, because the mutations appeared in different orders. Hence, their experiment yielded different messages about convergence, depending on whether they focused on convergence in state or convergence in path.

11.4.1 Experimental Adaptive Landscapes

The study of Wichman et al. (1999) demonstrates the feasibility of studying convergence using the approach of experimental evolution. Similar approaches have been useful in inferring properties of an adaptive landscape. Generally, these studies are of one of two types. In the first, the adaptive landscape is thoroughly described for a handful of mutations, which are then used to infer the number of possible paths on the landscape (Poelwijk et al. 2007; Kogenaru et al. 2009). In the second, empirical observations of convergence are used to infer some of the landscape's properties.

11.4.1.1 Reconstructing Adaptive Landscapes

Using site-directed mutagenesis, it is possible to reconstruct an adaptive landscape. With N observed mutations, a landscape is an N-dimensional hypercube, whose full characterization requires the construction and fitness determination of 2^N mutants (assuming two states per mutated site). To date, the experiments that have constructed adaptive landscapes have focused on four to nine mutations that were found together in an adapted genotype. Pioneering work by Lee (Lee et al. 1997), Lunzer (Lunzer et al. 2005), de Visser (de Visser et al. 1997, 2009) and Weinreich (Weinreich et al. 2006) have used this approach. However, de Visser, rather than using beneficial mutations that co-evolved together, generated a landscape from unrelated deleterious mutations with selectable phenotypes (Fig. 11.4a). Several similar studies have been published more recently (Lozovsky et al. 2009; O'Maille et al. 2008; da Silva et al. 2010; Tan et al. 2011; Khan et al. 2011; Chou et al. 2011). Many of the recent studies have focused on mutations that provide antibiotic resistance or adaptation to new culture conditions.

Interestingly, this experimental approach has revealed that the pattern of epistasis seems to vary as to whether mutations are found within a single gene or across a series of genes. Mutations within a single gene or enzyme frequently result in sign epistasis (Lee et al. 1997; Lunzer

Fig. 11.4 Three experimentally resolved adaptive landscapes. All proxies for 'fitness' values are represented here in log-scale. The landscape from de Visser et al. (1997) is a combination of deleterious mutation held in different genes of *Aspergillus niger*. Although there is a general tendency to have smaller fitness with more mutations, the landscape harbors several peaks and sinks; it is rugged. The landscape of Weinreich et al. (2006) explores the effect of mutations that have coevolved together in *E. coli* toward a high resistance to cefotaxime. The landscape has one peak and one sink but shows pervasive sign epistasis. From the sink to the peak, only 18 direct paths are possible (out of 120). The landscape of Chou et al. (2011) reports how 4 mutations that have coevolved to allow *Methylobacterium extorquens* to grow on methanol interact together. It shows no sign epistasis and is close to a model with no epistasis

This model requires only three meta-parameters:

- The number of phenotypic dimensions (or the phenotypic complexity).
- The fitness function, which represents the decay of fitness as a function of the distance to the optimum in the simplest case but that can also be extended to include some asymmetry.
- The way a mutation affects an organism. Here again, the simplest model will assume isotropy of mutations (mutation directions is random), but some forms of anisotropy (some preferential directions for mutations exist) can be added.

Despite its high level of abstraction, this model has been very valuable for describing various properties of the adaptive landscapes revealed by experimental evolution. This model has shed light into the distribution of mutational effects, epistatic interactions, fitness trajectories and even the existence of fitness equilibria depending on population sizes (Hartl and Taubes 1996, 1998; Burch and Chao 1999; Orr 1998, 2000, 2005, 2006; Martin and Lenormand 2008, 2006a, b; Martin et al. 2007; Poon and Otto 2000; Waxman and Welch 2005; Welch and Waxman 2003; Wang et al. 2010; Silander et al. 2007; Lourenco et al. 2011; Chevin et al. 2010; Gros et al. 2009; Gros and Tenaillon 2009; Tenaillon et al. 2007; Le Nagard et al. 2011; Trindade et al. 2012; Sousa et al. 2012). For instance, if the number of phenotypic dimensions is **n**, the log-fitness of populations having an effective size of $\mathbf{N_e}$ will converge towards an equilibrium value of $-n/(2\ Q\ N_e)$. In other words, when complexity is high, it is difficult to optimize all the traits simultaneously and a high population size is required for the population to reach a high fitness.

It is unclear why this simple model seems to be a reasonable approximation, and indeed few authors have addressed this question. However, one apparent limitation of the model is its continuity. As long as the vision of adaptation was dominated by the idea that there were only a very few potential beneficial mutations, the idea of a continuum seemed inaccurate; discrete landscapes appeared to be the best description of the adaptive landscape with only a few beneficial mutations. However, we have shown that in many environments at the gene or genome level the number of beneficial mutations is large – so large, in fact, that it prevents reconstruction of discrete landscapes. Even within highly constrained essential gene in the *E. coli* genome, a large diversity of alleles with slightly different fitness effects and phenotypic effects appear to be beneficial (Tenaillon et al. 2012). This suggests that a continuous phenotypic space is appropriate and may help explain some of the success of the model.

Yet, Fisher's model of phenotypic evolution does not explicitly suppose an underlying genetic model. In its classical form, mutations within the model are assumed to affect all traits in what is called 'universal pleiotropy', pleiotropy referring to the action of a single mutation on several phenotypes (Wagner and Zhang 2011). However, loci with different levels of pleiotropy and varying mutational spectra may be included in the mutation process (Chevin et al. 2010; Lourenco et al. 2011). This considerably limits the mathematical tractability of the model but allows analyses that may lead to a better understanding of the factors that explain convergence at the gene level (Chevin et al. 2010; Lourenco et al. 2011). In fact, restricted pleiotropy appears to be an important determinant of convergence. If all genes can affect all phenotypes, then many alternative solutions exist; conversely if a gene affects only a subset of phenotypes, then some convergence at the genetic level may be observed.

A coupling between Fisher's model and more explicit genetic models will be required to go beyond these observations. There are several ways to do so. The first one, which is quite similar to the approach of Chevin et al. (2010), is to define genetic loci with precise effects in the phenotypic space. Rather than doing so at the gene level only, defining a finite number of possible mutations within each gene may allow a more precise and multi-scale analysis of convergence. The second approach is to define a realistic and complex genetic system, based for instance on mechanistic properties of real organisms (e.g., regulatory or metabolic networks). The aim is then to translate the properties of this model in terms of Fisher model and to study convergence in this

defined space. We have studied, for example, adaptive neural network mimicking signalization pathways (Le Nagard et al. 2011) and found that the adaptive properties of the model could be understood through the prism of Fisher's model.

11.4.1.7 Phenotypic Convergence

If a continuous approach may help to better describe adaptive landscapes, it makes the analysis of convergence more difficult because there is no such thing as whole-organism phenotyping. Thus, ideally the identification of mutations at a genetic level should be sufficient to uncover the phenotypes that are under selection. This may be easily achievable for the case of a well-known gene inactivated through stop or frameshift mutations, but in most cases genetic experiments have to be performed to infer the phenotypic consequences of mutations (Philippe et al. 2009; Conrad et al. 2009; Applebee et al. 2011). Part of this is due to the fact many mutations involved in the first steps of adaptation affect global regulators that are highly pleiotropic (Hindre et al. 2012). Unraveling the phenotypic effect of these mutation can be done, but may require several years of investigation (Conrad et al. 2010).

A few options have been used to study convergence at a phenotypic level. One initial strategy is to focus on a phenotype that is known to be under tight selection under the conditions of the experiment. For example, Reynolds (2000) studied the phenotypic effect of compensatory mutations in the RNA polymerase following the acquisition of a mutation providing rifampicin resistance. She convincingly showed that different mutants improved the efficiency of transcription that had been deteriorated by the resistance mutation. This kind of experiment is interesting but can only be applied to genes that have been identified as a target of adaptation. However, as only a specific phenotype is studied, this approach may hinder the understanding of collateral phenotypic effects and therefore provides only a partial vision of convergence.

Several phenotypes can also be studied simultaneously. Convergence or divergence is then analyzed in term of movements in a multidimensional space that can be visualized with principal component analysis or multidimensional scaling. Populations evolving in similar directions in that space will be said to converge. Tyerman et al. (2008) used different statistical properties of growth curves to infer convergence using principal component analysis. To have an even broader vision, "-omics" approaches can be used (Cooper et al. 2003; Fong et al. 2005). Cooper et al. (2003) used transcriptomic characterization of LTEE lineages to uncover some parallel changes in two different lineages. They showed that 59 genes had their regulation affected in similar ways in the two lineages and could retrieve the signature of a global regulator, *spoT*.

The most integrated analysis of convergence comes from the work of Lewis et al. (2010), in which they used proteomics and transcriptomics in concert with a genome scale metabolic network. Though their analysis does not study convergence between lines per se, it analyses the similarity between the phenotypic evolution of adapted lineages and the predicted optimal flux in metabolic networks. This approach suggests convincingly that the phenotypes of the evolved lineages converge quantitatively towards the predicted optimality, and that this convergence occurs despite differences at the genetic level. This appears to be a very promising line of research, though it is currently limited to specific adaptations dealing with metabolic changes.

11.5 Conclusion

Throughout this chapter we have shown that all the different facets of convergence and the reproducibility of adaptation can be thoroughly studied with experimental evolution. Reductionist approaches, in which a handful of beneficial mutations are studied, have shed light on the importance of epistasis in shaping the adaptive landscape and adaptive path. Some of the limitations of this approach have been overcome by studying the large diversity of beneficial mutations recovered in parallel adaption experiments by the application of whole genome sequencing.

Connecting discrete genetic models and continuous trait models may be an interesting venue

for future research. Indeed, the ultimate goal is to study convergence among genotypes, phenotypes and fitness and to assess the parameter space in which all three approaches provide a coherent picture. The fitness landscape maps genotype to fitness, Fisher's model relates phenotypes to fitness maps, and quantitative genetics study the mapping of genotype to phenotype. The most promising way to combine these three mappings will presumably be with systems biology approaches that integrate the predictions of mechanistic models with the genotypes, phenotypes and fitness improvements of evolved populations (Lewis et al. 2010).

Acknowledgments This work was supported by grants from the Agence Nationale de la Recherche (Grant ANR-12-JSV7-0007-01 to G.A., Grant ANR-08-GENM-023-001 to O.T.) and NSF grant DEB-0748903 to BS.G and A.RV.

References

Anderson JB, Funt J, Thompson DA, Prabhu S, Socha A, Sirjusingh C, Dettman JR, Parreiras L, Guttman DS, Regev A, Kohn LM (2010) Determinants of divergent adaptation and Dobzhansky-Muller interaction in experimental yeast populations. Curr Biol 20(15):1383–1388

Applebee MK, Joyce AR, Conrad TM, Pettigrew DW, Palsson BO (2011) Functional and metabolic effects of adaptive glycerol kinase (GLPK) mutants in *Escherichia coli*. J Biol Chem 286(26):23150–23159

Arendt J, Reznick D (2008) Convergence and parallelism reconsidered: what have we learned about the genetics of adaptation? Trends Ecol Evol 23(1):26–32

Barrick JE, Yu DS, Yoon SH, Jeong H, Oh TK, Schneider D, Lenski RE, Kim JF (2009) Genome evolution and adaptation in a long-term experiment with *Escherichia coli*. Nat Genetics 461(7268):1243–1247

Besnard G, Muasya AM, Russier F, Roalson EH, Salamin N, Christin PA (2009) Phylogenomics of C(4) photosynthesis in sedges (Cyperaceae): multiple appearances and genetic convergence. Mol Biol Evol 26(8):1909–1919

Brown KM, Depristo MA, Weinreich DM, Hartl DL (2009) Temporal constraints on the incorporation of regulatory mutants in evolutionary pathways. Mol Biol Evol 26(11):2455–2462

Burch CL, Chao L (1999) Evolution by small steps and rugged landscapes in the RNA virus phi6. Genetics 151(3):921–927

Burch CL, Chao L (2000) Evolvability of an RNA virus is determined by its mutational neighbourhood. Nat Genetics 406(6796):625–628

Charusanti P, Conrad TM, Knight EM, Venkataraman K, Fong NL, Xie B, Gao Y, Palsson BO (2010) Genetic basis of growth adaptation of *Escherichia coli* after deletion of pgi, a major metabolic gene. PLoS Genet 6(11):e1001186

Chevin LM, Martin G, Lenormand T (2010) Fisher's model and the genomics of adaptation: restricted pleiotropy, heterogeneous mutation, and parallel evolution. Evolution 64(11):3213–3231

Chou HH, Chiu HC, Delaney NF, Segre D, Marx CJ (2011) Diminishing returns epistasis among beneficial mutations decelerates adaptation. Science 332(6034):1190–1192

Conrad TM, Joyce AR, Applebee MK, Barrett CL, Xie B, Gao Y, Palsson BO (2009) Whole-genome resequencing of *Escherichia coli* K-12 MG1655 undergoing short-term laboratory evolution in lactate minimal media reveals flexible selection of adaptive mutations. Genome Biol 10(10):R118

Conrad TM, Frazier M, Joyce AR, Cho BK, Knight EM, Lewis NE, Landick R, Palsson BO (2010) RNA polymerase mutants found through adaptive evolution reprogram *Escherichia coli* for optimal growth in minimal media. Proc Natl Acad Sci USA 107(47):20500–20505

Cooper TF, Rozen DE, Lenski RE (2003) Parallel changes in gene expression after 20,000 generations of evolution in *Escherichia coli*. Proc Natl Acad Sci USA 100(3):1072–1077

Costanzo MS, Brown KM, Hartl DL (2011) Fitness tradeoffs in the evolution of dihydrofolate reductase and drug resistance in *Plasmodium falciparum*. PLoS One 6(5):e19636

da Silva J, Coetzer M, Nedellec R, Pastore C, Mosier DE (2010) Fitness epistasis and constraints on adaptation in a human immunodeficiency virus type 1 protein region. Genetics 185(1):293–303

de Visser JAGM, Hoekstra RF, van den Ende H (1997) Test of interaction between genetic markers that affect fitness in *Aspergillus niger*. Evolution 51:1499–1505

de Visser JA, Park SC, Krug J (2009) Exploring the effect of sex on empirical fitness landscapes. Am Nat 174(Suppl 1):S15–S30

DePristo MA, Hartl DL, Weinreich DM (2007) Mutational reversions during adaptive protein evolution. Mol Biol Evol 24(8):1608–1610

Desai MM, Fisher DS (2007) Beneficial mutation selection balance and the effect of linkage on positive selection. Genetics 176(3):1759–1798

Fisher RA (1930) The genetical theory of natural selection. Oxford University Press, Oxford

Fong SS, Joyce AR, Palsson BO (2005) Parallel adaptive evolution cultures of *Escherichia coli* lead to convergent growth phenotypes with different gene expression states. Genome Res 15(10):1365–1372

Freddolino PL, Goodarzi H, Tavazoie S (2012) Fitness landscape transformation through a single amino acid change in the rho terminator. PLoS Genet 8(5):e1002744
Gerrish PJ, Lenski RE (1998) The fate of competing beneficial mutations in an asexual population. Genetica 103(1–6):127–144
Gerstein AC, Lo DS, Otto SP (2012) Parallel genetic changes and nonparallel gene-environment interactions characterize the evolution of drug resistance in yeast. Genetics 192(1):241–252
Gould SJ (1989) Wonderful life: the burgess shale and the nature of history. Norton, New York
Gros PA, Tenaillon O (2009) Selection for chaperone-like mediated genetic robustness at low mutation rate: impact of drift, epistasis and complexity. Genetics 182(2):555–564
Gros PA, Le Nagard H, Tenaillon O (2009) The evolution of epistasis and its links with genetic robustness, complexity and drift in a phenotypic model of adaptation. Genetics 182(1):277–293
Haas O, Simpson GG (1946) Analysis of some phylogenetic terms, with attempts at redefinition. Proc Am Philos Soc 90(319–49)
Hartl DL, Taubes CH (1996) Compensatory nearly neutral mutations: selection without adaptation. J Theor Biol 182(3):303–309
Hartl DL, Taubes CH (1998) Towards a theory of evolutionary adaptation. Genetica 102–103(1–6):525–533
Herring CD, Raghunathan A, Honisch C, Patel T, Applebee MK, Joyce AR, Albert TJ, Blattner FR, van den Boom D, Cantor CR, Palsson BO (2006) Comparative genome sequencing of *Escherichia coli* allows observation of bacterial evolution on a laboratory timescale. Nat Genet 38(12):1406–1412
Hindre T, Knibbe C, Beslon G, Schneider D (2012) New insights into bacterial adaptation through in vivo and in silico experimental evolution. Nat Rev Microbiol 10(5):352–365
Hoekstra HE, Hirschmann RJ, Bundey RA, Insel PA, Crossland JP (2006) A single amino acid mutation contributes to adaptive beach mouse color pattern. Science 313(5783):101–104
Kauffman S (1993) The origin of order. Oxford University Press, New York
Khan AI, Dinh DM, Schneider D, Lenski RE, Cooper TF (2011) Negative epistasis between beneficial mutations in an evolving bacterial population. Science 332(6034):1193–1196
Kogenaru M, de Vos MG, Tans SJ (2009) Revealing evolutionary pathways by fitness landscape reconstruction. Crit Rev Biochem Mol Biol 44(4):169–174
Kvitek DJ, Sherlock G (2011) Reciprocal sign epistasis between frequently experimentally evolved adaptive mutations causes a rugged fitness landscape. PLoS Genet 7(4):e1002056
Le Gac M, Brazas MD, Bertrand M, Tyerman JG, Spencer CC, Hancock RE, Doebeli M (2008) Metabolic changes associated with adaptive diversification in *Escherichia coli*. Genetics 178(2):1049–1060
Le Gac M, Plucain J, Hindre T, Lenski RE, Schneider D (2012) Ecological and evolutionary dynamics of coexisting lineages during a long-term experiment with *Escherichia coli*. Proc Natl Acad Sci USA 109(24):9487–9492
Le Nagard H, Chao L, Tenaillon O (2011) The emergence of complexity and restricted pleiotropy in adapting networks. BMC Evol Biol. In press 11:326
Lee YH, LM DS, Fox GE (1997) Equally parsimonious pathways through an RNA sequence space are not equally likely. J Mol Evol 45(3):278–284
Lenski RE, Rose MR, Simpson SC, Tadler SC (1991) Long-term experimental evolution in *Escherichia coli*. I. Adaptation and divergence during 2,000 generations. Am Nat 138:1315–1341
Lewis NE, Hixson KK, Conrad TM, Lerman JA, Charusanti P, Polpitiya AD, Adkins JN, Schramm G, Purvine SO, Lopez-Ferrer D, Weitz KK, Eils R, Konig R, Smith RD, Palsson BO (2010) Omic data from evolved *E. coli* are consistent with computed optimal growth from genome-scale models. Mol Syst Biol 6:390
Lobkovsky AE, Koonin EV (2012) Replaying the tape of life: quantification of the predictability of evolution. Front Genet 3:246
Lobkovsky AE, Wolf YI, Koonin EV (2011) Predictability of evolutionary trajectories in fitness landscapes. PLoS Comput Biol 7(12):e1002302
Lourenco J, Galtier N, Glemin S (2011) Complexity, pleiotropy, and the fitness effect of mutations. Evolution 65(6):1559–1571
Lozovsky ER, Chookajorn T, Brown KM, Imwong M, Shaw PJ, Kamchonwongpaisan S, Neafsey DE, Weinreich DM, Hartl DL (2009) Stepwise acquisition of pyrimethamine resistance in the malaria parasite. Proc Natl Acad Sci USA 106(29): 12025–12030
Lunzer M, Miller SP, Felsheim R, Dean AM (2005) The biochemical architecture of an ancient adaptive landscape. Science 310(5747):499–501
Martin G, Lenormand T (2006a) The fitness effect of mutations across environments: a survey in light of fitness landscape models. Evolution 60(12): 2413–2427
Martin G, Lenormand T (2006b) A general multivariate extension of Fisher's geometrical model and the distribution of mutation fitness effects across species. Evolution 60(5):893–907
Martin G, Lenormand T (2008) The distribution of beneficial and fixed mutation fitness effects close to an optimum. Genetics 179(2):907–916
Martin G, Elena SF, Lenormand T (2007) Distributions of epistasis in microbes fit predictions from a fitness landscape model. Nat Genet 39(4):555–560
O'Maille PE, Malone A, Dellas N, Andes Hess B Jr, Smentek L, Sheehan I, Greenhagen BT, Chappell J, Manning G, Noel JP (2008) Quantitative exploration of the catalytic landscape separating divergent plant sesquiterpene synthases. Nat Chem Biol 4(10):617–623

Orr HA (1998) The population genetics of adaptation: the distribution of factors fixed during adaptive evolution. Evolution 52(4):935–949

Orr HA (2000) Adaptation and the cost of complexity. Evolution 54(1):13–20

Orr HA (2005) The genetic theory of adaptation: a brief history. Nat Rev Genet 6(2):119–127

Orr H (2006) The distribution of fitness effects among beneficial mutations in Fisher's geometric model of adaptation. J Theor Biol 238(2):279–285

Ostrowski EA, Rozen DE, Lenski RE (2005) Pleiotropic effects of beneficial mutations in *Escherichia coli*. Evolution 59(11):2343–2352

Ostrowski EA, Woods RJ, Lenski RE (2008) The genetic basis of parallel and divergent phenotypic responses in evolving populations of *Escherichia coli*. Proc Biol Sci 275(1632):277–284

Pelosi L, Kuhn L, Guetta D, Garin J, Geiselmann J, Lenski RE, Schneider D (2006) Parallel changes in global protein profiles during long-term experimental evolution in *Escherichia coli*. Genetics 173(4):1851–1869

Philippe N, Pelosi L, Lenski RE, Schneider D (2009) Evolution of penicillin-binding protein 2 concentration and cell shape during a long-term experiment with *Escherichia coli*. J Bacteriol 191(3):909–921

Poelwijk FJ, Kiviet DJ, Weinreich DM, Tans SJ (2007) Empirical fitness landscapes reveal accessible evolutionary paths. Nat Genetics 445(7126):383–386

Poelwijk FJ, Tanase-Nicola S, Kiviet DJ, Tans SJ (2011) Reciprocal sign epistasis is a necessary condition for multi-peaked fitness landscapes. J Theor Biol 272(1):141–144

Poon A, Otto SP (2000) Compensating for our load of mutations: freezing the meltdown of small populations. Evolution 54(5):1467–1479

Rainey PB, Travisano M (1998) Adaptive radiation in a heterogeneous environment. Nat Genetics 394(6688):69–72

Reynolds MG (2000) Compensatory evolution in rifampin-resistant *Escherichia coli*. Genetics 156(4):1471–1481

Ritland K, Newton C, Marshall HD (2001) Inheritance and population structure of the white-phased "Kermode" black bear. Curr Biol 11(18):1468–1472

Rodriguez-Verdugo A, Gaut BS, Tenaillon O (2013) Evolution of *Escherichia coli* rifampicin resistance in an antibiotic-free environment during thermal stress. BMC Evol Biol 13:50

Rose MR, Passananti HB, Chippindale AK, Phelan JP, Matos M, Teotonio H, Mueller LD (2005) The effects of evolution are local: evidence from experimental evolution in *Drosophila*. Integr Comp Biol 45(3):486–491

Salverda ML, Dellus E, Gorter FA, Debets AJ, van der Oost J, Hoekstra RF, Tawfik DS, de Visser JA (2011) Initial mutations direct alternative pathways of protein evolution. PLoS Genet 7(3):e1001321

Silander OK, Tenaillon O, Chao L (2007) Understanding the evolutionary fate of finite populations: the dynamics of mutational effects. PLoS Biol 5(4):e94

Sousa A, Magalhaes S, Gordo I (2012) Cost of antibiotic resistance and the geometry of adaptation. Mol Biol Evol 29(5):1417–1428

Steiner CC, Weber JN, Hoekstra HE (2007) Adaptive variation in beach mice produced by two interacting pigmentation genes. PLoS Biol 5(9):e219

Tan L, Serene S, Chao HX, Gore J (2011) Hidden randomness between fitness landscapes limits reverse evolution. Phys Rev Lett 106(19):198102

Tenaillon O, Silander OK, Uzan JP, Chao L (2007) Quantifying organismal complexity using a population genetic approach. PLoS One 2:e217

Tenaillon O, Rodriguez-Verdugo A, Gaut RL, McDonald P, Bennett AF, Long AD, Gaut BS (2012) The molecular diversity of adaptive convergence. Science 335(6067):457–461

Toprak E, Veres A, Michel JB, Chait R, Hartl DL, Kishony R (2012) Evolutionary paths to antibiotic resistance under dynamically sustained drug selection. Nat Genet 44(1):101–105

Trindade S, Sousa A, Gordo I (2012) Antibiotic resistance and stress in the light of Fisher's model. Evolution 66(12):3815–3824

Tyerman JG, Bertrand M, Spencer CC, Doebeli M (2008) Experimental demonstration of ecological character displacement. BMC Evol Biol 8:34

Wagner GP (2010) The measurement theory of fitness. Evolution 64(5):1358–1376

Wagner GP, Zhang J (2011) The pleiotropic structure of the genotype-phenotype map: the evolvability of complex organisms. Nat Rev Genet 12(3):204–213

Wang Z, Liao BY, Zhang J (2010) Genomic patterns of pleiotropy and the evolution of complexity. Proc Natl Acad Sci USA 107(42):18034–18039

Waxman D, Welch JJ (2005) Fisher's microscope and Haldane's ellipse. Am Nat 166(4):447–457

Weinreich DM (2005) The rank ordering of genotypic fitness values predicts genetic constraint on natural selection on landscapes lacking sign epistasis. Genetics 171(3):1397–1405

Weinreich DM, Delaney NF, Depristo MA, Hartl DL (2006) Darwinian evolution can follow only very few mutational paths to fitter proteins. Science 312(5770):111–114

Welch JJ, Waxman D (2003) Modularity and the cost of complexity. Evolution 57(8):1723–1734

Wichman HA, Badgett MR, Scott LA, Boulianne CM, Bull JJ (1999) Different trajectories of parallel evolution during viral adaptation. Science 285(5426): 422–424

Woods R, Schneider D, Winkworth CL, Riley MA, Lenski RE (2006) Tests of parallel molecular evolution in a long-term experiment with *Escherichia coli*. Proc Natl Acad Sci USA 103(24):9107–9112

Wright S (1932) The roles of mutation, inbreeding, cross breeding and selection in evolution. In: Proceeding of the sixth international congress of genetics. Ithaca, New York pp 356–366

Ecological Genomics of Host Shifts in *Drosophila mojavensis*

12

Luciano M. Matzkin

Abstract

Advances in next-generation sequencing technologies have liberated our dependency on model laboratory species for answering genomic and transcriptomic level questions. These new techniques have dramatically expanded our breadth of study organisms and have allowed the analysis of species from diverse ecological environments. One such species is the cactophilic *Drosophila mojavensis* that inhabits the deserts of western North America. These insects feed and develop in the necrotic cacti, feeding largely on the microflora of the necrotic plant tissues. *Drosophila mojavensis* is composed of four geographically and ecologically separated populations. Each population (Baja California peninsula, mainland Sonoran Desert, Mojave Desert and Santa Catalina Island) utilizes the necrotic tissues of distinct cactus species. The differences in the nutritional and chemical composition of the necroses include a set of toxic compounds to which resident population must adapt. These ecological differences have facilitated many of the life history, behavior, physiological and genetic differences between the cactus host populations. Genomic resources have allowed investigators to examine the genomic and transcriptional level changes associated with the local adaptation of the four *D. mojavensis* populations, thereby providing further understanding of the genetic mechanism of adaptation and its role in the divergence of ecologically distinct populations.

Keywords

Ecological genomics • Cactophilic drosophila • Adaptation • *Drosophila mojavensis* • Transcriptomics

12.1 Introduction

One of the great advantages in studying model laboratory species (e.g. *Drosophila melanogaster*, *Caenorhabditis elegans*,

L.M. Matzkin (✉)
Department of Biological Sciences, University of Alabama in Huntsville, Huntsville, AL 35899, USA
e-mail: lmm0015@uah.edu

C.R. Landry and N. Aubin-Horth (eds.), *Ecological Genomics: Ecology and the Evolution of Genes and Genomes*, Advances in Experimental Medicine and Biology 781,
DOI 10.1007/978-94-007-7347-9_12,

Saccharomyces cerevisiae, *Arabidopsis thaliana* and *Escherichia coli*) is the presence of a completely sequenced and annotated genome (Adams et al. 2000; *C. elegans* Sequencing Consortium 1998; Blattner et al. 1997; Goffeau et al. 1996; Arabidopsis Genome Initiative 2000). These model systems offer the availability of many mutant and transgenic stocks, which allows for the manipulation and genetic dissection of the traits of interest. The lack of much ecological information about these species prevents in many instances the correlation of patterns of genomic variation to the ecological factors influencing its variance. The recent feasibility of large-scale sequencing, due to the development of next-generation platforms, has allowed the genomic and transcriptomic investigation of a greater number of ecologically diverse species. The placing of genomic data in an ecological context has been an instrumental approach in our understanding of the adaptation and evolution of species (Feder and Mitchell-Olds 2003). This ecological genomic approach has been embraced in many fascinating and ecologically well-studied systems such as the water flea (*Daphnia pulex*) and stickleback fish (*Gasterosteus aculeatus*) (Colbourne et al. 2011; Jones et al. 2012).

The ecological characteristic of certain species facilitates the change or shift in resource use. This is no more evident that in the many species of phytophagous insects (Agrawal et al. 2009; Janz 2011; Berenbaum 2002). Genetic and genomic analysis of host shifts in phytophagous insects has allowed investigators to peer into the mechanism of the adaptation process. In general, such as observed in the apple maggot fly (*Rhagoletis pomonella*) and the pea aphid (*Acyrthosiphon pisum*), host shifts involve not only changes associated with the use of an alternative host, but as well in several correlated life history, physiological and behavioral traits (Caillaud and Via 2012; Dambroski and Feder 2007; Linn et al. 2003; Via 1999). Genome level studies have begun to identify the loci and genomic regions associated many of these host shift related adaptations (Michel et al. 2010; Richards et al. 2010; Schwarz et al. 2009; Smadja et al. 2012) and in the case of *Drosophila pachea* investigators have been able to identify the genetic loci responsible for obligate host use (Lang et al. 2012). Furthermore, when correlated with reproductive isolation between host populations, the changes associated with host shifts could eventually lead to ecological speciation (reviewed in Nosil 2012). Among the many insect groups, the ecologically diverse and specious genus *Drosophila* offers a vast number of study systems to assist in the understanding of the genetic basis of host adaptation and its relationship to speciation.

The genus *Drosophila* is comprised of over 2,000 described species inhabiting a wide variety of ecological habitats from tropical rainforests to deserts (Markow and O'Grady 2006). The vast majority of these species are saprophytic, mainly feeding as larvae and adults on yeasts and bacteria growing in a variety of tissues, such as fruits, tree sap fluxes, leaves, cactus and mushrooms (Throckmorton 1975; Sturtevant 1921; Jaenike 1978; Kaneshiro et al. 1973; Heed 1978). A few species are known to utilize live flowers, nutrient-soaked soils and even land crabs for feeding (Brncic 1983; Carson 1974; Kaneshiro et al. 1973). Although yeasts are a major source of the *Drosophila's* nutrition, they are also exposed to chemical compounds found in the host. In certain systems these compounds are toxic and resident *Drosophila* species must adapt to their presence, such as in the case of α-amanitin tolerance in the mycophagous species *D. putrida*, *D. recens* and *D. tripunctata* (Jaenike et al. 1983) or octanoic acid in the *Morinda citrifolia* fruit, the host of *D. sechellia* (Legal et al. 1994).

Another example of *Drosophila* exposed to the toxic chemical profile of its host, are those species inhabiting cactus necroses. Cactophilic *Drosophila* species feed as adults, oviposit and develop in necrotic stems and/or fruits of cacti, with some fascinating exceptions such as *D. mettleri* which feeds as an adult in the necrotic tissues of cardón (*Pachycereus pringlei*), saguaro (*Carnegiea gigantean*) and occasionally senita

(*Lophocereus schottii*) cactus, but oviposits, develops and pupates in the necrotic exudate soaked soil (Heed 1977). These Drosophilids, including specialists and generalists, are found throughout the Americas utilizing a wide variety of plants within the family Cactaceae (Hasson et al. 1992; Heed 1978; Oliveira et al. 2012; Ruiz et al. 1990; Carson and Wasserman 1965; Heed and Kircher 1965; Markow and O'Grady 2008; Fontdevila et al. 1988). One exception to the New World distribution of cactophilic *Drosophila* is *D. buzzatti*, which has been introduced to the Mediterranean region and Australia (Carson and Wasserman 1965).

With the exception of the *D. nannoptera* species group, cactophilic *Drosophila* are members of the *D. repleta* species group, which is comprised of approximately 100 described species (Throckmorton 1975; Oliveira et al. 2012). Both of these groups are part of a larger species radiation (virilis-repleta) that occurred within the genus approximately 36 MYA (Throckmorton 1975). Among the species within the *D. repleta* species group is *D. mojavensis,* which has proven to be a powerful system for understanding the ecological genomics of adaptation.

12.2 The *Drosophila mojavensis* Study System

12.2.1 Evolutionary History

Drosophila mojavensis is one of four cactophilic species endemic to the Sonoran Desert of western North America (Heed 1978). Its distribution includes four geographically separated populations, or host races, each utilizing a distinct necrotic cactus host for both oviposition and adult feeding. *Drosophila mojavensis* utilizes the agria cactus (*Stenocereus gummosus*) in the Baja California peninsula, organ pipe cactus (*S. thurberi*) in the mainland Sonoran Desert, Red Barrel cactus (*Ferocactus cylindraceus*) in the Mojave Desert and Coastal Prickly Pear (*Opuntia littoralis*) in Santa Catalina Island(hereafter Catalina Island) (Fig. 12.1) (Heed 1978; Ruiz et al. 1990). In the mainland Sonoran Desert and in Baja California *D. mojavensis* is sympatric with its sister species, *D. arizonae*, a generalist cactophile known to utilize the same hosts as *D. mojavensis* in addition to the cina cactus (*S. alamosensis*) (Fellows and Heed 1972). In fact, the presence of *D. arizonae* across Baja California appears to be a relative recent observation. Field collections in the

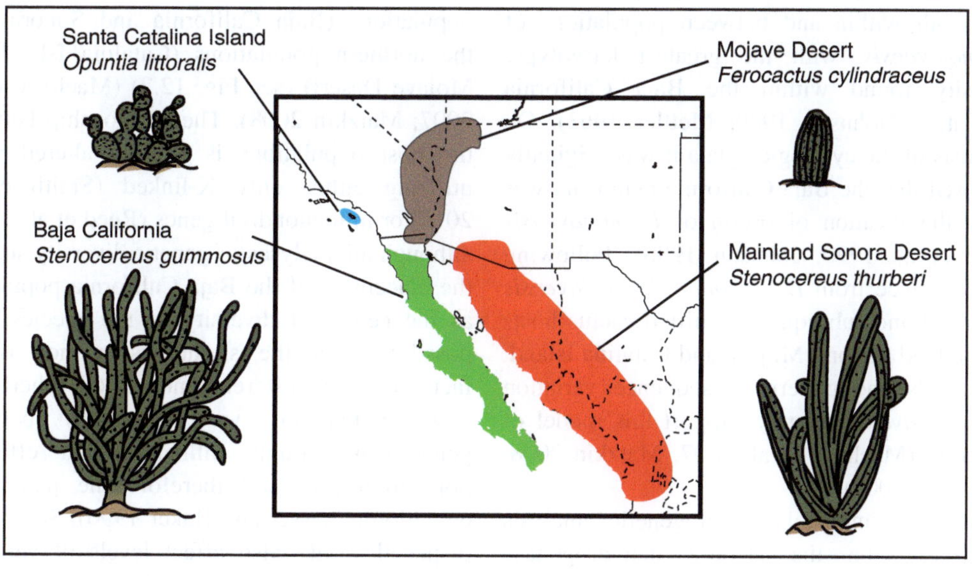

Fig. 12.1 Distribution and cactus host use of the four *D. mojavensis* host races

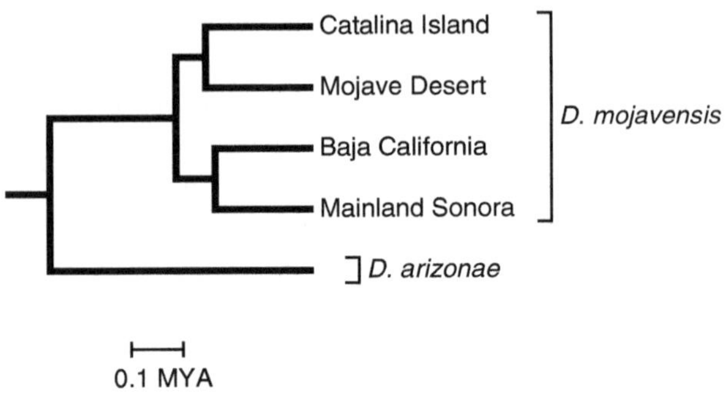

Fig. 12.2 Phylogenetic relationship of the four *D. mojavensis* host races. Population and species relationship and divergence times estimates based on previous molecular studies (Smith et al. 2012; Matzkin 2008; Machado et al. 2007; Reed et al. 2007)

1970s and 1980s rarely observed *D. arizonae* in Baja California, and when collected, all localities where in the Cape region of the peninsula (Heed 1982). Currently, *D. arizonae* can be collected across the peninsula and into Southern California, USA (Reed et al. 2007; Matzkin pers. obs.).

Drosophila mojavensis was originally collected by Warren Spencer from necrotic barrel cactus in the Mojave Desert of California (Patterson and Crow 1940; Spencer 1941). *Drosophila mojavensis*, *D. arizonae* and their sibling species *D. navojoa*, also a cactophile, are all members of the mulleri complex, which spans South and North America (Wasserman 1962, 1982). Cytologically, *D. mojavensis* and *D. arizonae* differ by inversions in the X, second and third chromosomes (Muller elements A, E and B, respectively) (Wasserman 1962). Polymorphic inversions in the second and third chromosome exist both within and between populations of *D. mojavensis*, with the greatest karyotypic diversity found within the Baja California population (Johnson 1980; Mettler 1963). On the basis of the cytological data it was originally proposed that the Baja California peninsula was likely the location of origin of *D. mojavensis* (Ruiz et al. 1990; Johnson 1980). Following its divergence from *D. arizonae*, *D. mojavensis* colonized and subsequently shifted cactus hosts in mainland Sonora, Mojave and Catalina Island. The levels and pattern of sequence variation in *D. mojavensis* largely support this model of evolution (Machado et al. 2007; Matzkin 2008; Reed et al. 2007).

The first molecular phylogenetic analysis was based on the alcohol dehydrogenase (*Adh*) paralogs and estimated the divergence time of *D. mojavensis* and *D. arizonae* to be approximately 4 MYA (Russo et al. 1995), subsequent population genetics analysis of the same loci demonstrated a more recent divergence time of approximately 1 MYA (Matzkin 2004; Matzkin and Eanes 2003). Further population genetic analysis of nuclear and mitochondrial genes suggest an even more recent divergence time of less than 0.5 MYA (Matzkin 2008; Reed et al. 2007). The relationship of the four host populations has been examined using both cytological and molecular data. Recently, given the level of morphological and molecular differences, the host races of *D. mojavensis* have been described as subspecies (Pfeiler et al. 2009). Evidence from multiple nuclear markers spanning all chromosomes suggests that soon after the establishment of *D. mojavensis*, the species diverged into two clades, the southern populations (Baja California and Sonora) and the northern populations (Catalina Island and Mojave Desert) (see Fig. 12.2) (Machado et al. 2007; Matzkin 2008). The relationship between the host populations is slightly altered when utilizing either only X-linked (Smith et al. 2012) or mitochondrial genes (Reed et al. 2007), although all analyses of genetic diversity support the placement of the Baja California population as the center of diversity of the species. One possibility for the slight discrepancy is the fact that the different modes of inheritance (e.g. uniparental vs. diparental) between genes can strongly influence their effective population size and therefore the pattern of evolution (Chesser and Baker 1996). Sex-biased dispersal would also affect levels of variation between uniparental and diparental inherited

loci (Chesser and Baker 1996), although, in the population studied (Sonora) no difference in dispersal was observed to occur between the sexes (Markow and Castrezana 2000). The ongoing sequencing of the genomes from all *D. mojavensis* host populations will provide a better understanding of the history and relationship among them (Matzkin unpub.)

12.2.2 Ecology

The different cacti utilized by each of the host races offers distinct biotic and abiotic environments to the resident *D. mojavensis*. Much of the chemical composition of the necrotic cactus is a function of both the host plant as well as the resident microflora (Fogleman and Starmer 1985; Starmer 1982a, b; Starmer et al. 1990). It is the bacterial and yeast communities found in the cactus necrosis that are instrumental in setting up the chemical environment for the flies (Starmer et al. 1986). The greatest microflora and chemical similarities are between organ pipe and agria, both columnar cactus species (Kircher 1982; Starmer and Phaff 1983). Cactus hosts can differ in a variety of compounds such as triterpene glycosides, unhydrolyzed glycosides, sterol diols, free fatty acids, sugars, many volatiles and in the case of *Opuntia* sp., alkaloids (Fogleman and Abril 1990; Kircher 1982; Starmer and Phaff 1983; Meyer et al. 1980). The compounds associated with cactus necroses have been shown to be detrimental, even lethal to other non-cactophilic species and in certain cases deleterious to non-native cactophilic *Drosophila* (Fellows and Heed 1972; Kircher et al. 1967).

In addition to the chemical composition, there are distinct differences in the physical properties of the cacti and their necroses (Etges 1989; Mangan 1982). The total size of the plant is positively correlated with the persistence and biomass of the necrosis, but negatively correlated with the density of the necroses in the desert landscape (Breitmeyer and Markow 1998). This would suggest that individuals that utilize necroses from small plants would have an easier task in discovering oviposition sites, but such sites would be available to adults and developing larvae for a shorter period of time. In contrast, individuals utilizing larger host would need to travel longer distances to locate a potential oviposition site. With respect to *D. mojavensis*, *Opuntia* cladodes (cactus pads) are significantly smaller than the arms of an organ pipe or agria cactus, which would influence the evolution of life history characters such as developmental time, dispersal rate, starvation and desiccation resistance and chemosensory behavior. Furthermore, abiotic factors such as the thermal environment differ across the populations, and data suggests that the genetic mechanisms underlying resistance to thermal stress might be distinct in each of the populations (Krebs and Thompson 2005). The many biotic and abiotic differences in the ecology of the four *D. mojavensis* host populations have influenced their evolutionary trajectory.

12.3 Genetic Variation and Population Genetics

The environmental conditions experienced by the *D. mojavensis* host races have shaped many aspects of their biology. Genetic variation in life history, morphological, physiological and behavioral characteristics exist across the populations. Furthermore, for several of these characters, there exists a significant interaction with environmental variables.

Many life history comparisons have involved the Baja California and Sonora populations, showing differences in developmental time and adult size (thorax length) between the two populations (Etges 1990, 1998; Etges et al. 2010). Thorax length is a significant life history trait, given its correlation with other characteristics such as flight performance, stress resistance, ovariole number (which is correlated with lifetime fecundity) and mating success (Markow and Ricker 1992; Azevedo et al. 1998; Hoffmann et al. 2001; Mangan 1978). Several morphological and pigmentation differences have been identified across all four host races, most notably divergence in features of the male genitalia such as the shape of the aedeagus

(Richmond et al. 2012; Pfeiler et al. 2009). Overall, *D. mojavensis* is highly resistant to water stress relative to other Drosophila (Gibbs et al. 2003; Gibbs and Matzkin 2001; Matzkin et al. 2009), but yet significant differences in desiccation resistance exist between Catalina, Baja California and Sonora populations (Rajpurohit et al. 2013; Matzkin et al. 2007). Furthermore, interpopulation differences extend to the composition of the hydrocarbons in the fly cuticle. The composition of these hydrocarbons is distinct between Baja California and Sonora populations and, although composition could be influenced by host utilization and temperature, it also affects courtship behavior both within and between populations (Havens and Etges 2013; Markow and Toolson 1990; Etges and Jackson 2001). Courtship behavior differences exist between several of the host races, which contributes to a reduction in gene flow (Markow et al. 1983; Etges et al. 2006).

Utilization of alternative cactus hosts can elicit negative fitness effects. Development of flies from Baja California and Sonora populations on non-native hosts (necrotic agria or organ pipe, respectively) results in significant life history consequences, affecting such characters as thorax length and developmental time (Etges 1990, 1993, 1998). Larval viability can be drastically reduced when Sonora flies develop in necrotic agria or cina (Matzkin and Markow 2013; Bono and Markow 2009). These viability differences are amplified when *D. mojavensis* utilize more chemically distinct cactus hosts (Fellows and Heed 1972). For example, relative to organ pipe or agria, the larval viability of *D. mojavensis* is 2 % in the senita cactus (*L. schottii*), a Sonoran and Baja California plant with high levels of alkaloids (Fellows and Heed 1972).

12.3.1 Candidate Gene Studies

Fermentation by the resident yeast communities largely contributes to the volatile concentration variation across cactus hosts (Fogleman 1982; Heed 1982; Kircher 1982; Vacek 1979). The cactus-specific substrates used in the fermentation process affect the concentration and composition of many of the volatiles such as alcohols. Relative to the organ pipe cactus (Sonora), necroses of agria (Baja California) contain relatively greater levels of 2-propanol than 1-propanol (Heed 1978; Kircher 1982; Starmer et al. 1986; Vacek 1979). In *Drosophila*, Alcohol Dehydrogenase (ADH) is a major pathway for the metabolism of small alcohol molecules (e.g. ethanol) (Chambers 1988). In *D. mojavensis* the *Adh* locus is duplicated, having a larval and adult ovarian tissue expressed locus (*Adh-1*) and a late larval stage and adult (non-ovarian tissue) expressed locus (*Adh-2*) (Batterham et al. 1983; Atkinson et al. 1988). Population genetic analyses date the duplication event to approximately 4 MYA (Matzkin 2004). Earlier studies have shown the presence of two major allozyme alleles (Fast and Slow) at the *Adh-2* locus. While in the Sonora population the Slow allele is at high frequency (>90 %), in Baja California the Fast allele is most frequent (>90 %) (Heed 1978). Resistance to specific alcohols is associated with *Adh-2* genotype, with *Adh-2* Fast homozygotes having increased resistance to 2-propanol relative to *Adh-2* Slow flies (Heed 1978; Starmer et al. 1977). Subsequent studies have shown that the mutation responsible for the Fast/Slow allozyme class (serine to arginine change at residue 28) is associated with as many as four other amino acid substitutions (Matzkin 2004; Matzkin and Eanes 2003). These amino acid differences between allozyme class alleles confer significant substrate specificity differences, with the ADH-2 Fast allele having greater activity on 2-propanol relative to 1-propanol, matching the alcohol concentration of the cactus necrosis in which the *Adh-2* Fast allele is commonly found (Matzkin 2005).

The different metabolic environment experienced by a larval and adult expressed ADH paralogs has distinctly shaped their evolution. Studies using *D. melanogaster* have shown that the control of metabolic flux of ADH in larvae is significantly greater relative to when expressed in adult tissues (Freriksen et al. 1991, 1994; Middleton and Kacser 1983). In any given pathway, the effect of activity changes of a single

enzymatic step on the overall metabolic rate or flux through that pathway is known as the flux control coefficient (Kacser and Burns 1973). Activity perturbations of an enzyme with a high flux control coefficient will produce greater changes on the overall metabolic flux of that pathway relative to an enzyme with little control.

Interestingly, *D. melanogaster* lacks a similar *Adh* duplication, although it produces two distinct transcripts (larval and adult), which resemble the expression pattern of *D. mojavensis* paralogs (*Adh-1* and *Adh-2*, respectively) (Benyajati et al. 1983; Savakis et al. 1986). In the lineage leading to *D. mojavensis*, the larval/ovarian expressed paralog (*Adh-1*) is under positive selection, unlike what is observed for *Adh-2* (Matzkin 2004; Matzkin and Eanes 2003). Furthermore, several amino acid substitutions have occurred between the *Adh* paralogs and are responsible for the observed substrate specificity and kinetic differences between the genes (Matzkin 2005). The non-overlapping expression pattern and functional differences of the *Adh* paralogs in *D. mojavensis*, coupled with the expression pattern of *Adh* in species with a single copy, strongly supports a subfunctionalization model of evolution for the *D. mojavensis* paralogs (Force et al. 1999; Hughes 1994; Lynch and Force 2000).

12.4 *Drosophila mojavensis* in the Genomic Era

The sequencing, assembly and annotation of the *D. melanogaster* genome was a major leap in the understanding of the genetics and evolution of a species in which a tremendous amount of information was already known (Adams et al. 2000). This was later followed by the genome sequencing of a relatively distant species, *D. pseudoobscura* (Richards et al. 2005). These accomplishments lead to the subsequent genome sequencing and comparative analysis of ten additional *Drosophila* species (Drosophila 12 Genomes Consortium 2007). Together these 12 species encompassed a wide breadth of the genus, with nine members of the Sophophora subgenus, and three from the Drosophila subgenus, including *D. mojavensis*.

Although a cDNA microarray was developed for *D. mojavensis* prior to the genome sequencing (Matzkin et al. 2006), knowledge of the genome allowed the construction of complete transcriptome oligonucleotide microarrays (Bono et al. 2011; Matzkin 2012; Matzkin and Markow 2009, 2013; Rajpurohit et al. 2013; Smith et al. 2013). Additional tools for *D. mojavensis* include a BAC library and the availability of transgenic stocks (Song et al. 2011; Holtzman et al. 2010). Although not as extensive as those available for *D. melanogaster*, the *D. mojavensis* transgenic stocks could be used to create null alleles via RNA interference to begin to identify the role of candidate loci in host adaptation. The expanse of ecological information of *D. mojavensis* and the addition of genomic and transcriptomics tools vastly increased the power of this system to help answer many fundamental questions in biology such as the genomic basis of adaptation, the role of local ecological adaptation in speciation and the genomic basis of the evolution of reproductive incompatibilities.

12.4.1 Host Adaptation

The four chemically distinct hosts of *D. mojavensis* have influenced the evolution of the populations at many levels. The genome level changes include structural (e.g. chromosomal inversions), coding sequence, and transcriptional changes. One clear advantage of the *D. mojavensis* system is the vast ecological information that has been previously gathered, which allows for genomic analyses in the ecological context of the species. In the case of *D. mojavensis* the necrotic cactus is a major component of its ecology. The creation of ecologically realistic breeding substrates using lab-generated cactus necroses incorporating the natural microflora has been extensively used in many prior life history studies (Etges 1989, 1990, 1993; Etges and Heed 1987). Similarly, recent microarray studies in *D. mojavensis* allow flies to develop in lab-generated necrotic cactus also including the natural yeast and bacteria

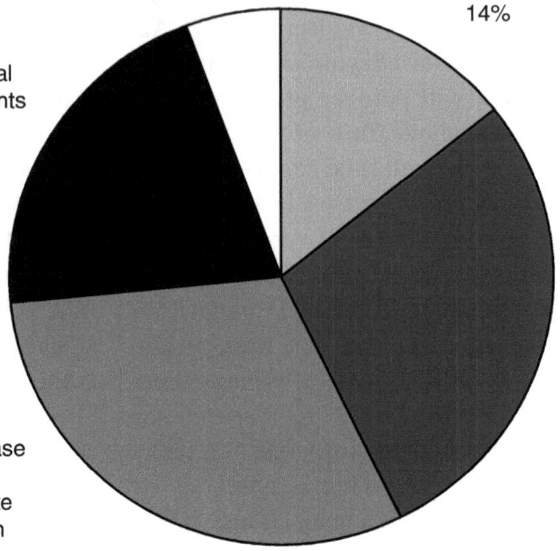

Fig. 12.3 Overrepresented (FDR < 0.05) molecular function and biological process gene ontology categories of differentially expressed genes (Data from Matzkin (2012) examining gene expression of third instar larva from nine isofemale lines from Sonora when reared in either necrotic organ pipe (native host) or agria)

microflora (Matzkin 2012; Matzkin et al. 2006; Rajpurohit et al. 2013; Smith et al. 2013).

In Matzkin et al. (2006) and Matzkin (2012) *D. mojavensis* larvae from either Baja California or Sonora were exposed to necrotic agria or organ pipe cactus. Development and exposure to necrotic cactus elicited a complex transcriptional response, with a variety of loci being differentially expressed (Matzkin 2012; Matzkin et al. 2006). The differentially expressed genes correspond to several gene ontology groups, including xenobiotic metabolism and detoxification (Fig. 12.3). Among the xenobiotic metabolism genes, Glutathione *S*-transferases, Cytochrome P450, and UDP-glycosyltransferase were modulated in response to cactus use (Table 12.1). In other insect systems, members of these three gene families have been known to play a central role in detoxification (Luque and O'Reilly 2002; Ranson and Hemingway 2005; Ranson et al. 2001; Feyereisen 2005; Li et al. 2007).

Although the ability to modulate gene expression in response to environmental change would be advantageous, over evolutionary time constant exposure to an environment, such as a host shift, may produce fixed expression differences (Waddington 1953; West-Eberhard 2003). There are a number of genes whose expression difference appear to be fixed when comparing across the *D. mojavensis* host races (Matzkin and Markow 2013). Recently, Matzkin and Markow (2013) examined the expression profile of third instar larvae reared in media lacking cactus compounds (i.e. standard banana media). In addition to detoxification genes, genes associated with metabolism were differentially expressed across the cactus host races. In fact, these metabolic genes included a large proportion of central metabolism enzymes located both at and outside branch points (Fig. 12.4). Branch enzymes are important control points of flux through pathways (LaPorte et al. 1984). For example, the activity of the enzyme Glucose-6-dehydrogenase (G6PD, see Fig. 12.4) not only influences flux through the pentose shunt, but given that its substrate (glucose-6-phosphate) is used by other enzymes (PGM, HEX and PGI), it also could modulate flux through those other pathways. Given the greater control of flux of branch point enzymes, it would be expected that these enzymes be involved in adaptation (Eanes 1999; Flowers et al. 2007; Rausher 2013). This suggests that the nutritional differences between hosts could have influenced flux through this pathway. Further analysis is needed to examine the functional consequence of these gene expression differences.

Table 12.1 Summary of known detoxification genes that are differentially expressed in response to cactus utilization

D. mojavensis annotation	*D. melanogaster* ortholog	Gene name	Gene family
GI16623	CG17523	GstE2	GST
GI16624	–	GstE2b	GST
GI19388	CG17522	GstE10	GST
GI20124	CG17534	GstE9	GST
GI24379	CG10045	GstD1	GST
GI23193	–	GstD1b	GST
GI23196	CG17639	CG17639	GST
GI10234	CG9716	Cyp313b1	P450
GI13002	CG33503	Cyp12d1-d	P450
GI16117	CG3656	Cyp4d1	P450
GI16990	CG9964	Cyp309a1	P450
GI18674	CG3540	Cyp4d14	P450
GI18951	CG8859	Cyp6g2	P450
GI20221	–	Cyp9h1b	P450
GI20230	CG13977	Cyp6a18	P450
GI20590	CG8453	Cyp6g1	P450
GI24047	CG14680	Cyp12e1	P450
GI10119	CG4739	Ugt86Dc	UGT
GI10120	CG18578	Ugt86Da	UGT
GI10122	CG4772	Ugt86Dh	UGT
GI14390	CG11289	CG11289	UGT
GI17058	CG13271	Ugt36Bb	UGT
GI17522	CG11012	Ugt37a1	UGT
GI22627	CG6644	Ugt35a	UGT
GI22628	CG6649	Ugt35b	UGT
GI22630	CG6633	Ugt86Dd	UGT

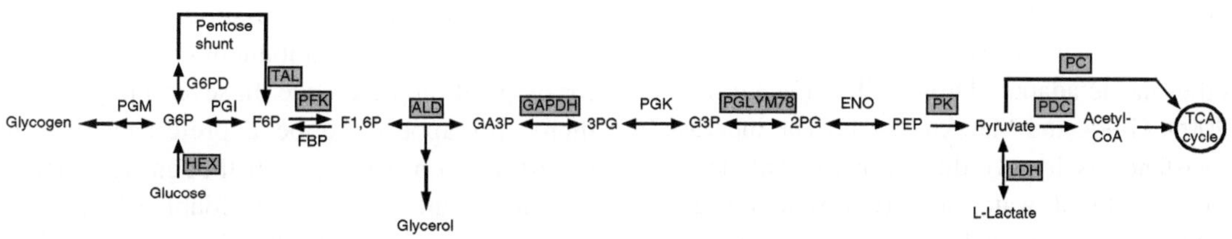

Fig. 12.4 Fixed expression differences in central metabolism genes across the four host races. The enzymes with fixed expression differences: Hexokinase (*HEX*), Transaldolase (*TAL*), Phosphofructose Kinase (*PFK*), Aldolase (*ALD*), Glyceraldehyde 3-phosphate Dehydrogenase (*GAPDH*), Phosphoglyceromutase (*PGLYM78*), Pyruvate Kinase (*PK*), Lactate Dehydrogenase (*LDH*) and Pyruvate Carboxylase (*PC*) are highlighted with a shaded box. The enzyme without fixed expression differences are: Phophoglucose Mutase (*PGM*), Glucose-6-phosphate Dehydrogenase (*G6PD*), Phosphoglucose Isomerase (*PGI*), Phophoglycerate Kinase (*PGK*) and Enolase (*ENO*) (Data from Matzkin and Markow (2013))

In conjunction with transcriptional differences, host adaptation has also been facilitated by changes in the coding region of detoxification genes. Interestingly, there appears to be an association between transcriptional modulations and coding sequence evolution. *Drosophila mojavensis* genes that lack an orthologous call to *D. melanogaster* were found disproportionally among the genes whose expression was significantly affected by cactus host use (Matzkin 2012). A specific example of a gene that fits this pattern is *Glutathione S-transferase D1*

(*GstD1*), a gene whose expression was observed to be affected by alternative host use (organ pipe cactus) in a population from Baja California (Matzkin et al. 2006). Population genetic analysis of *GstD1* suggests that this gene has been under positive selection in the lineage leading to the Baja California and Sonora populations (Matzkin 2008). Among the eight amino acid substitutions occurring in this lineage, two (Leu-7-Gln and His-39-Gln) occur in the active site pocket of the enzyme. Biochemical analysis suggests that these amino acid substitutions result in functional differences between the GSTD1 isoforms from Baja California/Sonora and Catalina Island/Mojave (Matzkin unpub.)

12.4.2 Genomics of Desiccation Resistance

Across the four *D. mojavensis* host races, the chemical composition of the host (including nutritional and toxic compounds) has shaped the pattern of variation at the genomic, transcriptional and functional levels. A major abiotic stress for *D. mojavensis* and other desert endemics is the adaptation to a desiccating environment. Desiccation resistance is significantly greater in *D. mojavensis* compared to other mesic adapted Drosophilids (Matzkin et al. 2009; Gibbs and Matzkin 2001). The increase in resistance is largely due to an overall decrease in the rate of water loss (Gibbs and Matzkin 2001), which is achieved via a decrease in respiratory rate (i.e. metabolic rate) (Gibbs et al. 2003). Analysis of gene expression differences during the desiccating process suggests that key points in central metabolism are modulated in a manner that would suggest a decrease in metabolic flux (Matzkin and Markow 2009). In a recent study, Rajpurohit et al. (2013) observed that in both Baja California and Sonora populations desiccating conditions were associated with the up-regulation of genes associated with the structure of the cuticle and sensory pathways.

12.4.3 Chemosensory Adaptation

Given the deleterious fitness consequences associated with developing in a non-native host, it is predicted that there would be strong selective pressure in *D. mojavensis* to correctly identify a cactus before oviposition. The chemosensory system in *Drosophila* is composed of a number of sensory neurons with transmembrane receptors distributed across the insect's body (Vosshall and Stocker 2007). In adults, odorant receptors are expressed in sensory neurons (ORN) housed in sensilla in the antenna and maxillary palps, while gustatory receptors are expressed in neurons (GRN) not only in the proboscis, but also the legs, wings and ovipositor (Stocker 1994; Vosshall and Stocker 2007). These transmembrane proteins initiate a signal cascade that leads to the perception of taste and smell. Unlike GRN, in ORN, a universal co-receptor (*Or83b*) is expressed which interacts with the neuron-specific odorant receptor (Benton et al. 2006). Overall, both odorant and gustatory receptors can specialize to interact with only a subset of ligands (Laissue and Vosshall 2008; Hallem and Carlson 2006).

Drosophila mojavensis females have the ability to assess host type prior to oviposition. Although there is some variation across studies, overall there is a large amount of genetic variation for oviposition preference. In several populations there still appears to be a preference for the ancestral agria cactus, even though agria is not present in all locations (Lofdahl 1985, 1986; Newby and Etges 1998). In contrast, there are also examples of adult preference to its native host, such as in the Mojave population (Newby and Etges 1998). Prior expression studies examining interpopulation and host-induced changes in *D. mojavensis* focused on the larval stage, and, interestingly, both odorant and gustatory receptors were significantly differentially expressed (Matzkin 2012; Matzkin and Markow 2013). *Drosophila mojavensis* larvae have been shown to selectively feed on certain cactophilic yeast species while ignoring others (Fogleman et al. 1981). Therefore,

it is quite possible that larvae, using their chemosensory system (including odorant and gustatory receptors), are selecting microhabitats within individual cactus necroses. Preliminary evidence shows that some *D. mojavensis* odorant receptors have been under positive selection (Matzkin unpub.), and thus might be candidates for involvement in the location of host-specific microhabitats. Further functional and behavioral studies are needed to fully understand the consequence and role of these receptors in cactus host adaptation.

12.5 Conclusion

The technological and computational advances of recent years have not only revolutionized the study of model laboratory organisms but have dramatically expanded our choice of organisms. These new methods have allowed for the investigation of ecologically defined species, those species in which ecological information is known and genomic information could be analyzed in an explicit ecological context.

In *D. mojavensis* genomic and transcriptomic tools have allowed us to peer into the genomic mechanisms of the adaptive process. We have seen how the transcriptome has been shaped by the various host shifts that have occurred in the history of *D. mojavensis*. These changes include the modulation and fixed expression pattern of a wide variety of genes, some of which we would have expected to be involved in the host adaptation process, such as detoxification, chemosensory and metabolic genes. The few analyses of candidate genes have shown how selection has shaped the pattern of genetic and functional variation, and its possible link to performance in the field. More studies are necessary to make the connection between the genetic, functional and life history variation in the ecological context of this fly. Furthermore, genome-wide surveys of sequence and structural variation will help us elucidate large-scale changes in the *D. mojavensis* populations and determine for example the presence of genomic islands of divergence between them (Turner et al. 2005, 2008). Several of these questions will begin to be answered using the previously sequenced Catalina Island genome (Drosophila 12 Genomes Consortium 2007) and the recently sequenced Baja California, Mojave and Sonora *D. mojavensis* genomes (Matzkin unpub.). Furthermore, meta-genomic analysis of the microflora of both the cactus necroses and the *D. mojavensis* gut, could shed light into the role of these interspecific interactions and host adaptation. Finally, a future aim is to examine links between local adaptation occurring in the four cactus host races, with their behavioral and genetic divergence and the ongoing pattern of incipient speciation (Nosil 2012).

Acknowledgments I would like to thank all the past and present investigators that have worked with cactophilic *Drosophila*, and whose work has made these ecologically interesting flies a powerful ecological genomics system of study. I thank Jeremy Bono for comments on the manuscript. Funding from NSF (DEB-1219387) and Science Foundation Arizona (in collaboration with T.A. Markow) has made much of my work possible.

References

Adams M, Celniker S, Holt R et al (2000) The genome sequence of *Drosophila melanogaster*. Science 287(5461):2185–2195

Agrawal AA, Fishbein M, Halitschke R et al (2009) Evidence for adaptive radiation from a phylogenetic study of plant defenses. Proc Natl Acad Sci USA 106(43):18067–18072

Arabidopsis Genome Initiative (2000) Analysis of the genome sequence of the flowering plant *Arabidopsis thaliana*. Nature 408(6814):796–815

Atkinson PW, Mills LE, Starmer WT, Sullivan DT (1988) Structure and evolution of the Adh genes of *Drosophila mojavensis*. Genetics 120(3):713–723

Azevedo RBR, James AC, McCabe J, Partridge L (1998) Latitudinal variation of wing: thorax size ratio and wing-aspect ratio in *Drosophila melanogaster*. Evolution 52(5):1353–1362

Batterham P, Lovett JA, Starmer WT, Sullivan DT (1983) Differential regulation of duplicate *alcohol dehydrogenase* genes in *Drosophila mojavensis*. Develop Biol 96:346–354

Benton R, Sachse S, Michnick SW, Vosshall LB (2006) Atypical membrane topology and heteromeric function of *Drosophila* odorant receptors in vivo. PLoS Biol 4(2):240–257

Benyajati C, Spoerel N, Haymerle H, Ashburner M (1983) The messenger RNA for alcohol dehydrogenase in *Drosophila melanogaster* differs in its 5′ end in different developmental stages. Cell 33:125–133

Berenbaum MR (2002) Postgenomic chemical ecology: from genetic code to ecological interactions. J Chem Ecol 28(5):873–896

Blattner FR, Plunkett G, Bloch CA et al (1997) The complete genome sequence of *Escherichia coli* k-12. Science 277(5331):1453–1462

Bono JM, Markow TA (2009) Post-zygotic isolation in cactophilic *Drosophila*: larval viability and adult life-history traits of *D. mojavensis*/*D. arizonae* hybrids. J Evol Biol 22(7):1387–1395

Bono JM, Matzkin LM, Kelleher ES, Markow TA (2011) Postmating transcriptional changes in reproductive tracts of con- and heterospecifically mated *Drosophila mojavensis* females. Proc Natl Acad Sci USA 108(19):7878–7883

Breitmeyer CM, Markow TA (1998) Resource availability and population size in cactophilic *Drosophila*. Funct Ecol 12(1):14–21

Brncic D (1983) Ecology of flower-breeding *Drosophila*. In: Ashburner M, Carson HL, Thompson JN (eds) The genetics and biology of *Drosophila*, vol 3d. Academic, New York, pp 333–382

C. elegans Sequencing Consortium (1998) Genome sequence of the nematode *C. elegans*: a platform for investigating biology. Science 282(5396): 2012–2018

Caillaud MC, Via S (2012) Quantitative genetics of feeding behavior in two ecological races of the pea aphid, *Acyrthosiphon pisum*. Heredity 108(3):211–218

Carson HL (1974) Three flies and three islands: parallel evolution in Drosophila. Proc Natl Acad Sci USA 71(9):3517–3521

Carson HL, Wasserman M (1965) A widespread chromosomal polymorphism in a widespread species, *Drosophila buzzatii*. Am Nat 99(905):111–116

Chambers GK (1988) The Drosophila alcohol dehydrogenase gene enzyme system. Adv Genet 25: 39–107

Chesser R, Baker R (1996) Effective sizes and dynamics of uniparentally and diparentally inherited genes. Genetics 144(3):1225–1235

Colbourne JK, Pfrender ME, Gilbert D et al (2011) The ecoresponsive genome of *Daphnia pulex*. Science 331(6017):555–561

Dambroski HR, Feder JL (2007) Host plant and latitude-related diapause variation in *Rhagoletis pomonella*: A test for multifaceted life history adaptation on different stages of diapause development. J Evol Biol 20(6):2101–2112

Drosophila 12 Genomes Consortium (2007) Evolution of genes and genomes on the Drosophila phylogeny. Nature 450(7167):203–218

Eanes WF (1999) Analysis of selection on enzyme polymorphisms. Annu Rev Ecol Syst 30:301–326

Etges WJ (1989) Evolution of developmental homeostasis in *Drosophila mojavensis*. Evol Ecol 3:189–201

Etges WJ (1990) Direction of life history evolution in *Drosophila mojavensis*. In: Barker JSF, Starmer WT, MacIntyre RJ (eds) Ecological and evolutionary genetics of Drosophila. Plenum Press, New York, pp 37–56

Etges WJ (1993) Genetics of host-cactus response and life-history evolution among ancestral and derived populations of cactophilic *Drosophila mojavensis*. Evolution 47:750–767

Etges WJ (1998) Premating isolation is determined by larval rearing substrates in cactophilic *Drosophila mojavensis*. IV. Correlated responses in behavioral isolation to artificial selection on a life-history trait. Am Nat 152(1):129–144

Etges WJ, Heed WB (1987) Sensitivity to larval density in populations of *Drosophila mojavensis*: influences of host plant variation on components fitness. Oecologia 71:375–381

Etges WJ, Jackson LL (2001) Epicuticular hydrocarbon variation in *Drosophila mojavensis* cluster species. J Chem Ecol 27(10):2125–2149

Etges WJ, Over KF, De Oliveira CC, Ritchie MG (2006) Inheritance of courtship song variation among geographically isolated populations of *Drosophila mojavensis*. Anim Behav 71:1205–1214

Etges WJ, de Oliveira CC, Noor MAF, Ritchie MG (2010) Genetics of incipient speciation in *Drosophila mojavensis*. III. Life-history divergence in allopatry and reproductive isolation. Evolution 64(12):3549–3569

Feder ME, Mitchell-Olds T (2003) Evolutionary and ecological functional genomics. Nat Rev Genet 4:649–655

Fellows DF, Heed WB (1972) Factors affecting host plant selection in desert-adapted cactiphilic *Drosophila*. Ecology 53:850–858

Feyereisen R (2005) Insect cytochrome P450. In: Gilbert LI, Iatrou K, Gill SS (eds) Comprehensive molecular insect science, vol 4. Elsevier, Amsterdam, pp 1–77

Flowers JM, Sezgin E, Kumagai S et al (2007) Adaptive evolution of metabolic pathways in *Drosophila*. Mol Biol Evol 24(6):1347–1354

Fogleman JC (1982) The role of volatiles in the ecology of cactophilic *Drosophila*. In: Barker JSF, Starmer WT (eds) Ecological genetics and evolution: the cactus-yeast-Drosophila model system. Academic, New York, pp 191–208

Fogleman JC, Abril JR (1990) Ecological and evolutionary importance of host plant chemistry. In: Barker JSF, Starmer WT, MacIntyre RJ (eds) Ecological and evolutionary genetics of *Drosophila*. Plenum Press, New York, pp 121–143

Fogleman JC, Starmer WT (1985) Analysis of the community structure of yeasts associated with the decaying stems of cactus. III. *Stenocereus thurberi*. Microb Ecol 11(2):165–173

Fogleman JC, Starmer WT, Heed WB (1981) Larval selectivity for yeast species by *Drosophila mojavensis* in natural substrates. Proc Natl Acad Sci USA 78(7):4435–4439

Fontdevila A, Pla C, Hasson E et al (1988) *Drosophila koepferae*: A new member of the *Drosophila serido*

(Diptera: Drosophilidae) superspecies taxon. Ann Entomol Soc Am 81(3):380–385

Force A, Lynch M, Pickett FB et al (1999) Preservation of duplicate genes by complementary, degenerative mutations. Genetics 151:1531–1545

Freriksen A, Seykens D, Scharloo W, Heinstra PWH (1991) Alcohol dehydrogenase controls the flux from ethanol into lipids in *Drosophila* larvae: A C-13 NMR-study. J Biol Chem 266(32):21399–21403

Freriksen A, Seykens D, Heinstra PWH (1994) Differences between larval and adult *Drosophila* in metabolic degradation of ethanol. Evolution 48(2): 504–508

Gibbs AG, Matzkin LM (2001) Evolution of water balance in the genus *Drosophila*. J Exp Biol 204(13):2331–2338

Gibbs AG, Fukuzato F, Matzkin LM (2003) Evolution of water conservation mechanisms in *Drosophila*. J Exp Biol 206(7):1183–1192

Goffeau A, Barrell BG, Bussey H et al (1996) Life with 6000 genes. Science 274(5287):546

Hallem EA, Carlson JR (2006) Coding of odors by a receptor repertoire. Cell 125(1):143–160

Hasson E, Naveira H, Fontdevila A (1992) The breeding sites of Argentinean cactophilic species of the *Drosophila mulleri* complex (subgenus *Drosophila repleta* group). Rev Chil Hist Nat 65(3):319–326

Havens JA, Etges WJ (2013) Premating isolation is determined by larval rearing substrates in cactophilic *Drosophila mojavensis*. IX. Host plant and population specific epicuticular hydrocarbon expression influences mate choice and sexual selection. J Evol Biol 26(3):562–576

Heed WB (1977) New cactus-feeding but soil-breeding species of *Drosophila* (Diptera: Drosophilidae). P Entomol Soc Wash 79(4):649–654

Heed WB (1978) Ecology and genetics of Sonoran desert *Drosophila*. In: Brussard PF (ed) Ecological genetics: the interface. Springer, New York, pp 109–126

Heed WB (1982) The origin of Drosophila in the sonoran desert. In: Barker JSF, Starmer WT (eds) Ecological genetics and evolution: the cactus-yeast-Drosophila model system. Academic, New York, pp 65–80

Heed WB, Kircher HW (1965) Unique sterol in ecology and nutrition of *Drosophila pachea*. Science 149(3685):758–761

Hoffmann AA, Hallas R, Sinclair C, Mitrovski P (2001) Levels of variation in stress resistance in *Drosophila* among strains, local populations, and geographic regions: patterns for desiccation, starvation, cold resistance, and associated traits. Evolution 55(8):1621–1630

Holtzman S, Miller D, Eisman RC et al (2010) Transgenic tools for members of the genus *Drosophila* with sequenced genomes. Fly 4(4):349–362

Hughes AL (1994) The evolution of functionally novel proteins after gene duplication. Proc R Soc B 256(1346):119–124

Jaenike J (1978) Resource predictability and niche breadth in *Drosophila quinaria* species group. Evolution 32(3):676–678

Jaenike J, Grimaldi DA, Sluder AE, Greenleaf AL (1983) A-amanitin tolerance in mycophagous *Drosophila*. Science 221(4606):165–167

Janz N (2011) Ehrlich and Raven revisited: mechanisms underlying codiversification of plants and enemies. Annu Rev Ecol Evol Syst 42:71–89

Johnson WR (1980) Chromosomal polymorphism in natural populations of the desert adapted species, *Drosophila mojavensis*. Ph.D. Thesis, University of Arizona, Tucson, AZ

Jones FC, Grabherr MG, Chan YF et al (2012) The genomic basis of adaptive evolution in threespine sticklebacks. Nature 484(7392):55–61

Kacser H, Burns JA (1973) The control of flux. Symp Soc Exp Biol 27:65–104

Kaneshiro KY, Carson HL, Clayton FE, Heed WB (1973) Niche separation in a pair of homosequential *Drosophila* species from island of Hawaii. Am Nat 107(958):766–774

Kircher HW (1982) Chemical composition of cacti and its relationship to Sonoran desert Drosophila. In: Barker JSF, Starmer WT (eds) Ecological genetics and evolution: the cactus-yeast-Drosophila model system. Academic, New York, pp 143–158

Kircher HW, Heed WB, Russell JS, Grove J (1967) Senita cactus alkaloids: their significance to Sonoran desert *Drosophila* ecology. J Insect Physiol 13(12):1869

Krebs RA, Thompson KA (2005) A genetic analysis of variation for the ability to fly after exposure to thermal stress in *Drosophila mojavensis*. J Therm Biol 30(4):335–342

Laissue PP, Vosshall LB (2008) The olfactory sensory map in Drosophila. Adv Exp Med Biol 628:102–114

Lang M, Murat S, Clark AG et al (2012) Mutations in the neverland gene turned *Drosophila pachea* into an obligate specialist species. Science 337(6102): 1658–1661

LaPorte DC, Walsh K, Koshland DE Jr (1984) The branch point effect: ultrasensitivity and subsensitivity to metabolic control. J Biol Chem 259:14068–14075

Legal L, Chappe B, Jallon JM (1994) Molecular basis of *Morinda citrifolia* (L.): toxicity on *Drosophila*. J Chem Ecol 20(8):1931–1943

Li XC, Schuler MA, Berenbaum MR (2007) Molecular mechanisms of metabolic resistance to synthetic and natural xenobiotics. Annu Rev Entomol 52:231–253

Linn C, Feder JL, Nojima S et al (2003) Fruit odor discrimination and sympatric host race formation in *Rhagoletis*. Proc Natl Acad Sci USA 100(20):11490–11493

Lofdahl KL (1985) A quantitative genetic analysis of habitat selection behavior in the cactus-breeding species *Drosophila mojavensis*. Ph.D. Thesis, University of Chicago, Chicago

Lofdahl KL (1986) A genetical analysis of habitat selection in the cactophilic species *Drosophila mojavensis*. In: Huettel MD (ed) Evolutionary genetics of invertebrate behavior: progress and prospects. Plenum, New York, pp 153–162

Luque T, O'Reilly DR (2002) Functional and phylogenetic analyses of a putative *Drosophila melanogaster*

UDP-glycosyltransferase gene. Insect Biochem Mol Biol 32(12):1597–1604
Lynch M, Force A (2000) The probability of duplicate gene preservation by subfunctionalization. Genetics 154(1):459–473
Machado CA, Matzkin LM, Reed LK, Markow TA (2007) Multilocus nuclear sequences reveal intra- and interspecific relationships among chromosomally polymorphic species of cactophilic *Drosophila*. Mol Ecol 16(14):3009–3024
Mangan RL (1978) Competitive interactions and host plant specific *Drosophila* species. University of Arizona, Tucson
Mangan RL (1982) Adaptations to competition in cactus breeding *Drosophila*. In: Barker JSF, Starmer WT (eds) Ecological genetics and evolution: the cactus-yeast-Drosophila model system. Academic, New York, pp 257–272
Markow TA, Castrezana S (2000) Dispersal in cactophilic *Drosophila*. Oikos 89(2):378–386
Markow TA, O'Grady PM (2006) Drosophila: a guide to species identification and use. Academic, London
Markow TA, O'Grady PM (2008) Reproductive ecology of *Drosophila*. Funct Ecol 22(5):747–759
Markow TA, Ricker JP (1992) Male size, developmental stability, and mating success in natural populations of 3 *Drosophila* species. Heredity 69:122–127
Markow TA, Toolson EC (1990) Temperature effects on epicuticular hydrocarbons and sexual isolation in *Drosophila mojavensis*. In: Barker JSF, Starmer WT, MacIntyre RJ (eds) Ecological and evolutionary genetics of *Drosophila*. Plenum Press, New York, pp 315–331
Markow TA, Fogleman JC, Heed WB (1983) Reproductive isolation in Sonoran desert *Drosophila*. Evolution 37(3):649–652
Matzkin LM (2004) Population genetics and geographic variation of alcohol dehydrogenase (*Adh*) paralogs and glucose-6-phosphate dehydrogenase (*G6pd*) in *Drosophila mojavensis*. Mol Biol Evol 21(2):276–285
Matzkin LM (2005) Activity variation in alcohol dehydrogenase paralogs is associated with adaptation to cactus host use in cactophilic *Drosophila*. Mol Ecol 14:2223–2231
Matzkin LM (2008) The molecular basis of host adaptation in cactophilic Drosophila: Molecular evolution of a glutathione s-transferase gene (*GstD1*) in *Drosophila mojavensis*. Genetics 178(2):1073–1083
Matzkin LM (2012) Population transcriptomics of cactus host shifts in *Drosophila mojavensis*. Mol Ecol 21(10):2428–2439
Matzkin LM, Eanes WF (2003) Sequence variation of alcohol dehydrogenase (*Adh*) paralogs in cactophilic *Drosophila*. Genetics 163:181–194
Matzkin LM, Markow TA (2009) Transcriptional regulation of metabolism associated with the increased desiccation resistance of the cactophilic *Drosophila mojavensis*. Genetics 182:1279–1288
Matzkin LM, Markow TA (2013) Transcriptional differentiation across the four cactus host races of *Drosophila mojavensis*. In: Michalak P (ed) Speciation: natural processes, genetics and biodiversity. Nova Science Publishers, Hauppauge
Matzkin LM, Watts TD, Bitler BG et al (2006) Functional genomics of cactus host shifts in *Drosophila mojavensis*. Mol Ecol 15:4635–4643
Matzkin LM, Watts TD, Markow TA (2007) Desiccation resistance in four *Drosophila* species: sex and population effects. Fly 1(5):268–273
Matzkin LM, Watts TD, Markow TA (2009) Evolution of stress resistance in *Drosophila*: Interspecific variation in tolerance to desiccation and starvation. Funct Ecol 23:521–527
Mettler LE (1963) *Drosophila mojavensis baja*, a new form in the mulleri complex. Drosophila Inf Serv 38:57–58
Meyer BN, Mohamed YAH, McLaughlin JL (1980) Beta-phenethylamines from the cactus genus *Opuntia*. Phytochemistry 19(4):719–720
Michel AP, Sim S, Powell THQ et al (2010) Widespread genomic divergence during sympatric speciation. Proc Natl Acad Sci USA 107(21):9724–9729
Middleton RJ, Kacser H (1983) Enzyme variation, metabolic flux and fitness: alcohol dehydrogenase in *Drosophila melanogaster*. Genetics 105(3):633–650
Newby BD, Etges WJ (1998) Host preference among populations of *Drosophila mojavensis* (Diptera: Drosophilidae) that use different host cacti. J Insect Behav 11(5):691–712
Nosil P (2012) Ecological speciation. Oxford series in ecology and evolution. Oxford University Press, Oxford
Oliveira DCSG, Almeida FC, O'Grady PM et al (2012) Monophyly, divergence times, and evolution of host plant use inferred from a revised phylogeny of the *Drosophila repleta* species group. Mol Phylogenet Evol 64(3):533–544
Patterson JT, Crow JF (1940) Hybrydization in the *mulleri* group of *Drosophila*. The University of Texas Publication 4032:251–256
Pfeiler E, Castrezana S, Reed LK (2009) Genetic, ecological and morphological differences among populations of the cactophilic *Drosophila mojavensis* from southwestern USA and northwestern Mexico, with descriptions of two new subspecies. J Nat Hist 43: 923–938
Rajpurohit S, Oliveira CC, Etges WJ, Gibbs AG (2013) Functional genomic and phenotypic responses to desiccation in natural populations of a desert drosophilid. Mol Ecol 22:2698–2715
Ranson H, Hemingway J (2005) Glutathione transferases. In: Gilbert LI, Iatrou K, Gill SS (eds) Comprehensive molecular insect science, vol 5. Elsevier, Amsterdam, pp 383–402
Ranson H, Rossiter L, Ortelli F et al (2001) Identification of a novel class of insect glutathione s-transferases involved in resistance to DDT in the malaria vector *Anopheles gambiae*. Biochem J 359:295–304
Rausher MD (2013) The evolution of genes in branched metabolic pathways. Evolution 67(1):34–48

Reed LK, Nyboer M, Markow TA (2007) Evolutionary relationships of *Drosophila mojavensis* geographic host races and their sister species *Drosophila arizonae*. Mol Ecol 16(5):1007–1022

Richards S, Liu Y, Bettencourt BR et al (2005) Comparative genome sequencing of *Drosophila pseudoobscura*: chromosomal, gene, and cis-element evolution. Genome Res 15(1):1–18

Richards S, Gibbs RA, Gerardo NM et al (2010) Genome sequence of the pea aphid *Acyrthosiphon pisum*. PLoS Biol 8(2)

Richmond MP, Johnson S, Markow TA (2012) Evolution of reproductive morphology among recently diverged taxa in the *Drosophila mojavensis* species cluster. Ecol Evol 2(2):397–408

Ruiz A, Heed WB, Wasserman M (1990) Evolution of the mojavensis cluster of cactophilic *Drosophila* with descriptions of two new species. J Hered 81:30–42

Russo CAM, Takezaki N, Nei M (1995) Molecular phylogeny and divergence times of Drosophilid species. Mol Biol Evol 12:391–404

Savakis C, Ashburner M, Willis JH (1986) The expression of the gene coding for alcohol dehydrogenase during the development of *Drosophila melanogaster*. Dev Biol 114(1):194–207

Schwarz D, Robertson HM, Feder JL et al (2009) Sympatric ecological speciation meets pyrosequencing: sampling the transcriptome of the apple maggot *Rhagoletis pomonella*. BMC Genomics 10

Smadja CM, Canback B, Vitalis R et al (2012) Large-scale candidate gene scan reveals the role of chemoreceptor genes in host plant specialization and speciation in the pea aphid. Evolution 66(9):2723–2738

Smith G, Lohse K, Etges WJ, Ritchie MG (2012) Model-based comparisons of phylogeographic scenarios resolve the intraspecific divergence of cactophilic *Drosophila mojavensis*. Mol Ecol 21(13):3293–3307

Smith G, Fang Y, Liu X et al (2013) Transcriptome-wide expression variation associated with environmental plasticity and mating success in cactophilic *Drosophila mojavensis*. Evolution 67:1950–1963

Song X, Goicoechea JL, Ammiraju JSS et al (2011) The 19 genomes of Drosophila: a bac library resource for genus-wide and genome-scale comparative evolutionary research. Genetics 187(4):1023–1030

Spencer WP (1941) Ecological factors and Drosophila speciation. Ohio J Sci 41:190–200

Starmer WT (1982a) Analysis of the community structure of yeasts associated with the decaying stems of cactus. I. *Stenocereus gummosus*. Microb Ecol 8(1):71–81

Starmer WT (1982b) Associations and interactions among yeasts, Drosophila and their habitats. In: Barker JSF, Starmer WT (eds) Ecological genetics and evolution: the cactus-yeast-Drosophila model system. Academic, New York, pp 159–174

Starmer WT, Phaff HJ (1983) Analysis of the community structure of yeasts associated with the decaying stems of cactus. II. *Opuntia* species. Microb Ecol 9(3):247–259

Starmer WT, Heed WB, Rockwood-Sluss ES (1977) Extension of longevity in *Drosophila mojavensis* by environmental ethanol: differences between subraces. Proc Natl Acad Sci USA 74:387–391

Starmer WT, Barker JSF, Phaff HJ, Fogleman JC (1986) Adaptations of *Drosophila* and yeasts: their interactions with the volatile 2-propanol in the cactus microorganism *Drosophila* model system. Aust J Biol Sci 39(1):69–77

Starmer WT, Lachance MA, Phaff HJ, Heed WB (1990) The biogeography of yeasts associated with decaying cactus tissue in North America, the Caribbean, and northern Venezuela. In: Hecht MK, Wallace B, Macintyre RJ (eds) Evolutionary biology, vol 24. Plenum, New York, pp 253–296

Stocker RF (1994) The organization of the chemosensory system in *Drosophila melanogaster*: a review. Cell Tissue Res 275(1):3–26

Sturtevant AH (1921) The North American species of *Drosophila*, vol 301. Carnegie Institution of Washington, Washington, DC

Throckmorton LH (1975) The phylogeny, ecology, and geography of *Drosophila*. In: King RC (ed) Handbook of genetics, vol 3. Plenum, New York, pp 421–469

Turner TL, Hahn MW, Nuzhdin SV (2005) Genomic islands of speciation in *Anopheles gambiae*. PLoS Biol 3(9):1572–1578

Turner TL, Levine MT, Eckert ML, Begun DJ (2008) Genomic analysis of adaptive differentiation in *Drosophila melanogaster*. Genetics 179(1):455–473

Vacek DC (1979) The microbial ecology of the host plants of *Drosophila mojavensis*. Ph.D. Thesis, University of Arizona, Tucson, AZ

Via S (1999) Reproductive isolation between sympatric races of pea aphids. I. Gene flow restriction and habitat choice. Evolution 53(5):1446–1457

Vosshall LB, Stocker RE (2007) Molecular architecture of smell and taste in *Drosophila*. Annu Rev Neurosci 30:505–533

Waddington CH (1953) Genetic assimilation of an acquired character. Evolution 7(2):118–126

Wasserman M (1962) Cytological studies of the *repleta* group of the genus *Drosophila*. V. The *mulleri* subgroup. University of Texas Publication 6205: 85–117

Wasserman M (1982) Evolution in the repleta group. In: Ashburner M, Carson HL, Thompson JN (eds) The genetics and biology of *Drosophila*, vol 3b. Academic, London, pp 61–139

West-Eberhard MJ (2003) Developmental plasticity and evolution. Oxford University Press, New York

The Genomics of an Adaptive Radiation: Insights Across the *Heliconius* Speciation Continuum

13

Megan Supple, Riccardo Papa, Brian Counterman, and W. Owen McMillan

Abstract

Fueled by new technologies that allow rapid and inexpensive assessment of fine scale individual genomic variation, researchers are making transformational discoveries at the interface between genomes and biological complexity. Here we review genomic research in *Heliconius* butterflies – a radiation characterized by extraordinary phenotypic diversity in warningly colored wing patterns and composed of a continuum of taxa across the stages of speciation. These characteristics, coupled with a 50-year legacy of ecological and behavioral research, offer exceptional prospects for genomic studies into the nature of adaptive differences and the formation of new species. Research in *Heliconius* provides clear connections between genotype, phenotype, and fitness of wing color patterns shown to underlie adaptation and speciation. This research is challenging our perceptions about how speciation occurs in the presence of gene flow and the role of hybridization in generating adaptive novelty. With the release of the first *Heliconius* genome assembly, emerging genomic studies are painting a dynamic picture of the evolving species boundary. As the field of speciation genomics moves beyond describing patterns, towards a more integrated understanding of the process of speciation, groups such as *Heliconius*, where there is a clear speciation continuum and the traits underlying adaptation and speciation are known, will provide a roadmap for identifying variation crucial in the origins of biodiversity.

M. Supple
Smithsonian Tropical Research Institute, Panama City, Panama

Biomathematics Program, North Carolina State University, Raleigh, NC 27695, USA

W.O. McMillan (✉)
Smithsonian Tropical Research Institute, Panama City, Panama
e-mail: McMillanO@si.edu

R. Papa
Department of Biology and Center for Applied Tropical Ecology and Conservation, University of Puerto Rico, Rio Piedras, San Juan 00921, Puerto Rico

B. Counterman
Department of Biological Sciences, Mississippi State University, Mississippi State, MS 39762, USA

C.R. Landry and N. Aubin-Horth (eds.), *Ecological Genomics: Ecology and the Evolution of Genes and Genomes*, Advances in Experimental Medicine and Biology 781,
DOI 10.1007/978-94-007-7347-9_13,

Keywords
Adaptation • Association mapping • Genomic divergence • Hybridization • Introgression • Phenotypic evolution • Speciation

13.1 Introduction

Over 150 years ago, Henry Walter Bates published his first observations of butterfly diversity in the New World Tropics (Bates 1862). Exploring deep within the Amazon basin for over a decade, Bates documented extraordinary cases of mimicry in the vivid wing patterns of distantly related butterfly species. His writings provided Darwin with some of the most visually stunning examples of evolution by natural selection and the best evidence of a link between natural selection and speciation. Thanks to Bates and Fritz Müller, who arrived in Brazil a few years after Bates, butterflies, arguably more than any other group, contributed to the early establishment and acceptance of evolutionary theory (Carroll et al. 2009). Research on butterflies continues to be as relevant today as it was 150 years ago. Using modern technologies, there is an active research community encompassing most areas of ecology and evolution, ranging from the molecular details of vision to the analysis of human impact on biodiversity (Boggs et al. 2003; Kotiaho et al. 2005; Briscoe et al. 2010). This includes a vibrant genomics research community working on a number of different species, including passion-vine butterflies (*Heliconius* spp., Heliconius Genome Consortium 2012), monarchs (*Danaus plexippus*, Zhan et al. 2011), swallowtails (*Papilio* spp., O'Neil et al. 2010), the Glanville fritillary (*Melitae acinxia*, Hanski 2011), *Bicyclus anynana* (Brakefield et al. 2009), and *Lycaeides* (Gompert et al. 2012). The past several years have seen remarkable progress in the development of genomic resources in these species, culminating in the publication of the first two butterfly reference genomes (Zhan et al. 2011; Heliconius Genome Consortium 2012), with a number of additional genomes scheduled to be released in the coming year.

In this review, we examine emerging ecological and evolutionary genomic research addressing adaptation and speciation in *Heliconius* butterflies. Genomic research is ongoing in several butterfly species, but genomic studies on *Heliconius* are arguably further developed. Research on *Heliconius* provides a foundation to discuss larger issues relating to the nature of adaptive differences and the formation of new species. We begin with an overview of the *Heliconius* radiation – a radiation that has produced an extraordinary evolutionary continuum composed of divergent races and species at different stages of speciation (Mallet et al. 1998; Mallet 2008; Merrill et al. 2011a). Using this continuum as a backdrop, we review recent progress to (i) identify functional variation in the group and reconstruct the history of adaptive alleles, (ii) understand the importance of hybridization in speciation, and (iii) explore the genomic architecture that allows speciation to proceed in the face of gene flow. We conclude with a discussion of how to move beyond patterns of genomic variation to gain a deeper understanding of the processes that drive divergence and speciation in nature.

13.2 The *Heliconius* Radiation: A Primer

The butterfly subtribe Heliconiina (Lepidoptera: Nymphalidae: Heliconiinae) is restricted to the New World tropics and has a host of life history, ecological, and phenotypic traits that have long fascinated biologists and naturalists. Heliconiines get their common name, passion-vine butterfly, from their strong association with the passion flower family (Passifloraceae). Passion vines are protected by a diverse arsenal of cyanogenic compounds that are likely a by-product of an evolutionary arms race with heliconiines (Spencer 1988). Heliconiines have adopted this defensive tactic–evolving the ability to make and, in some cases sequester, cyanogenic glycosides (Engler et al. 2000; Engler-Chaouat and Gilbert 2007). These compounds render the bearer highly

distasteful and avian predators quickly learn to associate a wing color pattern with unpalatability (Chai 1986).

Within the subtribe, the genus *Heliconius* is characterized by an ecological shift to pollen feeding. Unlike most butterflies, which feed only on fluids (e.g. nectar, decomposing animals and fruits, and dung), *Heliconius* actively collect and process pollen (Gilbert 1972). The origin of pollen feeding is associated with subtle changes in morphology (Gilbert 1972; Krenn and Penz 1998; Eberhard et al. 2009), coupled with more profound changes in a host of life history and behavioral traits. In particular, the transition to pollen feeding is hypothesized to be important in the butterflies' ability to synthesize toxic compounds and to enable a very long adult life – one of the longest recorded for a butterfly (Gilbert 1972). Pollen feeding is also associated with a rapid increase in brain size (Sivinski 1989), the evolution of slower and more maneuverable flight, the development of large eyes with accentuated ultraviolet color vision (Briscoe et al. 2010), and the evolution of a suite of complex behaviors, including trap-line feeding, gregarious roosting, and elaborate mating strategies (Brown 1981).

The *Heliconius* genus is best known for the extraordinary mimicry-related wing pattern diversity seen among its 43 species (Fig. 13.1). With over 400 distinct color pattern varieties, the group represents one of the most striking adaptive radiations in the animal kingdom. Repeated convergent and divergent evolution creates a colorful tapestry where distantly related species often look identical and closely related species or races can look strikingly different (Fig. 13.1). The complexity of this tapestry is exemplified by the parallel radiations of *H. erato* and *H. melpomene*, which, although phylogenetically distant and unable to hybridize, have converged on 25 different mimetic color patterns across the Neotropics (Fig. 13.2). Most of the diversity in these two species can be partitioned into two major phenotypic groups: (i) "postman" phenotypes, which have red on their forewing and either possess or lack a yellow hindwing bar; and (ii) "rayed" phenotypes, which have a yellow forewing band, a red patch on the proximal area of the forewing, and red rays on the hindwing. Variations on these themes generate the abundant pattern diversity we see in nature (Fig. 13.2b). In addition to the postman and rayed phenotypes, there are also a number of tiger striped *Heliconius*. For example, *H. numata* shows numerous sympatric color patterns races that are likely a result of strong selection pressure to mimic distantly related *Melinaea* species (Nymphalidae: Ithomiinae) (Brown and Benson 1974; Joron et al. 1999), which can vary dramatically in abundance over small spatial and temporal scales. Geographic variation and convergent evolution are common across *Heliconius* and the wing patterns of most species converge onto a handful of common color patterns, so called mimicry rings, which coexist locally (Mallet and Gilbert 1995). This convergence between species led to the original hypothesis of mimicry (Bates 1862) and *Heliconius* is now a classic example of Müllerian mimicry, where distantly related, but similarly distasteful, species converge on the same warningly colored pattern.

Divergence in wing color pattern is also associated with speciation due to the dual role of color pattern in signaling to predators and in mate selection. For example, *H. melpomene* and *H. cydno* are very closely related, broadly sympatric, and occasionally hybridize in nature. In this case, speciation is associated with, and reinforced by, divergence in mimetic color patterns. *Heliconius melpomene* is generally black with red and yellow markings and mimics *H. erato* (Flanagan et al. 2004) (Fig. 13.2); whereas, *H. cydno* is black with white or yellow markings and typically mimics *H. sapho* and *H. eleuchia* (see Fig. 13.1). Mate recognition involves visual attraction of males to females, which leads to strong color-pattern based assortative mating and disruptive sexual selection against hybrids (Jiggins et al. 2001b; Naisbit et al. 2001). Furthermore, there is ecological post-mating isolation that results from increased predation on hybrids due to their non-mimetic wing patterns (Merrill et al. 2012). Species boundaries are often associated with mimetic color pattern shifts, highlighting the pervasive role that color pattern evolution plays in reproductive isolation across the radiation (Mallet et al. 1998).

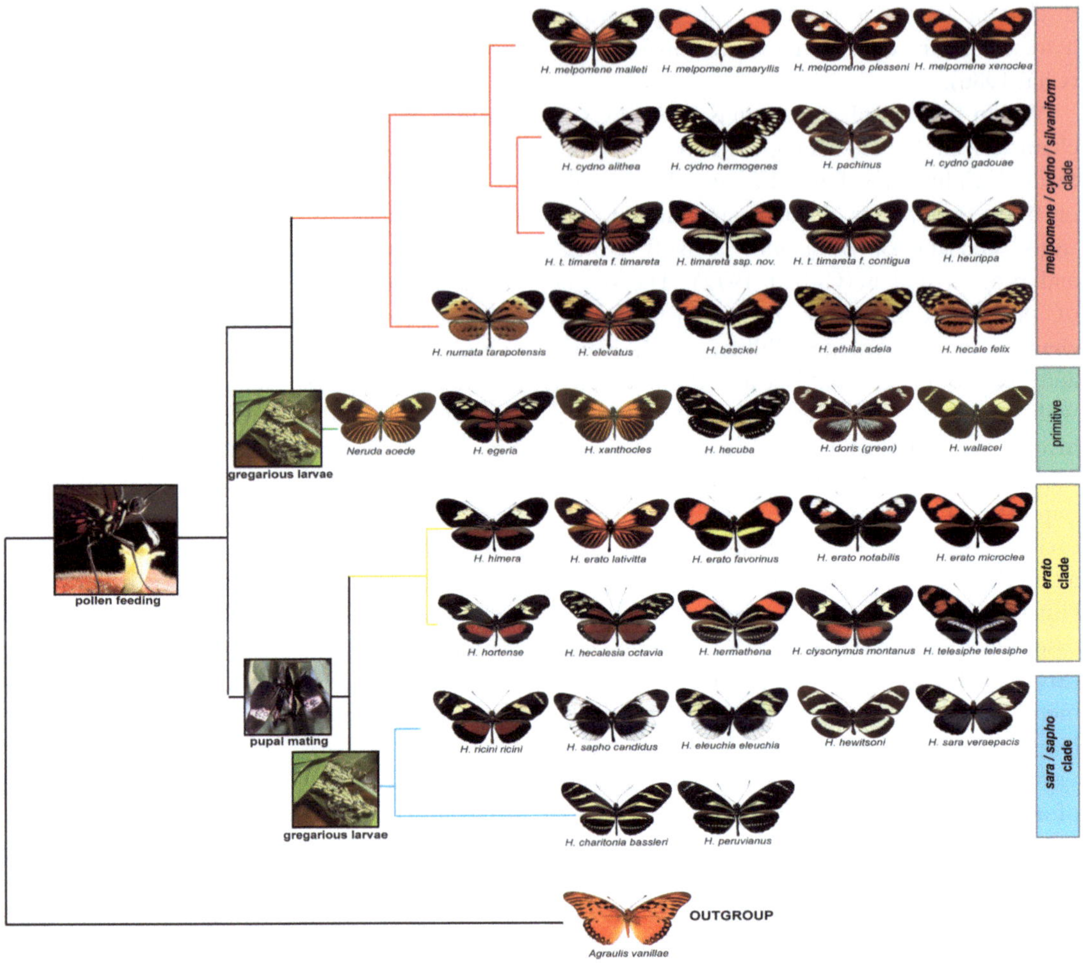

Fig. 13.1 **Nature's palette – the *Heliconius* radiation.** This representation of the major *Heliconius* clades incorporates over half of the 43 different *Heliconius* species and a large portion of wing pattern variation. Note the reuse of phenotypes across the phylogeny and the extreme phenotypic variation that exists within and between closely related species. The origins of several major ecological and behavioral innovations are indicated, including pollen feeding, pupal mating, and gregarious larvae. Pupal mating is the reproductive behavior where males search and guard pupae and mate with females as they eclose (Deinert et al. 1994). Pollen feeding and pupal mating likely evolved only once in the radiation. In contrast, gregarious aggregations of larvae evolved several times including the lineage containing the primitive *Heliconius*. Branch length does not necessarily reflect divergence time (Beltrán et al. 2007)

13.3 The Genetics of Color Pattern in *Heliconius*

Given the visually dynamic nature and clear importance of coloration in the *Heliconius* radiation, much effort has focused on characterizing the genetic basis of wing pattern variation in the group. The spectrum of pattern diversity in *Heliconius* is largely controlled by variation at a handful of loci of large phenotypic effect (Sheppard et al. 1985; Joron et al. 2006, 2011; Reed et al. 2011; Martin et al. 2012; Papa et al. 2013). These loci are phylogenetically conserved "hotspots" of phenotypic evolution, responsible for color pattern evolution in both convergent and divergent *Heliconius* species. Allelic variation at these loci produce complex phenotypic changes

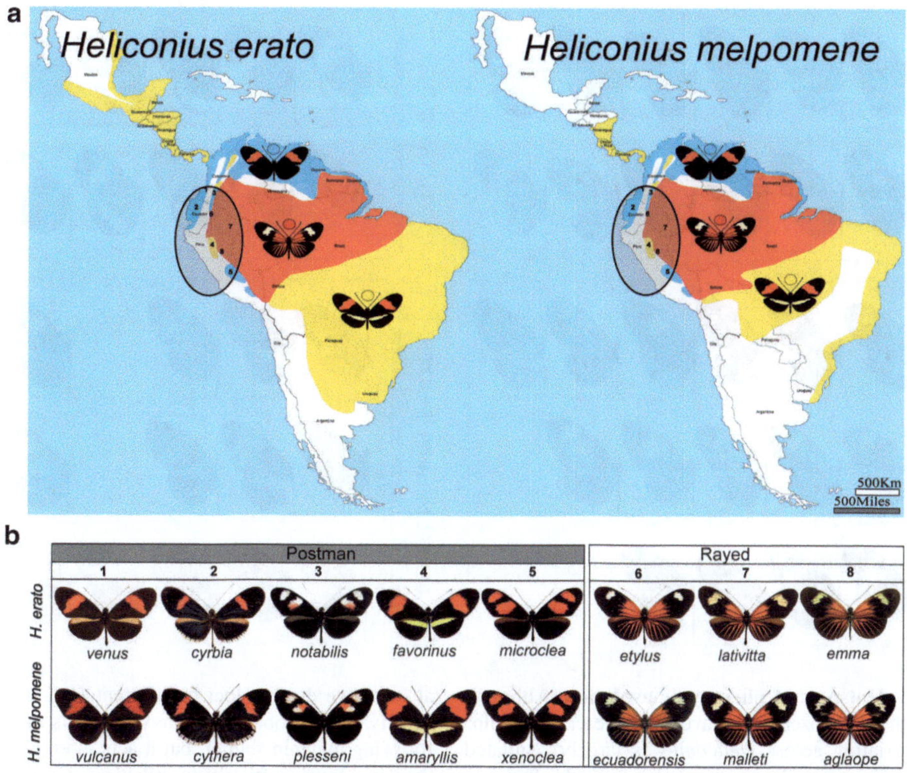

Fig. 13.2 Variations on a theme – parallel color pattern radiations in *Heliconius erato* and *Heliconius melpomene*. (**a**) Geographic distributions of the major phenotypes of *H. erato* and *H. melpomene*, showing the "rayed" phenotype throughout the Amazon basin and disjoint populations of the two variations of the "postman" phenotype. (**b**) Each row shows some of the color pattern divergence within the two concordant radiations, classified by the two major phenotypes ("postman" and "rayed"). Comparing color pattern phenotypes between the two species reflects the convergent color pattern evolution

that show spatial variation across wing surfaces (Fig. 13.3). The identification and characterization of these color pattern alleles has been a major step forward in understanding the evolution and development of lepidopteran wing patterns and provides a unique glimpse into the developmental genetic architecture of pattern evolution in nature.

The nomenclature surrounding the genetics of color pattern variation in *Heliconius* that has developed over the past 50 years is complicated; however, most described color pattern variation is due to the combined effects of four major loci. All four loci contain functional variation that affects distinct color pattern elements. For simplicity, we will refer to them as the "red switch", the "yellow switch", the "forewing melanin shutter", and the "global melanin shutter" (Fig. 13.3). The red switch controls the presence of several discrete red elements, including the forewing band, the proximal forewing "dennis" patch, and the hindwing rays (Fig. 13.3). The yellow switch causes a loss of yellow ommochrome pigments across both wing surfaces, resulting in a switch from yellow to white pattern elements (Linares 1997; Naisbit et al. 2003) (Fig. 13.3). This locus has largely been described from crosses of *H. cydno*, but probably underlies pattern differences in races of *H. sapho* and *H. eleuchia* as well. The two "shutters" (sensu Gilbert 2003) modulate the complex distribution of black/melanin across both wing surfaces. The forewing melanin shutter acts predominately in the central portion of the forewing, generating variation in the shape, position, and size of the forewing band (Fig. 13.3).

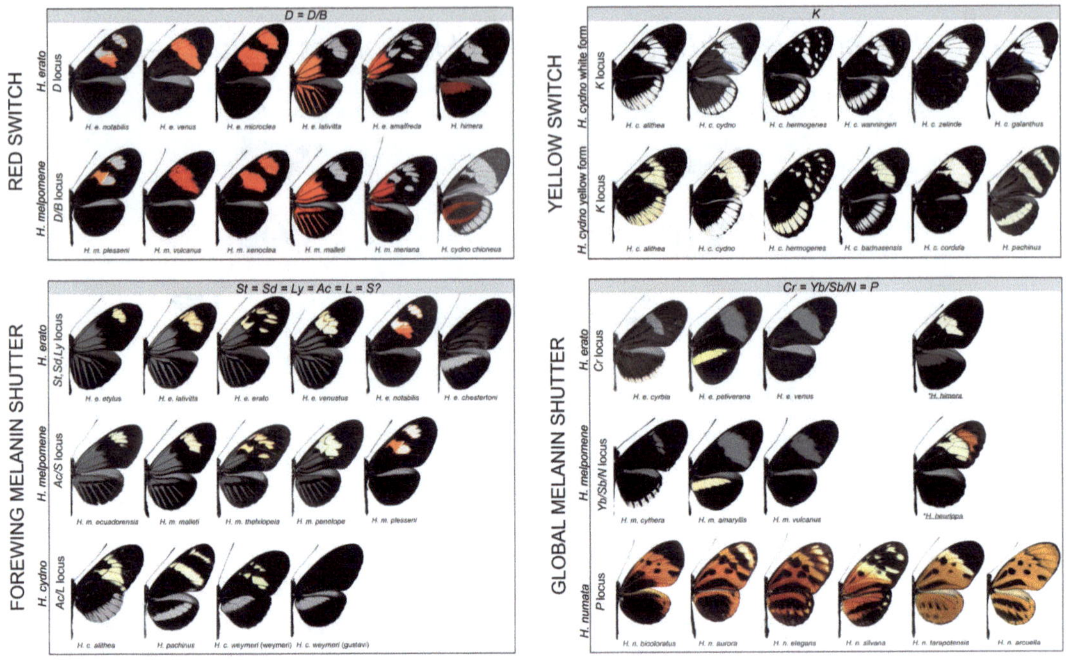

Fig. 13.3 Hotspots of phenotypic evolution. Although more than two dozen loci have been described, color pattern variation across *Heliconius* is largely regulated by four major effect loci – two color "switch" loci and two melanin "shutter" loci. These loci interact to create the natural diversity of *Heliconius* color patterns. The "red switch" controls various red elements across the wing surfaces and was originally described as multiple loci (Sheppard et al. 1985). The "yellow switch" changes forewing and hindwing pattern elements between white and yellow. The "forewing melanin shutter" controls the distribution of black/melanic wing scale cells on the forewing to either expose or cover white or yellow pattern elements and generates variation in forewing band shape, size and position. This locus also originally described as at least five distinct loci in different *Heliconius* species. Finally, the "global melanin shutter" is similar to the forewing melanin shutter, but it acts more broadly across the wing. It was similarly described as a number of distinct loci in different *Heliconius* species. Allelic variation at this locus can have multiple phenotypic effects across the wing including: (i) creating the white fringes on the fore wing and hindwing, (ii) causing the presence or absence of the yellow hindwing bar, and (iii) changing the shape and color of the forewing band in some species, including *H. himera* and *H. heurippa*. In addition, allelic variation at this locus, in the form of a series of localized inversions, controls all color and pattern variation in *H. numata* (see Joron et al. 2011)

The global melanin shutter affects the distribution of melanin across both wing surfaces to create variation in a number of red, yellow, and white pattern elements (Fig. 13.3). Due to the broad phenotypic effects of these color pattern loci, most were originally described as supergenes (Mallet 1989), or clusters of linked genes that facilitate co-segregation of adaptive variation (sensu Mather 1950).

The above synthesis is an oversimplification in several ways. First, there are other loci that have moderate size effects that contribute to pattern variation in important ways (Papa et al. 2013). This includes completely unexplored loci that alter the nanostructures of wing scale surfaces, causing light to scatter, which results in iridescent wing surfaces. Second, all of the major loci have pleiotropic effects that extend beyond the neat categorizations described above. Furthermore, most interact epistatically with each other to generate additional pattern diversity. For example, in *H. melpomene* the global melanin shutter interacts with the red switch and the forewing melanin shutter to finely control the color, size, and position of the forewing band. Finally, the magnitude of color pattern variation generated by allelic differences within these loci can be extreme. This is best exemplified in *H. numata*,

where the entire spectrum of wing pattern variation is controlled by variation at a single locus – the global melanin shutter (Fig. 13.3) (Joron et al. 2006).

13.4 Genomic Divergence Across the Speciation Continuum

A major strength of *Heliconius* as a genomic model system is that the radiation has produced a continuum of divergent races and species, which provide an excellent opportunity to study taxa at different stages of speciation (Fig. 13.4). At one end of the continuum, many divergent color pattern races freely hybridize, showing only weak reproductive isolation from each other. These racial boundaries are maintained primarily by selection on wing color patterns, with free gene flow across the rest of the genome. This heterogeneity in gene flow across the genome is ideal for identifying functional variation, as the genomes of divergent, hybridizing races should only differ at regions responsible for phenotypic differences. For some pairs of *Heliconius* taxa, speciation has progressed further. For example, the hybrid zone between *H. erato* and *H. himera* in Ecuador is characterized by the evolution of strong pre-mating isolation through assortative mating, but there is no evidence for hybrid inviability or sterility (McMillan et al. 1997). Both prezygotic and postzygotic isolation have been shown to contribute to isolation in

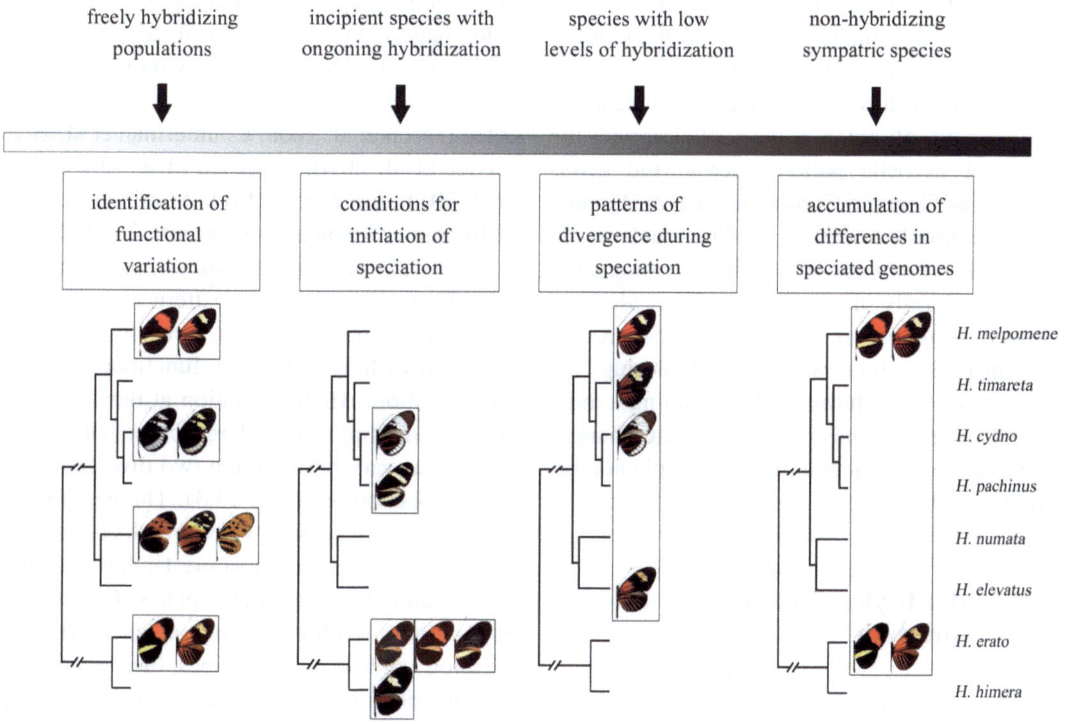

Fig. 13.4 Across the speciation continuum. The *Heliconius* radiation provides an exceptional opportunity to study taxa at different stages of speciation. The continuum, from freely hybridizing populations to non-hybridizing species, provides insight into a variety of key areas of research related to speciation. The genomes of freely hybridizing races should diverge only at genomic regions driving phenotypic divergence, allowing identification of important functional variation. Species pairs at the earliest stages of speciation give insights into how reproductive isolation is established. As reproductive isolation increases, but low levels of hybridization remain, patterns of divergence across the genome reflect the complex interplay between evolutionary forces and genome structure. Studying mimetic species pairs that do not hybridize gives insights into how genomes diverge with complete reproductive isolation and how distantly related species can converge on nearly identical phenotypes

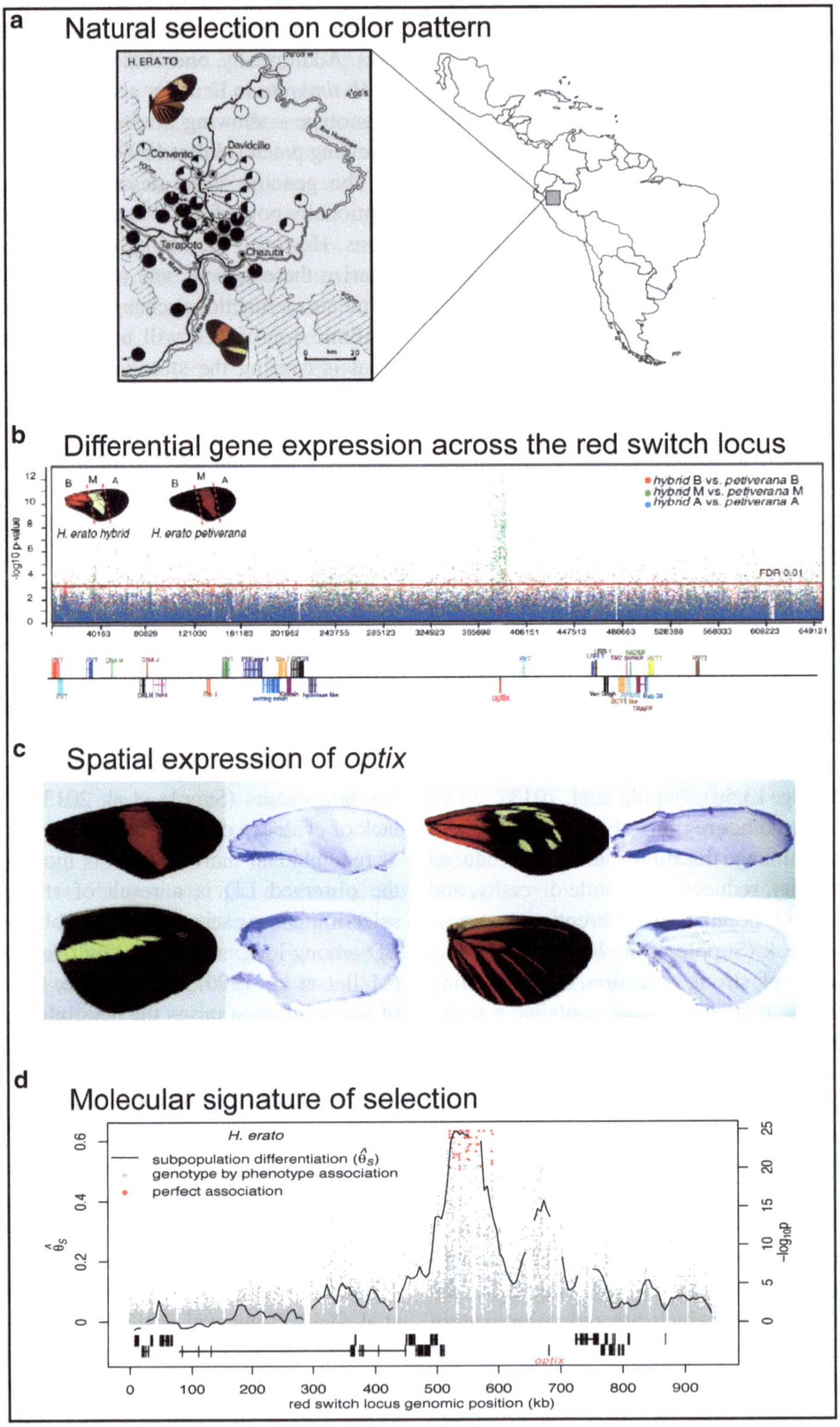
a
Natural selection on color pattern
H. ERATO
Convento
Davidcillo
Tarapoto
Chazuta
b
Differential gene expression across the red switch locus
H. erato hybrid
H. erato petiverana
hybrid B vs. petiverana B
hybrid M vs. petiverana M
hybrid A vs. petiverana A
-log10 p-value
FDR 0.01
optix
c
Spatial expression of optix
d
Molecular signature of selection
H. erato
subpopulation differentiation ($\hat{\theta}_S$)
genotype by phenotype association
perfect association
$\hat{\theta}_S$
$-\log_{10}P$
optix
red switch locus genomic position (kb)

pharmacological treatments all indicate that the *WntA* ligand modulates variation of the forewing band (Martin et al. 2012). The spatial pattern of *WntA* expression corresponds to the black forewing pattern in multiple species across the *Heliconius* radiation. As with *optix,* the *WntA* protein is highly conserved and variation in *cis*-regulatory regions is likely responsible for pattern diversity. *WntA* is a signaling ligand that creates a morphogen gradient across the developing wing tissue and is the type of molecule predicted to underlie pattern formation in theoretical models of wing color pattern development (Kondo and Miura 2010; Nijhout 1991). *WntA* is expressed earlier in pattern formation than *optix* and may act as a negative regulator of *optix,* with the interaction between the two genes being responsible for establishing black versus non-black wing pattern boundaries. This is only the second report of a morphogen involved in pattern generation (see Werner et al. 2010), but it is the first that directly links change in a patterning molecule to the evolution of a highly variable trait with clear adaptive significance (Martin et al. 2012).

The yellow switch and the global melanin shutter have similarly been positionally cloned (Joron et al. 2006; M. Kronforst, pers. comm.). Nonetheless, specific candidate genes and/or functional elements have yet to be identified. The global melanin shutter, in particular, has proven difficult to characterize. It was the first color pattern locus to be positionally cloned and it has the broadest range of phenotypic effects of any of the *Heliconius* color loci. Moreover, this region has recently been shown to underlie pattern change in several Lepidoptera species, including eyespot size in *Bicyclus* (Saenko et al. 2010) and the classic case of industrial melanism in the British peppermoth, *Biston betularia* (van't Hof et al. 2011), underscoring the flexibility and broad evolutionary importance of the *Heliconius* patterning loci. In *Heliconius,* this locus has been hypothesized to be a "supergene" composed of a number of distinct co-adapted protein coding loci. Support for this comes from recent work on *H. numata* that showed that allelic variants at this locus are actually a set of different chromosomal inversions across a region containing at least 18 genes (Joron et al. 2011). However, ongoing expression and genome scan studies in *H. erato* and *H. melpomene* indicate that, similar to the red switch and the forewing melanin shutter, only a single protein coding region at this locus may underlie the global variation in melanin pattern across *Heliconius* wings (C. Jiggins, pers. comm.).

13.4.2 The Origins of Novel Phenotypes

Natural selection can explain why wing patterns of different *Heliconius* species should converge – strong selection against rare color patterns promotes mimicry (Müller 1879). Natural selection can also explain the maintenance of existing wing pattern diversity – strong frequency-dependent selection removes non-mimetic individuals creating sharp transition zones between divergent phenotypes (Mallet et al. 1998). However, natural selection cannot easily explain the origin and spread of new phenotypes in *Heliconius*. This is a complex issue at the core of the mimicry paradox – the frequency – dependent selection that

Fig. 13.5 Linking genotype to phenotype to fitness in *Heliconius*. (**a**) Allele frequency variation at the red switch locus across the Peruvian hybrid zone reflects strong natural selection on color pattern. (**b**) *optix* is the only gene in the red switch locus showing strong and significant differential expression between the red and yellow forewing bands of divergent red phenotypes. (**c**) *In situ* hybridizations show that spatial expression of *optix* exactly prefigures red color pattern elements in pupal wings 60 hours after pupation. (**d**) Sliding window genomic divergence between two major *H. erato* phenotypes ("postman" and "rayed") from three hybrid zones across the red switch locus. There are two peaks of divergence, one near *optix* and one at a 65 kb region 3' of *optix*, indicating potentially functional regions. The 65 kb peak also contains numerous SNPs perfectly associated with phenotype. The divergent peak stands in marked contrast to the lack of differentiation at regions unlinked to color pattern $\left(\widehat{\theta}_s = -0.07\right)$ (Figure modified from Mallet and Barton 1989; Reed et al. 2011; Supple et al. 2013)

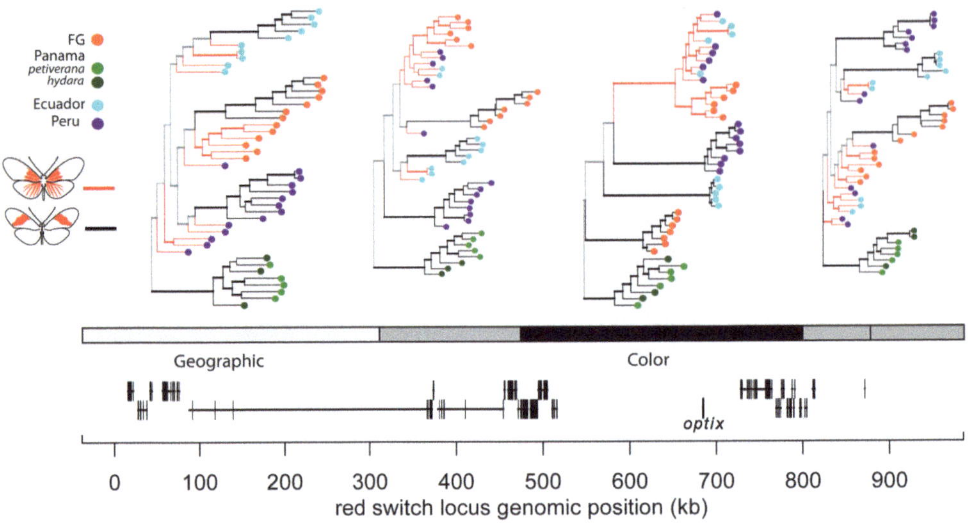

Fig. 13.6 The origins of an adaptive radiation – phylogenetic analysis of the red switch locus in *H. erato*. Phylogenetic analysis of optimal topological partitions highlight a region around the gene *optix* as clustering samples by phenotype, rather than geographic proximity. The tree topologies are shown, with phenotypes represented by branch color and geographic regions by terminal node color. Trees were generated from SNPs determined by aligning short sequencing reads to a reference genome. The grey scale bar is colored by the general history inferred. Around *optix*, samples are clustered by phenotype (*black bar*), while the farthest partition from *optix* clusters by geography (*light grey bar*), and the regions in between are intermediate, clustering by a mix of geography and phenotype (*dark grey bar*). Gene annotations, with the gene *optix* indicated, are shown below (Figure modified from Supple et al. 2013)

stabilizes existing patterns is the same force that eliminates novel forms, yet pattern divergence is rampant (Mallet and Gilbert 1995; Turner and Mallet 1996; Joron and Mallet 1998). To begin to explain this paradox, we first need to understand the evolutionary dynamics of the genomic regions that cause pattern change. An essential first question to ask is whether novel phenotypes arise once and spread within and between species or are there multiple, independent origins of the same phenotype? It has only been with the identification of the regions that modulate phenotypic differences that we can begin to address this question. The answer seems to be a bit of both – phenotypic evolution within races and species with even low levels of hybridization occurs by sharing uniquely derived color pattern alleles; while convergent phenotypes evolve independently in more distantly related co-mimics.

Analyses of the genomic region responsible for color pattern diversity support a single origin for major red color pattern phenotypes within species. For both *H. erato* and *H. melpomene*, variation around the red switch locus sorts by color pattern phenotype (Hines et al. 2011; Supple et al. 2013). In both species, individuals possessing a rayed phenotype cluster together to the exclusion of individuals possessing the postman phenotype (Fig. 13.6). Rayed patterns are found in the Amazon basin and are co-mimetic with several other *Heliconius* species and day flying moths; whereas, the postman phenotypes are largely unique to *H. erato* and *H. melpomene* and are found in multiple disjunct regions around the periphery of the Amazon and in Central America (Fig. 13.2). The pattern of genetic variation around the red switch locus supports the hypothesis that one rayed phenotype evolved in each species and spread quickly, fragmenting the geographic distribution of the older postman phenotypes. This phylogenetic signal is restricted to a region containing the 65 kb divergence peak identified in the hybrid zone comparisons (Fig. 13.5d). As you move away from this region, the phylogenetic signal increasingly reflects a history of recent gene flow, with variation

clustering by geographic proximity and biogeographic boundaries, regardless of color pattern (Fig. 13.6). This pattern of clustering by geography is the same pattern that is observed across regions unlinked to color pattern, which previously led to the incorrect conclusion that similar color pattern phenotypes evolved multiple times within each species (Brower 1994; Flanagan et al. 2004; Quek et al. 2010). This discordance demonstrates how inferences drawn from a specific subset of the genome can be misleading (Hines et al. 2011).

Mimetic convergence between distantly related species, in contrast, likely occurs by independent evolution. For example, population genetic and phylogenetic analyses of the *H. erato* and *H. melpomene* radiations, using variation within color pattern intervals, consistently clusters individuals by species designation, which is similar to the groupings obtained at loci unlinked to color pattern. Thus, although the same genomic region regulates mimetic color pattern variation, the changes responsible for mimetic convergence likely arose independently in the two species. The independent origin of similar color patterns in *H. erato* and *H. melpomene* is perhaps not unexpected, given that the two species diverged from each other over 15 million years ago (Pohl et al. 2009) and do not hybridize.

It is less clear how often more closely related species "borrow" color pattern alleles to acquire a mimetic wing color pattern. A cursory review of a phylogeny of the *Heliconius* radiation shows the high frequency that similar wing patterns are shared by species across the tree (Fig. 13.1). For example, within the melpomene/cydno/silvaniform (MCS) clade, the rayed and postman phenotypes occur within three of the four major lineages (Fig. 13.1). These species are all closely related and many are known to hybridize in the wild (Mallet et al. 2007) and in greenhouses (Gilbert 2003). These observations lead to a proposed model whereby *Heliconius* mimicry evolved by repeated interspecific transfer of color patterning alleles (Gilbert 2003). Adaptive introgression, which is the spread of beneficial variation through interspecific hybridization, may have provided the genetic raw material for both accelerated adaptation and speciation due to the dual role of color patterns in mimicry and mating behavior.

Compelling evidence for hybridization and introgression, particularly around color pattern loci, comes from analyses of closely related *Heliconius* species that share similar mimetic patterns. Phylogenetic analysis of genetic variation from the region identified as crucial to red color pattern differences, clusters populations and species by red pattern phenotype across species boundaries, rather than by phylogenetic relationship (Heliconius Genome Consortium 2012; Pardo-Diaz et al. 2012) (Fig. 13.7). This clustering by phenotype includes *H. elevatus*, which is the only rayed species in the more distantly related silvaniform clade, which usually have orange, black, and yellow tiger patterns and are known to hybridize, albeit rarely, with *H. melpomene*. Additional support comes from genome-wide tests that attempt to distinguish shared ancestral polymorphism from shared polymorphism resulting from recent gene flow (Green et al. 2010; Durand et al. 2011). This analysis shows a statistically significant excess of shared polymorphisms between sympatric co-mimics than would be expected from random sorting of ancestral polymorphisms, with a particularly strong signal at known color pattern loci. Although this test has been shown to be biased by population structure and to be sensitive to genotyping errors (Durand et al. 2011), the pattern of shared polymorphisms combined with the traditional phylogenetic analyses paints a dramatic picture of the adaptive spread of color pattern alleles across species boundaries.

13.4.3 Wing Color Patterns as a "Magic Trait" Promoting Speciation

As we are beginning to understand the role that adaptive introgression plays in the spread of mimetic color pattern alleles, it is becoming equally clear that divergent color pattern alleles in *Heliconius* likewise play a profound role in speciation. Disruptive selection on an ecological trait, such as *Heliconus* wing patterns, imposes

Heterogeneity is emerging as a common feature of genomic divergence in a number of recent studies. For example, studies of differentiation in *Anopheles* mosquitoes, *Ficedula* flycatchers, and *Coregonus* whitefish also showed highly heterogeneous patterns of differentiation (Lawniczak et al. 2010; Ellegren et al. 2012; Gagnaire et al. 2013). In *Ficedula* and *Coregonus,* the patterns are thought to be the result of recent admixture following allopatric divergence. In contrast, the *Heliconius* and *Anopheles* patterns are thought to have emerged as a result of speciation without periods of allopatry. Both speciation with gene flow and allopatric divergence with secondary contact can generate genomic heterogeneity. However, finer analysis of the patterns should allow one to distinguish the two scenarios. A commonly used measure of the extent and timing of gene flow is the number and distribution of shared polymorphisms between species. The basic principle is that (i) longer periods of gene flow should result in a greater proportion of shared polymorphic alleles between older species, (ii) recent gene flow will reduce differentiation and increase the proportion of shared alleles among sympatric populations, as compared to allopatric populations, and (iii) recent gene flow should initially result in strong linkage disequilibrium between shared alleles at linked sites, which would breakdown over time if gene flow was ongoing for longer periods. Various methods for comparing the numbers and distribution of shared polymorphisms are being developed and have recently been used to study the role of gene flow during speciation in the handful of organisms with population genomic data available (Kulathinal et al. 2009; Ellegren et al. 2012), including humans and neandertals (Green et al. 2010).

The melpomene/cydno/silvaniform clade provides an ideal opportunity to explore the genomics of speciation with gene flow versus allopatric divergence with secondary contact in a rich comparative framework. This is because the clade includes many allopatric and sympatric/parapatric races and species with varying degrees of known phylogenetic relatedness that can be used to compare patterns of shared polymorphisms and test different speciation models. For example, the observed increase of shared polymorphisms across the genome at increasing phylogenetic depths is suggestive of a long history of gene flow during speciation (Martin et al. 2013). Interestingly, a similar conclusion was reached using a completely different approach that involved modeling introgression rates in a community assessment of genomic differentiation among Costa Rican *Heliconius* species (Kronforst et al. 2013). However, extreme caution in interpreting these patterns is warranted. High heterogeneity in genomic divergence is indicative of the complex interplay between a diverse array of ecological and demographic factors, including selection, gene flow, and population history, as well as intrinsic genomic features such as variation in recombination rate.

With these new data, we are beginning to appreciate the complexity and the challenges of identifying genomic regions responsible for adaptive divergence and reproductive isolation and understanding how they affect genome-wide patterns of divergence throughout the speciation process. This is a serious challenge, yet systems with replicated examples of adaptation or speciation, such as *Heliconius*, sticklebacks, and whitefish, can be extremely powerful for inferring the functional importance of regions of divergence and understanding the history of gene flow between species.

13.5 From Patterns to Process

Moving forward requires a much better understanding of the how genomes diverge. As genomic technologies advance, empirical descriptions of genomic divergence will be layered onto one another to describe how the genomes of species change through space and time. The accumulating genomic data are already revealing extremely heterogeneous patterns of divergence that result from complex interactions between selection and gene flow. The research is quickly transitioning to identifying systems that have the most promise to provide insights into the process of genomic divergence. In this respect, new model systems, such as *Heliconius*, which (i) have replicated examples of adaptation, (ii) are

14.3.1 Genetic Architecture and Evolution of Hybrid Sterility

Postzygotic reproductive isolating barriers, expressed as reduced viability and fertility of hybrids, are most often due to deleterious genetic interactions between loci that have differentiated independently in diverging lineages. The precise molecular changes responsible for these sterility barriers have been identified in fewer than a handful of species pairs (Presgraves 2010). Identifying these genes can provide direct insight into the raw material and evolutionary dynamics that drive the formation of new species. For example, the genetic architecture (e.g., the number of contributing genes, their degree of dominance, and their mode of inheritance) of isolating barriers can fundamentally influence the magnitude and permeability of these barriers, and the time course, pace, and ease of evolution of reproductive isolation (Orr 1995; Turelli and Orr 2000; Turelli and Moyle 2007).

Solanum species are known to be isolated by a range of substantial but incomplete postzygotic barriers, ranging from seed inviability to pollen sterility (e.g., Rick and Lamm 1995; Rick 1963, 1979, and reviewed in Moyle 2007). Similarly, experimental hybridizations among *Jaltomata* species (J. L. Kostyun, unpubl. data) and among *Capsicum* (Walsh and Hoot 2001) indicate that species within each of these genera are also crossable to some degree, but show a variety of readily diagnosed, postzygotic reproductive barriers. Therefore, these plant groups have features that make them ideal for examining the genetics and evolution of postzygotic isolation: numerous, closely-related crossable species, with variable expression of postzygotic barriers in interspecific hybrids.

To date, we have examined the genetic architecture of hybrid sterility between five pairs of *Solanum* species, using QTL (quantitative trait locus) mapping and other methods (Moyle and Graham 2005, 2006; Moyle and Nakazato 2008, 2009; Posto 2009). By doing so, we now have a well developed understanding of both the number and chromosomal location of loci that contribute to the expression of hybrid sterility between particular species (Moyle and Graham 2005, 2006; Moyle 2007; Moyle and Nakazato 2008, 2009) (Fig. 14.2a), but also an emerging understanding of the generality of these findings across a range of genetic divergence within the group as a whole. As such these data provide an interesting contrast with general patterns detected in other well-studied groups, as well as the opportunity to test long-standing theoretical predictions about the evolutionary accumulation of isolation loci.

For example, our analyses in *Solanum* indicate that most hybrid incompatibility QTL act recessively (or, at most, additively), that a relatively modest number of QTL underlie hybrid incompatibility between any two species, and that roughly comparable numbers of male- and female-acting sterility QTL act to isolate species (Moyle and Nakazato 2008). The latter two patterns contrast strongly with equivalent observations in other model systems such as *Drosophila* (e.g., Hollocher and Wu 1996; True et al. 1996; Tao et al. 2003a, b), and suggest that different reproductive processes and evolutionary dynamics predominate during the accumulation of hybrid incompatibility barriers in these two systems (Moyle and Nakazato 2008). For example, our observation of roughly equal numbers of sex-specific isolation loci in *Solanum* is notably different from *Drosophila* – in which most assayed species pairs are isolated by tens of male sterility loci before any other postzygotic isolation phenotypes are observed (Masly and Presgraves 2007, and references therein). Many biological differences between these two groups could contribute to this disparity, but differences in genetic sex determination (e.g. the presence/absence of sex chromosomes) and the relative strength and efficacy of sexual selection are most intriguing. Both these factors are known to play a special role in the evolution of hybrid sterility in groups where they occur (e.g., Presgraves 2008), including accelerating the rate of change in male reproductive loci. The emergence of equivalent genome-wide data in multiple species pairs in other diverse systems (e.g. mice; Payseur and Place 2007), will help with interpreting whether these broad

three QTL involved in retarded pollen growth within interspecific pistils (Posto 2009). In parallel, emerging work in *Solanum* examines the connection between genetically-determined self-incompatibility (SI) mechanisms (molecular interactions in the female reproductive tract that prevent hermaphrodites from fertilizing with their own pollen) and the expression of post-mating prezygotic isolation barriers between species. These phenotypes are qualitatively similar and one long-standing model proposes that the molecular machinery responsible for SI is also directly involved in the expression pollen-pistil barriers between species (de Nettancourt 2000). By examining the genetics of pollen-pistil barriers between wild tomato lineages with different SI status, using QTL mapping and candidate gene studies, this work aims to clarify the mechanistic links between known molecular players in SI and the presence and strength of such barriers (e.g., Covey et al. 2010; Li and Chetelat 2010; Bedinger et al. 2011). Already it is clear that the SRNase responsible for female-side (pistil) expression of SI reactions is not individually necessary to express pollen-pistil barriers between species; however at least two other loci (HT-A and HT-B) also involved in SI might work individually, or epistatically with SRNase, to bring about this barrier phenotype (reviewed in Bedinger et al. 2011). If there are shared mechanistic underpinnings between SI and isolating barriers, the persistence of pollen-pistil species barriers could be dependent upon mating system transitions that involve the maintenance or loss of SI. Therefore, this work has the potential to draw interesting connections between the ecological (mating system dynamics) and genetic (molecular players in SI) factors that can contribute to such barriers.

Finally, in addition to physiological interactions in the style, other morphological traits can influence the expression of post-mating prezygotic barriers, including species differences in the size of flower parts (i.e., mechanical gametic isolation) (Ramsey et al. 2003; Dell'Olivo et al. 2011). Relevant traits, including style length and flower size, are equally amenable to genetic analysis (reviewed in Sicard and Lenhard 2011, and see below).

14.3.3 Ecological and Developmental Genetics of Floral Variation and Divergence

Although above we have focused on reproductive traits that influence sexual interactions or postzygotic fecundity, angiosperms – 'flowering plants' – are best known for their remarkable diversity in floral traits (Darwin 1876; Endress 2011). These traits play multiple roles, including as sexual advertisements and arenas for selecting sexual partners, and as organs for producing and protecting gametes. Variation in floral traits also plays a role in reproductive isolation among angiosperm species, especially via pre-mating sexual isolation due to differential pollinator attraction (Kay and Sargent 2009, and see below) or variable fit between pollinator and floral structures (Grant 1949), and postzygotic sexual isolation due to floral intermediacy (and therefore reduced attractiveness) in hybrids (e.g., Schemske and Bradshaw 1999). Although relatively poorly addressed, floral differentiation could also potentially influence postzygotic species barriers if divergent floral development programs have pleiotropic effects on hybrid viability and fecundity (Walia et al. 2009). Therefore a complete understanding of reproductive trait variation within and between species, and their isolating barriers, requires an examination of the ecology and genetics of floral variation.

Within our study groups, there is considerable variation in floral traits associated with mating system variation or with pollinator differences. For instance, *Solanum* encompasses variation that is associated with mating system transitions, including in corolla size, and size of other floral structures – especially variation in stigma exertion (see above, and reviewed in Bedinger et al. 2011). In *Capsicum*, flowers exhibit color variation between and among species. However in comparison to both *Solanum* and *Capsicum*, which are considered bee pollinated and lack

Fig. 14.3 Comparison of floral diversity between representative species in *Jaltomata* (*left*) and *Solanum* (*right*). The figure includes a stylized cladogram illustrating species relationships (Based on Miller et al. 2011 and Haak et al. (unpubl.)). *R* rotate, *C* campanulate, *ST* short tubular, *T* tubular. *Jaltomata* photos courtesy of Dr. Thomas Mione and colleagues, Dr. Eric J. Tepe and Victor Quipuscoa, where indicated

appreciable variation in traits associated with differential pollinator attraction (Knapp 2010), the flowers of *Jaltomata* show tremendous variation in many floral characteristics (Fig. 14.3), and associated variation in service by different pollinators. In general, pollinator attraction is known to be differentially affected by variation in multiple floral traits, including color (Bradshaw and Schemske 2003; Hoballah et al. 2007; Hopkins and Rausher 2012), shape (Owen and Bradshaw 2011), scent (Xu et al. 2011), and nectar characteristics (Medina-Tapia et al. 2012) as well as combinations of particular floral traits (Hodges et al. 2002; Campbell 2009). *Jaltomata* species span these critical axes of floral variation, with species that vary in corolla color (including white, green, deep purple, and fuchsia), corolla shape (including rotate, campanulate, and tubular

forms), and nectar quality, quantity, and color. Indeed, several shifts from generalist bees to hummingbirds have been directly observed in *Jaltomata*, and multiple additional shifts are suspected (T. Mione, pers. comm.). Thus, identifying underlying mechanisms of floral variation involved in pollinator interactions in this clade offers an exciting opportunity to link genetics to traits with clear ecological functions, and with potential collateral effects on pollinator-mediated reproductive barriers among species.

In addition to their evident ecological importance, floral traits are also attractive as targets for ecological genomics work because they are increasingly amenable to genetic and developmental analysis. For example, several studies have identified QTL associated with phenotypic changes involved in transitions from outcrossing to selfing (e.g. Lin and Ritland 1997; Fishman et al. 2002; Goodwillie et al. 2006), including in *Solanum* (e.g., Bernacchi and Tanksley 1997). Between species of *Solanum*, we have identified between two and six QTL for traits related to flower size and shape, that could influence pollinator attraction and/or selfing rates (Moyle 2007) indicating the genetics of these traits is not highly complex. Similarly, QTL underlying pollinator-influencing floral traits have been identified in several plant systems (e.g., Fulton and Hodges 1999; Hodges et al. 2002; Bradshaw and Schemske 2003), and in some cases the underlying genes have been localized (e.g. Hopkins and Rausher 2012). These examples provide clear models for connecting both genetic and ecological approaches to understand ecologically important floral trait variation.

In parallel, substantial work has been devoted to identifying the molecular genetic and developmental mechanisms responsible for producing floral structures (Heijmans et al. 2012), especially the role of MADS-box genes in specifying the development of distinct floral organs (sepals, petals, stamens, carpels) (Causier et al. 2010). Recent research has identified upstream regulators as well as likely targets of MAD-Box genes (Liu and Mara 2010; Sablowski 2010), and characterized duplications and molecular evolution within MADS-box lineages (Hileman et al. 2006; Irish 2006). MADS-box genes are being identified and characterized in a number of angiosperm lineages – including domestic tomato (Hileman et al. 2006), as well as close relative *Petunia* (Rijpkema et al. 2006). This groundwork provides an excellent set of candidate genes and developmental mechanisms for understanding floral diversity observed in our groups. By leveraging these techniques and data, we aim to investigate how variation in floral development mechanisms and genes are connected to the broad diversity of floral forms among our numerous closely related species. Given the remarkable diversity of floral form within *Jaltomata*, including four major categories of corolla shape, multiple transitions among shape categories, and multiple independent origins of convergent shapes (Fig. 14.3), this system is particularly promising for studying the developmental mechanisms responsible for rapid, ecologically-important, floral diversification.

14.3.4 Synthesis: Comparative Ecological Genomics of Reproductive Traits and Reproductive Isolation

Reproductive traits and reproductive isolation are both ecologically important and genetically tractable in our study groups, and in other close relatives (Smith et al. 2008; Jewell et al. 2012). With the expanding development of tools and genomic information in these groups (see below), we expect ultimately to identify the ecological importance, evolutionary origin, and mechanistic basis of these traits, in numerous individual cases. In addition, by taking an integrated approach, we also aim to assess the importance of mechanistic and evolutionary connections between classes of traits. For example, the potential connection between differentiation in floral morphology and the expression of postzygotic hybrid incompatibility is currently poorly understood. Because many genes involved in floral development and patterning (e.g., MADS-box genes and their upstream regulators) also function in developmental processes throughout the plant life cycle (Smaczniak et al. 2012), incompatibilities

between these gene products could induce strong, or even complete, intrinsic postzygotic isolation between parents with different floral structures; therefore postzygotic isolation could evolve as the pleiotropic consequence of divergent floral development. By examining reproductive divergence from floral variation through to postzygotic hybrid incompatibility, we aim to assess evidence for such mechanistic connections.

We expect that examining such questions in parallel across closely related species and genera will also open up exciting opportunities to evaluate new comparative questions. For example, the large disparity in quantitative and qualitative floral variation between *Jaltomata* and *Solanum* (e.g., Fig. 14.3) provides an excellent comparative opportunity to examine the genetic mechanisms and ecological forces that might be responsible for affecting the rates of evolutionary origin of novel, ecologically-important, floral variants. Given their differences in pollinator associations (via floral variation) and mating system diversity, *Jaltomata* and *Solanum* also likely experience quite different drivers of reproductive isolation, with potentially different consequences for the rate at which isolation accumulates at pre-mating, post-mating prezygotic, and postzygotic stages. Therefore, comparisons between these groups in the genetic and molecular basis of such barriers, promises to provide insight into the importance of these differences for the strength and accumulation of isolating barriers.

14.4 The Genetics and Evolution of Adaptive Ecological Traits

Ecological adaptations are the second major class of traits of interest in our ecological genomics studies. Darwin presumed that phenotypic differentiation among populations resulted from differential adaptation in response to environmental heterogeneity (Kawecki and Ebert 2004). Although we now recognize that adaptation is abundant, it is also clear that not all differences between populations are adaptive (Futuyma and Moreno 1988). Rather, adaptation must be demonstrated with empirical data that connects the physiological, genetic, and ecological mechanisms of variation with evolutionary (fitness) consequences of this variation. As with much 'ecological genomics' research, this is a long-term goal of our work in this area. To achieve this, our strategy is to focus on populations and species with physiological variation, to investigate the relationship between this variation and ecological/environmental factors (i.e., to understand the adaptive significance of physiological variation), and to dissect the physiological pathways and genetic factors that underlie these physiological responses.

Our interests lie in two general kinds of ecological adaptation – abiotic and biotic – and the mechanistic and ecological interactions between these classes. Much of the discussion below focuses on trait variation related to drought (abiotic) and defense (biotic) responses, as these are among the most important variants differentiating populations and species in our study systems (see below). An important emerging perspective from this work, however, is that there are intimate mechanistic and ecological connections between these adaptations to abiotic and biotic ecological interactions. Therefore, to address most of the fundamental questions within ecological genomics, it is essential that they be studied in tandem rather than in isolation.

14.4.1 Ecological Differentiation in Abiotic Environmental Adaptations

Responses to variation in the physical environment are particularly relevant to organisms, such as plants, that are unable to escape their immediate surroundings. Accordingly, studies of ecological differentiation in plants have most frequently focused on adaptive trait responses to abiotic environmental variation (McKay et al. 2008). Still, an integrated understanding of the molecular genetic basis of such environmental adaptations remains remarkably rare (Mitchell-Olds and Schmitt 2006). Wild Solanaceous species are known to span an enormous range of geographical and environmental variation. For instance, our

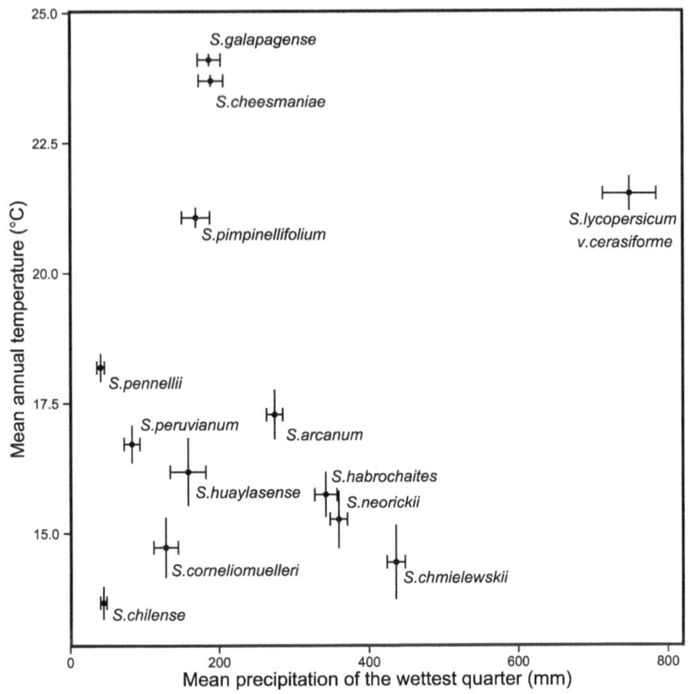

Fig. 14.4 Wild tomato species differentiation along climatological gradients (Adapted from Moyle 2008). Species are expected to differ in relevant functional traits along these gradients of water and temperature

focal groups span over 60° of latitude and the breadth of two continents (Table 14.1; and see, *Systems*, above). Accordingly, species in each of these groups are expected to be adapted to a broad range of environmental conditions, including extremes in drought, salt, altitude and thermal tolerance, making them very attractive as systems to study the ecological genomics of abiotic environmental adaptation.

14.4.1.1 Ecological Importance of Natural Ecophysiological Trait Differentiation

To understand adaptive responses to environment, it is important to understand which environmental factors are most influential in shaping species distributions and their ecophysiological traits (Turesson 1925). While many approaches are possible (Guisan and Zimmermann 2000) we have used Geographical Information Systems (GIS) to reveal the broad-scale environmental preferences of individual species (their 'bioclimate envelopes'), and therefore the specific abiotic factors that characterize these species ecological ranges and amplitudes (Nakazato et al. 2010). These analyses also identify the most important environmental differences between species. Among wild tomatoes, for example, the greatest axes of differentiation between species are average annual rainfall and temperature (Nakazato et al. 2010; Fig. 14.4).

Identifying the climatic variables most strongly associated with species differences generates hypotheses about the most important abiotic selective agents shaping this variation. It can also be used as a filter for identifying important candidate traits associated with each axis of environmental variation, when combined with studies of natural phenotypic variation. For example, in common garden investigations with wild tomatoes, we found that natural quantitative genetic variation in specific leaf area (SLA) and time to wilting (TW; under water limited conditions) are both significantly associated with natural variation in native water availability among wild populations (Nakazato et al. 2008). These traits are therefore among our primary targets for further mechanistic physiological and genetic studies.

By identifying the traits most strongly associated with climatic variation, trait-environment associations also provide a point

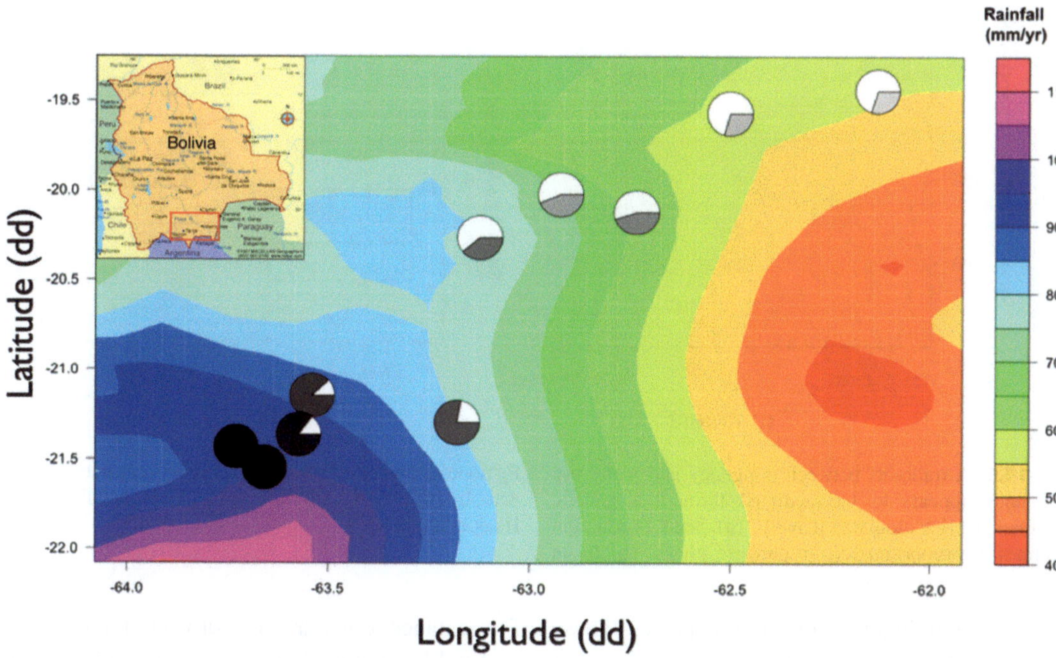

Fig. 14.5 Natural rainfall gradient across south-eastern Bolivia where polymorphic populations occur; pie charts indicate, the proportion of non-pungent plants (*white*), capsaicinoid concentration (intensity of *grey*), filled pie charts indicate populations monomorphic for pungency (Modified from Haak et al. 2012)

of departure for more directed studies of the ecological mechanisms maintaining this variation. For example, the wild chili *C. chacoense* is polymorphic both for pungent versus non-pungent genotypes (presence versus absence of capsaicinoids) and for degree of pungency (capsaicinoid concentration). In both field and common garden studies we have demonstrated that this physiological variation coincides with a natural gradient in moisture across south-eastern Bolivia (Haak et al. 2012; Fig. 14.5), and experiments confirm a connection between pungency and adaptation to moisture gradients. For example, in a common garden experiment, non-pungent plants produced 100 % more seed than pungent plants, under water-stressed conditions (Haak et al. 2012; Fig. 14.6a). This suggests that non-pungent plants have greater water use efficiency (WUE; defined as the amount of carbon assimilated per unit water loss) and therefore increased drought tolerance relative to pungent plants. In follow-up experiments using an F_2 population segregating for pungency, we found that stomatal density (a negative correlate of WUE) was 40 % higher in pungent plants than non-pungent plants (Fig. 14.6b; Haak et al. 2012). Our results indicate that a genetic correlation between pungency and stomatal density is likely responsible for the increased frequency of non-pungent plants under water limited conditions (Haak et al. 2012). This finding is especially intriguing, as it points to an ecological trade-off between lower WUE (due to stomatal density) versus higher plant defense in pungent genotypes, as we discuss further below.

Interestingly, data from both wild chilies and wild tomatoes suggest that trait responses to natural gradients in water availability are among the most critical for determining adaptive abiotic differences. Whether this is equally true in *Jaltomata* awaits similar analyses. Provisionally, however, traits that influence plant water relations (e.g., stomatal density and conductance, time to wilting, leaf morphology) including responses to both acute (seasonal) and chronic drought stress, are likely to be very important for adaptive

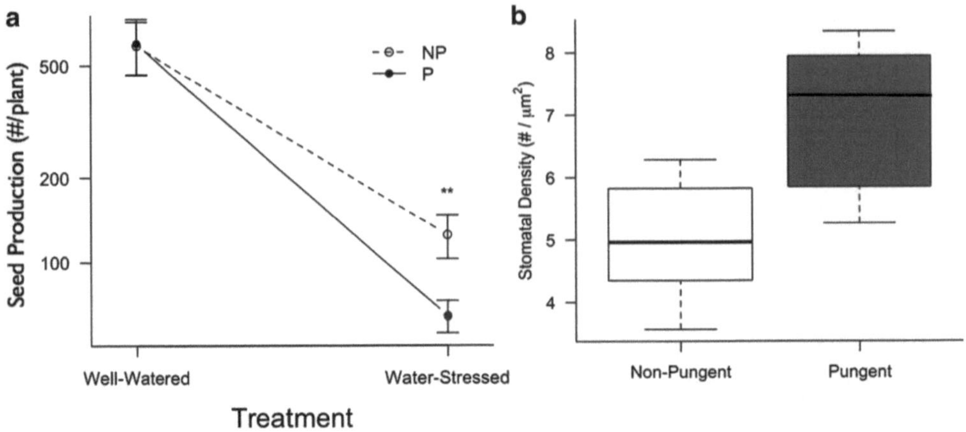

Fig. 14.6 A trade-off between pungency and water-use efficiency appears to be mediated through a genetic correlation with stomatal density. (**a**) Seed production declines disproportionately in pungent plants. (**b**) In an F_2 population segregating for pungency, stomatal density was significantly higher in pungent plants (Modified from Haak et al. 2012)

responses within and between our species. These traits are therefore among our primary targets for further mechanistic physiological and genetic studies.

14.4.1.2 Physiological Mechanisms and Genetics of Natural Ecophysiological Trait Variation

Studies across a broad range of plant species – including domesticated tomato cultivars (e.g. Hsiao 1973) – have consistently shown that ecophysiological adaptation, especially to drought, can be complex (Juenger et al. 2005; Collins et al. 2008). Though limited, studies in wild tomato species reveal substantial variation in drought avoidance strategies, in their associated ecophysiological traits, and in underlying mechanisms (reviewed in Moyle and Muir 2010). For example, the xeric wild tomatoes, *S. chilense* and *S. pennellii* have distinctly different strategies for avoiding dehydration: an extensive root system (>6 m) in *S. chilense* presumably allows greater access to available soil moisture (Rick 1973), whereas *S. pennellii* copes with water limitation via increased WUE (Moyle and Muir 2010).

In spite of this mechanistic complexity, identifying the genetic basis of ecophysiological traits still appears to be tractable in these systems. For instance, comparisons between domesticated and wild tomatoes have identified QTL directly associated with ecophysiological trait differences (Martin et al. 1989; Foolad et al. 2003; Xu et al. 2008; Muir and Moyle 2009). Interestingly, these studies have revealed both main effect QTL and transgressive QTL (interactions among QTL which drive a phenotypic distribution outside that of the parents). For instance, *Lycopersicon* species differences in WUE (Xu et al. 2008), TW and SLA (Muir and Moyle 2009) are all associated with both main effect QTL and transgressive QTL. This complexity may echo the alternative strategies seen in mechanistic responses.

Ecophysiological traits in chillies have received considerably less attention, however the developing picture reveals parallels to the tomato system. Responses to limited water in domesticated chilies involve similar dehydration avoidance mechanisms as those found in cultivated tomato (i.e., osmotic, developmental, and morphological shifts) (Choi et al. 2002; De Pascale et al. 2003). Also similar, SLA in *Capsicum annuum* was associated with both a main effect QTL and transgressive QTL (Alimi et al. 2012). Together these studies in tomato and chili support the hypothesis (Malmberg and Mauricio 2005) that epistatic QTL contribute to ecologically relevant variation, and suggests a role for genomic

interactions in ecotypic differences among wild tomatoes and wild chilies.

Our goal is to expand these kinds of analyses to diverse natural variation across ecogeographic environmental gradients. Ultimately we aim to describe and compare drought response pathways across tomatoes, *Jaltomata*, and chilies, and to connect the underlying molecular changes with shifts in ecophysiological traits within each of these clades (see final section). An important aspect in this work is to understand how and when the same, or different, pathways are used to affect similar ecological responses.

14.4.2 Ecological Differentiation in Adaptations for Biotic Interactions

Traits that mediate biotic interactions have long been implicated as drivers of population differentiation and evolutionary diversification (Darwin 1862; Ehrlich and Raven 1964; Dawkins and Krebs 1979; Schemske et al. 2009). In plant-animal interactions, fitness is effected either directly by altering reproductive success (e.g., via variation in traits mediating pollinator service), or indirectly, through herbivory or pathogen resistance (e.g., physical and chemical resistance) (De Bodt et al. 2005). In terms of biotic defense, members of the family Solanaceae (a.k.a. "nightshades") are well known for producing a diverse array of toxic compounds (e.g., scopolamine, atropine, solanine), and physical deterrents (e.g., spines and trichomes (plant hairs)) that are thought to be critical in plant defense. Therefore, this group is exceptionally rich in opportunities to examine the ecological genomics of natural defense.

14.4.2.1 Ecological Importance of Natural Defense Trait Differentiation

Chilies arguably exemplify the potential of ecological genomics approaches to understand adaptive plant defense. As outlined above, the pungent (hot) flavor of chili fruit is due to capsaicinoids, which are produced exclusively in this genus (Levey et al. 2007). In addition to capsaicinoids, chilies are also defended by other chemical (Harborne 1986; Andersson 1999) and structural traits (Drummond 1986; Barboza 2011), which likely play adaptive defensive roles. However, of these defenses, the ecological significance of pungency is currently the best understood so we focus on it here.

Capsaicinoids provide defenses against biotic enemies. For example, the fungus *Fusarium* is a necrotrophic seed pathogen in natural *Capsicum* populations; capsaicinoids in ripening fruit reduce the growth rate of *Fusarium spp.* in a dose dependent manner (Tewksbury et al. 2008). Patterns of trait variation also suggest that the inhibition of this fungal pathogen is an adaptive benefit of capsaicinoids. For example, although fruit of both pungent and non-pungent plants have *Fusarium* infection in natural *C. chacoense* populations, non-pungent seed carries a higher fungal load, a difference that translates into a 15 % reduction in germination success (Tewksbury et al. 2008). Importantly, the intensity of this selective cost for non-pungent plants is also likely to be associated with native rainfall, via indirect effects on the density of a fruit feeding insect, *Acroleucus coxalis* [Stäl]. *A. coxalis* provides points of entry for the fungus into the chili fruits via proboscis holes; its density is higher in wetter areas (Carlo et al., unpublished data), and the number of proboscis holes is also correlated with rainfall, suggesting that the opportunity for fungal infection is higher in higher moisture regions (Tewksbury et al. 2008). Thus, pungency appears to offer an adaptive benefit in response to a seed pathogen, although this benefit comes at a cost in drier environments where non-pungent plants appear to be favored (Fig. 14.5).

Although less dramatic than capsaicinoids in their culinary effects, wild tomatoes are also known to produce a range of toxic compounds (e.g., α-tomatine, sesquiterpenes), either throughout the plant or isolated to particular tissues (e.g., fruit, leaves, glandular trichomes). Many of these compounds target specific classes of insect pests (Kennedy 2007), either through chemical or physical defense strategies. For example, the methyl ketone,

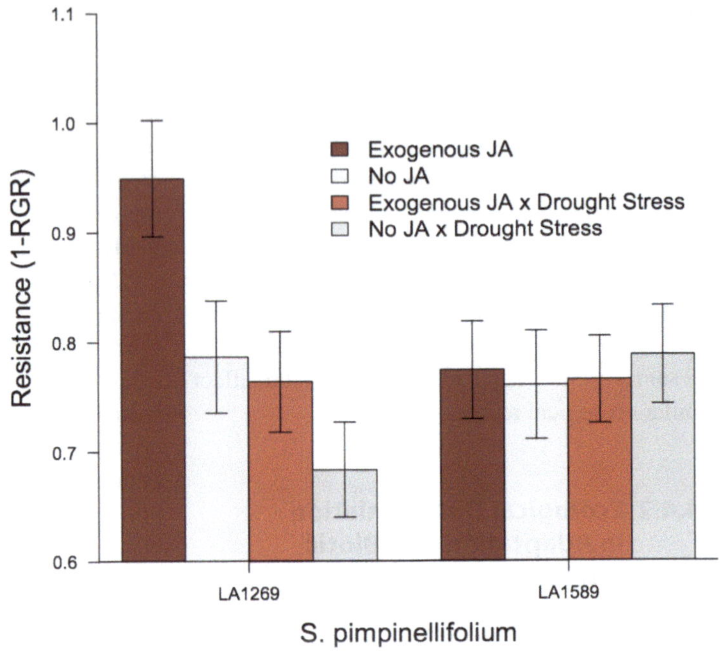

Fig. 14.7 Variation in the crosstalk between the induced defense pathway and the drought stress response pathway. The X-axis shows four treatments for each of two populations (LA1269, and LA1589) of the wild species *S. pimpinellifolium*. Under drought stress, the JA induced defense response in LA1269 is attenuated

2-tridecanone, is produced in the glandular trichomes of *S. habrochaites* and, like other compounds, is often lethal to small arthropods at natural concentrations (Kennedy and Sorenson 1985). Conversely, acylsugars produced in the glandular trichomes of *S. pennellii* form a sticky layer on the leaf surface where insect pests become entangled and die from physical entrapment (Kennedy 2007). In addition to variation in trichome chemistry (McDowell et al. 2011), wild tomatoes are also differentiated for the production of independent leaf and fruit chemical defenses (Courtney and Lambeth 1977; Juvik et al. 1982), and for structural defenses like trichome type (McDowell et al. 2011).

Just as with chilies, patterns of defense trait variation in tomatoes can potentially reveal important ecological information about the adaptive significance of this variation. Recent studies of glandular trichome mediated defense traits in *S. habrochaites* identified within species variation in the production of terpenes and acylsugars (Gonzales-Vigil et al. 2012; Kim et al. 2012). We have also found that wild populations and species show large differences in their levels of natural resistance, as quantified by growth rates in a herbivore bioassay using larvae of the Solanaceous specialist *Manduca sexta* (Haak et al., unpubl. data.). Moreover, as found in cultivated tomato varieties (reviewed in Karban and Baldwin 1997), this array of defenses can be expressed constitutively (i.e., 'always on'), or induced in response to specific elicitors that indicate prior herbivore attack (Fig. 14.7 and see above). Induced resistance is known to be an important systemic defense response for both physical and chemical resistance to insects and pathogens (Schilmiller and Howe 2005). We find natural variation in the degree of induction among wild tomatoes (Haak et al. unpubl. data, and Fig. 14.7), suggesting different ecological (selective) conditions might be acting upon this induced defense strategy among populations and species.

14.4.2.2 Physiological Mechanisms and Genetics of Natural Defense Trait Differentiation

As with ecophysiological traits, relatively few studies of the physiological and genetic basis of defense responses have been done in wild Solanaceous species (Sánchez-Peña et al. 2006; Kessler et al. 2011). Nonetheless, results from cultivated tomato (Kennedy 2003; Thaler et al. 2004) and chili pepper (Harborne 1986; Römer

et al. 2007), can provide insights into the physiological mechanisms and genetics of resistance in our wild systems. For example, in chilies there is extensive variation in the constitutive production of capsaicinoids (Tewksbury et al. 2006) and in the induced production of capsidiols in response to fungal pathogens (Egea et al. 1996). Non-pungency (no capsaicinoid accumulation) in cultivated pepper appears to be controlled by a single locus (Stewart et al. 2007); similarly, among cultivars that do produce capsaicinoids, a single QTL accounts for 35 % of the variation in total concentration (Blum et al. 2003). While the genetic basis of non-pungency in wild chilies is not resolved, and differs from cultivated peppers, it also appears to be relatively simple (Stellari et al. 2009). In contrast, 13 QTL were associated with induced resistance to a fungal pathogen in *C. annuum* (Lefebvre and Palloix 1996); interestingly, epistatic effects between these QTL accounted for most of the variation in resistance (Lefebvre and Palloix 1996).

More attention has been devoted to the physiology and genetics of biotic traits in domesticated and wild tomatoes, including studies of the expression of chemical and physical defense traits (constitutive and induced) across tissues (e.g., Kennedy 2007). In addition, the signaling cascades associated with induced resistance are well understood in domesticated tomato and other cultivated relatives (Karban and Baldwin 1997; Thaler et al. 2012). For example, the plant hormone Jasmonic Acid (JA; Farmer and Ryan 1992; Schilmiller and Howe 2005) is central to signaling cascades associated with herbivore resistance in cultivated tomatoes and chilies (Rodriguez-Saona et al. 2010). Among other functions, JAs are associated with wound responses (Li 2003), trichome plasticity (Kang et al. 2010), and defense of reproductive tissue (Hause et al. 2000). In our experiments, natural variation among populations in the degree of JA-induced defense indicates variation in the JA-mediated signal cascade (Haak et al. unpubl. data, and Fig. 14.7), possibly as the result of differing local selective forces.

From a genetic and developmental perspective, mutant screens have uncovered variants for chemical defense and identified steps in the induced resistance signal transduction pathway (Thaler et al. 2004; Schilmiller and Howe 2005). Several of these mutants in tomato indicate intriguing pleiotropic effects on multiple defense-related traits. For instance, a mutant deficient in the production of terpenes and flavonoids from glandular trichomes also exhibits a reduction in trichome density (Kang et al. 2010; see Rick et al. 1976 for another example). These resources promise to provide useful candidate loci for defense trait variation in wild tomatoes. Finally, there is increasing interest in identifying the genetic basis of defense trait variation directly in wild tomato species. For example, among natural populations of the wild species *S. habrochaites* there is allelic variation associated with biochemical variation in terpene composition (Gonzalez-Vigil et al. 2012) and acyl sugar content (Kim et al. 2012). QTL for resistance to specific herbivores have also been identified, for example spider mite resistance in *S. pimpinellifolium* (Salinas et al. 2012). This work, as well as our own emerging data, suggests that wild tomatoes are a fruitful system for uncovering the ecological genomics of specific defense traits and for understanding the evolution of, and interactions between, primary defense strategies (i.e., constitutive versus induced).

14.4.3 Synthesis: Comparative Ecological Genomics of Abiotic and Biotic Traits

One advantage of an ecological genomics approach is its ability to accommodate analyses of both individual traits and complex, integrated interactions among traits.

Above we have largely discussed adaptive abiotic and biotic responses as if these classes of traits were ecologically and physiologically independent. However, there is ample evidence that abiotic and biotic responses are mechanistically associated under many ecological and physiological conditions. One of our goals is to understand the nature, and evolutionary consequences, of these mechanistic associations.

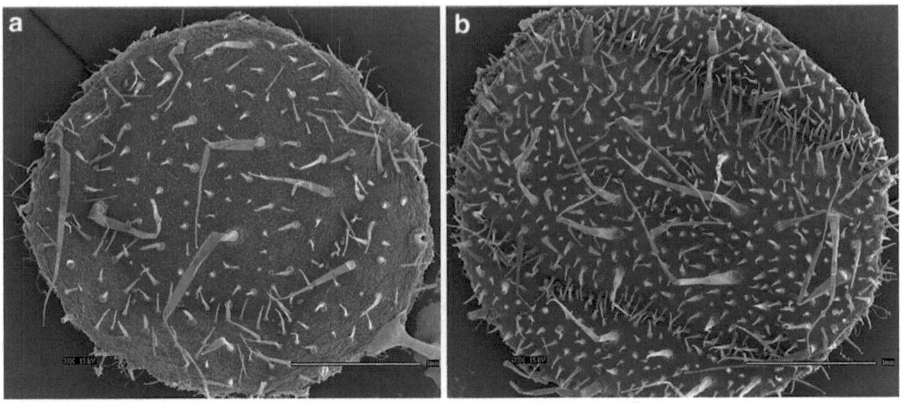

Fig. 14.8 Application of exogenous jasmonic acid increases trichome density as well as the elongation of particular trichome types in *S. habrochaites,* consistent with induced defense. Scanning electron micrographs of leaf discs from *S. habrochaites* ecotype LA1353, (**a**) expanded leaf prior to application of methyljasmonate (meJA) and (**b**) expanded leaf emerged after application of meJA

Pungency in wild chilies provides a particularly compelling example of interdependence between adaptive responses to abiotic and biotic interactions. As shown above, pungency is clearly beneficial for fungal pathogen resistance, but non-pungency is associated with increased fitness under water-limited conditions, most likely via a genetic association between pungency and stomatal density (therefore reduced WUE). Thus, a trade-off between a biotic trait (pungency) and an abiotic trait (stomatal density) results in an adaptive constraint (Haak et al. 2012). In wild tomatoes we also see evidence that abiotic and biotic drivers can have correlated effects on the same key traits, suggesting a similar potential for ecological interactions to reciprocally shape these classes of traits. For example, exposure to drought stress or exogenous JA (a mimic of herbivory) both precipitate a shift in trichome density and trichome morphology (Fig. 14.8). This suggests that there are shared downstream targets and that trichomes are a key trait that could act to integrate different classes of ecological input.

Indeed, one clear way that alternative adaptive responses can be mechanistically linked is via 'crosstalk' – the convergence on shared targets between alternative physiological pathways (Harrison 2012). Phytohormones, especially JA and abscisic acid (ABA), are known to be critical mediators of crosstalk in plants (Wasternack and Hause 2002). In response to multiple stressors, this crosstalk can be synergistic or antagonistic, within and between biotic and abiotic responses (Thaler et al. 2012). In wild tomatoes, we find evidence of natural variation in crosstalk between abiotic and biotic signal cascades. For example, in one ecotype of *S. pimpinellifolium* we find that JA induced resistance is attenuated when plants are also under water limited conditions (Fig. 14.7). This suggests that the induced resistance response is constrained under drought stress.

The consequences of physiological and molecular crosstalk for understanding the ecological genomics of adaptive trait variation should be evident: if pathways responsible for adaptive responses to abiotic and biotic environmental factors are physiologically linked, then responses to one class of factors will inevitably affect responses to the other class. To address this in both *Capsicum* and *Solanum* we are using multiple approaches including measuring whole-genome transcriptional responses to factorially imposed environmental manipulations (see below). One key goal of our ongoing research is to understand the extent to which physiological responses to alternative ecological drivers can facilitate or constrain other adaptive responses.

14.5 Perspective and New Avenues

14.5.1 Emerging Themes

While ecological genomics undoubtedly offers many potential insights, we find two themes emerging from our efforts to understand trait diversification, adaptation, and speciation in our developing systems:

First, it is increasingly clear that traits often interact to shape organismal fitness, because of underlying mechanistic and ecological connections between them. Therefore, in our work it will be important to simultaneously examine variation in suites of traits in order to understand the interplay between developmental, physiological, and genetic mechanisms, and ecological and evolutionary responses.

For instance, we have emphasized physiological links between adaptive responses to abiotic and biotic stressors such as drought and herbivory, and the pathways (e.g., JA) and traits (e.g., trichomes) that are most likely to integrate and link these responses. This mechanistic crosstalk has the potential to reach beyond interactions governing adaptive responses to environmental factors. For example, JAs are not only associated with biotic and abiotic stress, but are also known to play a role in seed and pollen viability (Li et al. 2004; Balbi and Devoto 2008). An unresolved question is whether these mechanisms could therefore also be responsible for the critical link between adaptive environmental responses and the pleiotropic expression of species isolating barriers.

We already recognize the potential fitness consequences of such interactions in individual cases. In wild chilies, for example, biotic and abiotic trait interactions appear to constrain adaptation because of the pleiotropic action of a single trait, pungency, on loci associated with stomatal density. A better understanding of these connections will allow us to determine whether the mechanisms underpinning one trait could act to enhance or constrain evolutionary responses in associated traits. We aim to dissect the nature of these interactions, by simultaneously examining the phenotypic and mechanistic basis of trait variation across suites of traits, ecological conditions, and taxa.

Second, studies situated within an explicitly comparative framework have the potential to answer new and exciting questions about the mechanisms and ecology of adaptation and speciation. Indeed, while there will always be a place for careful, detailed studies in individual systems, unique information can emerge from comparisons between groups. For example, comparative studies of accumulation of hybrid incompatibilities between multiple *Solanum* species have enabled us to explicitly test predictions from evolutionary theory. Alternatively, investigations of the environmental factors driving adaptive trait variation reveal the parallel importance of water availability in shaping whole organism adaptations (morphology to physiology) in both *Solanum* and *Capsicum*.

We are particularly excited about the insights that might emerge from comparisons between groups that differ in their evolutionary drivers of diversification. For example, given their large differences in quantitative and qualitative patterns of floral diversity, *Jaltomata* and *Solanum* likely experience quite different ecological drivers of floral diversification, and thereby potentially of reproductive isolation. A comparison between these groups in the genetic architecture and molecular basis of such traits, and their associated barriers, could provide unique insight into the importance of these differences for the strength and accumulation of isolating barriers.

Ultimately, one of our goals is to use phylogenetic comparative methods to more directly examine such associations between ecological transitions and changes at the phenotypic and molecular levels within and across our study systems. This will provide a more powerful investigation of the generality of underlying mechanisms and evolutionary patterns of trait variation, including whether this variation is derived from shared or novel molecular changes. Especially given the close relationships between

Solanum, *Jaltomata* and *Capsicum*, these systems are strongly suited for studies integrating genetics and ecology within a comparative context.

14.5.2 New Avenues: Unification Through Comparative Genomic Analyses

One current limitation on our ability to integrate analyses between systems (across populations, species, or genera), within individual cases (across ecological, physiological, and genetic levels), and/or among traits, is an incomplete genomic framework within which to place these studies. The last several years have seen great strides in the development and availability of resources to fill such a gap (Song and Mitchell-Olds 2011), including advances in sequencing technologies (e.g., deep sequencing with Illumina Hi-Seq, 'third-generation' long read platforms) as well as cost saving, reduced representation sequencing methods (e.g., RNA-seq, ChIP-seq), and the marriage of high-throughput sequencing with traditional quantitative genetic processes (e.g., bulk segregant analysis, RAD-seq). These approaches are already beginning to yield cost effective, *de novo* genomic resources in non-model systems (Anderson et al. 2011; Ellegren et al. 2012). To further 'genomicise' our study systems, we are employing several of these strategies, including comparative transcriptomics, experimental transcriptomics, and genotyping by sequencing.

Comparative transcriptomics: To construct a phylogenomic framework for our target genera, we are using RNA-seq to generate genome-wide transcriptomic data from multiple ecotypes encompassing the entire clade of wild tomatoes as well as ecotypes and species from *Capsicum* and *Jaltomata*. RNA-seq uses deep sequencing technologies (e.g., Illumina) to generate base pair resolution of the transcriptome, without requiring a priori knowledge of gene models in any particular species. By focusing on expressed sequences, RNA-seq can generate a large common set of sequences (i.e., gene models or unigenes) across the genome from all target species, without expending sequencing and assembly effort on repetitive, non-coding regions. Because there is a high degree of synteny observed in the Solanaceae (Wang et al. 2008), the two published tomato reference genomes (The Tomato Genome Consortium 2012) provide a natural framework around which we can assemble and organize these genomic resources across wild tomatoes, *Capsicum*, and *Jaltomata*.

In the first instance, these data can reveal the phylogenetic relationships among taxa across the entire genome, and provide insight into the degree and distribution of such phenomena as incomplete lineage sorting and historical introgression (McCormack et al. 2012). They can also be used to test for genome-wide patterns of molecular variation, such as the pattern of relationship between polymorphism and divergence across the genome (e.g., Eisen and Fraser 2003). Second, by revealing genome-wide patterns of molecular evolutionary rates, these analyses can also identify classes of genes and individual loci that have experienced rapid molecular evolutionary change (i.e., via a 'reverse ecology' approach to identifying candidate loci (Li et al. 2008)), including identifying the specific branch(es) where this rapid change is concentrated. Third, by focusing on known candidate genes involved in specific functional pathways and traits, such data can also be used to evaluate molecular evolutionary changes in loci previously implicated in relevant trait variation. Finally, in combination with phylogenetic inferences about the location of important trait transitions, these data can point to specific loci whose patterns of molecular evolution are associated with functional abiotic, biotic, and reproductive trait variation. These analyses will reveal genome-wide patterns of molecular evolution within and among species, and identify a series of candidates for further examining underlying mechanisms of ecologically relevant phenotypic variation.

Experimental transcriptomics: Molecular sequence variation alone is unlikely to provide a complete picture of the underlying molecular mechanisms of trait variation within and

between species. Therefore, paired with our comparative framework, we are using next generation transcriptomics (RNA-seq) as a tool for understanding genome-wide quantitative gene expression responses that are associated with trait variation. These 'experimental transcriptomics' are complementary to, and can build upon, our comparative genomic framework, by providing a link between DNA sequence and phenotypic variation, at a whole-genome scale.

For example, to examine the complex physiological interactions between stress response pathways, we are measuring the gene expression responses elicited by biotic (JA), abiotic (drought), and the combination of biotic and abiotic stressors, across ecotypes and tissues. This manipulative approach allows us to evaluate the degree and strength of shared molecular responses among stressors. In addition, comparing tissue specific gene expression patterns, we can investigate correlated molecular responses between relevant tissues, e.g., leaves and trichomes. For example, in wild tomato exposure to drought stress or exogenous JA precipitates a shift in both trichome density and morphology (Fig. 14.8) as well as leaf stomatal density. Gene expression patterns can be used to evaluate the degree of molecular overlap in these ecological responses, including whether such responses are more highly integrated in some tissues versus others. Finally, by looking within tissue types across ecotypes and species, we can also potentially gain insight into the role of shared molecular mechanisms (parallel evolution) across multiple instances of convergent ecological adaptation to the same environmental conditions.

Integrating other NGS and genomic information: Similar Next Generation Sequencing (NGS) approaches can be used to speed the identification of loci underlying our traits of interest. For example, fine-mapping and positional cloning of QTL has become increasingly efficient by incorporating NGS into approaches such as bulked segregant analysis. In this case, high-throughput sequencing of pools of individuals at the tails of the phenotypic distribution allows cost-effective, rapid mapping of ecological traits (Magwene et al. 2011). Leveraging on genomic information, we can use such approaches to identify small genomic regions and a narrow list of candidate genes for each QTL, to target with further functional validation.

Finally, the availability of complementary data from other genetic approaches, and physiological and mechanistic studies, will be critical for making sense of the wealth of genomic data generated from high throughput sequencing. For instance, the transcriptomic approaches we are taking (above) will yield a deep list of putatively important loci. Winnowing this list to a tractable set of strong candidates for further functional analyses is a challenging but necessary next step. In our groups, we can pair high throughput sequence data with traditional approaches (e.g., existing QTL) to gain rapid insight into the genetic underpinning of known traits (Stinchcombe and Hoekstra 2007). For instance, of the >2,000 QTL that have been identified for trait variation in tomato, a smaller set (ca. 140) are associated with biotic and abiotic adaptations (Foolad 2007; Li 2010). Combining data from multiple genetic approaches thus provides a phenotypically-informed method of identifying a tractable set of candidate loci for the specific traits most relevant to ecological responses.

14.6 Synopsis

By linking ecological phenotype to genotype, ecological genomics seeks to understand how ecological forces act on physiological and genetic mechanisms to shape the evolution of biodiversity. By approaching these questions in a multifaceted framework, we aim to investigate variation in diverse ecological traits using pre-existing and emerging tools necessary to uncover the associated molecular variation. As we have emphasized here, this work requires a detailed understanding of natural trait variation, and its adaptive significance. In ongoing work, we are using next generation sequencing approaches to pair underlying molecular evolutionary shifts with abiotic and biotic trait transitions. In doing so, we hope to understand

ecologically-significant trait variation – from molecular to ecological levels – in numerous individual cases. Moreover, by making comparisons among traits, ecotypes, species, and genera, in an explicitly comparative framework, we also aim to generate a robust picture of adaptation and trait diversification as general processes. We hope such data can illuminate broad factors that influence rates and patterns of adaptation and speciation, and identify common mechanistic and ecological features of these processes.

Acknowledgments Members of the Moyle lab, and two anonymous reviewers, provided helpful feedback. T. Mione kindly shared *Jaltomata* seed material and unpublished data, and T. Carlo kindly shared unpublished *Capsicum* data. The C.M. Rick Tomato Genetics Resource Center (tgrc.ucdavis.edu) provided access to their collections and collection information database. This work was supported by National Science Foundation (Division of Environmental Biology/Dimensions of Biodiversity) award 1136707 (to L.C.M. and D.C.H). J.L.K. was supported by the IU Biology Floyd Plant and Fungal Biology Summer Fellowship, Amherst College Graduate Fellowship, and an NIH Genetics, Cellular, and Molecular Sciences training grant, award 5T32GM007757.

References

Alimi NA, Bink MCAM, Dieleman JA et al (2012) Genetic and QTL analyses of yield and a set of physiological traits in pepper. Euphytica 100(2):181–201

Andersson C (1999) Glycoalkaloids in tomatoes, eggplants, pepper and two Solanum species growing wild in the Nordic countries. Nordic Council of Ministers, Copenhagen

Anderson JT, Willis JH, Mitchell-Olds T (2011) Evolutionary genetics of plant adaptation. Trends Genet 27:258–266

Balbi V, Devoto A (2008) Jasmonate signaling network in Arabidopsis thaliana: crucial regulatory nodes and new physiological scenarios. New Phytol 177:301–318

Barboza G (2011) Lectotypifications, synonymy, and a new name in *Capsicum* (Solanoideae, Solanaceae). Phytokeys 2:91

Barrett SCH (2002) Sexual interference of the floral kind. Heredity 88:154–159

Bedinger PA, Chetelat RT, McClure B et al (2011) Interspecific reproductive barriers in the tomato clade: opportunities to decipher mechanisms of reproductive isolation. Sex Plant Reprod 24:171–187

Bernacchi D, Tanksley SD (1997) An interspecific backcross of *Lycopersicon esculentum* x *L. hirsutum*: linkage analysis and a QTL study of sexual compatibility factors and floral traits. Genetics 147:861–877

Blum E, Mazourek M, O'Connell M, Curry J, Thorup T, Liu K, Paran I (2003) Molecular mapping of capsaicinoid biosynthesis genes and quantitative trait loci analysis for capsaicinoid content in *Capsicum*. Theor Appl Genet 108(1):79–86

Boughman JW, Rundle HD, Schluter D (2005) Parallel evolution of sexual isolation in sticklebacks. Evolution 59:361–373

Bradshaw H, Schemske D (2003) Allele substitution at a flower colour locus produces a pollinator shift in monkeyflowers. Nature 426:176–178

Campbell DR (2009) Using phenotypic manipulations to study multivariate selection of floral trait associations. Ann Bot 103:1557–1566

Causier B, Schwarz-Sommer Z, Davies B (2010) Floral organ identity: 20 years of ABCs. Semin Cell Dev Biol 21:73–79

Choi KH, Hong CB, Kim WT (2002) Isolation and characterization of drought-induced cDNA clones from hot pepper (*Capsicum annuum*). J Plant Biol 45:212–218

Collins NC, Tardieu F, Tuberosa R (2008) Quantitative trait loci and crop performance under abiotic stress: where do we stand? Plant Physiol 147:469–486

Courtney WH III, Lambeth VN (1977) Glycoalkaloid content of mature green fruit of *Lycopersicon* species (tomatoes, interspecific breeding). Hort Sci 12(6): 550–551

Covey PA, Kondo K, Welch L et al (2010) Multiple features that distinguish unilateral incongruity and self-incompatibility in the tomato clade. Plant J 64:367–378

Coyne JA, Orr HA (2004) Speciation. Sinauer Associates, Sunderland

Darwin C (1862) On the contrivances by which British and foreign orchids are fertilized by insects, and on the good effects of intercrossing. John Murray, London

Darwin C (1876) The effects of cross- and self-fertilization in the vegetable kingdom. John Murray, London

Darwin SC, Knapp S, Peralta IE (2003) Taxonomy of tomatoes in the Galápagos Islands: native and introduced species of *Solanum* section *Lycopersicon* (Solanaceae). Syst Biodivers 1:29–53

Dawkins R, Krebs JR (1979) Arms races between and within species. Proc R Soc B 205:489–511

De Bodt S, Maere S, Van de Peer Y (2005) Genome duplication and the origin of angiosperms. Trends Ecol Evol 20:591–597

de Nettancourt (2000) Incompatibility and incongruity in wild and cultivated plants, 2nd edn. Springer, Berlin/Heidelberg/New York

De Pascale S, Ruggiero C, Barbieri G, Maggio A (2003) Physiological responses of pepper to salinity and drought. J Am Soc Hortic Sci 128:48–54

Dell'Olivo A, Hoballah ME, Gübitz T, Kuhlemeier C (2011) Isolation barriers between *Petunia axillaris* and *Petunia integrifolia* (Solanaceae). Evolution 65:1979–1991

Drummond (1986) Coevolution of ithomiine butterflies and solanaceous plants. In: D'Arcy WG (ed) The Solanaceae: biology and systematics. Columbia University Press, New York, pp 303–327

Egea C, Perez MDG, Candela ME (1996) Capsidiol accumulation in *Capsicum annuum* stems during the hypersensitive reaction to *Phytophthora capsici*. J Plant Physiol 149(6):762–764

Ehrlich P, Raven P (1964) Butterflies and plants – a study in coevolution. Evolution 18:586–608

Eisen JA, Fraser CM (2003) Phylogenomics: intersection of evolution and genomics. Science 300: 1706–1707

Ellegren H, Smeds L, Burri R et al (2012) The genomic landscape of species divergence in *Ficedula* flycatchers. Nature 491:756–760

Endress PK (2011) Evolutionary diversification of the flowers in angiosperms. Am J Bot 98:370–396

Farmer EE, Ryan CA (1992) Octadecanoid precursors of jasmonic acid activate the synthesis of wound-inducible proteinase inhibitors. Plant Cell 4: 129–134

Feder M, Mitchell-Olds T (2003) Evolutionary and ecological functional genomics. Nat Rev Genet 4:651–657

Fishman L, Kelly JA, Willis JH (2002) Minor quantitative trait loci underlie floral traits associated with mating system divergence in *Mimulus*. Evolution 56:2138–2155

Foolad M (2007) Genome mapping and molecular breeding of tomato. Int J Plant Genom 2007:52

Foolad MR, Zhang LP, Subbiah P (2003) Genetics of drought tolerance during seed germination in tomato: inheritance and QTL mapping. Genome 46:536–545

Fulton M, Hodges SA (1999) Floral isolation between *Aquilegia formosa* and *Aquilegia pubescens*. Proc R Soc B 266:2247–2252

Futuyma DJ, Moreno G (1988) The evolution of ecological specialization. Annu Rev Ecol Evol Syst 19:207–233

Gentry AH (1982) Patterns of neotropical plant species diversity. Evol Biol 15:1–84

Gonzalez-Vigil E, Hufnagel DE, Kim J et al (2012) Evolution of TPS20-related terpene synthases influences chemical diversity in the glandular trichomes of the wild tomato relative *Solanum habrochaites*. Plant J 71:921–935

Goodwillie C, Ritland C, Ritland K (2006) The genetic basis of floral traits associated with mating system evolution in *Leptosiphon* (Polemoniaceae): an analysis of quantitative trait loci. Evolution 60:491–504

Grandillo S, Chetelat R, Knapp S et al (2011) *Solanum* sect. *Lycopersicon*. In: Wild crop relatives: genomic and breeding resources. Springer, New York, pp 129–215

Grant V (1949) Pollination systems as isolating mechanisms in angiosperms. Evolution 3:82–97

Guisan A, Zimmermann NE (2000) Predictive habitat distribution models in ecology. Ecol Model 135:147–186

Haak DC, McGinnis LA, Levey DJ, Tewksbury JJ (2012) Why are not all chilies hot? A trade-off limits pungency. Proc R Soc B 279:2012–2017

Hansen DM, Olesen JM, Mione T et al (2007) Coloured nectar: distribution, ecology, and evolution of an enigmatic floral trait. Biol Rev 82:83–111

Harborne JB (1986) Systematic significance of variations in defense chemistry in the Solanaceae. In: Solanaceae: biology and systematics. Columbia University Press, New York, pp 328–344

Harrison MA (2012) Cross-talk between phytohormone signaling pathways under both optimal and stressful environmental conditions. In: Phytohormones and abiotic stress tolerance in plants. Springer, Berlin/Heidelberg/New York, pp 49–76

Hause B, Stenzel I, Miersch O et al (2000) Tissue-specific oxylipin signature of tomato flowers: allene oxide cyclase is highly expressed in distinct flower organs and vascular bundles. Plant J 24:113–126

Heijmans K, Morel P, Vandenbussche M (2012) MADS-box genes and floral development: the dark side. J Exp Bot 63:5397–5404

Heiser CB (1987) The fascinating world of the nightshades: tobacco, mandrake, potato, tomato, pepper, eggplant, etc. Dover publications, New York

Hileman LC, Sundstrom JF, Litt A, Chen M, Shumba T, Irish VF (2006) Molecular and phylogenetic analyses of the MADS-Box gene family in tomato. Mol Biol Evol 23:2245–2258

Hoballah ME, Gubitz T, Stuurman J, Broger L, Barone M, Mandel T, Dell'Olivo A, Arnold M, Kuhlemeier C (2007) Single gene-mediated shift in pollinator attraction in *Petunia*. Plant Cell 19:779–790

Hodges SA, Whittall JB, Fulton M, Yang JY (2002) Genetics of floral traits influencing reproductive isolation between *Aquilegia formosa* and *Aquilegia pubescens*. Am Nat 159:S51–S60

Holle M, Rick CM, Hunt DG (1979) Catalog of collections of green-fruited Lycopersicon species and *Solanum pennellii* found in watersheds of Peru. Report Tomato Genetics Cooperative

Hollocher H, Wu CI (1996) The genetics of reproductive isolation in the *Drosophila simulans* clade: X vs autosomal effects and male vs female effects. Genetics 143:1243–1255

Hopkins R, Rausher MD (2011) Identification of two genes causing reinforcement in the Texas wildflower *Phlox drummondii*. Nature 469:411–414

Hopkins R, Rausher MD (2012) Pollinator-mediated selection on flower color allele drives reinforcement. Science 335:1090–1092

Hsiao TC (1973) Plant responses to water stress. Annu Rev Plant Physiol 24(1):519–570

Irish VF (2006) Duplication, diversification, and comparative genetics of angiosperm MADS-box genes. In: Soltis DE, LeebensMack JH, Soltis PS, Callow JA (eds) Advances in botanical research: incorporating advances in plant pathology, vol 44, Developmental genetics of the flower. Elsevier, Boston, pp 129–161

Jewell C, Papineau AD, Freyre R, Moyle LC (2012) Patterns of reproductive isolation in *Nolana* (Chilean Bellflower). Evolution 66:2628–2636

Juenger TE, McKay JK, Hausmann N et al (2005) Identification and characterization of QTL underlying whole-plant physiology in *Arabidopsis thaliana*: delta13C, stomatal conductance and transpiration efficiency. Plant Cell Environ 28:697–708

Juvik JA, Stevens MA, Rick CM (1982) Survey of the genus *Lycopersicon* for variability in α-tomatine content. HortScience 17(5):764–766

Kang J, Shi F, Jones A et al (2010) Distortion of trichome morphology by the hairless mutation of tomato affects leaf surface chemistry. J Exp Bot 61:1053

Karban R, Baldwin IT (1997) Induced responses to herbivory. University of Chicago Press, Chicago

Kawecki TJ, Ebert D (2004) Conceptual issues in local adaptation. Ecol Lett 7:1225–1241

Kay KM, Sargent RD (2009) The role of animal pollination in plant speciation: integrating ecology, geography, and genetics. Ann Rev Ecol Evolut Syst 40:637–656

Kennedy GG (2003) Tomato, pests, parasitoids, and predators: tritrophic interactions involving the genus *Lycopersicon*. Annu Rev Entomol 48:51–72

Kennedy GG (2007) Resistance in tomato and other *Lycopersicon* species to insect and mite pests. Genet Improv Solanaceous Crops 2:488–519

Kennedy GG, Sorenson CF (1985) Role of glandular trichomes in the resistance of *Lycopersicon hirsutum f. glabratum* to Colorado potato beetle (Coleoptera: Chrysomelidae). J Econ Entomol 78(3): 547–551

Kessler A, Halitschke R, Poveda K (2011) Herbivory-mediated pollinator limitation: negative impacts of induced volatiles on plant-pollinator interactions. Ecology 92:1769–1780

Kim H-J, Baek K-H, Lee S-W et al (2008) Pepper EST database: comprehensive in silico tool for analyzing the chili pepper (*Capsicum annuum*) transcriptome. BMC Plant Biol 8:101–101

Kim J, Kang K, Gonzales-Vigil E et al (2012) Striking natural diversity in glandular trichome acylsugar composition is shaped by variation at the Acyltransferase2 locus in the wild tomato *Solanum habrochaites*. Plant Physiol 160:1854–1870

Knapp S (2010) On "various contrivances": pollination, phylogeny and flower form in the Solanaceae. Philos Trans R Soc B 365:449–460

Lefebvre V, Palloix A (1996) Both epistatic and additive effects of QTLs are involved in polygenic induced resistance to disease: a case study, the interaction pepper – Phytophthora capsici Leonian. Theor Appl Genet 93:503–511

Levey DJ, Tewksbury JJ, Izhaki I, Tsahar E, Haak DC (2007) Evolutionary ecology of secondary compounds in ripe fruit: case studies with capsaicin and emodin. In: Dennis AJ, Schupp EW, Green RJ, Westcott DA (eds) Seed dispersal: theory and its application in a changing world. CABI, Wallingford, pp 37–58

Li C (2003) The tomato suppressor of prosystemin-mediated responses: gene encodes a fatty acid desaturase required for the biosynthesis of jasmonic acid and the production of a systemic wound signal for defense gene expression. Plant Cell 15:1646–1661

Li (2010) Exploration of wild relatives of tomato for enhanced stress tolerance. Thesis, Wageningen University, Wageningen

Li W, Chetelat RT (2010) A pollen factor linking inter- and intraspecific pollen rejection in tomato. Science 330:1827–1830

Li L, Zhao Y, McCaig B et al (2004) The tomato homolog of CORONATINE-INSENSITIVE1 is required for the maternal control of seed maturation, jasmonate-signaled defense responses, and glandular trichome development. Plant Cell 16:126

Li Y, Costello J, Holloway A, Hahn M (2008) "Reverse ecology" and the power of population genomics. Evolution 62:2984–2994

Lin JZ, Ritland K (1997) Quantitative trait loci differentiating the outbreeding *Mimulus guttatus* from the inbreeding *M. platycalyx*. Genetics 146:1115–1121

Liu Z, Mara C (2010) Regulatory mechanisms for floral homeotic gene expression. Semin Cell Dev Biol 21:80–86

Magwene PM, Willis JH, Kelly JK (2011) The statistics of bulk segregant analysis using next generation sequencing. PLoS Comput Biol 7(11):e1002255

Malmberg RL, Mauricio R (2005) QTL-based evidence for the role of epistasis in evolution. Genet Res 86:89–95

Martin B, Nienhuis J, King G, Schaefer A (1989) Restriction fragment length polymorphisms associated with water use efficiency in tomato. Science 243:1725–1728

Masly JP, Presgraves DC (2007) High-resolution genome-wide dissection of the two rules of speciation in *Drosophila*. PLoS Biol 5:e243

McCormack JE, Faircloth BC, Crawford NG et al (2012) Ultra-conserved elements are novel phylogenomic markers that resolve placental mammal phylogeny when combined with species-tree analysis. Genome Res 22:746–754

McDowell ET, Kapteyn J, Schmidt A et al (2011) Comparative functional genomic analysis of *Solanum* glandular trichome types. Plant Physiol 155:524–539

McKay J, Richards J, Nemali K et al (2008) Genetics of drought adaptation in *Arabidopsis thaliana* II. QTL analysis of a new mapping population KAS-1x TSU-1. Evolution 62:3014–3026

Medina-Tapia N, Ayala-Berdon J, Morales-Perez L, Miron Melo L, Schondube JE (2012) Do hummingbirds have a sweet-tooth? Gustatory sugar thresholds and sugar selection in the broad-billed hummingbird Cynanthus latirostris. Comp Biochem Physiol A Mol Integr Physiol 161:307–314

Miller RJ, Mione T, Phan H-L, Olmstead RG (2011) Color by numbers: nuclear gene phylogeny of *Jaltomata* (Solanaceae), sister genus to *Solanum*, supports three clades differing in fruit color. Syst Bot 36:153–162

Mitchell-Olds T, Schmitt J (2006) Genetic mechanisms and evolutionary significance of natural variation in *Arabidopsis*. Nature 441:947

Mitchell-Olds T, Willis JH, Goldstein DB (2007) Which evolutionary processes influence natural genetic variation for phenotypic traits? Nat Rev Genet 8:845–856

Moyle LC (2007) Comparative genetics of potential prezygotic and postzygotic isolating barriers in a *Lycopersicon* species cross. J Hered 98:123–135

Moyle LC (2008) Ecological and evolutionary genomics in the wild tomatoes (*Solanum sect. Lycopersicon*). Evolution 62(12):2995–3013

Moyle LC, Graham EB (2005) Genetics of hybrid incompatibility between *Lycopersicon esculentum* and *L. hirsutum*. Genetics 169:355–373

Moyle LC, Graham EB (2006) Genome-wide associations between hybrid sterility QTL and marker transmission ratio distortion. Mol Biol Evol 23:973–980

Moyle LC, Muir CD (2010) Reciprocal insights into adaptation from agricultural and evolutionary studies in tomato. Evol Appl 3:409–421

Moyle LC, Nakazato T (2008) Comparative genetics of hybrid incompatibility: sterility in two *Solanum* species crosses. Genetics 179:1437–1453

Moyle LC, Nakazato T (2009) Complex epistasis for Dobzhansky-Muller hybrid incompatibility in *Solanum*. Genetics 181:347–351

Moyle LC, Nakazato T (2010) Hybrid incompatibility "snowballs" between *Solanum* species. Science 329:1521–1523

Moyle LC, Payseur BA (2009) Reproductive isolation grows on trees. Trends Ecol Evol 24:591–598

Muir CD, Moyle LC (2009) Antagonistic epistasis for ecophysiological trait differences between *Solanum* species. New Phytol 183:789–802

Nakazato T, Bogonovich M, Moyle LC (2008) Environmental factors predict adaptive phenotypic differentiation within and between two wild Andean tomatoes. Evolution 62(4):774–792

Nakazato T, Warren DL, Moyle LC (2010) Ecological and geographic modes of species divergence in wild tomatoes. Am J Bot 97:680–693

Nesbitt TC, Tanksley SD (2002) Comparative sequencing in the genus *Lycopersicon*: implications for the evolution of fruit size in the domestication of cultivated tomatoes. Genetics 162(1):365–379

Olmstead RG, Bohs L, Migid HA, Santiago-Valentin E, Garcia VF, Collier SM (2008) A molecular phylogeny of the Solanaceae. Taxon 57(4):1159–1181

Orr HA (1995) The population genetics of speciation-the evolution of hybrid incompatibilities. Genetics 139:1805–1813

Owen CR, Bradshaw HD (2011) Induced mutations affecting pollinator choice in *Mimulus lewisii* (Phrymaceae). Arthropod-Plant Interact 5:235–244

Payseur BA, Place M (2007) Searching the genomes of inbred mouse strains for incompatibilities that reproductively isolate their wild relatives. J Hered 98:115–122

Peralta I, Knapp S, Spooner D (2007) The taxonomy of tomatoes: a revision of wild tomatoes (*Solanum* L. section *Lycopersicon* (Mill.) Wettst.) and their outgroup relatives (*Solanum* sections *Juglandifolium* (Rydb.) Child and *Lycopersicoides* (Child) Peralta). Syst Bot Monogr 84:1–186

Posto AL (2009) The genetics of pre-and postzygotic reproductive isolation in *Solanum* (Masters of Science Thesis, Indiana University, Department of Biology)

Presgraves DC (2008) Sex chromosomes and speciation in *Drosophila*. Trends Genet 24:336–343

Presgraves DC (2010) The molecular evolutionary basis of species formation. Nat Rev Genet 11:175–180

Ramsey J, Bradshaw HD, Schemske DW (2003) Components of reproductive isolation between the monkeyflowers *Mimulus lewisii* and *M. cardinalis* (Phrymaceae). Evolution 57(7):1520–1534

Rick CM (1963) Barriers to interbreeding in *Lycopersicon peruvianum*. Evolution 17:216–232

Rick CM (1973) Potential genetic resources in tomato species: clues from observations in native habitats. Basic Life Sci 2:255–269

Rick CM (1979) Biosystematic studies in *Lycopersicon* and closely related species of *Solanum*. In: Hawkes JG, Lester RN, Skelding AD (eds) The biology and taxonomy of the Solanaceae. Academic Press for the Linnean Society, London, pp 667–678

Rick, Lamm (1995) Biosystematic studies on the status of *Lycopersicon chilense*. Am J Bot 42(7):663–675

Rick CM, Quiros CF, Harry Lange W, Allen Stevens M (1976) Monogenic control of resistance in the tomato to the tobacco flea beetle: probable repellance by foliage volatiles. Euphytica 25:521–530

Rijpkema A, Gerats T, Vandenbussche M (2006) Genetics of floral development in Petunia. In: Soltis DE, LeebensMack JH, Soltis PS, Callow JA (eds) Advances in botanical research: incorporating advances in plant pathology, vol 44: developmental genetics of the flower., pp 237–278

Rodriguez-Saona CR, Musser RO, Vogel H et al (2010) Molecular, biochemical, and organismal analyses of tomato plants simultaneously attacked by herbivores from two feeding guilds. J Chem Ecol 36:1043–1057

Römer P, Hahn S, Jordan T et al (2007) Plant pathogen recognition mediated by promoter activation of the pepper Bs3 resistance gene. Science 318:645–648

Sablowski R (2010) Genes and functions controlled by floral organ identity genes. Semin Cell Dev Biol 21:94–99

Salinas M, Capel C, Alba JM et al (2012) Genetic mapping of two QTL from the wild tomato *Solanum pimpinellifolium* L. controlling resistance against two-spotted spider mite (*Tetranychus urticae* Koch). Theor Appl Genet 126:83–92

Sánchez-Peña P, Oyama K, Núñez-Farfán J, Fornoni J, Hernandez-Verdugo S, Marquez-Guzman J, Garzón-Tiznado JA (2006) Sources of resistance to whitefly (*Bemisia* spp.) in wild populations of *Solanum lycopersicum var.cerasiforme* (Dunal). Spooner GJ

Anderson et RK Jansen in Northwestern Mexico. Genet Resour Crop Evol 53(4):711–719

Schemske DW, Bradshaw HD (1999) Pollinator preference and the evolution of floral traits in monkeyflowers (*Mimulus*). Proc Natl Acad Sci USA 96:11910–11915

Schemske DW, Mittelbach GG, Cornell HV, Sobel JM, Roy K (2009) Is there a latitudinal gradient in the importance of biotic interactions? Annu Rev Ecol Evol Syst 40:245–269

Schilmiller AL, Howe GA (2005) Systemic signaling in the wound response. Curr Opin Plant Biol 8:369–377

Sicard A, Lenhard M (2011) The selfing syndrome: a model for studying the genetic and evolutionary basis of morphological adaptation in plants. Ann Bot 107:1433–1443

Smaczniak C, Immink RGH, Angenent GC, Kaufmann K (2012) Developmental and evolutionary diversity of plant MADS-domain factors: insights from recent studies. Development 139:3081–3098

Smith SD, Rausher MD (2011) Gene loss and parallel evolution contribute to species difference in flower color. Mol Biol Evol 28:2799–2810

Smith SD, Ané C, Baum DA (2008) The role of pollinator shifts in the floral diversification of *Iochroma* (Solanaceae). Evolution 62:793–806

Song B, Mitchell Olds T (2011) Evolutionary and ecological genomics of non-model plants. J Syst Evol 49:17–24

Stebbins GL (1957) Self fertilization and population variability in the higher plants. Am Nat 91:337–352

Stellari GM, Mazourek M, Jahn MM (2009) Contrasting modes for loss of pungency between cultivated and wild species of *Capsicum*. Heredity 104(5):460–471

Stewart C, Mazourek M, Stellari GM, O'Connell M, Jahn M (2007) Genetic control of pungency in *C. chinense* via the Pun1 locus. J Exp Bot 58(5):979–991

Stinchcombe JR, Hoekstra HE (2007) Combining population genomics and quantitative genetics: finding the genes underlying ecologically important traits. Heredity 100(2):158–170

Tao Y, Xhen SN, Hartl DL, Laurie CC (2003a) Genetic dissection of hybrid incompatibilities between *Drosophila simulans* and *D. mauritiana*. I. Differential accumulation of hybrid male sterility effects on the X and autosomes. Genetics 164:1383–1397

Tao Y, Zeng ZB, Li J, Hartl DL, Laurie CC (2003b) Genetic dissection of hybrid Incompatibilities between *Drosophila simulans* and *D. mauritiana*. II. Mapping hybrid male sterility loci on the third chromosome. Genetics 164:1399–1418

Tewksbury J, Manchego C, Haak DC, Levey D (2006) Where did the chili get its spice? Biogeography of capsaicinoid production in ancestral wild chili species. J Chem Ecol 32:547–564

Tewksbury JJ, Reagan KM, Machnicki NJ, Carlo TA, Haak DC, Peñaloza ALC, Levey DJ (2008) Evolutionary ecology of pungency in wild chilies. Proc Natl Acad Sci USA 105(33):11808–11811

Thaler JS, Owen B, Higgins VJ (2004) The role of the jasmonate response in plant susceptibility to diverse pathogens with a range of lifestyles. Plant Physiol 135:530–538

Thaler JS, Humphrey PT, Whiteman NK (2012) Evolution of jasmonate and salicylate signal crosstalk. Trends Plant Sci 17:260–270

Tomato Genome Consortium (2012) The tomato genome sequence provides insights into fleshy fruit evolution. Nature 485:635–641

True JR, Weir BS, Laurie CC (1996) A genome-wide survey of hybrid incompatibility factors by the introgression of marked segments of *Drosophila mauritiana* chromosomes into *Drosophila simulans*. Genetics 142:819–837

Turelli M, Moyle LC (2007) Asymmetric postmating isolation: Darwin's corollary to Haldane's rule. Genetics 176:1059–1088

Turelli M, Orr HA (2000) Dominance, epistasis and the genetics of postzygotic isolation. Genetics 154:1663–1679

Turesson G (1925) The plant species in relation to habitat and climate. Hereditas 6:147–236

Turner LM, Hoekstra HE (2006) Adaptive evolution of fertilization proteins within a genus: variation in ZP2 and ZP3 in deer mice (*Peromyscus*). Mol Biol Evol 23:1656–1669

Walia H, Josefsson C, Dilkes B, Kirkbride R, Harada J, Comai L (2009) Dosage-dependent deregulation of an AGAMOUS-LIKE gene cluster contributes to interspecific incompatibility. Curr Biol 19:1128–1132

Walsh BM, Hoot SB (2001) Phylogenetic relationships of *Capsicum* (Solanaceae) using DNA sequences from two noncoding regions: the chloroplast atpB-rbcL spacer region and nuclear waxy introns. Int J Plant Sci 162:1409–1418

Wang Y, Diehl A, Wu F et al (2008) Sequencing and comparative analysis of a conserved syntenic segment in the solanaceae. Genetics 180:391–408

Wasternack C, Hause B (2002) Jasmonates and octadecanoids: signals in plant stress responses and development. Prog Nucleic Acid Res Mol Biol 72:165–221

Xu X, Martin B, Comstock JP et al (2008) Fine mapping a QTL for carbon isotope composition in tomato. Theor Appl Genet 117:221–233

Xu S, Schlueter PM, Scopece G, Breitkopf H, Gross K, Cozzolino S, Schiestl FP (2011) Floral isolation is the main reproductive barrier among closely related sexually deceptive orchids. Evolution 65:2606–2620

Integrated Genomics Approaches in Evolutionary and Ecological Endocrinology

15

Jun Kitano, Asano Ishikawa, and Sean C. Lema

Abstract

Hormones can act on a variety of target tissues to regulate the expression of multiple phenotypic traits. Therefore, phenotypes regulated by the same hormones can be genetically correlated due to their common regulatory mechanism. Such genetic correlations may either facilitate or constrain adaptive evolution. In addition, hormone signaling pathways are regulated by external environmental factors, so hormones can mediate *phenotypic plasticity* and *polyphenism*. When different responses to environmental signals are favored, hormone signaling pathways can vary between populations and species exploiting dissimilar environments and thus mediate genotype-by-environment interactions. A complete understanding of the evolutionary causes and ecological implications of hormone signal variation requires examining several components of hormone signaling pathways across multiple individuals, populations, and species. Genomic technologies are excellent tools for undertaking genetic studies of naturally occurring variation in hormone signals. In this chapter, we review how genomic approaches can help to answer major questions in evolutionary endocrinology, including how environmental cues can be translated into phenotypic development through hormone pathways, how multiple hormone-mediated phenotypic traits are coupled and decoupled, how gene functions in hormone pathways influence the evolutionary rate of genes, and how divergence in hormone pathways

J. Kitano (✉)
Ecological Genetics Laboratory, National Institute of Genetics, Yata 1111, Mishima, Shizuoka 411-8540, Japan

Precursory Research for Embryonic Science and Technology, Japan Science and Technology Agency, Kawaguchi, Saitama, Japan
e-mail: jkitano@lab.nig.ac.jp

A. Ishikawa
Ecological Genetics Laboratory, National Institute of Genetics, Yata 1111, Mishima, Shizuoka 411-8540, Japan

S.C. Lema
Biological Sciences Department, Center for Coastal Marine Sciences, California Polytechnic State University, San Luis Obispo, CA 93407, USA

C.R. Landry and N. Aubin-Horth (eds.), *Ecological Genomics: Ecology and the Evolution of Genes and Genomes*, Advances in Experimental Medicine and Biology 781,
DOI 10.1007/978-94-007-7347-9_15, © Springer Science+Business Media Dordrecht 2014

can contribute to phenotypic diversification and speciation in non-model organisms. We also discuss how emerging analytical and experimental technologies in genomics and hormone measurement can provide valuable new insights into the roles of hormone signal variation in adaptive evolution and phenotypic diversification.

Keywords
Hormone • Pleiotropy • Behavior • Speciation • Next generation sequencer • Transcriptome

15.1 Introduction

Hormones play key roles in phenotype expression, and a single hormone can regulate a variety of morphological, behavioral, and physiological traits. As chemical messenger molecules, *hormones* are produced by specialized cells in endocrine organs and released into systemic circulation (Fig. 15.1; see the glossary for definitions of italicized terms). Once in circulation, hormones affect peripheral target tissues by binding to intracellular or cell membrane receptors, which subsequently triggers a transduction cascade of intracellular phosphorylation or dephosphorylation, changes in ion channel permeability, or an increase or decrease in gene transcription. *Endocrine pathways* are typically organized into an 'axis' with multiple levels of signaling control (Fig. 15.2), including upstream releasing and tropic hormones that regulate the production and secretion of hormones from other organs down-axis, and several feedback pathways (i.e., 'feedback loops') where a hormone down-axis regulates the release of hormones up-axis.

Variation in hormone signals has been found across disparate taxa (Adkins-Regan 2005; Bradshaw 2007). For example, animal populations inhabiting different latitudes often differ in circulating levels of thyroid hormones, which are important for regulating metabolism and body temperature in the homoiotherms (Ishikawa and Kitano 2012). Males from species exhibiting different mating systems (e.g., polygamy vs. monogamy) have been shown to differ in androgen hormone responses to attacks by competitive males (Wingfield et al. 1990; Hirschenhauser and Oliveira 2006). Such examples where naturally occurring variation in hormone signaling has been identified between individuals, populations, or species are now numerous and have been selectively reviewed elsewhere (e.g., Adkins-Regan 2005; Bradshaw 2007). Rather, in this chapter, we consider the genetic mechanisms underlying variation in endocrine pathway signaling.

Genetic variation in hormone signaling could arise from evolutionary changes in the structure or regulation of genes at several levels of the endocrine pathways from the gene that encodes the hormone itself to the genes for other components of the pathways such as blood carrier proteins, conversion enzymes, cell membrane transporters, and membrane or intracellular receptors (Fig. 15.1). A complete understanding of mechanisms by which hormone signaling variation evolves thereby requires examining simultaneously the production, transport, and metabolism of hormones, as well as both the hormone reception in target tissues and the cellular responses of gene expression. The study of such multidimensional endocrine phenotypes will be facilitated by integrating techniques from classical physiological and biochemical endocrinology with so-called "omics" technologies, such as genomics, transcriptomics, proteomics, metabolomics, and epigenomics.

In this chapter, we discuss how the study of endocrine signaling from an ecologically relevant, genomics perspective is enlightening our

understanding of the genetic and physiological bases for individual, population, and species-level phenotypic variation. We provide a brief conceptual overview of the questions currently being addressed in evolutionary endocrinology using genomic approaches, and outline a framework for how genomic technologies are being employed in endocrine research to advance our understanding of how environmental cues influence phenotypic development through hormone pathways, how hormone-mediated phenotypic traits are coupled and decoupled, how a gene's function within a hormone pathway influences the evolutionary rate of that gene and how divergence in hormone pathways can contribute to speciation. Lastly, we explore how emerging analytical and experimental approaches in genomics and hormone measurement can provide valuable new insights into the roles of hormone signal variation in phenotypic diversification and adaptive evolution.

15.2 Key Questions in the Evolution of Hormone-Signal Variation

15.2.1 How Do Correlations Among Hormone-Mediated Phenotypes Facilitate or Constrain Evolutionary Change?

Hormones can act on a variety of target tissues to regulate the expression of multiple phenotypic traits (Figs.15.1 and 15.2). For instance, the androgen testosterone regulates spermatogenesis in most vertebrates, and can also regulate social behaviors, secondary sexual characteristics, and immune function (Wingfield et al. 1990; Ketterson and Nolan 1992; Hau 2007; Mank 2007). AVT regulates not only aggression and courtship, but also hydromineral balance and

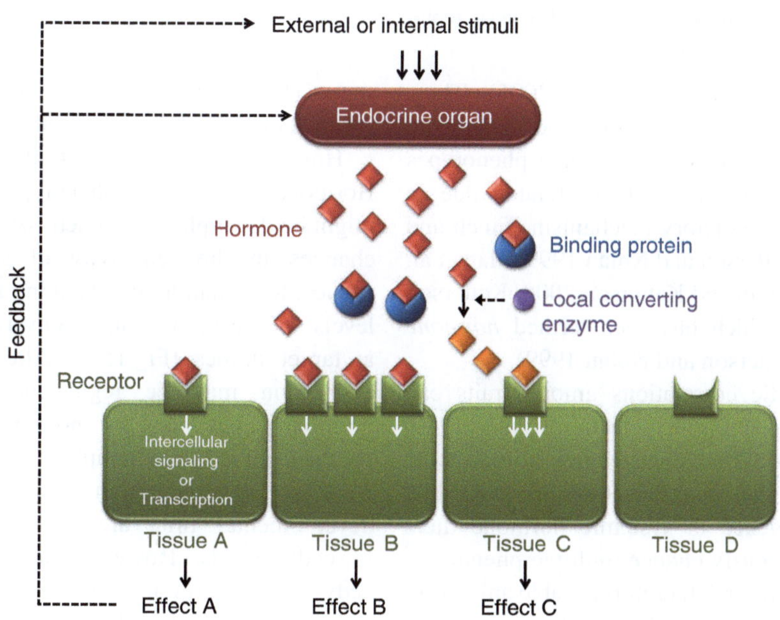

Fig. 15.1 The typical endocrine signaling axis is composed of multiple steps, each of which provides an opportunity for regulation of hormone secretion or hormone effects. Hormones are secreted from endocrine glands or organs in response to environmental or internal stimuli. Secreted hormones usually bind to extracellular carrier proteins and are then either stored in an extracellular matrix or carried through the circulatory system to multiple target tissues. Hormones might then metabolized by local converting enzymes at local target tissues to more active or inactive forms. Finally, hormones bind to receptors (membrane or intercellular) expressed at the target cells to exert their functions by changing intracellular signaling and/or transcriptomes

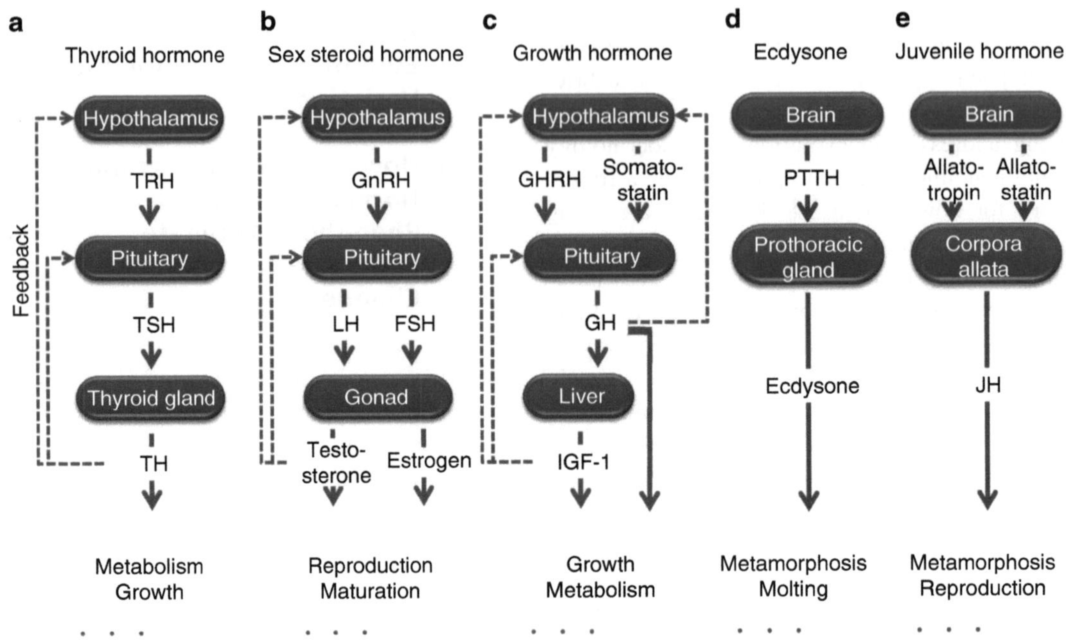

Fig. 15.2 Several examples of hormone axes. *Red axes* indicate examples from vertebrates, while *blue* ones indicate examples from insects. Abbreviations: *TRH* thyrotropin-releasing hormone, *TSH* thyroid-stimulating hormone, *TH* thyroid hormone, *GnRH* gonadotropin-releasing hormone, *LH* lutenizing hormone, *FSH* follicle-stimulating hormone, *GHRH* growth hormone-releasing hormone, *IGF-1* insulin like growth factor-1, *PTTH* Prothoracicotropic hormone, *JH* juvenile hormone

stress responses (Balment et al. 2006). Thus, it is common for a single hormone to regulate a suite of traits. Hormone-mediated phenotypes can therefore be genetically correlated due to their common regulatory mechanism (Finch and Rose 1995; Ketterson and Nolan 1999; Flatt et al. 2005; McGlothlin and Ketterson 2008; Ketterson et al. 2009), which has been termed *hormone pleiotropy* (Ketterson and Nolan 1999).

Such genetic correlations among traits can influence the response to selection (Lande 1979; Schluter 2000). When changes in a suite of traits controlled by the same hormones are favorable, genetic divergence in systemic hormone titers may simultaneously change multiple phenotypic traits at one time and, therefore, enable animals to rapidly adapt to new environments. By contrast, when independent evolution of different phenotypic traits regulated by the same hormones is favorable, adaptive evolution may become slow because of the presence of hormone-based genetic correlations. However, experimental tests of this idea in an ecologically-relevant context are sparse (Ketterson and Nolan 1999; Zijlstra et al. 2004; Hau 2007; Ketterson et al. 2009).

How can such correlated traits be decoupled? Hormonally-mediated phenotypic correlations might be decoupled via genetic or environmental changes in the sensitivity of specific target tissues, for example, by changing the expression levels of receptors and converting enzymes at target tissues (Fig.15.1). Alternatively, trait expression may be regulated not only by hormones, but also by hormone-independent mechanisms. For example, in the butterfly *Bicyclus anynana*, both wing eyespots and developmental time are under the influence of ecdysteroids. However, in addition to the ecdysteroid hormone system, other developmental mechanisms also play important roles in determining the eyespot size. Therefore, in an artificial selection experiment, eyespot size could change independently of ecdysteroid titer and developmental time (Zijlstra et al. 2004). Genomic studies that identify the network of genes regulated by a given hormone and examine

variation in these networks between individuals inhabiting dissimilar environments will provide valuable insights into the mechanisms underlying the developmental and genetic decoupling of hormonally-mediated multiple traits (see Sect. 15.4).

Endocrine studies can also inform our understanding of the genetic correlations between males and females. Males and females differ in optimal values of several phenotypic traits (Darwin 1874; Andersson 1994). For example, exaggerated male ornaments may be advantageous for males, because they increase male mating success, but disadvantageous for females as they can increase predation risk and, ultimately, decrease fitness (Arnqvist and Rowe 2005). In vertebrates, the development and expression of male ornaments is often controlled by sex steroids (Adkins-Regan 2005; Hau 2007). Therefore, sex steroid levels can be a target of sexually antagonistic selection (i.e., favorable for one sex, but detrimental for another sex); high androgen levels may be favorable for males, but detrimental for females (Roberts et al. 2004; Mank 2007). To fully understand the evolutionary mechanisms whereby sexually dimorphic traits evolved, it is crucial to understand how the genetic correlations of sex steroid signals are resolved in the evolution of male and female phenotypes (Rice 1984; Arnqvist and Rowe 2005; Cox and Calsbeek 2009; Williams and Carroll 2009; Kitano et al. 2011).

15.2.2 What Are the Roles of Hormones in Phenotypic Plasticity and Polyphenism?

Hormone signaling pathways are regulated not only by genetic factors, but also by external environmental factors. Because hormones act on a variety of tissues to regulate multiple functions in response to environmental stimuli, hormones can mediate *phenotypic plasticity*, the ability of a single genotype to produce different phenotypes (Zera and Harshman 2001; Nijhout 2003; Bradshaw 2007; Lema 2008). For example, metamorphosis in amphibians involves changes in morphology, physiology and behavior, and metamorphic timing is both plastic and hormonally controlled (Denver 1997, 1998). In several amphibian species such as the Western spadefoot toad (*Spea hammondii*), metamorphosis is accelerated by environmental cues indicative of a desiccating habitat, such as low water and reduced food. These cues trigger increases in neural corticotropin-releasing hormone (CRH) content and systemic thyroid hormone (T_3) levels, which elicit metamorphosis (Boorse and Denver 2004). Facultative metamorphosis caused by a variety of abiotic and biotic conditions also exists in some amphibian populations and species (Sexton and Bizer 1978). Similar scenarios where hormone-mediated plasticity contributes to population-level phenotypic variation also occur in other taxa, such as pupfish where specific environmental conditions (e.g., high temperature, low food) alter thyroid signaling to contribute to a developmentally neoteonic phenotype (Lema and Nevitt 2006).

One extreme form of phenotypic plasticity is *polyphenism*, in which organisms of the same genotype can develop into discrete phenotypes (Nijhout 2003). In most cases, hormone signaling pathways regulate the expression of polyphenism (Nijhout 2003; West-Eberhard 2003). For example, most aphids can switch their reproductive strategies from asexual to sexual reproduction in response to short photoperiods (reproductive polyphenism). Reproductive polyphenism in aphids is regulated by juvenile hormone (JH), whose titer decreases in response to short photoperiods via up-regulation of a JH degradation pathway (Ishikawa et al. 2012). In a second example, environmental temperature during development induces distinct adult eyespot morphs of the butterfly *B. anynana* (Oostra et al. 2011). Linear variation in developmental temperature generates a binary response in the timing – but not maximum level – of peak ecdysteroid hormone levels in this butterfly species, suggesting that ecdysteroid signaling has evolved to translate continuous environmental gradients into discrete responses of hormonal titers, ultimately generating a polyphenic pattern of phenotypic expression (Oostra et al. 2011).

Importantly, phenotypic plasticity is an evolvable trait. Organisms tend to evolve to an

increased capacity for phenotypic plasticity under unstable environments, but reduced capacity under stable environments (DeWitt and Scheiner 2004). In some cases, phenotypic plasticity first allows populations to survive when faced with environmental changes, but the subsequent genetic evolution can also occur, leading to the genetic fixation of phenotypic traits (Price et al. 2003), a process that has been called *genetic assimilation* (Waddington 1961) (Glossary). The genetic modification of hormone signaling pathway can underlie evolutionary changes in plasticity. For example, rapid development can be favored in environments with high predation pressures or high desiccation risk (Buchholz and Hayes 2005), whereas stable, low food environments favor slower development and even lead to sexual maturation without metamorphosis (paedomorphosis) (Werner 1986; Richter-Boix et al. 2011). The absence of metamorphosis in the axolotl (*Ambystoma mexicanum*), for example, is caused by fixation of low thyroid hormone production (Shaffer 1993) and a low tissue capacity to respond to thyroid hormones (Voss et al. 2012). In the case of reproductive polyphenism of aphids, several aphid populations have lost the sexual phase (Moran 1992), and in these populations, neither JH titer nor JH-degradation enzymes show photoperiodic changes (Ishikawa et al. unpublished). Thus, modification of hormone pathways can lead to variation in phenotypic plasticity and polyphenism between species and populations.

15.2.3 Does Divergence in Hormone Signaling Pathways Promote Speciation?

Ecological adaptation to divergent environments can lead to the establishment of reproductive isolation (Schluter 2000; Nosil 2012). For instance, contrasting environments can select for different hormonal phenotypes (Bradshaw 2007; Kitano et al. 2010). Once two populations exploiting contrasting environments acquire divergent hormone signaling pathways, migrants from different habitats or inter-population hybrids may lack optimal hormone responses and, therefore, have lower fitness than native pure populations, thereby resulting in reduced gene flow between divergent populations. Thus, studies on adaptive divergence in hormone signals are important for ecological speciation research.

In addition, sex steroids regulate the expression of male display signals (Adkins-Regan 2005; Hau 2007; Mank 2007) and female mate choice behaviors (McGlothlin et al. 2004; Alvergne and Lummaa 2009; Gabor and Grober 2010; Ramsey et al. 2011). Therefore, divergence in sex steroid signaling can affect the patterns and magnitude of sexual isolation. Incipient sympatric species of three-spined stickleback (*Gasterosteus aculeatus*) were found to differ in sex steroid levels both in males and females (Kitano et al. 2011), although the functional roles of sex steroid divergence remain elusive in this sympatric pair. It will be crucial to identify the hormonally-regulated gene networks that shape variation both in ecologically important traits and in species recognition traits among populations to better understand speciation mechanisms (Crews and Williams 1977).

15.3 How Genomic Technologies Can Be Employed in Endocrinology

The application of genomic methods to the study of endocrinology in an ecologically relevant context is still in its early stages. Here, we describe current genomic approaches that can be applied to the genetic study of phenotypic variation resulting from hormone signaling variation in the wild, and use several studies as examples to illustrate how these techniques will advance our understanding of endocrine network evolution and evolutionary constraints on adaptive evolution.

15.3.1 Sequence Comparison Among Species or Populations

Next-generation sequencing (NGS) methods greatly facilitate the study of protein sequences for genes encoding polypeptide hormones,

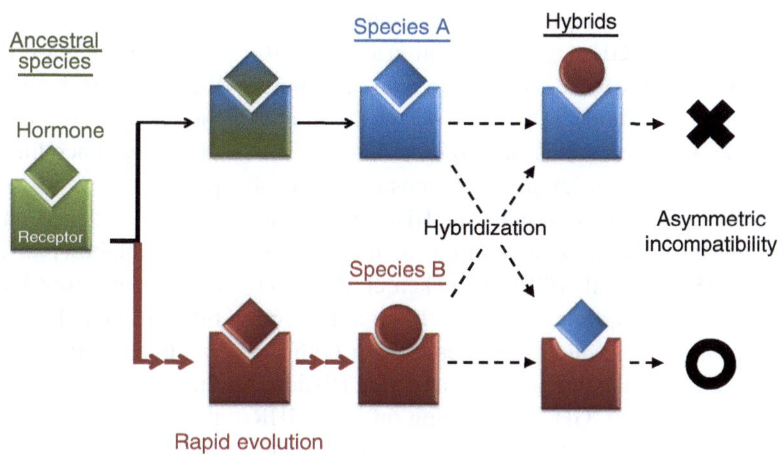

Fig. 15.3 Evolution of hormone-receptor incompatibities. Rapid evolution of the ligand and the receptor in the lineage leading to Species B (*red*) cause hybrid incompatibility between the ligand of Species A (*blue*) and the receptor of Species B (*red*)

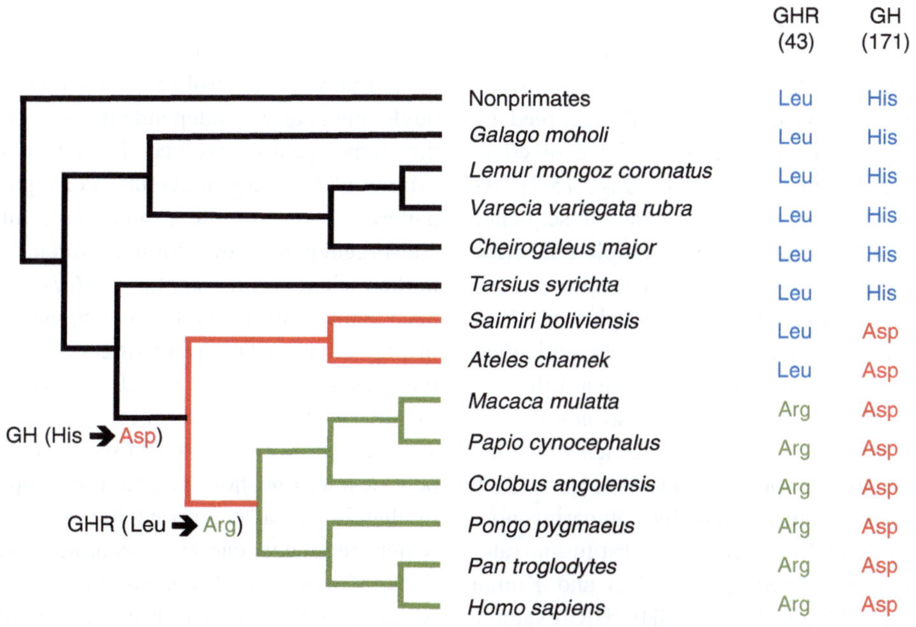

Fig. 15.4 Amino acid residue at position 43 of growth hormone receptor (GHR) and position 171 of GH are shown next to the phylogenetic tree of primates. The tree shows only topology of phylogenetic relationships, so branch lengths do not correspond to divergence time. The lineages that acquired amino acid changes are shown on the branches. The tree is based on Liu et al. 2001

hormone-binding carrier proteins, membrane transporters, and target tissue converting enzymes and receptors as well as the non-coding sequences that regulate their expression. NGS is easily applied to non-model organisms (Gilad et al. 2009; Johansson 2009; Stapley et al. 2010; Ekblom and Galindo 2011). Comparison of sequences of endocrine-related genes across diverse taxa should provide valuable insights into many important aspects of evolutionary processes, such as co-evolution between interacting proteins, selective pressures on hormone genes, and the genetic basis for convergent evolution. For example, the study of growth hormone (GH) signaling in primates suggests that rapid amino acid changes of hormones and receptors may cause hybrid incompatibility (Figs. 15.3 and 15.4). Non-primate GH cannot activate the

human (*Homo sapiens*) GH receptor (GHR), while human GH can activate non-primate GHRs (Souza et al. 1995). This differential specificity is due to variation at position 43 of the human GHR, which contains an Arginine residue that interacts with an Asparate at position 171 of human GH, but will repulse the Histidine residue found at the equivalent position of non-primate GHs (Souza et al. 1995; Behncken et al. 1997). Non-primate GHRs, however, have a Leucine at position 43, which can interact with both the Asparate in the human GH and the Histidine in non-primate GH. Sequencing of the GHR gene from multiple primate species revealed that position 43 evolved from a Leucine to an Arginine in the common ancestor of Old World primates (Liu et al. 2001). Interestingly, New World monkeys have the ancestral state, where position 43 of GHR is still occupied by a Leucine, but position 171 of GH evolved to an Asparate (Liu et al. 2001) (Fig. 15.4). As this example illustrates, when genomic sequence information is obtained from multiple endocrine genes from several closely-related species, that information can provide novel insights into how evolutionary changes in hormones and their receptors contribute to hybrid incompatibility.

When cDNA sequence data are available from multiple species, one can also investigate whether genes were under positive selection or relaxed selection in certain lineages by comparing non-synonymous and synonymous substitution rates across closely-related species (Nei and Kumar 2000). In addition to the GH/GHR example described above, prolactin (PRL) genes are also known to have evolved rapidly in primates (Ohta 1993; Wallis 2000; Wallis et al. 2005). The patterns of rapid evolution seen in both GH and PRL protein sequences may have resulted from parent-offspring conflict (Haig 2008). During pregnancy in primates, pituitary GH production in the mother is suppressed, and the placenta produces both PRL and a GH form variant (GH-V) that are bioavailable to act on the receptors of mothers (Handwerger and Freemark 2000). The structure of the GH-V and PRL hormones or the regulatory mechanisms for production of these hormones from the placenta may thus evolve to increase the fitness of the fetus, while the hormone receptors may evolve to increase the fitness of the mother. Rapid evolution of genes encoding growth hormone/prolactin pathways was also found in two other lineages, muroid rodents and ruminants (Haig 2008). The availability of genomic or cDNA sequence data from several closely-related primates and other placental mammals should help elucidate whether selective pressures acted differently on these hormone signal genes and receptor genes in the past.

The large coverage of gene sequences obtained by NGS can also help address another question in evolutionary endocrinology: when the same patterns of convergent evolution of hormonally controlled characters occur in phylogenetically independent lineages, are the same genes involved in that convergent evolution? Convergent evolution of pigmentation patterns is one of the prominent examples of convergent genetic evolution. Different mutations in the melanocortin receptor-1 (*Mc1r*) gene have been shown to generate convergent evolution in pigmentation among phylogenetically distant lineages (Manceau et al. 2010). The primary ligand of the *Mc1r* receptor is α-melanocyte stimulating hormone (α-MSH). The α-MSH peptide hormone, however, has pleiotropic effects in that it can also activate the *Mc3r* receptor, which regulates energy expenditure and food intake (Chen et al. 2000), and the *Mc5r* receptor, which influences exocrine gland activity (Chen et al. 1997). The convergent evolution seen in *Mc1r* genes might therefore be indicative of constraints on the evolution of α-MSH, since any evolutionary changes in α-MSH peptide structure might generate correlated changes in other traits as well (Ducrest et al. 2008). Future studies that use NGS to examine genetic variation in other hormone pathways promise to enlighten whether convergent evolution of receptors may be a general feature of hormone-controlled trait evolution and whether such convergence might occur due to evolutionary constraints associated with the pleiotropic effects of hormones.

15.3.2 Identification of Genomic Regions Under Selection

Genomic loci under recent natural selection are expected to show reduced genetic diversity, long linkage disequilibrium, and increased divergence between populations (Nielsen 2005; Sabeti et al. 2006; Oleksyk et al. 2010). Genome-wide single nucleotide polymorphisms (SNPs) can be identified using NGS (Davey et al. 2011), and then used to detect genes or genomic loci under natural selection. Whole genome re-sequencing (Jones et al. 2012b), transcriptome sequencing of multiple individuals (Sauvage et al. 2012b), and restriction site-associated DNA (RAD) tag sequencing (Miller et al. 2007; Baird et al. 2008; Hohenlohe et al. 2010; King et al. 2012) are options for identifying SNP markers. These methods can also be used directly for genome scan analysis (Hohenlohe et al. 2010; Ellegren et al. 2012; Jones et al. 2012b). Alternatively, identified sets of SNPs can be used for designing custom SNP assay systems and then applied in genome scanning (Deagle et al. 2011; Jones et al. 2012a).

Genes involved in hormone signals or genes expressed in endocrine organs are often under selection (Kitano et al. 2010; López Herráez et al. 2010; Ishikawa and Kitano 2012; Jarvis et al. 2012). For example, in African Pygmies, two genes involved in thyroid hormone signaling, thyroid hormone receptor interactor 4 (TRIP4) and iodotyrosine deiodinase (IYD), are found at the peak of the signature of selection (López Herráez et al. 2010). Because African Pygmies rarely show goiter, despite the fact that they usually inhabit iodine-deficient geographic regions (Dormitzer et al. 1989), these changes in TRIP4 and IYD might reflect genetic adaptations of Pygmies to iodine-deficient diets. In another set of studies addressing the evolution of thyroid hormone signaling, Kitano and co-workers (2010) found signatures of divergent selection at the thyroid-stimulating hormone-β2 subunit (*tshβ2*) locus between marine and stream-resident ecotypes of three-spined stickleback. Populations of marine and stream-resident sticklebacks show repeated evolution of systemic thyroid hormone levels and pituitary *tshβ2* mRNA expression (Kitano et al. 2010; Kitano and Lema 2013). Hormonal manipulations demonstrated that thyroid hormones increase swimming activity and metabolic rate in marine sticklebacks (Kitano et al. 2010). High thyroid hormone levels may be adaptive for marine ecotypes because their long distance migrations demand a lot of energy, whereas high thyroid hormone levels may be deleterious for stream-resident ecotypes because thyroid hormones increase oxygen consumption rate and oxygen is often scarce in stream habitats. And, as a third example, thyroid-stimulating hormone (TSH) receptor loci (*tshr*) were found to be under strong selection during domestication in chickens (Rubin et al. 2010) and sheep (Kijas et al. 2012). Because TSH regulates photoperiodic control of reproduction (Hanon et al. 2008; Nakao et al. 2008; Dardente et al. 2010), artificial selection for the independency of reproduction from natural photoperiods might act on the *tshr* locus. These examples of evolutionary variation in thyroid hormone signaling illustrate how genome scan analysis can be used to identify hormone signaling pathways under selection.

15.3.3 Quantitative Trait Loci (QTL) Mapping and Genome-Wide Association Studies (GWAS)

Quantitative trait loci (QTL) mapping and *genome-wide association studies (GWAS)* are able to detect genomic loci associated with particular phenotypes. As described above, NGS can facilitate identification of large numbers of SNPs. These SNP-detection methods, such as RAD tag sequencing, can then be directly used for genotyping multiple individuals of hybrid progeny for QTL mapping and GWAS (Baird et al. 2008; Chutimanitsakun et al. 2011; King et al. 2012) or used for making custom SNP assay systems (Kitano et al. 2009; Greenwood et al. 2011; Sauvage et al. 2012b; Wark et al. 2012).

By conducting QTL mapping, we can not only identify the genomic loci significantly

associated with phenotypic variation, but also estimate the number and the effect size of causal genes (Lynch and Walsh 1998). QTL mapping of hormone levels have been conducted in a number of domesticated and laboratory animals. Most studies investigate QTL for hormones related to growth and stress responses. Examples include QTL mapping of variation in insulin-like growth factor 1 (IGF1) levels in mice (Rosen et al. 2000; Harper et al. 2003; Leduc et al. 2010), levels of 17β-estradiol and IGF1 levels and stress-induced cortisol levels in salmonids (Sauvage et al. 2012a, b), and thyroid hormone status in rats (Harper et al. 2003; Baum et al. 2005). QTL mapping of the response to thyroid hormone has also been conducted in non-model organisms; the developmental response to thyroid hormones was investigated in the backcross hybrids between a paedomorphic *A. mexicanum* and an obligatory metamorphic *A. tigrinum tigrinum*, and a few major QTLs were found, suggesting that the loss of phenotypic plasticity in *A. mexicanum* might be caused by changes in a small number of major effect genes (Voss et al. 2012). Further QTL mapping with more species will lead to a better understanding of the general features of genetic architectures underlying hormonal variation and hormone-mediated phenotypic variation.

GWAS is used to find common genetic variants associated with particular phenotypic traits, such as susceptibility to diseases (Hindorff et al. 2009; Barsh et al. 2012). GWAS is a good option to find candidate loci associated with particular traits in cases where laboratory crosses are difficult to obtain, but a number of individuals can be sampled from natural populations. For example, the National Human Genome Research Institute, USA, developed an online database of human GWAS (http://www.genome.gov/gwastudies) (Hindorff et al. 2009). On this database (as of Feb 2, 2013), there are 8,387 SNPs found associated with hormone-related traits. Loci with these SNPs will be good candidate genes for potential associations with hormone-related traits in ecologically important non-model organisms.

15.3.4 Analysis of Variation in Transcriptomes: Mechanisms of Transcriptional Regulation

By using microarray approaches and RNA sequencing, transcript levels of multiple endocrine-related genes can now be assayed at one time (Aubin-Horth and Renn 2009). Studies using these transcriptome approaches have found that divergence in transcript levels can occur at each level of the hormone axis illustrated in Fig. 15.1. For example, microarray screening studies of genes responding to photoperiodic cues in quail (*Coturnix japonica*) (Nakao et al. 2008), rats (Ross et al. 2011), and sticklebacks (Ishikawa et al. unpublished), each found transcription of mRNAs encoding the β-subunit of TSH (*tshβ*) to be under strong photoperiodic control. However, evolutionary variation in the transcriptional response of *tshβ* to photoperiod has also been demonstrated (Dardente et al. 2010; Kitano et al. 2010).

Transcriptome differences between populations and species can be caused by *cis*-regulatory changes, *trans*-regulatory changes, and their interactions (Wittkopp et al. 2004; Landry et al. 2005; McManus et al. 2010). Allele-specific expression analysis of F1 hybrids between populations or species are possible using pyrosequencing technologies or NGS when the transcripts have SNPs that can be used to distinguish between cDNAs derived from two parental populations/species (Wittkopp et al. 2004; Landry et al. 2005; McManus et al. 2010). For example, RNA-sequencing of *Drosophila melanogaster, D. sechellia*, and their F1 hybrids revealed that *cis*- and *trans*-regulatory divergence influences 51 % and 66 % of expressed genes, respectively, with 35 % of genes being affected by both (McManus et al. 2010).

Cis-regulatory mutations in hormones and hormone receptors have also been identified. For example, in the case of transcriptional regulation of *tshβ2* in mammals, variation in a conserved D-element in the *tshβ* gene promoter, which binds to transcription factor complexes, has been found to underlie variation in photoperiodic

responses among species (Dardente et al. 2010). Variation in a microsatellite located in the *cis*-regulatory region of the gene encoding the V1a-type receptor (*avpr1a*) for arginine vasopressin (AVP) has been tied to species differences both in brain expression pattern of the V1a receptor and in social attachment and sexual fidelity behaviors between monogamous prairie voles (*Microtrus ochrogaster*) and nonmonogamous montane voles (*M. montanus*) (Young et al. 1999; Insel and Young 2000; Lim et al. 2004). In a rather elegant application of genetic techniques, insertion of the monogamous prairie vole *avpr1a* gene into polygamous mice resulted in a change in V1a receptor distribution in the mouse brain and a corresponding shift in its behavior toward monogamy (Young et al. 1999), showing that the genetic variation underlying these receptor expression differences occurs in the 5′ flanking region of the *avpr1a* gene, and not in the coding region itself. Along similar lines, the *H. sapiens*-specific loss of a *cis*-regulatory region of the androgen receptor gene may underlie differential androgen receptor expression and sexual dimorphism between humans and apes (McLean et al. 2011).

Taken together, these studies demonstrate that integration of genome sequence analysis and transcriptome analysis can be used to identify *cis*-regulatory regions responsible for variation in endocrine-associated gene expression and, ultimately, in hormone-mediated traits. In addition, by using QTL and association mapping, it should also be possible to find *trans*-regulatory regions responsible for differential expression of hormone-related genes (Gilad et al. 2008; Majewski and Pastinen 2011). Finally, differences in transcript expression patterns can result not only from changes in nucleotide sequences but also by *epigenetic* modification of *cis*-regulatory regions, which we will discuss in the next section.

15.3.5 A Role for Epigenetic Regulation

Epigenetic regulation includes DNA methylation and histone modifications (Allis et al. 2007), which can persist through mitotic or meiotic cell division (see Glossary for definition of epigenetics used here). Such mitotically or meiotically heritable epigenetic modifications can alter gene expression levels without the need of nucleotide changes, so will be overlooked by comparisons of nucleotide sequences alone. There are several methods of genome-wide DNA methylation and histone modification analysis, details of which are available elsewhere (Moss and Leblanc 2009; Tost 2009). Importantly, recent studies have demonstrated that there is substantial variation in epigenome between *Arabidopsis* populations (Becker et al. 2011; Schmitz et al. 2013), suggesting the possibility that epigenetic modifications may play important roles in producing phenotypic diversity. At present, however, the application of epigenetic analyses to endocrinology is still in its infancy, and our understanding of the endocrine epigenome comes almost entirely from controlled laboratory studies.

Even so, a number of genes involved in hormone signaling have already been found to be under epigenetic control (Zhang and Ho 2011). For example, in neonatal rodents, low levels of maternal licking and grooming have been shown to induce CpG hyper-methylation in the 5′ promoter regions of a glucocorticoid receptor gene and an estrogen receptor gene (Weaver et al. 2004; Champagne et al. 2006). Socially-induced changes in DNA methylation of these receptor genes may lead to altered hypothalamic-pituitary-adrenal (HPA) axis stress reactivity and reproductive behaviors, when the pups become adults (Champagne and Curley 2009).

Hormones themselves can alter the epigenetic regulation of other genes (Fowden and Forhead 2009). For example, in the bed nucleus of the stria terminalis of rats, exogenous testosterone treatment increases, while castration decreases, CpG methylation in the promoter region of the gene encoding AVP, but has the opposing effects on promoter methylation of the gene encoding estrogen receptor α (Auger et al. 2011). Likewise, glucocorticoids, which become elevated in response to isolation stress during adolescence, modulate CpG methylation in the promoter of the tyrosine hydroxylase gene in dopaminergic

neurons in mice (Niwa et al. 2013). Hormones regulate not only CpG methylation, but also histone modification. For instance, estrogen has been shown to promote histone acetylation in the promoter region of the *Kiss1* gene encoding kisspeptin, effectively down regulating *Kiss1* transcription (Tomikawa et al. 2012). Histone modification also may play an important role in the negative feedback regulation present in most endocrine axes (Sasaki et al. 1999; Wang et al. 2010).

Epigenetic modification is often induced by *small RNAs*. Small RNAs, such as microRNAs and small interfering RNAs, are non-coding RNAs that range from 20 to 30 nucleotides in length. Small RNAs are best known for silencing gene expression at the transcriptional level by inhibiting translation of mRNAs, but small RNAs can also interact with DNA to induce DNA methylation and histone modifications (Filipowicz et al. 2008; Ghildiyal and Zamore 2009; Kim et al. 2009). There are now several studies suggesting that hormone-induced transcriptional changes are regulated by small RNAs (Kuokkanen et al. 2010), and that hormones themselves can alter small RNA transcriptomes (Bethke et al. 2009). Recent genomic studies have also revealed divergence in small RNA transcriptomes between closely related species of cichlids (Loh et al. 2011) and sticklebacks (Kitano et al. 2013), suggesting that evolutionary changes to the small RNA transcriptome may play a role in phenotypic differentiation in the wild.

Importantly, some epigenetic modifications show transgenerational inheritance, although we know at present little about the importance of transgenerational epigenetic inheritance in evolutionary changes (Grossniklaus et al. 2013). For example, prenatal exposure of humans to famine reduced DNA methylation at the insulin-like growth factor-2 (*igf2*) locus, and this environmentally-induced epigenomic change was inherited across generations (Heijmans et al. 2008). Endocrine disruptors have also been shown to alter DNA methylation in the germ line, ultimately resulting in reduced male fertility (Anway et al. 2005), altered stress responses (Crews et al. 2012), and changes in mate preferences (Crews et al. 2007) in the progeny. These DNA methylation patterns and impaired phenotypes generated by endocrine disruptor exposure can be maintained for generations (Anway et al. 2005; Crews et al. 2007, 2012).

Although still limited, such evidence suggests that studies that examine epigenetic variation in hormone signaling genes in an ecologically relevant context could lead to a better understanding of the contribution of epigenetic regulation to phenotypic diversification in natural populations. Recent advances in genomic technologies are making it possible to conduct whole-genome methylome analysis and chromatin immunoprecipitation (ChIP) sequencing on populations of many non-model species (Moss and Leblanc 2009; Tost 2009). These epigenetic analyses enable us to find epigenetically modified genomic loci that might explain phenotypic diversity, but would be overlooked by comparing nucleotide sequences only.

15.4 Network Analysis of Hormone Pathway Evolution

Because of the comprehensive data that genomic technologies can produce, we will be able to address questions about the evolutionary constraints on hormone-mediated phenotypes that were intractable just a few years ago. For example, when hormone pathways evolutionarily diverge between species, which components of these pathways – the hormone signal, conversion enzyme, receptor, etc. (see Fig. 15.1) – evolve most rapidly? In other words, how does the functional role of a gene product in a hormone pathway influence the impacts of selection on that gene?

Mutations that change circulating hormone levels may affect multiple traits given the *pleiotropic effect* of many hormones. Amino acid changes in hormone receptors, however, might also have pleiotropic effects, since most receptors are expressed in multiple tissues. Therefore, nucleotide changes in the structure of hormones and their receptors may be constrained and evolutionarily conserved. In contrast, changes

in the sensitivity of specific target tissues by shifts in the activity of local hormone-converting enzymes, the abundance of membrane transporters, or the type and relative density of receptors may alter only a portion of a hormone's biological functions, so these changes may occur more frequently. Studies examining the effects of exogenous hormones (e.g., thyroid hormone) on endocrine gene (e.g., deiodinase enzyme, thyroid hormone receptors) transcription in several tissues have revealed that the same endocrine gene can have distinct patterns of expressional regulation in different target tissues (Soma et al. 1999; Johnson and Lema 2011). The genetic or epigenetic mechanisms underlying these tissue-specific responses may allow for evolutionary decoupling of hormone-mediated traits. These predictions can now be empirically tested more easily by looking at several components of the hormone signaling pathways simultaneously with the employment of the genomic technologies.

Comparisons of evolutionary rates between different components in a hormone axis will require the application of network analysis approaches. For instance, in the analysis of non-synonymous and synonymous mutation rates of genes involved in the plant anthocyanin biosynthetic pathway, genes encoding downstream enzymes were found to have evolved faster than genes encoding upstream genes (Rausher et al. 1999; Lu and Rausher 2003), suggesting that there may be evolutionary constraints on the upstream genes. Similar results where down-axis genes evolved faster than up-axis genes were found in studies of the terpenoid synthesis pathway of angiosperms (Ramsay et al. 2009; Yang et al. 2009), the high osmolarity glycerol (HOG) pathway of yeast (Wu et al. 2010), and both the insulin/TOR pathway and Ras signaling pathway of *Drosophila* (Alvarez-Ponce et al. 2009). However, there are also cases where the opposite relationship between evolutionary rate and pathway position have been found (Olsen et al. 2002; Cork and Purugganan 2004). For example, in vertebrate sex determination pathways, genes involved in signal cascades in the downstream are relatively conserved, while master sex determination genes often differ between closely related species (Graves and Peichel 2010). The rapid changes of sex determination genes might be linked to the rapid turnover of sex chromosomes, which could be promoted by sexually antagonistic selection (van Doorn and Kirkpatrick 2007).

A negative correlation between connectivity and the rate of protein evolution was also found in the protein-protein interaction network of yeast. In this study, the most highly connected proteins were found to evolve more slowly than proteins with fewer direct network connections (Fraser et al. 2002). Interestingly, however, this difference in evolutionary rate may not have been due to the number of network connections, but rather due to a greater proportion of the structure of highly connected proteins being involved in functional protein interactions. There is also evidence that genes expressed by many tissues may evolve more slowly than genes expressed in only one or a few tissues (Duret and Mouchiroud 2000). In the case of hormones, might this mean that the structure and regulation of versatile hormones that act on multiple tissues evolve more slowly than hormones that act on few tissues? What about hormone receptors or conversion enzymes? Might genome duplication relax any such genetic constraints (Ohno 1970; Zhang 2003)? Because the network structures of hormone pathways are relatively simple (Figs. 15.1 and 15.2), hormone pathways may be a good model system for studying gene network evolution.

15.5 The Future of Ecological and Endocrine Genomics

15.5.1 Genetic Engineering

One of the advantages of focusing on hormone-mediated traits is that we can experimentally manipulate hormone levels by injecting or implanting exogenous hormones or receptor antagonists and directly study hormone function in vivo (Ketterson et al. 1996). However, in order to understand the effects of particular genetic changes, it is also essential to manipulate genes. Especially in invertebrates, RNAi seems to work well to suppress the functions of particular genes

(Martin et al. 2006; Zhou et al. 2007; Minakuchi et al. 2008; Sim and Denlinger 2008; Pamuru et al. 2012). For the manipulation of nuclear sequences, zinc-finger nucleases (ZFNs), which recognize arrays of nucleotide triplets and induce double-strand DNA breaks, were developed (Carroll 2005). Although it is theoretically possible to manipulate any genes in any organisms with ZFNs, context-dependent effects have limited their widespread use (Ramirez et al. 2008). Recent improvement of transcription activator-like effector nucleases (TALEN) technologies has overcome this shortcoming (Bogdanove and Voytas 2011). These enzymes consist of a fusion between a transcription activator-like (TAL) effector DNA recognition domain and a FokI nuclease domain. Because the amino acid repeats of TAL effectors independently recognize a single nucleotide of a target DNA, the modularity of DNA recognition allows customization of the TAL domain to target any specific sequences. The TALEN technology has become a powerful method for genomic editing in a lot of model organisms (Li et al. 2011; Ansai et al. 2012; Lei et al. 2012; Ma et al. 2012; Ochiai et al. 2012; Watanabe et al. 2012). Thus, technologies like TALEN are very promising for gene manipulation in non-model organisms. In the next decades, we expect to see more progress in the technologies of genetic engineering of non-model organisms.

15.5.2 Quantifying the "Hormonome"

One of the remaining challenges in endocrinology is the quantification of systemic or tissue levels of multiple hormones simultaneously, especially when studying peptide hormones from non-model organisms. While transcriptome analyses such as real-time quantitative RT-PCR (qRT-PCR) or digital PCR can be used to quantify relative or absolute mRNA levels of many peptide hormones from a single endocrine tissue, the relationships between mRNA levels and protein expression are not consistently coupled under all conditions (Vogel and Marcotte 2012), and quantification of protein hormone levels needs to be conducted separately. Analyses of circulating or tissue levels of hormones usually require hormone-specific antibodies that can be used for enzyme-linked immunosorbent assays (ELISAs) or radioimmunoassays (RIAs). Such antibodies are rarely available for peptide hormone measurement in non-model species, and even antibodies for thyroid hormones or steroid hormones may not work effectively in all species depending on a given species' composition of hormone binding proteins.

However, new approaches like liquid chromatography-tandem mass spectrometry (LC-MS/MS) are emerging that will enable analysis of multiple hormones at one time (Soldin and Soldin 2009, 2011; Faupel-Badger et al. 2010). For example, in a LC-MS/MS analysis of feces of a New World monkey *Cebus capusinus*, 10–20 steroids could be simultaneously measured to confirm that adult males have higher androgens than juvenile males and lactating females (Weltring et al. 2012). Such parallel analyses of a suite of hormones – of what might be termed the "hormonome" – will make it possible to get a comprehensive picture of hormone signal divergence between ecologically important species, even those from which large amounts of blood may be difficult to obtain. As the application of these new, more comprehensive methods for hormone quantification becomes increasingly common and integrated with transcriptomic or genomic data, we can expect the emergence of a clearer picture of how the multidimensional endocrine phenotype is shaped by environmental conditions and evolutionary history, and of how those hormone signals shape ecologically relevant patterns of phenotypic variation among individuals, populations and species.

15.6 Conclusions

Genomic approaches now offer many new tools for identifying how genetic variation in hormone signaling pathways relates to variation in behavioral, physiological and life history

phenotypes. As genomic technologies accelerate the discovery of genetic mechanisms underlying hormone signal variation, we expect that interdisciplinary studies that use genomic and physiological methods to address questions in ecological and evolutionary contexts will provide valuable insights into the adaptive evolution of endocrine signaling pathways, hormone-mediated variation in phenotypic plasticity, and the role of hormone pleiotropy and evolutionary constraints in adaptation and speciation. The further integration of methods from endocrinology and ecological genomics promises an exciting time for evolutionary endocrinology in the coming years, as our expanding understanding of hormone signaling contributes a better appreciation for the processes generating the evolutionary diversity of life.

Glossary

Endocrine signaling pathway The collection of physiologically-linked hormones, binding proteins, membrane transporters, conversion enzymes and receptors that jointly mediate the effects of a hormone signal on a target cell

Endocrine system The system of ductless glands or organs that secrete hormone molecules into systemic circulation

Epigenetics There exists several definitions of epigenetics. Here, we define epigenetic modification as mitotically heritable functional modification of gene expression, including DNA methylation and histone modification, which can remain through cell divisions for the reminder of the cell's life. Some of them are meiotically heritable and show transgenerational inheritance

Genetic assimilation Environmentally induced phenotypic variation that becomes constitutively produced without environmental cues

Genome-wide association studies (GWAS) Identification of common genetic variants associated with a particular phenotypic trait, such as disease, by genotyping many individuals

Hormone Chemical messenger produced by specialized populations of cells located in an endocrine organ, released into systemic circulation, and carried throughout the organism's body. Examples of hormones include polypeptides (e.g., growth hormone [GH], arginine vasotocin [AVT], gonadotropin-releasing hormone [GnRH]), modified amino acids (e.g., thyroid hormones, epinephrine, serotonin), and cholesterol-derived steroids (e.g., androgens, estrogens, glucocorticoids)

Hormone pleiotropy The phenomenon in which a single hormone affects two or more different phenotypic traits

Phenotypic plasticity The ability of a single genotype to produce different phenotypes in response to environmental stimuli

Polyphenism The ability of a single genotype to produce different discrete phenotypes in response to environmental stimuli

Quantitative trait loci (QTL) mapping Identifying genomic regions that explain substantial variation of a particular quantitative trait in experimental crosses

Small RNA Non-coding RNAs, ranging from 20 to 30 nucleotides in length, which contribute to the regulation of gene expression

References

Adkins-Regan E (2005) Hormones and animal social behavior. Princeton University Press, Princeton

Allis CD, Jenuwein T, Reinberg D, Caparros M-L (2007) Epigenetics. Cold Spring Harbor Laboratory Press, New York

Alvarez-Ponce D, Aguade M, Rozas J (2009) Network-level molecular evolutionary analysis of the insulin/TOR signal transduction pathway across 12 *Drosophila* genomes. Genome Res 19:234–242

Alvergne A, Lummaa V (2009) Does the contraceptive pill alter mate choice in humans? Trends Ecol Evol 25:171–179

Andersson M (1994) Sexual selection. Princeton University Press, Princeton

Ansai S, Ochiai H, Kanie Y, Kamei Y, Gou Y, Kitano T, Yamamoto T, Kinoshita M (2012) Targeted disruption of exogenous EGFP gene in medaka using zinc-finger nucleases. Dev Growth Differ 54:546–556

Anway MD, Cupp AS, Uzumcu M, Skinner MK (2005) Epigenetic transgenerational actions of endocrine disruptors and male fertility. Science 308:1466–1469

Arnqvist G, Rowe L (2005) Sexual conflict. Princeton University Press, Princeton
Aubin-Horth N, Renn SCP (2009) Genomic reaction norms: using integrative biology to understand molecular mechanisms of phenotypic plasticity. Mol Ecol 18:3763–3780
Auger CJ, Coss D, Auger AP, Forbes-Lorman RM (2011) Epigenetic control of vasopressin expression is maintained by steroid hormones in the adult male rat brain. Proc Natl Acad Sci USA 108:4242–4247
Baird NA, Etter PD, Atwood TS, Currey MC, Shiver AL, Lewis ZA, Selker EU, Cresko WA, Johnson EA (2008) Rapid SNP discovery and genetic mapping using sequenced RAD markers. PLoS One 3:e3376
Balment RJ, Lu W, Weybourne E, Warne JM (2006) Arginine vasotocin a key hormone in fish physiology and behaviour: a review with insights from mammalian models. Gen Comp Endocrinol 147:9–16
Barsh GS, Copenhaver GP, Gibson G, Williams SM (2012) Guidelines for genome-wide association studies. PLoS Genet 8:e1002812
Baum AE, Solberg LC, Kopp P, Ahmadiyeh N, Churchill G, Takahashi JS, Jameson JL, Redei EE (2005) Quantitative trait loci associated with elevated thyroid-stimulating hormone in the Wistar-Kyoto rat. Endocrinology 146:870–878
Becker C, Hagmann J, Muller J, Koenig D, Stegle O, Borgwardt K, Weigel D (2011) Spontaneous epigenetic variation in the *Arabidopsis thaliana* methylome. Nature 480:245–249
Behncken SN, Rowlinson SW, Rowland JE, Conway-Campbell BL, Monks TA, Waters MJ (1997) Aspartate 171 Is the major primate-specific determinant of human growth hormone: engineering porcine growth hormone to activate the human receptor. J Biol Chem 272:27077–27083
Bethke A, Fielenbach N, Wang Z, Mangelsdorf DJ, Antebi A (2009) Nuclear hormone receptor regulation of microRNAs controls developmental progression. Science 324:95–98
Bogdanove AJ, Voytas DF (2011) TAL effectors: customizable proteins for DNA targeting. Science 333:1843–1846
Boorse GC, Denver RJ (2004) Endocrine mechanisms underlying plasticity in metamorphic timing in spadefoot toads. Inetgr Comp Biol 43:646–657
Bradshaw D (2007) Environmental endocrinology. Gen Comp Endocrinol 152:125–141
Buchholz DR, Hayes TB (2005) Variation in thyroid hormone action and tissue content underlies species differences in the timing of metamorphosis in desert frogs. Evol Dev 7:458–467
Carroll SB (2005) Evolution at two levels: on genes and form. PLoS Biol 3:e245
Champagne FA, Curley JP (2009) Epigenetic mechanisms mediating the long-term effects of maternal care on development. Neurosci Biobehav Rev 33: 593–600
Champagne FA, Weaver ICG, Diorio J, Dymov S, Szyf M, Meaney MJ (2006) Maternal care associated with methylation of the estrogen receptor-α1b promoter and estrogen receptor-α expression in the medial preoptic area of female offspring. Endocrinology 147:2909–2915
Chen W, Kelly MA, Opitz-Araya X, Thomas RE, Low MJ, Cone RD (1997) Exocrine gland dysfunction in MC5-R-deficient mice: evidence for coordinated regulation of exocrine gland function by melanocortin peptides. Cell 91:789–798
Chen AS, Marsh DJ, Trumbauer ME, Frazier EG, Guan X-M, Yu H, Rosenblum CI, Vongs A, Feng Y, Cao L, Metzger JM, Strack AM, Camacho RE, Mellin TN, Nunes CN, Min W, Fisher J, Gopal-Truter S, MacIntyre DE, Chen HY, Van der Ploeg LHT (2000) Inactivation of the mouse melanocortin-3 receptor results in increased fat mass and reduced lean body mass. Nat Genet 26:97–102
Chutimanitsakun Y, Nipper R, Cuesta-Marcos A, Cistue L, Corey A, Filichkina T, Johnson E, Hayes P (2011) Construction and application for QTL analysis of a Restriction Site Associated DNA (RAD) linkage map in barley. BMC Genomics 12:4
Cork JM, Purugganan MD (2004) The evolution of molecular genetic pathways and networks. Bioessays 26:479–484
Cox RM, Calsbeek R (2009) Sexually antagonistic selection, sexual dimorphism, and the resolution of intralocus sexual conflict. Am Nat 173:176–187
Crews D, Williams EE (1977) Hormones, reproductive behavior, and speciation. Am Zool 17:271–286
Crews D, Gore AC, Hsu TS, Dangleben NL, Spinetta M, Schallert T, Anway MD, Skinner MK (2007) Transgenerational epigenetic imprints on mate preference. Proc Natl Acad Sci USA 104:5942–5946
Crews D, Gillette R, Scarpino SV, Manikkam M, Savenkova MI, Skinner MK (2012) Epigenetic transgenerational inheritance of altered stress responses. Proc Natl Acad Sci USA 109:9143–9148
Dardente H, Wyse CA, Birnie MJ, Dupré SM, Loudon ASI, Lincoln GA, Hazlerigg DG (2010) A molecular switch for photoperiod responsiveness in mammals. Curr Biol 20:2193–2198
Darwin C (1874) The decent of man; and selection in relation to sex. Humbolt, New York
Davey JW, Hohenlohe PA, Etter PD, Boone JQ, Catchen JM, Blaxter ML (2011) Genome-wide genetic marker discovery and genotyping using next-generation sequencing. Nat Rev Genet 12:499–510
Deagle BE, Jones FC, Chan YF, Absher DM, Kingsley DM, Reimchen TE (2011) Population genomics of parallel phenotypic evolution in stickleback across stream-lake ecological transitions. Proc R Soc B 279:1277–1286
Denver RJ (1997) Environmental stress as a developmental cue: corticotropin-releasing hormone is a proximate mediator of adaptive phenotypic plasticity in amphibian metamorphosis. Horm Behav 31:169–179
Denver RJ (1998) Hormonal correlates of environmentally induced metamorphosis in the western spadefoot toad, *Scaphiopus hammondii*. Gen Comp Endocrinol 110:326–336

DeWitt TJ, Scheiner SM (2004) Phenotypic plasticity: functional and conceptual approaches. Oxford University Press, Oxford

Dormitzer PR, Ellison PT, Bode HH (1989) Anomalously low endemic goiter prevalence among Efe pygmies. Am J Phys Anthropol 78:527–531

Ducrest A-L, Keller L, Roulin A (2008) Pleiotropy in the melanocortin system, coloration and behavioural syndromes. Trends Ecol Evol 23:502–510

Duret L, Mouchiroud D (2000) Determinants of substitution rates in mammalian genes: expression pattern affects selection intensity but not mutation rate. Mol Biol Evol 17:68–070

Ekblom R, Galindo J (2011) Applications of next generation sequencing in molecular ecology of non-model organisms. Heredity 107:1–15

Ellegren H, Smeds L, Burri R, Olason PI, Backstrom N, Kawakami T, Kunstner A, Makinen H, Nadachowska-Brzyska K, Qvarnstrom A, Uebbing S, Wolf JBW (2012) The genomic landscape of species divergence in *Ficedula* flycatchers. Nature 491:756–760

Faupel-Badger JM, Fuhrman BJ, Xu X, Falk RT, Keefer LK, Veenstra TD, Hoover RN, Ziegler RG (2010) Comparison of liquid chromatography-tandem mass spectrometry, RIA, and ELISA methods for measurement of urinary estrogens. Cancer Epidemiol Biomark Prev 19:292–300

Filipowicz W, Bhattacharyya SN, Sonenberg N (2008) Mechanisms of post-transcriptional regulation by microRNAs: are the answers in sight? Nat Rev Genet 9:102–114

Finch CE, Rose MR (1995) Hormones and the physiological architecture of life history evolution. Q Rev Biol 70:1–52

Flatt T, Tu MP, Tatar M (2005) Hormonal pleiotropy and the juvenile hormone regulation of Drosophila development and life history. Bioessays 27:999–1010

Fowden AL, Forhead AJ (2009) Hormones as epigenetic signals in developmental programming. Exp Physiol 94:607–625

Fraser HB, Hirsh AE, Steinmetz LM, Scharfe C, Feldman MW (2002) Evolutionary rate in the protein interaction network. Science 296:750–752

Gabor CR, Grober MS (2010) A potential role of male and female androgen in species recognition in a unisexual-bisexual mating complex. Horm Behav 57:427–433

Ghildiyal M, Zamore PD (2009) Small silencing RNAs: an expanding universe. Nat Rev Genet 10:94–108

Gilad Y, Rifkin SA, Pritchard JK (2008) Revealing the architecture of gene regulation: the promise of eQTL studies. Trends Genet 24:408–415

Gilad Y, Pritchard JK, Thornton K (2009) Characterizing natural variation using next-generation sequencing technologies. Trends Genet 25:463–471

Graves JAM, Peichel CL (2010) Are homologies in vertebrate sex determination due to shared ancestry or to limited options? Genome Biol 11:205

Greenwood AK, Jones FC, Chan YF, Brady SD, Absher DM, Grimwood J, Schmutz J, Myers RM, Kingsley DM, Peichel CL (2011) The genetic basis of divergent pigment patterns in juvenile threespine sticklebacks. Heredity 107:155–166

Grossniklaus U, Kelly B, Ferguson-Smith AC, Pembrey M, Lindquist S (2013) Transgenerational epigenetic inheritance: how important is it? Nat Rev Genet 14:228–235

Haig D (2008) Placental growth hormone-related proteins and prolactin-related proteins. Placenta 29(Suppl A):S36–S41

Handwerger S, Freemark M (2000) The roles of placental growth hormone and placental lactogen in the regulation of human fetal growth and development. J Pediatr Endocrinol Metab 13:343–356

Hanon EA, Lincoln GA, Fustin JM, Dardente H, Masson-Pevet M, Morgan PJ, Hazlerigg DG (2008) Ancestral TSH mechanism signals summer in a photoperiodic mammal. Curr Biol 18:1147–1152

Harper JM, Galecki AT, Burke DT, Pinkosky SL, Miller RA (2003) Quantitative trait loci for insulin-like growth factor I, leptin, thyroxine, and corticosterone in genetically heterogeneous mice. Physiol Genomics 15:44–51

Hau M (2007) Regulation of male traits by testosterone: implications for the evolution of vertebrate life histories. Bioessays 29:133–144

Heijmans BT, Tobi EW, Stein AD, Putter H, Blauw GJ, Susser ES, Slagboom PE, Lumey LH (2008) Persistent epigenetic differences associated with prenatal exposure to famine in humans. Proc Natl Acad Sci USA 105:17046–17049

Hindorff LA, Sethupathy P, Junkins HA, Ramos EM, Mehta JP, Collins FS, Manolio TA (2009) Potential etiologic and functional implications of genome-wide association loci for human diseases and traits. Proc Natl Acad Sci USA 106:9362–9367

Hirschenhauser K, Oliveira RF (2006) Social modulation of androgens in male vertebrates: meta-analyses of the challenge hypothesis. Anim Behav 71:265–277

Hohenlohe PA, Bassham S, Etter PD, Stiffler N, Johnson EA, Cresko WA (2010) Population genomics of parallel adaptation in threespine stickleback using sequenced RAD tags. PLoS Genet 6:e1000862

Insel TR, Young LJ (2000) Neuropeptides and the evolution of social behavior. Curr Opin Neurobiol 10:784–789

Ishikawa A, Kitano J (2012) Ecological genetics of thyroid hormone physiology in humans and animals. In: Agrawal NK (ed) Thyroid hormone. InTech, Rijeka, pp 37–50

Ishikawa A, Ogawa K, Gotoh H, Walsh TK, Tagu D, Brisson JA, Rispe C, Jaubert-Possamai S, Kanbe T, Shiotsuki T, Miura T (2012) Juvenile hormone titer and gene expression during the change of reproductive modes in the pea aphid. Insect Mol Biol 12:49–60

Jarvis JP, Scheinfeldt LB, Soi S, Lambert C, Omberg L, Ferwerda B, Froment A, Bodo J-M, Beggs W, Hoffman G, Mezey J, Tishkoff SA (2012) Patterns of

ancestry, signatures of natural selection, and genetic association with stature in Western African pygmies. PLoS Genet 8:e1002641

Johansson ML (2009) Next generation sequencing in nonmodel organisms: has the future arrived? J Hered 100:807–807

Johnson KM, Lema SC (2011) Tissue-specific thyroid hormone regulation of gene transcripts encoding iodothyronine deiodinases and thyroid hormone receptors in striped parrotfish (*Scarus iseri*). Gen Comp Endocrinol 172:505–517

Jones FC, Chan YF, Schmutz J, Grimwood J, Brady SD, Southwick AM, Absher DM, Myers RM, Reimchen TE, Deagle BE, Schluter D, Kingsley DM (2012a) A genome-wide SNP genotyping array reveals patterns of global and repeated species-pair divergence in sticklebacks. Curr Biol 22:83–90

Jones FC, Grabherr MG, Chan YF, Russell P, Mauceli E, Johnson J, Swofford R, Pirun M, Zody MC, White S, Birney E, Searle S, Schmutz J, Grimwood J, Dickson MC, Myers RM, Miller CT, Summers BR, Knecht AK, Brady SD, Zhang H, Pollen AA, Howes T, Amemiya C, Lander ES, Di Palma F, Lindblad-Toh K, Kingsley DM (2012b) The genomic basis of adaptive evolution in threespine sticklebacks. Nature 484:55–61

Ketterson ED, Nolan VJ (1992) Hormones and life histories: an integrative approach. Am Nat 140:S33–S62

Ketterson ED, Nolan V Jr (1999) Adaptation, exaptation, and constraint: a hormonal perspective. Am Nat 154:S4–S24

Ketterson ED, Nolan VJ, Cawthorn MJ, Parker PG, Ziegenfus C (1996) Phenotypic engineering: using hormones to explore the mechanistic and functional bases of phenotypic variation in nature. Ibis 138:70–86

Ketterson ED, Atwell JW, McGlothlin JW (2009) Phenotypic integration and independence: hormones, performance, and response to environmental change. Integr Comp Biol 49:365–379

Kijas JW, Lenstra JA, Hayes B, Boitard S, Porto Neto LR, San Cristobal M, Servin B, McCulloch R, Whan V, Gietzen K, Paiva S, Barendse W, Ciani E, Raadsma H, McEwan J, Dalrymple B, Members ISGC (2012) Genome-wide analysis of the world's sheep breeds reveals high levels of historic mixture and strong recent selection. PLoS Biol 10:e1001258

Kim VN, Han J, Siomi MC (2009) Biogenesis of small RNAs in animals. Nat Rev Mol Cell Biol 10: 126–139

King EG, Merkes CM, McNeil CL, Hoofer SR, Sen S, Broman KW, Long AD, Macdonald SJ (2012) Genetic dissection of a model complex trait using the *Drosophila* synthetic population resource. Genome Res 22:1558–1566

Kitano J, Lema S (2013) Divergence in thyroid hormone levels between juveniles of marine and stream ecotypes of the threespine stickleback (*Gasterosteus aculeatus*). Curr Zool 15:143–153

Kitano J, Ross JA, Mori S, Kume M, Jones FC, Chan YF, Absher DM, Grimwood J, Schmutz J, Myers RM, Kingsley DM, Peichel CL (2009) A role for a neo-sex chromosome in stickleback speciation. Nature 461:1079–1083

Kitano J, Lema SC, Luckenbach JA, Mori S, Kawagishi Y, Kusakabe M, Swanson P, Peichel CL (2010) Adaptive divergence in the thyroid hormone signaling pathway in the stickleback radiation. Curr Biol 20:2124–2130

Kitano J, Kawagishi Y, Mori S, Peichel CL, Makino T, Kawata M, Kusakabe M (2011) Divergence in sex steroid hormone signaling between sympatric species of Japanese threespine stickleback. PLoS ONE 6:e29253

Kitano J, Yoshida K, Suzuki Y (2013) RNA sequencing reveals small RNAs differentially expressed between incipient Japanese threespine sticklebacks. BMC Genomics 14:214

Kuokkanen S, Chen B, Ojalvo L, Benard L, Santoro N, Pollard JW (2010) Genomic profiling of microRNAs and messenger RNAs reveals hormonal regulation in microRNA expression in human endometrium. Biol Reprod 82:791–801

Lande R (1979) Quantitative genetic analysis of multivariate evolution, applied to brain: body size allometry. Evolution 33:402–416

Landry CR, Wittkopp PJ, Taubes CH, Ranz JM, Clark AG, Hartl DL (2005) Compensatory cis-trans evolution and the dysregulation of gene expression in interspecific hybrids of *Drosophila*. Genetics 171:1813–1822

Leduc MS, Hageman RS, Meng Q, Verdugo RA, Tsaih S-W, Churchill GA, Paigen B, Yuan R (2010) Identification of genetic determinants of IGF-1 levels and longevity among mouse inbred strains. Aging Cell 9:823–836

Lei Y, Guo X, Liu Y, Cao Y, Deng Y, Chen X, Cheng CH, Dawid IB, Chen Y, Zhao H (2012) Efficient targeted gene disruption in *Xenopus* embryos using engineered transcription activator-like effector nucleases (TALENs). Proc Natl Acad Sci USA 109:17484–17489

Lema S (2008) The phenotypic plasticity of Death Valley's pupfishes. Am Sci 96:28–36

Lema SC, Nevitt GA (2006) Testing an ecophysiological mechanism of morphological plasticity in pupfish and its relevance to conservation efforts for endangered Devils Hole pupfish. J Exp Biol 209:3499–3509

Li T, Huang S, Jiang WZ, Wright D, Spalding MH, Weeks DP, Yang B (2011) TAL nucleases (TALNs): hybrid proteins composed of TAL effectors and FokI DNA-cleavage domain. Nucleic Acids Res 39:359–372

Lim MM, Wang Z, Olazabal DE, Ren X, Terwilliger EF, Young LJ (2004) Enhanced partner preference in a promiscuous species by manipulating the expression of a single gene. Nature 429:754–757

Liu JC, Makova KD, Adkins RM, Gibson S, Li WH (2001) Episodic evolution of growth hormone in primates and emergence of the species specificity of human growth hormone receptor. Mol Biol Evol 18:945–953

Loh Y-HE, Yi SV, Streelman JT (2011) Evolution of microRNAs and the diversification of species. Genome Biol Evol 3:55–65

López Herráez D, Bauchet M, Tang K, Theunert C, Pugach I, Li J, Nandineni MR, Gross A, Scholz M, Stoneking M (2010) Genetic variation and recent positive selection in worldwide human populations: evidence from nearly 1 million SNPs. PLoS One 4:e7888

Lu Y, Rausher MD (2003) Evolutionary rate variation in anthocyanin pathway genes. Mol Biol Evol 20:1844–1853

Lynch M, Walsh B (1998) Genetics and analysis of quantitative traits. Sinauer, Sunderland

Ma S, Zhang S, Wang F, Liu Y, Liu Y, Xu H, Liu C, Lin Y, Zhao P, Xia Q (2012) Highly efficient and specific genome editing in silkworm using custom TALENs. PLoS One 7:e45035

Majewski J, Pastinen T (2011) The study of eQTL variations by RNA-seq: from SNPs to phenotypes. Trends Genet 27:72–79

Manceau M, Domingues VS, Linnen CR, Rosenblum EB, Hoekstra HE (2010) Convergence in pigmentation at multiple levels: mutations, genes and function. Philos Trans R Soc B 365:2439–2450

Mank JE (2007) The evolution of sexually selected traits and antagonistic androgen expression in Actinopterygiian fishes. Am Nat 169:142–149

Martin D, Maestro O, Cruz J, Mane-Padros D, Belles X (2006) RNAi studies reveal a conserved role for RXR in molting in the cockroach *Blattella germanica*. J Insect Physiol 52:410–416

McGlothlin JW, Ketterson ED (2008) Hormone-mediated suites as adaptations and evolutionary constraints. Philos Trans R Soc B 363:1611–1620

McGlothlin JW, Neudorf DL, Casto JM, Nolan JV, Ketterson ED (2004) Elevated testosterone reduces choosiness in female dark-eyed juncos (*Junco hyemalis*): evidence for a hormonal constraint on sexual selection? Proc R Soc B 271:1377–1384

McLean CY, Reno PL, Pollen AA, Bassan AI, Capellini TD, Guenther C, Indjeian VB, Lim X, Menke DB, Schaar BT, Wenger AM, Bejerano G, Kingsley DM (2011) Human-specific loss of regulatory DNA and the evolution of human-specific traits. Nature 471:216–219

McManus CJ, Coolon JD, Duff MO, Eipper-Mains J, Graveley BR, Wittkopp PJ (2010) Regulatory divergence in *Drosophila* revealed by mRNA-seq. Genome Res 20:816–825

Miller MR, Dunham JP, Amores A, Cresko WA, Johnson EA (2007) Rapid and cost-effective polymorphism identification and genotyping using restriction site associated DNA (RAD) markers. Genome Res 17:240–248

Minakuchi C, Namiki T, Yoshiyama M, Shinoda T (2008) RNAi-mediated knockdown of juvenile hormone acid O-methyltransferase gene causes precocious metamorphosis in the red flour beetle *Tribolium castaneum*. FEBS J 275:2919–2931

Moran NA (1992) The evolution of aphid life cycles. Annu Rev Ecol Syst 37:321–348

Moss T, Leblanc B (2009) DNA-protein interactions: principles and protocols. Humana Press, New York

Nakao N, Ono H, Yamamura T, Anraku T, Takagi T, Higashi K, Yasuo S, Katou Y, Kageyama S, Uno Y, Kasukawa T, Iigo M, Sharp PJ, Iwasawa A, Suzuki Y, Sugano S, Niimi T, Mizutani M, Namikawa T, Ebihara S, Ueda HR, Yoshimura T (2008) Thyrotrophin in the pars tuberalis triggers photoperiodic response. Nature 452:317–322

Nei M, Kumar S (2000) Molecular evolution and phylogenetics. Oxford University Press, Oxford

Nielsen R (2005) Molecular signatures of natural selection. Annu Rev Genet 39:197–218

Nijhout HF (2003) Development and evolution of adaptive polyphenism. Evol Dev 5:9–18

Niwa M, Jaaro-Peled H, Tankou S, Seshadri S, Hikida T, Matsumoto Y, Cascella NG, Kano S-i, Ozaki N, Nabeshima T, Sawa A (2013) Adolescent stress-induced epigenetic control of dopaminergic neurons via glucocorticoids. Science 339:335–339

Nosil P (2012) Ecological speciation. Oxford University Press, New York

Ochiai H, Sakamoto N, Fujita K, Nishikawa M, Suzuki K-i, Matsuura S, Miyamoto T, Sakuma T, Shibata T, Yamamoto T (2012) Zinc-finger nuclease-mediated targeted insertion of reporter genes for quantitative imaging of gene expression in sea urchin embryos. Proc Natl Acad Sci USA 109:10915–10920

Ohno S (1970) Evolution by gene duplication. Springer, New York

Ohta T (1993) Pattern of nucleotide substitutions in growth hormone-prolactin gene family: a paradigm for evolution by gene duplication. Genetics 134: 1271–1276

Oleksyk TK, Smith MW, O'Brien SJ (2010) Genome-wide scans for footprints of natural selection. Philos Trans R Soc B 365:185–205

Olsen KM, Womack A, Garrett AR, Suddith JI, Purugganan MD (2002) Contrasting evolutionary forces in the *Arabidopsis thaliana* floral developmental pathway. Genetics 160:1641–1650

Oostra V, de Jong MA, Invergo BM, Kesbeke F, Wende F, Brakefield PM, Zwaan BJ (2011) Translating environmental gradients into discontinuous reaction norms via hormone signalling in a polyphenic butterfly. Proc R Soc B 278:789–797

Pamuru RR, Rosen O, Manor R, Chung JS, Zmora N, Glazer L, Aflalo ED, Weil S, Tamone L, Sagi A (2012) Stimulation of molt by RNA interference of the molt-inhibiting hormone in the crayfish *Cherax quadricarinatus*. Gen Comp Endocrinol 178:227–236

Price TD, Qvarnström A, Irwin DE (2003) The role of phenotypic plasticity in driving genetic evolution. Proc R Soc B 270:1433–1440

Ramirez CL, Foley JE, Wright DA, Muller-Lerch F, Rahman SH, Cornu TI, Winfrey RJ, Sander JD, Fu F, Townsend JA, Cathomen T, Voytas DF, Joung JK (2008) Unexpected failure rates for modular assembly of engineered zinc fingers. Nat Methods 5:374–375

Ramsay H, Rieseberg LH, Ritland K (2009) The correlation of evolutionary rate with pathway position in plant terpenoid biosynthesis. Mol Biol Evol 26: 1045–1053
Ramsey ME, Wong RY, Cummings ME (2011) Estradiol, reproductive cycle and preference behavior in a northern swordtail. Horm Behav 170:381–390
Rausher MD, Miller RE, Tiffin P (1999) Patterns of evolutionary rate variation among genes of the anthocyanin biosynthetic pathway. Mol Biol Evol 16: 266–274
Rice WR (1984) Sex chromosomes and the evolution of sexual dimorphism. Evolution 38:735–742
Richter-Boix A, Tejedo M, Rezende EL (2011) Evolution and plasticity of anuran larval development in response to desiccation. A comparative analysis. Ecol Evol 1:15–25
Roberts ML, Buchanan KL, Evans MR (2004) Testing the immunocompetence handicap hypothesis: a review of the evidence. Anim Behav 68:227–239
Rosen CJ, Churchill GA, Donahue LR, Shultz KL, Burgess JK, Powell DR, Beamer WG (2000) Mapping quantitative trait loci for serum insulin-like growth factor-1 levels in mice. Bone 27:521–528
Ross AW, Helfer G, Russell L, Darras VM, Morgan PJ (2011) Thyroid hormone signalling genes are regulated by photoperiod in the hypothalamus of F344 rats. PLoS ONE 6:e21351
Rubin CJ, Zody MC, Eriksson J, Meadows JR, Sherwood E, Webster MT, Jiang L, Ingman M, Sharpe T, Ka S, Hallböök F, Besnier F, Carlborg O, Bed'hom B, Tixier-Boichard M, Jensen P, Siegel P, Lindblad-Toh K, Andersson L (2010) Whole-genome resequencing reveals loci under selection during chicken domestication. Nature 464:587–591
Sabeti PC, Schaffner SF, Fry B, Lohmueller J, Varilly P, Shamovsky O, Palma A, Mikkelsen TS, Altshuler D, Lander ES (2006) Positive natural selection in the human lineage. Science 312:1614–1620
Sasaki S, Lesoon-Wood LA, Dey A, Kuwata T, Weintraub BD, Humphrey G, Yang W-M, Seto E, Yen PM, Howard BH, Ozato K (1999) Ligand-induced recruitment of a histone deacetylase in the negative-feedback regulation of the thyrotropin β gene. EMBO J 18:5389–5398
Sauvage C, Vagner M, Derôme N, Audet C, Bernatchez L (2012a) Coding gene SNP mapping reveals QTL linked to growth and stress response in brook charr (*Salvelinus fontinalis*). G3 2:707–720
Sauvage C, Vagner M, Derôme N, Audet C, Bernatchez L (2012b) Coding gene single nucleotide polymorphism mapping and quantitative trait loci detection for physiological reproductive traits in brook charr, *Salvelinus fontinalis*. G3 2:379–392
Schluter D (2000) The ecology of adaptive radiation. Oxford University Press, New York
Schmitz RJ, Schultz MD, Urich MA, Nery JR, Pelizzola M, Libiger O, Alix A, McCosh RB, Chen H, Schork NJ, Ecker JR (2013) Patterns of population epigenomic diversity. Nature 495:193–198
Sexton OJ, Bizer JR (1978) Life history patterns of Ambystoma tigrinum in Montane Colorado. Am Midl Nat 99:101–118
Shaffer HB (1993) Phylogenetics of model organisms: the laboratory axolotl, *Ambystoma mexicanum*. Syst Biol 42:508–522
Sim C, Denlinger DL (2008) Insulin signaling and FOXO regulate the overwintering diapause of the mosquito *Culex pipiens*. Proc Natl Acad Sci USA 105:6777–6781
Soldin SJ, Soldin OP (2009) Steroid hormone analysis by tandem mass spectrometry. Clin Chem 55:1061–1066
Soldin OP, Soldin SJ (2011) Thyroid hormone testing by tandem mass spectrometry. Clin Biochem 44:89–94
Soma KK, Bindra RK, Gee J, Wingfield JC, Schlinger BA (1999) Androgen-metabolizing enzymes show region-specific changes across the breeding season in the brain of a wild songbird. J Neurobiol 41:176–188
Souza SC, Frick GP, Wang X, Kopchick JJ, Lobo RB, Goodman HM (1995) A single arginine residue determines species specificity of the human growth hormone receptor. Proc Natl Acad Sci USA 92: 959–963
Stapley J, Reger J, Feulner PGD, Smadja C, Galindo J, Ekblom R, Bennison C, Ball AD, Beckerman AP, Slate J (2010) Adaptation genomics: the next generation. Trends Ecol Evol 25:705–712
Tomikawa J, Uenoyama Y, Ozawa M, Fukanuma T, Takase K, Goto T, Abe H, Ieda N, Minabe S, Deura C, Inoue N, Sanbo M, Tomita K, Hirabayashi M, Tanaka S, Imamura T, Okamura H, Maeda K, Tsukamura H (2012) Epigenetic regulation of Kiss1 gene expression mediating estrogen-positive feedback action in the mouse brain. Proc Natl Acad Sci USA 109:E1294–E1301
Tost J (2009) DNA methylation: methods and protocols. Humana Press, New York
van Doorn GS, Kirkpatrick M (2007) Turnover of sex chromosomes induced by sexual conflict. Nature 449:909–912
Vogel C, Marcotte EM (2012) Insights into the regulation of protein abundance from proteomic and transcriptomic analyses. Nat Rev Genet 13:227–232
Voss SR, Kump DK, Walker JA, Shaffer HB, Voss GJ (2012) Thyroid hormone responsive QTL and the evolution of paedomorphic salamanders. Heredity 109:293–298
Waddington CH (1961) Genetic assimilation. Adv Genet 10:257–293
Wallis M (2000) Episodic evolution of protein hormones: molecular evolution of pituitary prolactin. J Mol Evol 50:465–473
Wallis OC, Mac-Kwashie A, Makri G, Wallis M (2005) Molecular evolution of prolactin in primates. J Mol Evol 60:606–614
Wang D, Xia X, Weiss RE, Refetoff S, Yen PM (2010) Distinct and histone-specific modifications mediate positive versus negative transcriptional regulation of TSHβ promoter. PLoS ONE 5:e9853

Wark AR, Mills MG, Dang L-H, Chan YF, Jones FC, Brady SD, Absher DM, Grimwood J, Schmutz J, Myers RM, Kingsley DM, Peichel CL (2012) Genetic architecture of variation in the lateral line sensory system of threespine sticklebacks. G3 2:1047–1056

Watanabe T, Ochiai H, Sakuma T, Horch HW, Hamaguchi N, Nakamura T, Bando T, Ohuchi H, Yamamoto T, Noji S, Mito T (2012) Non-transgenic genome modifications in a hemimetabolous insect using zinc-finger and TAL effector nucleases. Nat Commun 3:1017

Weaver ICG, Cervoni N, Champagne FA, D'Alessio AC, Sharma S, Seckl JR, Dymov S, Szyf M, Meaney MJ (2004) Epigenetic programming by maternal behavior. Nat Neurosci 7:847–854

Weltring A, Schaebs FS, Perry SE, Deschner T (2012) Simultaneous measurement of endogenous steroid hormones and their metabolites with LC–MS/MS in faeces of a New World primate species, *Cebus capucinus*. Physiol Behav 105:510–521

Werner EE (1986) Amphibian metamorphosis: growth rate, predation risk, and the optimal size at transformation. Am Nat 128:319–341

West-Eberhard MJ (2003) Developmental plasticity and evolution. Oxford University Press, Oxford

Williams TM, Carroll SB (2009) Genetic and molecular insights into the development and evolution of sexual dimorphism. Nat Rev Genet 10:797–804

Wingfield JC, Hegner RE, Dufty AM, Ball GF (1990) The 'challenge hypothesis': theoretical implications for patterns of testosterone secretion, mating systems and breeding strategies. Am Nat 136:829–846

Wittkopp PJ, Haerum BK, Clark AG (2004) Evolutionary changes in cis and trans gene regulation. Nature 430:85–88

Wu X, Chi X, Wang P, Zheng D, Ding R, Li Y (2010) The evolutionary rate variation among genes of HOG-signaling pathway in yeast genomes. Biol Direct 5:46

Yang Y-h, Zhang F-m, Ge S (2009) Evolutionary rate patterns of the Gibberellin pathway genes. BMC Evol Biol 9:206

Young LJ, Nilsen R, Waymire KG, MacGregor GR, Insel TR (1999) Increased affiliative response to vasopressin in mice expressing the V1a receptor from a monogamous vole. Nature 400:766–768

Zera AJ, Harshman LG (2001) The physiology of life history trade-offs in animals. Annu Rev Ecol Syst 32:95–126

Zhang J (2003) Evolution by gene duplication: an update. Trends Ecol Evol 18:292–298

Zhang X, Ho S-M (2011) Epigenetics meets endocrinology. J Mol Endocrinol 46:R11–R32

Zhou X, Tarver MR, Scharf ME (2007) Hexamerin-based regulation of juvenile hormone-dependent gene expression underlies phenotypic plasticity in a social insect. Development 134:601–610

Zijlstra W, Steigenga MJ, Koch PB, Zwaan BJ, Brakefield PM (2004) Butterfly selected lines explore the hormonal basis of interactions between life histories and morphology. Am Nat 163:E76–E87

Evolutionary Genomics of Environmental Pollution

16

Andrew Whitehead

Abstract

Chemical toxins have been a persistent source of evolutionary challenges throughout the history of life, and deep within the genomic storehouse of evolutionary history lay ancient adaptations to diverse chemical poisons. However, the rate of change of contemporary environments mediated by human-introduced pollutants is rapidly screening this storehouse and severely testing the adaptive potential of many species. In this chapter, we briefly review the deep history of evolutionary adaptation to environmental toxins, and then proceed to describe the attributes of stressors and populations that may facilitate contemporary adaptation to pollutants introduced by humans. We highlight that phenotypes derived to enable persistence in polluted habitats may be multi-dimensional, requiring global genome-scale tools and approaches to uncover their mechanistic basis, and include examples of recent progress in the field. The modern tools of genomics offer promise for discovering how pollutants interact with genomes on physiological timescales, and also for discovering what genomic attributes of populations may enable resistance to pollutants over evolutionary timescales. Through integration of these sophisticated genomics tools and approaches with an understanding of the deep historical forces that shaped current populations, a more mature understanding of the mechanistic basis of contemporary ecological-evolutionary dynamics should emerge.

Keywords

Contemporary evolution • Evolutionary-ecological dynamics • Evolutionary genomics • Ecological genomics • Comparative genomics • Toxicogenomics • Pollution • Adaptation • Evolutionary toxicology

16.1 Introduction

Humans are increasingly promoting rapid and global environmental change (Vitousek et al. 1997) that can serve as a vehicle for driving

A. Whitehead (✉)
Department of Environmental Toxicology, University of California, Davis, CA 95616, USA
e-mail: awhitehead@ucdavis.edu

C.R. Landry and N. Aubin-Horth (eds.), *Ecological Genomics: Ecology and the Evolution of Genes and Genomes*, Advances in Experimental Medicine and Biology 781,
DOI 10.1007/978-94-007-7347-9_16,

evolutionary change in diverse taxa (Palumbi 2001). The population-level consequences of such pressures can include decline leading to extinction, persistence mediated by physiological or behavioral plasticity, or persistence mediated by adaptation through natural selection (Chevin et al. 2010). Evolutionary adaptation has traditionally been perceived as a gradual process, but that natural selection can quickly derive solutions for challenges imposed by rapidly changing contemporary environments is gaining in recognition (Hendry and Kinnison 1999; Kinnison and Hendry 2001; Hairston et al. 2005). The likelihood of adaptation as an evolutionary solution to human-altered environments depends broadly on demographic characteristics of species and the genomic architecture that underlies adaptive variation. The purpose of this chapter is to explore intersections between historical and *contemporary adaptation* to environmental poisons, and how ecological and evolutionary genomics tools may be exploited to offer insight into the mechanisms whereby contemporary life may rapidly evolve adaptations to the current and growing environmental challenges posed by *anthropogenic pollutants* (see the glossary for definitions of italicized terms).

16.2 Evolutionary Solutions to Toxic Challenges Are Ancient

Toxic chemicals in the environment are typically considered as stressors that have *recently* imposed physiological and evolutionary challenges to species unlucky enough to be exposed. In contrast, life has been perpetually challenged with toxic stress since the beginning, and has persisted through evolution of diverse mechanisms to evade, tolerate, and compensate for such challenges, facilitated by sorting of genetic variation by natural selection. By viewing adaptation to *toxins* through the lens of deep evolutionary time, we can achieve insight into the mechanisms through which novel poisons exert their toxic effects, into the reasons for species variation in their sensitivity to poisons, and into the landscape of mechanisms whereby contemporary species may derive tolerance to modern human-derived *toxicants*. This historical perspective has been nicely and thoroughly summarized by Emily Monosson (2012).

Toxin defense systems are ancient. Growing evidence indicates that very early life inhabited niches near deep-ocean hydrothermal vents, where extremes in many environmental parameters such as temperature, sulfides, radioactivity, and heavy metals, are common (La Duc et al. 2007). Comparative genomics is offering insights into the mechanisms that have evolved to enable tolerance to such environmental extremes (Campbell et al. 2009) – mechanisms that must have evolved early in the history of life, for example in response to fluctuating redox potentials and temperatures, to high sulfur and CO_2, and to low oxygen. Similarly, early life had to contend with severe UV radiation prior to establishment of atmospheric ozone protection, especially when life emerged close to ocean surfaces where photosynthesis was possible but UV penetrance was high. Even once the ozone layer was established, sufficient DNA-damaging UV still penetrated to the planet surface. Among the most ancient of enzyme systems is DNA photolyase, which catalyzes the repair of UV-induced DNA lesions. DNA photolyase, in its various forms, is shared among all domains of life (Kiontke et al. 2011), and represents a defensive system that was likely derived in the last universal common ancestor (Ouzounis et al. 2006). Variation in these enzyme systems contributes to variable tolerance to UV radiation. For example, frog populations and species that occupy high-UV niches tend to be more tolerant to the damaging effects of UV radiation, which correlates with elevated photolyase activity (Blaustein and Belden 2003). Furthermore, frogs from high-UV niches are also more resistant to the DNA-damaging effects of pollutants (Marquis et al. 2009), thus emphasizing that evolutionary history that imparts tolerance to natural stressors can inadvertently confer cross-tolerance to anthropogenic pollutants.

Following the rise of cyanobacteria and the origin of oxygenic photosynthesis came the "Great Oxygenation Event" in the early paleoproterozoic era (~2.5 BYA), which represented a radical environmental change from an anoxic to an oxygenated world (Kump 2008). The availability of free oxygen in the atmosphere posed a series of major challenges to early life. Oxygen is a double-edged sword; it is clearly essential for much of life, especially multi-cellular life, but oxygen radicals rank among the most biologically destructive of molecules. As with defenses to UV, defense systems for oxygen free radicals, such as catalase and superoxide dismutase, are ancient (Ouzounis et al. 2006), and their emergence likely contributed to evolutionary solutions to this radically new and toxic oxygenic atmosphere. The rise in atmospheric oxygen also caused changes in the oxidation state of many metals, thereby affecting their bioavailability to aquatic organisms. Some biologically essential metals, such as iron, became less bioavailable, whereas some toxic metals, such as copper and zinc, became more water-soluble and therefore more bioavailable (Williams 2007). This radical change in the distribution of metals in the aquatic environment likely posed another key challenge to early life (Williams and Rickaby 2012). Indeed, the derivation of metabolic systems for managing transport and sequestration of metals is ancient. For example, the metallothioneins (MTs) are a large and functionally diversified protein family that harbor diverse metal-binding and redox capabilities, and are shared among all domains of life (Coyle et al. 2002). Variation in these metal binding and transport mechanisms contributes to evolved variation in tolerance to metals. For example, increase in the *copy number* of genes that mediate efflux of citrate (citrate chelates aluminum) contributes to variation in aluminum tolerance in maize (Maron et al. 2013). Also, heavy metal tolerance in *Arabidopsis halleri*, which predates anthropogenic spread of heavy-metal pollution of terrestrial habitats (Roux et al. 2011), is attributable to variation in copy number and regulation of metal-transporting proteins (Hanikenne et al. 2008). Similarly, MT copy number variation can be widespread in natural insect populations, and is correlated with tolerance to heavy metal toxicity (Maroni et al. 1987).

Another major milestone in this history of life was the invasion of terrestrial and aerial habitats in the Silurean era first by plants then soon followed by animals. Subsequent evolution of rooted sessile plants, susceptible to animal predation, recruited a diverse suite of chemical metabolites to serve as defensive poisons, and reciprocal evolution in animals has derived suites of metabolic strategies to compensate for these dietary poisons (Despres et al. 2007). Genetic innovations, for example achieved through recurrent duplication of the ancient cytochrome P450 (CYP) protein family and subsequent *subfunctionalization* (Goldstone et al. 2007), have been important enablers of this evolutionary arms race (Gonzalez and Nebert 1990). The CYPs are a hugely diversified (Bernhardt 2006) and ancient (Werck-Reichhart and Feyereisen 2000) protein family that catalyzes the oxidation of a broad suite of organic substrates, thereby contributing to metabolism of endogenous and *xenobiotic* chemicals. Indeed, the xenobiotic detoxifying family members are much diversified, and duplication followed by subfunctionalization is associated with derived tolerances to toxic plant allelochemicals. For example, unique CYP duplicates that exhibit high metabolic activities toward toxic plant allelochemicals are derived in some butterflies that have evolved specialized exploitation of the plants that produce these chemical deterrents (Wen et al. 2006). The diverse suite of CYPs that has emerged from the deep history of plant-herbivore chemical warfare (Gonzalez and Nebert 1990) is serving important contemporary roles in detoxification of human-derived pollutants. In several cases, evolution of CYPs has enabled adaptive tolerance to contemporary pesticides (Feyereisen 1999; Li et al. 2007; Puinean et al. 2010), and the pathway that governs CYP-mediated metabolism of industrial pollutants has been the target of natural selection in fish (Nacci et al. 2010; Wirgin et al. 2011).

The few examples introduced above serve to emphasize that the history of life is replete with

examples of evolved solutions to environmental poisons. Given the fact that many human-derived pollutants are structurally similar to more ancient environmental chemicals (pollutants such as metals, pesticides, organic industrial chemicals, and hormone mimics), key questions are (1) May heritable variation and natural selection provide a buffer for some species in contemporary polluted habitats?, (2) Under what circumstances might derived tolerance be most likely?, and (3) What might be the costs of evolved tolerance to populations and communities?

16.3 Contemporary Evolution

Clearly, life has a long history of deriving adaptive solutions to extreme environmental change. A growing recognition that evolutionary change can arise rapidly within contemporary time-scales, then feed back on ecological processes, is accelerating research in "eco-evolutionary dynamics"(Pelletier et al. 2009). Under what circumstances may evolution be sufficiently nimble to compensate for the extreme rates of environmental change, and the extreme selection pressures, imposed by human alteration of environments? A growing body of research is documenting adaptive responses to human-altered environments, and these adaptive responses are manifest through plasticity, natural selection, or a combination of the two. Plastic responses (developmental, physiological, behavioral) often appear important for maintaining persistence in newly changed environments, but evolved genetic change in populations can also be crucial for long-term resilience of at-risk populations (Hendry et al. 2008). Indeed, evolutionary adaptation has been detected for example in response to alterations in climate (Hoffmann and Sgro 2011), introduction of invasive species (Shine 2012), harvest pressure (Allendorf and Hard 2009), and environmental pollution (Van Veld and Nacci 2008a). What are the traits of populations and genomes that might facilitate adaptation to rapid environmental change?

Demographic traits are important, insofar as large populations enable more efficient natural selection and can harbor more potentially adaptive genetic variation (Graur and Li 2000), and short generation times can facilitate quicker adaptation to rapidly changing environments. Genomic context is important insofar as the number of genes that contribute to a trait, the penetrance of underlying genes and the degree of their effects, the standing frequency of putatively adaptive alleles in populations, and the nature of interactions among contributing genes (*pleiotropy* and *epistasis*), can all influence the efficiency of natural selection (Orr 2005). Given the importance of these demographic and genomic contexts for adaptation in general, we may ask more specifically what kinds of genomic variation and demographic contexts might enable rapid evolution of pollution tolerance?

16.3.1 Contemporary Adaptation to Poisons

Human-derived chemicals can impact natural populations either intentionally, for example as poisons introduced to kill pests, or unintentionally, for example as by-products of industrial processes. More is known about the evolutionary consequences of intentionally introduced poisons than unintentional pollution. Chemical pesticides are designed and formulated to target very specific biomolecules and are introduced into the environment in extremely large amounts and often as single chemicals. Evolved adaptation to such chemicals is common (Wright 2007; Whalon et al. 2008; Laxminarayan and Heymann 2012). What is the genetic architecture of such adaptations? One prediction from the pesticide resistance literature is that relatively weak selection on standing variation is likely to promote polygenic adaptation, whereas very strong selection is more likely to promote adaptation underlain by few genes of major effect (Mckenzie and Batterham 1994; ffrench-Constant et al. 2004).

This prediction is supported by laboratory studies, where incremental selection pressure for tolerance to pesticides involves polygenic adaptation, whereas field cases that involved intense targeted poisoning of insects typically results in adaptation underpinned by few genes of major effect (Mckenzie and Batterham 1994; ffrench-Constant et al. 2004). Furthermore, evolutionary adaptation of pests and pathogens to poisons can be swift when challenged with single chemicals, but treatment with multiple drugs or pesticides, thereby increasing the *dimensionality* of natural selection (Hendry et al. 2011), tends to slow the pace of adaptation (Barbaro et al. 2005; Beckie and Reboud 2009).

The pace of adaptation to environmental chemicals can be rapid if the molecular mechanisms of action are specific, the dimensionality of selection is small, and the intensity of selection is strong, but may also be much accelerated if adaptive variation pre-exists as standing variation within populations. If putatively adaptive alleles pre-exist within populations, then natural selection gains a head start, which may be especially important for adaptation to the rapid pace of human-induced environmental change and for populations with relatively long generation times. Though pathogen adaptation to pharmaceuticals is commonly thought to involve selection on new mutations, recent studies indicate that variation relevant for resistance to modern pharmaceuticals long pre-dates the modern era of clinical antibiotic use (D'Costa et al. 2011). Similarly, adaptive variation for pesticides exists in natural populations for example in *C. elegans* (Ghosh et al. 2012), brown rats (Pelz et al. 2005), and blowflies (Hartley et al. 2006). The pre-existence of such variants at appreciable frequencies in populations suggests their adaptive importance for functions other than resistance to human-imposed stressors. In the cases of evolved resistance introduced above, species likely benefited from the virtues of large population size, including greater efficiency of balancing selection and capacity to harbor greater genetic variation (Lacy 1987).

Agricultural pesticides enter the environment intentionally, where single chemicals designed for interaction with highly specific biochemical targets are applied at high concentrations, often quickly followed by rapid adaptation underlain by a relatively simple genetic architecture. In contrast, industrial pollution is unintentional, insofar as poisons are not delivered to the environment for strategic ends. Unintentionally introduced poisons can be found at concentrations that span the continuum from trace to extremely high, and pollution may be patchy over space and time. This kind of pollution is often characterized by complex mixtures of chemicals with diverse mechanisms of biological activity, especially in non-agricultural ecosystems where pollutants from multiple sources may converge and mix (Lyman 1984). Much is known about the nature of adaptation to intentionally-introduced poisons such as pesticides and drugs. Comparatively little is known of the evolutionary consequences of unintentional environmental pollution. Perhaps this is because adaptation to intentionally administered biocides often has direct economic impacts, thus promoting a discovery bias. Given the broad differences in the nature of intentional versus unintentionally introduced poisons, might one anticipate differences in the scope, phenotypic nature, and genetic basis of evolutionary solutions to unintentional environmental pollution?

16.4 Genomics Approaches for Studying the Ecological-Evolutionary Dynamics of Environmental Pollution

Compared to intentionally introduced poisons such as pesticides and antibiotics, relatively little is known of the potential for, the occurrence of, or the nature of, adaptation to most environmental pollutants. Given the exploratory nature of global and high-throughput technologies that enable genome-scale research, and their growing accessibility for deployment in non-traditional

model species, there exists much opportunity for discovery. Though much knowledge of the biochemical and molecular targets of many pollutants provides context for interpreting signatures of natural selection, the universe of potential genomic targets is too great to risk not adopting a global approach for discovery of adaptive targets. Some of the adaptive targets for pollutant resistance identified to date include few genes of large effect. However, there often remains adaptive variation that is unexplained, and other evolved traits that are important for survival in the novel environment likely remain undiscovered. For example, mutations in single genes of major effect explain most, but not all, of the variation in avermectin resistance in *C. elegans* (Ghosh et al. 2012) and TTX resistance in garter snakes (Geffeney et al. 2005). Repeated diversification of stickleback fishes from marine into freshwater habitats involved many evolutionary innovations, where single genes of major effect have been discovered for various morphological variants (Shapiro et al. 2004; Colosimo et al. 2005). However, *genome-scale scans* for adaptive variation identified not only those genes, but also many additional regions of the genome under strong diversifying selection in freshwater-adapted populations (Hohenlohe et al. 2010). Accordingly, global genome-scale approaches, that cast a wide net, hold great promise for not only testing hypotheses about the genetic basis of particular traits, but also for discovering other traits under selection, or discovering important compensatory mutations that may be missed with more targeted methods. This may be especially relevant in ecotoxicology, where pollutants are often encountered as complex mixtures, and adaptive solutions may require change in multiple phenotypes (Wirgin and Waldman 2004).

Few studies to date have exploited genome-scale tools for discovering the mechanistic basis of adaptation to unintentional anthropogenic pollution. Yet, these few studies serve to highlight how genome-scale tools may enable insight into the underlying mechanisms that facilitated persistence of populations in the face of rapid and sometimes extreme environmental change.

16.4.1 Peppered Moth Adaptation to Industrial Air Pollution

Widespread air pollution following the industrial revolution in England caused darkening of trees by soot, and subsequent evolution of darkened melanic forms of peppered moth is one of the best-known examples of contemporary evolution (Kettlewell 1973; Majerus 1998). Many genes from diverse taxa (including insects) are well-known to encode melanic variation, but the candidate-gene approach for discovering adaptive variation in peppered moths was unsuccessful (van't Hof and Saccheri 2010). Further exploration adopted a linkage mapping approach, where the melanic phenotype was mapped to a single core sequence that carried the signature of recent and strong natural selection (van't Hof et al. 2011). Comparative genomics showed that the melanic core region overlapped with regions in other insects that harbor genes known to encode color-patterning loci, and thus allowed scope for interpreting the functional targets of selection in the peppered moth. Since few genes of large effect often underpin variation in color patterning (e.g., (King 2003; Wittkopp et al. 2003; Reed et al. 2011)), it is perhaps not surprising that this is also the case for peppered moths. Though the evolution of cryptic coloration was adaptive for reducing predation risk in a novel polluted habitat (Cook et al. 2012), this evolutionary challenge and adaptive solution is clearly different from those pollution scenarios where the fitness challenge is imposed by the toxic effects of ingested environmental poisons. This serves to highlight that adaptive persistence in polluted habitats may in some cases involve many phenotypic (and therefore genotypic) dimensions.

16.4.2 Adaptation to Metal Pollution in Diverse Species

Exposures to metals in the environment are not an exclusively modern phenomenon, as metals have been persistent sources of physiological stress through deep time, and evolved mechanisms to move and sequester metals are ancient. Though

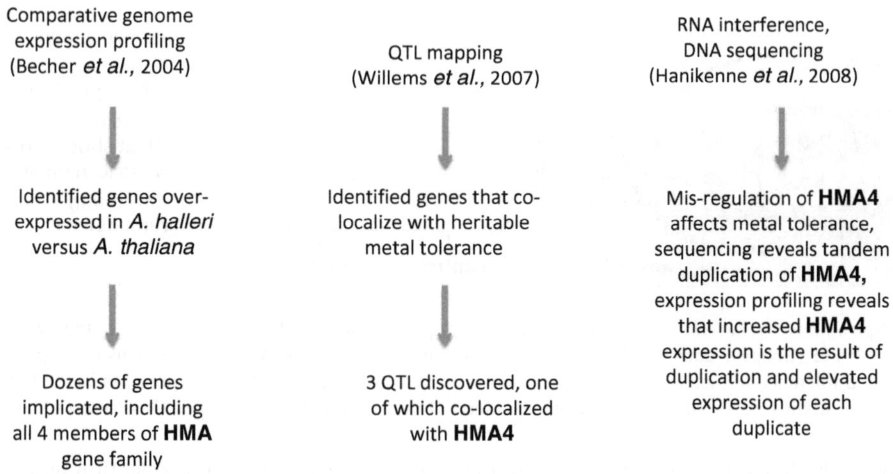

Fig. 16.1 A summary of the various lines of genomic inference used to discover the importance of the HMA4 gene for enabling extreme heavy metal tolerance in *Arabidopsis halleri*

modern massive re-distribution of metals in nature by humans poses severe physiological challenges to resident species, recently evolved tolerance to toxic metals has been reported for diverse terrestrial and aquatic taxa (Antonovics et al. 1971; Klerks and Weis 1987; Posthuma and van Straalen 1993). In many cases variation in ancient mechanisms of metal metabolism has been exploited by natural selection to enable tolerance, including for example elevated expression of metal binding proteins such as metallothioneins (MTs). Elevated cellular levels of a particular protein may be achieved by increasing the expression at a particular locus, or by duplicating many copies of the same gene, thereby enhancing protein dosage-related functions such as metal chelation. In fruit flies, increased copy number of MT genes is found naturally within populations and confers increased tolerance to metals (Maroni et al. 1987). Heavy metal tolerance can evolve quickly in fruit flies under strong selection, where adaptation is associated with inheritance of MT duplications and corresponding increase in MT protein expression (Otto et al. 1986). Heavy metal tolerance in an aquatic oligochaete involves up-regulation of MT-like proteins (Klerks and Bartholomew 1991), where tolerance appears to have a simple underlying genetic architecture (Martinez and Levinton 1996), and laboratory experiments indicate potential for very rapid evolution (Klerks and Levinton 1989). Similarly, elevated MT expression in springtails (*Orchesella cincta*) correlates with metal resistance that recently evolved in polluted soils (Sterenborg and Roelofs 2003).

In plants, evolved metal tolerance often involves a very simple genetic architecture involving one or few genes, though there is much evidence that additional genetic changes that enhance or modify tolerance are important in natural populations (reviewed in (Macnair 1993)). In the plant *Arabidopsis halleri*, *QTL mapping* (Willems et al. 2007), comparative transcriptomics (Becher et al. 2004), and genetic manipulation studies (Hanikenne et al. 2008) show that functional variation for metal tolerance lies in promoter substitutions and copy number variation of metal-transporting proteins (Fig. 16.1).

Though metal tolerance appears to often be underlain by relatively simple genetic architecture, additional genetic changes that enhance or modify tolerance may be important. One of the few comparative genome-scale studies of adaptive field-selected metal tolerance implicated a complex gene regulatory phenotype associated with metal resistance in springtails that expanded far beyond simple variation in MT expression (Roelofs et al. 2009). One intriguing feature of this variation was

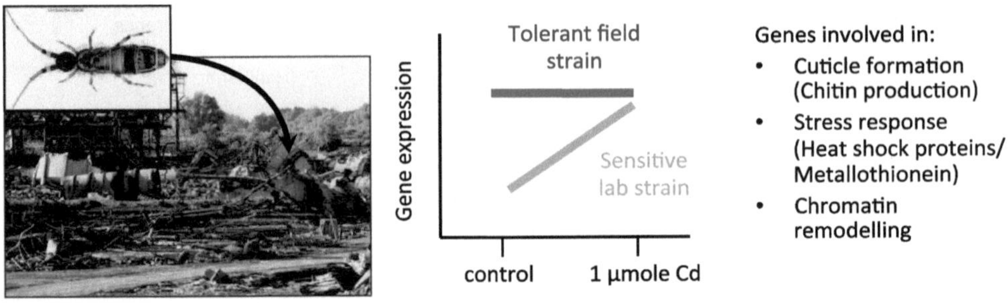

Fig. 16.2 Springtails *Orchesellacincta* collected from a metal-polluted site (Images; courtesy of Dick Roelofs) exhibit elevated tolerance to metals toxicity. *Comparative transcriptomics* studies revealed that many genes (see gene functions listed on the right) that were inducible by cadmium exposure in the sensitive springtails were constitutively elevated in expression in the field-collected tolerant animals (*line graph*) (Roelofs et al. 2009)

constitutively high expression of genes in the tolerant population that were transcriptionally up-regulated in response to metal exposures in the sensitive population (Fig. 16.2), indicating that canalization of environment-dependent gene expression plasticity contributes to adaptive tolerance in derived populations.

16.4.3 Killifish Adaptation to Industrially-Polluted Estuaries

Atlantic killifish represent a rare case of vertebrate animal adaptation to complex mixtures of environmental pollutants. Atlantic killifish (*Fundulus heteroclitus*) are abundant in shallow saltmarsh habitats along the entire United States Atlantic coast. Apparently healthy populations of killifish also occupy habitats that are polluted with persistent industrial chemicals, where sediments harbor chemical burdens of dioxin-like compounds (DLCs) that are orders of magnitude greater than at clean reference sites and that far exceed the lethal limit of typical killifish (Nacci et al. 2010). Though *F. heteroclitus* are relatively sensitive among fishes to the toxic effects of these chemicals (Van Veld and Nacci 2008a), population sizes are very large and animals are non-migratory and exhibit high site-fidelity (Lotrich 1975; Sweeney et al. 1998; Yozzo and Smith 1998; Teo and Able 2003). These attributes maximize the opportunity for local adaptation, insofar as large populations maximize the efficiency of natural selection and can harbor much potentially adaptive genetic variation, low migration rates offset the homogenizing effects of gene flow, and the high chemical stability of the industrial chemicals in the polluted habitats impose a consistent selection pressure across generations. Indeed, populations inhabiting polluted sites exhibit extreme heritable tolerance to the toxic effects of contaminating chemicals, where the degree of tolerance matches the degree of contamination. That is, sediment loads and chemical body burdens of pollutants at contaminated sites are two to three orders of magnitude greater than at reference sites, and resistance to the toxic effects of model chemicals is two to three orders of magnitude greater for fish from polluted sites than reference sites (Nacci et al. 2010). Intriguingly, this resistance has repeatedly evolved multiple times in *F. heteroclitus* in various polluted estuaries. What is the mechanism of evolved tolerance? What are the genomic targets of natural selection? Are the genomic targets of selection common in parallel-evolved populations? Are there tradeoffs or costs associated with resistance? Beyond direct toxicant resistance, have additional adaptations evolved in polluted habitats?

Prior to the genomics era, much was discovered about the genetic, biochemical, and physiological differences between tolerant and sensitive populations (reviewed in (Nacci et al. 2002; Van Veld and Nacci 2008b)). Several key discoveries of the resistant populations provided context and guidance for recent and future genomics studies:

First, pollutant tolerance involves a variety of endpoints, including embryo and adult survival, teratogenicity, and altered gene expression (Van Veld and Nacci 2008b). Second, each of the tolerant populations shows cross-tolerance to classes of compounds not abundant at their native sites; Dioxin-resistant Newark fish are also resistant to *PCBs* (Elskus et al. 1999), PCB-resistant New Bedford fish are also resistant to *PAHs* (Nacci et al. 1999; Bello et al. 2001), and PAH-resistant Elizabeth River fish are also resistant to PCBs (Meyer and Di Giulio 2002). Third, the tolerance appears to be largely heritable (Nacci et al. 1999; Meyer and Di Giulio 2002; Meyer et al. 2002; Ownby et al. 2002; Van Veld and Nacci 2008b). Fourth, the tolerant populations are genetically distinct from nearby sensitive populations and do show some evidence for selection at specific loci (Cohen 2002; Mulvey et al. 2003; Hahn et al. 2004, 2005; Williams and Oleksiak 2011), but genetic diversity is not eroded in tolerant populations (Mulvey et al. 2002, 2003; Roark et al. 2005; McMillan et al. 2006). Finally, although the fitness costs or "trade-offs" of pollutant tolerance in *F. heteroclitus* are not yet well understood, there is evidence that such costs accompany some pollutant tolerant phenotypes (Meyer and Di Giulio 2003; Clark and Di Giulio 2012; Harbeitner et al. 2013).

To more globally explore the mechanisms that enable persistence in these extreme niches, comparative transcriptomics approaches have been deployed. Remarkably few evolved differences in constitutive gene expression were observed during development between tolerant and sensitive populations raised in a common clean environment (Bozinovic and Oleksiak 2010). In contrast, exposures to model toxicants revealed global as well as specific gene expression patterns that were profoundly different between tolerant and sensitive populations (Whitehead et al. 2010; Oleksiak et al. 2011), indicating that key evolved differences in gene expression were environment-dependent. In a sensitive population chemical exposures produced dramatic, dose-dependent toxic effects, concurrent with the alterations in the expression of many genes. However, genome-wide expression was comparatively refractory to chemical induction in a tolerant population (Whitehead et al. 2010). Tolerance was associated with the global blockade of the aryl hydrocarbon receptor (*AHR*) signaling pathway, the key mediator of PCB toxicity, in contrast to the strong dose-dependent up-regulation of AHR pathway elements observed in the sensitive population (Whitehead et al. 2010). The AHR is a cytosolic receptor, that when bound by a ligand translocates to the nucleus, associates with other proteins, binds to response elements in the genome, and activates the transcription of a battery of genes, some of which are important for metabolism, and others that contribute to toxicity. Altered regulation of signaling pathways that cross-talk with AHR was implicated as one candidate mechanism for the adaptive AHR signaling repression and the pollution tolerance that it affords (Whitehead et al. 2010).

Expansion of comparative studies to include multiple pairs of tolerant and sensitive populations indicated that tolerant populations have each converged on a common molecular mechanism of resistance (Whitehead et al. 2012) (Fig. 16.3). Though functional convergence appears to have evolved in tolerant populations, genome scans with ~300 AFLP markers indicated that different loci were outliers in different tolerant-sensitive population pairs, implying different mechanisms of convergence (Williams and Oleksiak 2008). However, if selection sorted among pre-existing variation, one might expect that different alleles would be in linkage disequilibrium with adaptive loci in different ancestral populations. Genome-scale tools, and a reference genome assembly available for mapping population genetic markers, is a key tool for more precise inference about the identity and parallelism of causative loci. Indeed, such studies are underway.

A reference genome has been completed for *F. heteroclitus* (www.fundulus.org), and full-genome scans are underway for tolerant-sensitive population pairs to detect signatures of selection in tolerant populations (Fig. 16.3).

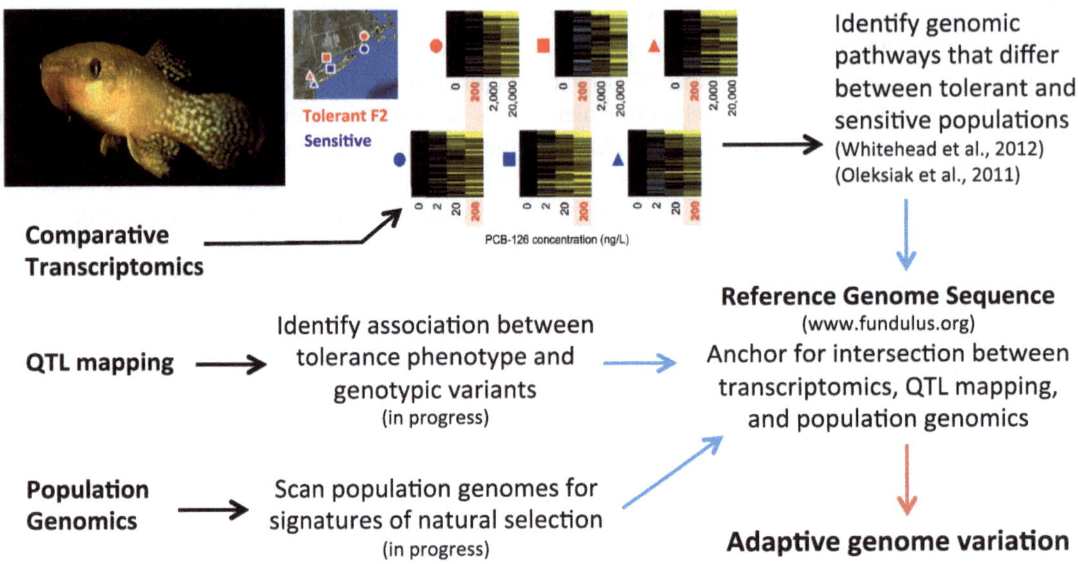

Fig. 16.3 Summary of the various lines of evidence deployed to discover the genomic basis of rapid and repeatedly evolved extreme pollution tolerance in the killifish *Fundulus heteroclitus* (Photo courtesy of Andrew Whitehead). The map of the northeast Atlantic coast of the United States shows the source of six populations used in comparative studies, where *red* symbols indicate tolerant populations collected from polluted sites and raised under clean conditions for two generations, and *blue* symbols indicate reference sensitive populations collected from relatively clean sites. Common symbols (*circles, squares, triangles*) indicate pairs of populations collected from the same geographical area. The heatmap summarizes the expression of genes differentially expressed between populations from comparative transcriptomics studies where developing embryos were exposed to increasing concentrations of the model toxicant PCB-126 (Whitehead et al. 2012). Each column represents a treatment, each row represents a different gene, and the color of a cell represents mean expression where *yellow* is up-regulated and *blue* is down-regulated relative to the control. The group of genes shown is enriched for transcriptional targets of the pollutant-activated aryl hydrocarbon receptor (AHR). At a common exposure concentration (200 ng/L), the transcriptional responses of the tolerant populations differ greatly compared to sensitive populations, but converge when concentrations are increased by two orders of magnitude, indicating that the mechanism of evolved tolerance involves profound desensitization of the AHR signaling pathway. To discover the genetic changes responsible for this divergence, QTL mapping and whole-genome scans of multiple tolerant and reference populations are being deployed, and allelic variants anchored to a reference genome sequence (www.fundulus.org). The integration of resulting data (*blue arrows*) will facilitate inference of the genomic architecture of the pollutant-adaptive phenotype

Concurrently, QTL mapping studies seek to link markers associated with the tolerant phenotypes with loci showing signatures of selection in population genome scans (Fig. 16.3). Candidate adaptation genes can be verified using a variety of approaches in follow-up functional studies. For example, pharmacological manipulation confirmed candidates governing craniofacial divergence in cichlid fish (Roberts et al. 2011), mRNA expression manipulation verified candidates governing eye degeneration in blind cavefish (Yamamoto et al. 2004), and the creation of transgenic mutants verified candidates conferring avermectin resistance in nematodes (Ghosh et al. 2012). Integrating candidate genes implicated by these studies with knowledge of pathways that mediate toxicity (e.g., (Clark et al. 2010)), and with knowledge of genes differentially expressed between populations (e.g., (Whitehead et al. 2010, 2012; Oleksiak et al. 2011)), should afford considerable insight into the genomic landscape that facilitated rapid repeated and dramatic adaptation to contemporary and complex pollutant mixtures in nature (Fig. 16.3).

Upon discovery of evolutionarily modified genes and pathways, one may pose more informed hypotheses about predicted pleiotropic

effects, including fitness costs, of adaptation. For example, elements of the AHR signaling pathway are involved in immune function, hypoxia responses, and metabolism, such that functional divergence in AHR for pollution tolerance may compromise these other functions. Long-term persistence of resistant populations may be enabled by additional adaptations to offset physiological tradeoffs possibly imposed by altered AHR. Furthermore, ecological feedbacks such as altered pathogen loads in polluted sites may impose additional selective pressures that required adaptive solutions (Cohen 2002). Genome scans and QTL mapping may also enable discovery of whether adaptation was underpinned by single genes of large effect as appears typical for much rapid adaptation, whether the targets of selection were protein-encoding variants or regulatory variants, whether the targets of selection were common among parallel-evolved populations, and whether the adaptive alleles pre-existed in normal sensitive populations. If adaptive alleles did exist in populations prior to pollution, maintained for example by balancing selection, then what function might these alternate alleles support in the absence of twentieth century pollutants? Genome-scale data afford the opportunity to discover the genomic basis of ecologically important traits, but also data to provoke additional hypotheses about the causes and consequences of the adaptive phenotype at the individual and population levels.

16.5 Conclusions and Future Directions

Adaptation to environmental pollution may be multi-dimensional, depending on species and habitat. For example, evolution of an adapted phenotype may involve selection on morphological variation (e.g., peppered moths), avoidance behavior, metabolism of ingested toxicants (e.g., springtails, killifish), and selection on compensatory phenotypes that perhaps evolved to offset the fitness costs of initial resistance. Furthermore, recent studies have shown that exposure to xenobiotics can alter the genome expression, physiology, and structure of the microbiome within exposed animal hosts (Maurice et al. 2013), and that natural selection on symbiont communities can facilitate adaptive resistance of the host to insecticides (Kikuchi et al. 2012). It is important to recognize that specific pollutants often co-occur with other pollutants and other stressors in human-altered environments, thereby increasing the complexity of the evolutionary challenge. Within the context of this organismal and environmental complexity, one may anticipate that phenotypes that are adaptive for polluted environments may be underlain by a complex landscape of genomic changes. If this is true, then discovery of single genes of large effect should not represent the terminus of investigations. This is because such findings may represent an illusion that emerges from insufficient experimental and statistical power (Rockman 2012), and also because exclusive focus on the tallest peak in the range may disable a nuanced appreciation of the entire landscape.

Genome-scale data may also be exploited to generate hypotheses about mechanisms beyond specific resistance to the toxic effects of chemicals that may contribute in important ways to the adaptive phenotype, just as genome scans in freshwater sticklebacks identified many genomic signatures of selection beyond those known to govern the most obvious morphological changes (Hohenlohe et al. 2010). Though toxicant-resistance genes may in fact be few of large effect, genetic conflicts and physiological tradeoffs may emerge from strong selection thereby promoting natural selection for compensatory traits. Costs of rapid adaptation to environmental pollutants are generally considered in the context of compromised ability to accommodate other stressors, or compromised life history traits (van Straalen and Hoffmann 2000; Medina et al. 2007; De Schamphelaere et al. 2011; Jansen et al. 2011; Klerks et al. 2011), that may manifest as fitness tradeoffs in contemporary environments or limited adaptability to future environments (e.g., (Ward and Robinson 2005)). In *Daphnia*,

for example, pesticide tolerance evolved at the cost of increased pathogen sensitivity (Jansen et al. 2011), and metals tolerance evolved at the cost of decreased feeding rates that were genetically correlated with survival (Agra et al. 2010). Negative effects of pollutant adaptation could extend beyond costs imposed by genetic conflicts, physiological tradeoffs, or reduced evolutionary potential, if adaptation of one species feeds back to affect community and ecological dynamics in polluted niches. These feedbacks may impact the tolerant animal's extended phenotype (sensu (Dawkins 1983)), or could pose indirect selective constraints on other resident species. For example, adaptation in some community members but not others may disrupt symbiotic or trophic relationships, or resistant species may serve as vectors for trophic transfer of *contaminants* (Wirgin and Waldman 2004). Upon cleanup of polluted sites, reversion of tolerant to sensitive phenotypes can rapidly evolve (Levinton et al. 2003), thereby reducing the potential for trophic transfer of contaminants. Population genomics studies could be useful for tracing the evolutionary forces (selection, migration, hybridization) that contribute to population change following remedial cleanup of polluted sites. Given the potential complexity outlined here, global systems-level approaches should be powerful for revealing the nature of ecological-evolutionary dynamics within polluted environments.

Deep within the storehouse of evolutionary history lies ancient adaptations to diverse chemical poisons, and this adaptive variation provides substrate for adaptive fine-tuning to suit contemporary and future environments. But humans are imposing chemical challenges within ecosystems that may be unprecedented in their nature and scale, requiring rapid and dramatic adaptations that may exceed the scope of adaptive potential for many species. The modern tools of genomics offer promise for discovering how pollutants interact with genomes on physiological timescales, and also for discovering what genomic attributes of populations may enable resistance to pollutants over evolutionary timescales. Perhaps by integrating these sophisticated genomics tools and approaches with an understanding of the deep historical dynamics and contemporary ecological-evolutionary dynamics of environmental pollution, a more mature study of environmental toxicology may emerge, akin to the recent emergence of Evolutionary Medicine (Gluckman et al. 2009).

Glossary

AHR The aryl hydrocarbon receptor (AHR) is a cytosolic transcription factor. Pollutants such as dioxins, PCBs, and PAHs, act as ligands for this receptor. Once activated, the receptor-ligand complex migrates to the nucleus, and acts as a transcription factor to activate the transcription of a battery of genes. Some of the activated genes are responsible for metabolism of the ligand and for the emergence of toxicity

Anthropogenic Anthropogenic refers to effects or objects that are of human origin, or that are influenced by humans

Comparative transcriptomics Comparative transcriptomics is an experimental design that includes the comparison of transcriptomic responses to some experimental variable (e.g., environmental or pollutant challenge) between strains, populations, or species

Contaminant A contaminant is a compound that is released into the environment as a result of human activities, but that may or may not be toxic to exposed organisms

Contemporary evolution Contemporary evolution refers to evolutionary change in response to recent changes in the environment

Copy number variation Specific regions of the genome may increase in copy number from duplication of chromosome segments, duplication of whole chromosomes, or duplication of entire genomes. Protein family expansion (e.g., globins, HOX genes) is often from duplication of chromosome segments that include an entire protein sequence. The fate of duplicate copies may include pseudogenization, maintenance of function, neofunctionalization, or subfunctionalization

Epistasis Epistasis is a phenomenon that refers to the non-additive effects of multiple genes, where the effects of different loci are not independent – where the effects of one gene are dependent on other genes

Genome scan Given genome-scale sequence data for many individuals for two or more populations, genome scans are a suite of exploratory methods that are designed to detect genetic signatures of natural selection in populations. Genetic markers that exhibit signatures of selection could be the causative genetic variants, but are more likely to be physically linked to causative variants

Neofunctionalization Neofunctionalization is a result of functional divergence of gene paralogs following duplication, where the duplicate copy acquires an entirely new function that is distinct from the original gene's functions

PAH Polycyclic aromatic hydrocarbons (PAHs) are a class of chemicals that are common environmental pollutants, where major sources are from spilled crude oil and from fuel combustion. PAHs act as ligands for the AHR, and exert at least part of their toxicity through the activation of the AHR signaling pathway

PCB Polychlorinated biphenyls (PCBs) are a class of chemicals that are persistent environmental pollutants, where major sources are from wastes from industrial processes where they were used as coolant fluids, hydraulic fluids, and insulation fluids. PCBs act as ligands for the AHR, and exert at least part of their toxicity through the activation of the AHR signaling pathway

Pleiotropy A gene product is considered pleiotropic if it influences more than one phenotypic trait

Pollutant A pollutant is a compound that is released into the environment as a result of human activities, and that has negative effects in exposed organisms

QTL mapping Quantitative trait locus (QTL) mapping involves matching genotype to phenotype from experimental crosses to detect genetic variants that are physically linked to phenotypes of interest. Genetic markers that are associated with the phenotype are not necessarily the causative loci, but are presumed to be physically linked with causative loci

Subfunctionalization Subfunctionalization is a result of functional divergence of gene paralogs following duplication, where the duplicate copies retain different parts of the original gene's functions

Toxicant Toxicants are chemicals that exert toxic effects at sufficiently high doses, and that are not natural products, but rather products of human activity

Toxin Toxins are chemicals that exert toxic effects at sufficiently high doses, and that are natural products of biosynthesis by organisms.

Xenobiotic A compound that may be found within organisms, but that is foreign to those organisms, usually in reference to manufactured chemicals

References

Agra AR, Guilhermino L, Soares AMVM, Barata C (2010) Genetic costs of tolerance to metals in *Daphnia longispina* populations historically exposed to a copper mine drainage. Environ Toxicol Chem 29: 939–946

Allendorf FW, Hard JJ (2009) Human-induced evolution caused by unnatural selection through harvest of wild animals. Proc Natl Acad Sci USA 106:9987–9994

Antonovics J, Bradshaw AD, Turner RG (1971) Heavy metal tolerance in plants. Adv Ecol Res 7:1–85

Barbaro G, Scozzafava A, Mastrolorenzo A, Supuran CT (2005) Highly active antiretroviral therapy: current state of the art, new agents and their pharmacological interactions useful for improving therapeutic outcome. Curr Pharm Design 11:1805–1843

Becher M, Talke IN, Krall L, Kramer U (2004) Cross-species microarray transcript profiling reveals high constitutive expression of metal homeostasis genes in shoots of the zinc hyperaccumulator *Arabidopsis halleri*. Plant J 37:251–268

Beckie HJ, Reboud X (2009) Selecting for weed resistance: herbicide rotation and mixture. Weed Technol 23:363–370

Bello SM, Franks DG, Stegeman JJ, Hahn ME (2001) Acquired resistance to aryl hydrocarbon receptor agonists in a population of *Fundulus heteroclitus* from a marine superfund site: in vivo and in vitro studies on the induction of xenobiotic-metabolizing enzymes. Toxicol Sci 60:77–91

Bernhardt R (2006) Cytochromes P450 as versatile biocatalysts. J Biotechnol 124:128–145

Blaustein AR, Belden LK (2003) Amphibian defenses against ultraviolet-B radiation. Evol Dev 5:89–97

Bozinovic G, Oleksiak MF (2010) Embryonic gene expression among pollutant resistant and sensitive *Fundulus heteroclitus* populations. Aquat Toxicol 98:221–229

Campbell BJ, Smith JL, Hanson TE, Klotz MG, Stein LY, Lee CK, Wu DY, Robinson JM, Khouri HM, Eisen JA, Cary SC (2009) Adaptations to submarine hydrothermal environments exemplified by the genome of *Nautilia profundicola*. Plos Genet 5(2):e1000362

Chevin LM, Lande R, Mace GM (2010) Adaptation, plasticity, and extinction in a changing environment: towards a predictive theory. Plos Biol 8(4):e1000357

Clark BW, Di Giulio RT (2012) *Fundulus heteroclitus* adapted to PAHs are cross-resistant to multiple insecticides. Ecotoxicology 21:465–474

Clark BW, Matson CW, Jung D, Di Giulio RT (2010) AHR2 mediates cardiac teratogenesis of polycyclic aromatic hydrocarbons and PCB-126 in Atlantic killifish (*Fundulus heteroclitus*). Aquat Toxicol 99:232–240

Cohen S (2002) Strong positive selection and habitat-specific amino acid substitution patterns in MHC from an estuarine fish under intense pollution stress. Mol Biol Evol 19:1870–1880

Colosimo PF, Hosemann KE, Balabhadra S, Villarreal G, Dickson M, Grimwood J, Schmutz J, Myers RM, Schluter D, Kingsley DM (2005) Widespread parallel evolution in sticklebacks by repeated fixation of ectodysplasin alleles. Science 307:1928–1933

Cook LM, Grant BS, Saccheri IJ, Mallet J (2012) Selective bird predation on the peppered moth: the last experiment of Michael Majerus. Biol Lett 8:609–612

Coyle P, Philcox JC, Carey LC, Rofe AM (2002) Metallothionein: the multipurpose protein. Cell Mol Life Sci 59:627–647

Dawkins R (1983) The extended phenotype: the long reach of the gene. Oxford University Press, Oxford

D'Costa VM, King CE, Kalan L, Morar M, Sung WWL, Schwarz C, Froese D, Zazula G, Calmels F, Debruyne R, Golding GB, Poinar HN, Wright GD (2011) Antibiotic resistance is ancient. Nature 477:457–461

De Schamphelaere KA, Glaholt S, Asselman J, Messiaen M, De Coninck D, Janssen CR, Colbourne JK, Shaw JR (2011) Will genetic adaptation of natural populations to chemical pollution result in lower or higher tolerance to future climate change? Integr Environ Assess Manag 7:141–143

Despres L, David JP, Gallet C (2007) The evolutionary ecology of insect resistance to plant chemicals. Trends Ecol Evol 22:298–307

Elskus AA, Monosson E, McElroy AE, Stegeman JJ, Woltering DS (1999) Altered CYP1A expression in *Fundulus heteroclitus* adults and larvae: a sign of pollutant resistance? Aquat Toxicol 45:99–113

Feyereisen R (1999) Insect P450 enzymes. Annu Rev Entomol 44:507–533

Ffrench-Constant RH, Daborn PJ, Le Goff G (2004) The genetics and genomics of insecticide resistance. Trends Genet 20:163–170

Geffeney SL, Fujimoto E, Brodie ED, Brodie ED, Ruben PC (2005) Evolutionary diversification of TTX-resistant sodium channels in a predator–prey interaction. Nature 434:759–763

Ghosh R, Andersen EC, Shapiro JA, Gerke JP, Kruglyak L (2012) Natural variation in a chloride channel subunit confers avermectin resistance in *C. elegans*. Science 335:574–578

Gluckman PD, Beedle A, Hanson MA (2009) Principles of evolutionary medicine. Oxford University Press, Oxford

Goldstone JV, Goldstone HMH, Morrison AM, Tarrant A, Kern SE, Woodin BR, Stegeman JJ (2007) Cytochrome p450 1 genes in early deuterostomes (tunicates and sea urchins) and vertebrates (chicken and frog): origin and diversification of the CYP1 gene family. Mol Biol Evol 24:2619–2631

Gonzalez FJ, Nebert DW (1990) Evolution of the P450-gene superfamily – animal plant warfare, molecular drive and human genetic-differences in drug oxidation. Trends Genet 6:182–186

Graur D, Li W-H (2000) Fundamentals of molecular evolution. Sinauer Associates, Sunderland

Hahn ME, Karchner SI, Franks DG, Merson RR (2004) Aryl hydrocarbon receptor polymorphisms and dioxin resistance in Atlantic killifish (*Fundulus heteroclitus*). Pharmacogenetics 14:131–143

Hahn ME, Karchner SI, Franks DG, Evans BR, Nacci D, Champlin D, Cohen S (2005) Mechanism of PCB – and dioxin-resistance in fish in the Hudson River Estuary: role of receptor polymorphisms. Final report, Hudson River Foundation Grant 004/02A

Hairston NG, Ellner SP, Geber MA, Yoshida T, Fox JA (2005) Rapid evolution and the convergence of ecological and evolutionary time. Ecol Lett 8:1114–1127

Hanikenne M, Talke IN, Haydon MJ, Lanz C, Nolte A, Motte P, Kroymann J, Weigel D, Kramer U (2008) Evolution of metal hyperaccumulation required cis-regulatory changes and triplication of HMA4. Nature 453:391–395, U344

Harbeitner RC, Hahn ME, Timme-Laragy AR (2013) Differential sensitivity to pro-oxidant exposure in two populations of killifish (*Fundulus heteroclitus*). Ecotoxicology 22:387–401

Hartley CJ, Newcomb RD, Russell RJ, Yong CG, Stevens JR, Yeates DK, La Salle J, Oakeshott JG (2006) Amplification of DNA from preserved specimens shows blowflies were preadapted for the rapid evolution of insecticide resistance. Proc Natl Acad Sci USA 103:8757–8762

Hendry AP, Kinnison MT (1999) Perspective: the pace of modern life: measuring rates of contemporary microevolution. Evolution 53:1637–1653

Hendry AP, Farrugia TJ, Kinnison MT (2008) Human influences on rates of phenotypic change in wild animal populations. Mol Ecol 17:20–29

Hendry AP, Kinnison MT, Heino M, Day T, Smith TB, Fitt G, Bergstrom CT, Oakeshott J, Jorgensen PS, Zalucki MP, Gilchrist G, Southerton S, Sih A, Strauss S, Denison RF, Carroll SP (2011) Evolutionary principles and their practical application. Evol Appl 4:159–183

Hoffmann AA, Sgro CM (2011) Climate change and evolutionary adaptation. Nature 470:479–485

Hohenlohe PA, Bassham S, Etter PD, Stiffler N, Johnson EA, Cresko WA (2010) Population genomics of parallel adaptation in threespine stickleback using sequenced RAD tags. Plos Genet 6(2):e1000862

Jansen M, Coors A, Stoks R, De Meester L (2011) Evolutionary ecotoxicology of pesticide resistance: a case study in Daphnia. Ecotoxicology 20:543–551

Kettlewell B (1973) The evolution of melanism. The study of a recurring necessity; with special reference to industrial melanism in the Lepidoptera. Clarendon, Oxford

Kikuchi Y, Hayatsu M, Hosokawa T, Nagayama A, Tago K, Fukatsu T (2012) Symbiont-mediated insecticide resistance. Proc Natl Acad Sci USA 109:8618–8622

King RB (2003) Mendelian inheritance of melanism in the garter snake *Thamnophis sirtalis*. Herpetologica 59:484–489

Kinnison MT, Hendry AP (2001) The pace of modern life II: from rates of contemporary microevolution to pattern and process. Genetica 112:145–164

Kiontke S, Geisselbrecht Y, Pokorny R, Carell T, Batschauer A, Essen LO (2011) Crystal structures of an archaeal class II DNA photolyase and its complex with UV-damaged duplex DNA. Embo J 30:4437–4449

Klerks PL, Bartholomew PR (1991) Cadmium accumulation and detoxification in a Cd-resistant population of the oligochaete *Limnodrilus-hoffmeisteri*. Aquat Toxicol 19:97–112

Klerks PL, Levinton JS (1989) Rapid evolution of metal resistance in a benthic oligochaete inhabiting a metal-polluted site. Biol Bull 176:135–141

Klerks PL, Weis JS (1987) Genetic adaptation to heavy-metals in aquatic organisms – a review. Environ Pollut 45:173–205

Klerks PL, Xie LT, Levinton JS (2011) Quantitative genetics approaches to study evolutionary processes in ecotoxicology; a perspective from research on the evolution of resistance. Ecotoxicology 20:513–523

Kump LR (2008) The rise of atmospheric oxygen. Nature 451:277–278

La Duc MT, Benardini JN, Kempf MJ, Newcombe DA, Lubarsky M, Venkateswaran K (2007) Microbial diversity of Indian Ocean hydrothermal vent plumes: microbes tolerant of desiccation, peroxide exposure, and ultraviolet and gamma-irradiation. Astrobiology 7:416–431

Lacy RC (1987) Loss of genetic diversity from managed populations: interacting effects of drift, mutation, immigration, selection, and population subdivision. Conserv Biol 1:143–158

Laxminarayan R, Heymann DL (2012) Challenges of drug resistance in the developing world. Br Med J 344

Levinton JS, Suatoni E, Wallace W, Junkins R, Kelaher B, Allen BJ (2003) Rapid loss of genetically based resistance to metals after the cleanup of a Superfund site. Proc Natl Acad Sci USA 100:9889–9891

Li XC, Schuler MA, Berenbaum MR (2007) Molecular mechanisms of metabolic resistance to synthetic and natural xenobiotics. Annu Rev Entomol 52:231–253

Lotrich VA (1975) Summer home range and movements of *Fundulus heteroclitus* (Pisces: Cyprinodontidae) in a tidal creek. Ecology 56:191–198

Lyman WJ (1984) Establishing sediment criteria for chemicals – industrial perspectives. In: Dickson KL, Maki AW, Brungs WA (eds) Fate and effects of sediment-bound chemicals in aquatic systems. Permagon Press, New York, pp 378–387

Macnair MR (1993) The genetics of metal tolerance in vascular plants. New Phytol 124:541–559

Majerus MEN (1998) Melanism: evolution in action. Oxford University Press, Oxford

Maron LG, Guimaraes CT, Kirst M, Albert PS, Birchler JA, Bradbury PJ, Buckler ES, Coluccio AE, Danilova TV, Kudrna D, Magalhaes JV, Pineros MA, Schatz MC, Wing RA, Kochian LV (2013) Aluminum tolerance in maize is associated with higher MATE1 gene copy number. Proc Natl Acad Sci USA 110:5241–5246

Maroni G, Wise J, Young JE, Otto E (1987) Metallothionein gene duplications and metal tolerance in natural-populations of *Drosophila-melanogaster*. Genetics 117:739–744

Marquis O, Miaud C, Ficetola GF, Bocher A, Mouchet F, Guittonneau S, Devaux A (2009) Variation in genotoxic stress tolerance among frog populations exposed to UV and pollutant gradients. Aquat Toxicol 95:152–161

Martinez DE, Levinton J (1996) Adaptation to heavy metals in the aquatic oligochaete *Limnodrilus hoffmeisteri*: evidence for control by one gene. Evolution 50:1339–1343

Maurice CF, Haiser HJ, Turnbaugh PJ (2013) Xenobiotics shape the physiology and gene expression of the active human gut microbiome. Cell 152:39–50

Mckenzie JA, Batterham P (1994) The genetic, molecular and phenotypic consequences of selection for insecticide resistance. Trends Ecol Evol 9:166–169

McMillan AM, Bagley MJ, Jackson SA, Nacci DE (2006) Genetic diversity and structure of an estuarine fish (*Fundulus heteroclitus*) indigenous to sites associated with a highly contaminated urban harbor. Ecotoxicology 15:539–548

Medina MH, Correa JA, Barata C (2007) Micro-evolution due to pollution: possible consequences for ecosystem responses to toxic stress. Chemosphere 67:2105–2114

Meyer J, Di Giulio R (2002) Patterns of heritability of decreased EROD activity and resistance to PCB 126-induced teratogenesis in laboratory-reared offspring of killifish (*Fundulus heteroclitus*) from a creosote-contaminated site in the Elizabeth River, VA, USA. Mar Environ Res 54:621–626

Meyer JN, Di Giulio RT (2003) Heritable adaptation and fitness costs in killifish (*Fundulus heteroclitus*) inhabiting a polluted estuary. Ecol Appl 13:490–503

Meyer JN, Nacci DE, Di Giulio RT (2002) Cytochrome P4501A (CYP1A) in killifish (*Fundulus heteroclitus*): heritability of altered expression and relationship to survival in contaminated sediments. Toxicol Sci 68:69–81

Monosson E (2012) Evolution in a toxic world: how life responds to chemical threats. Island Press, Washington, DC

Mulvey M, Newman MC, Vogelbein W, Unger MA (2002) Genetic structure of *Fundulus heteroclitus* from PAH-contaminated and neighboring sites in the Elizabeth and York Rivers. Aquat Toxicol 61:195–209

Mulvey M, Newman MC, Vogelbein WK, Unger MA, Ownby DR (2003) Genetic structure and mtDNA diversity of *Fundulus heteroclitus* populations from polycyclic aromatic hydrocarbon-contaminated sites. Environ Toxicol Chem 22:671–677

Nacci D, Coiro L, Champlin D, Jayaraman S, McKinney R, Gleason TR, Munns WR, Specker JL, Cooper KR (1999) Adaptations of wild populations of the estuarine fish *Fundulus heteroclitus* to persistent environmental contaminants. Mar Biol 134:9–17

Nacci DE, Gleason T, Gutjahr-Gobell R, Huber M, Munns WRJ (2002) Effects of environmental stressors on wildlife populations. In: Newman MC (ed) Coastal and Estuarine Risk Assessment: risk on the edge. CRC Press/Lewis Publishers, Washington, DC

Nacci D, Champlin D, Jayaraman S (2010) Adaptation of the estuarine fish *Fundulus heteroclitus* (Atlantic Killifish) to polychlorinated biphenyls (PCBs). Estuar Coast Shelf Sci 33:853–864

Oleksiak MF, Karchner SI, Jenny MJ, Franks DG, Welch DBM, Hahn ME (2011) Transcriptomic assessment of resistance to effects of an aryl hydrocarbon receptor (AHR) agonist in embryos of Atlantic killifish (*Fundulus heteroclitus*) from a marine Superfund site. BMC Genomics 12:263

Orr HA (2005) The genetic theory of adaptation: a brief history. Nat Rev Genet 6:119–127

Otto E, Young JE, Maroni G (1986) Structure and expression of a tandem duplication of the *Drosophila* metallothionein gene. Proc Natl Acad Sci USA 83:6025–6029

Ouzounis CA, Kunin V, Darzentas N, Goldovsky L (2006) A minimal estimate for the gene content of the last universal common ancestor – exobiology from a terrestrial perspective. Res Microbiol 157:57–68

Ownby DR, Newman MC, Mulvey M, Vogelbein WK, Unger MA, Arzayus LF (2002) Fish (*Fundulus heteroclitus*) populations with different exposure histories differ in tolerance of creosote-contaminated sediments. Environ Toxicol Chem 21:1897–1902

Palumbi SR (2001) Evolution – humans as the world's greatest evolutionary force. Science 293:1786–1790

Pelletier F, Garant D, Hendry AP (2009) Eco-evolutionary dynamics. Philos Trans R Soc B 364:1483–1489

Pelz HJ, Rost S, Hunerberg M, Fregin A, Heiberg AC, Baert K, MacNicoll AD, Prescott CV, Walker AS, Oldenburg J, Muller CR (2005) The genetic basis of resistance to anticoagulants in rodents. Genetics 170:1839–1847

Posthuma L, van Straalen NM (1993) Heavy-metal adaptation in terrestrial invertebrates – a review of occurrence, genetics, physiology and ecological consequences. Comp Biochem Phys C 106:11–38

Puinean AM, Foster SP, Oliphant L, Denholm I, Field LM, Millar NS, Williamson MS, Bass C (2010) Amplification of a cytochrome P450 gene is associated with resistance to neonicotinoid insecticides in the aphid Myzus persicae. Plos Genet 6:1–11

Reed RD, Papa R, Martin A, Hines HM, Counterman BA, Pardo-Diaz C, Jiggins CD, Chamberlain NL, Kronforst MR, Chen R, Halder G, Nijhout HF, McMillan WO (2011) Optix drives the repeated convergent evolution of butterfly wing pattern mimicry. Science 333:1137–1141

Roark SA, Nacci D, Coiro L, Champlin D, Guttman SI (2005) Population genetic structure of a nonmigratory estuarine fish (*Fundulus heteroclitus*) across a strong gradient of polychlorinated biphenyl contamination. Environ Toxicol Chem 24:717–725

Roberts RB, Hu YN, Albertson RC, Kocher TD (2011) Craniofacial divergence and ongoing adaptation via the hedgehog pathway. Proc Natl Acad Sci USA 108:13194–13199

Rockman MV (2012) The QTN program and the alleles that matter for evolution: all that's gold does not glitter. Evolution 66:1–17

Roelofs D, Janssens TKS, Timmermans MJTN, Nota B, Marien J, Bochdanovits Z, Ylstra B, Van Straalen NM (2009) Adaptive differences in gene expression associated with heavy metal tolerance in the soil arthropod *Orchesella cincta*. Mol Ecol 18:3227–3239

Roux C, Castric V, Pauwels M, Wright SI, Saumitou-Laprade P, Vekemans X (2011) Does speciation between *Arabidopsis halleri* and *Arabidopsis lyrata* coincide with major changes in a molecular target of adaptation? Plos One 6:e26872

Shapiro MD, Marks ME, Peichel CL, Blackman BK, Nereng KS, Jonsson B, Schluter D, Kingsley DM (2004) Genetic and developmental basis of evolutionary pelvic reduction in threespine sticklebacks. Nature 428:717–723

Shine R (2012) Invasive species as drivers of evolutionary change: cane toads in tropical Australia. Evol Appl 5:107–116

Sterenborg I, Roelofs D (2003) Field-selected cadmium tolerance in the springtail *Orchesella cincta* is correlated with increased metallothionein mRNA expression. Insect Biochem Molec 33:741–747

Sweeney J, Deegan L, Garritt R (1998) Population size and site fidelity of *Fundulus heteroclitus* in a macrotidal saltmarsh creek. Biol Bull 195:238–239

Teo SLH, Able KW (2003) Habitat use and movement of the mummichog (*Fundulus heteroclitus*) in a restored salt marsh. Estuar Coast Shelf Sci 26:720–730

van Straalen N, Hoffmann A (2000) Review of experimental evidence for physiological costs of tolerance to toxicants. In: Kammenga J, Laskowski R (eds) Demography in ecotoxicology. Wiley, Chichester, p xix, 297 p

Van Veld PA, Nacci DE (2008a) Toxicity resistance. In: Di Giulio RT, Hinton DE (eds) The toxicology of fishes. Taylor and Francis, Boca Raton, pp 597–641

Van Veld PA, Nacci DE (2008b) Toxicity resistance. In: Di Giulio RT, Hinton DE (eds) The toxicology of fishes. Taylor and Francis, Boca Raton

van't Hof AE, Saccheri IJ (2010) Industrial melanism in the peppered moth is not associated with genetic variation in canonical melanisation gene candidates. PLoS One 5:e10889

van't Hof AE, Edmonds N, Dalikova M, Marec F, Saccheri IJ (2011) Industrial melanism in British peppered moths has a singular and recent mutational origin. Science 332:958–960

Vitousek PM, Mooney HA, Lubchenco J, Melillo JM (1997) Human domination of earth's ecosystems. Science 277:494–499

Ward TJ, Robinson WE (2005) Evolution of cadmium resistance in *Daphnia magna*. Environ Toxicol Chem 24:2341–2349

Wen ZM, Rupasinghe S, Niu GD, Berenbaum MR, Schuler MA (2006) CYP6B1 and CYP6B3 of the black swallowtail (*Papilio polyxenes*): Adaptive evolution through subfunctionalization. Mol Biol Evol 23:2434–2443

Werck-Reichhart D, Feyereisen R (2000) Cytochromes P450: a success story. Genome Biol 1: REVIEWS3003

Whalon ME, Mota-Sanchez D, Hollingworth RM (2008) Global pesticide resistance in arthropods. CABI, Wallingford

Whitehead A, Triant DA, Champlin D, Nacci D (2010) Comparative transcriptomics implicates mechanisms of evolved pollution tolerance in a killifish population. Mol Ecol 19:5186–5203

Whitehead A, Pilcher W, Champlin D, Nacci D (2012) Common mechanism underlies repeated evolution of extreme pollution tolerance. Proc R Soc B 279: 427–433

Willems G, Drager DB, Courbot M, Gode C, Verbruggen N, Saumitou-Laprade P (2007) The genetic basis of zinc tolerance in the metallophyte *Arabidopsis halleri* ssp halleri (Brassicaceae): an analysis of quantitative trait loci. Genetics 176:659–674

Williams RJP (2007) A system's view of the evolution of life. J R Soc Interface 4:1049–1070

Williams LM, Oleksiak MF (2008) Signatures of selection in natural populations adapted to chronic pollution. BMC Evol Biol 8:282

Williams LM, Oleksiak MF (2011) Ecologically and evolutionarily important SNPs identified in natural populations. Mol Biol Evol 28:1817–1826

Williams RJP, Rickaby R (2012) Evolution's destiny: co-evolving chemistry of the environment and life. Royal Society of Chemistry, London

Wirgin I, Waldman JR (2004) Resistance to contaminants in North American fish populations. Mutat Res Fundam Mol Mech Mutagenesis 552:73–100

Wirgin I, Roy NK, Loftus M, Chambers RC, Franks DG, Hahn ME (2011) Mechanistic basis of resistance to PCBs in Atlantic tomcod from the Hudson River. Science 331:1322–1325

Wittkopp PJ, Carroll SB, Kopp A (2003) Evolution in black and white: genetic control of pigment patterns in Drosophila. Trends Genet 19:495–504

Wright GD (2007) The antibiotic resistome: the nexus of chemical and genetic diversity. Nat Rev Microbiol 5:175–186

Yamamoto Y, Stock DW, Jeffery WR (2004) Hedgehog signalling controls eye degeneration in blind cavefish. Nature 431:844–847

Yozzo DJ, Smith DE (1998) Composition and abundance of resident marsh-surface nekton: comparison between tidal freshwater and salt marshes in Virginia, USA. Hydrobiologia 362:9–19

Signatures of Natural Selection and Ecological Differentiation in Microbial Genomes

17

B. Jesse Shapiro

Abstract

We live in a microbial world. Most of the genetic and metabolic diversity that exists on earth – and has existed for billions of years – is microbial. Making sense of this vast diversity is a daunting task, but one that can be approached systematically by analyzing microbial genome sequences. This chapter explores how the evolutionary forces of recombination and selection act to shape microbial genome sequences, leaving signatures that can be detected using comparative genomics and population-genetic tests for selection. I describe the major classes of tests, paying special attention to their relative strengths and weaknesses when applied to microbes. Specifically, I apply a suite of tests for selection to a set of closely-related bacterial genomes with different microhabitat preferences within the marine water column, shedding light on the genomic mechanisms of ecological differentiation in the wild. I will focus on the joint problem of simultaneously inferring the boundaries between microbial populations, and the selective forces operating within and between populations.

Keywords

Microbial genomics • Natural selection • Recombination • Reverse ecology • Evolution • Bacteria • Convergent evolution • McDonald-Kreitman test • Speciation • Ecological differentiation • Adaptive divergence • Vibrio • Long-range haplotype test

17.1 Introduction

Microbes are key players in global biogeochemical cycles, human health and disease; yet the microbial world is largely hidden from view. Even with the best microscopes and experimental techniques, it is exceedingly difficult to know the predominant selective

B.J. Shapiro (✉)
Département de sciences biologiques, Université de Montréal, Montréal, QC, Canada
e-mail: jesse.shapiro@umontreal.ca

C.R. Landry and N. Aubin-Horth (eds.), *Ecological Genomics: Ecology and the Evolution of Genes and Genomes*, Advances in Experimental Medicine and Biology 781,
DOI 10.1007/978-94-007-7347-9_17, © Springer Science+Business Media Dordrecht 2013

pressures and ecological interactions at play in the wild. Microbial genome sequences provide a comprehensive and accessible record of the forces that drive microbial evolution. Using a *reverse ecology* approach (Li et al. 2008; Whitaker and Banfield 2006), we can analyze genome sequences – for example, by deploying sequence-based statistical tests to identify genes under positive selection – in order to discover ecologically distinct populations and how they adapt to different niches. Our motivation for this line of research could be driven by basic curiosity about the microbial world, but could also have practical goals in both environmental (e.g. linking microbial populations to nutrient cycles) and clinical spheres (e.g. understanding mechanisms of pathogenesis). The long-term goal of reverse ecology is to gain a mechanistic understanding of ecological processes, but short-term benefits can be expected along the way. For example, genes or mutations that are consistently associated with niches or phenotypes of interest (e.g. antibiotic resistance) can serve as diagnostic biomarkers, helping to predict environmental or clinical outcomes and suggesting effective interventions.

Gaining biological insight from microbial genome sequences and tests for selection poses several challenges. First, there are challenges arising from the enormous range of microbial evolutionary time scales: we may be interested in comparing species that diverged hundreds of millions to billions of years ago, or that diverged so recently that it is unclear if they constitute separate species or not. Second, while it was once thought that microbes do not form species in the classical sense because they reproduce clonally and do not recombine their DNA through sex, the idea is now gaining popularity that they do not form proper species because they have *too much* promiscuous sex, due to their ability to exchange genes by horizontal transfer spanning great genetic distances (Doolittle and Papke 2006).

In this chapter, I will begin by explaining how the problem of defining bacterial species is inextricably bound to the process of natural selection. I will then describe how genomic sequence data, analyzed with appropriate statistical and computational methods, can distinguish among evolutionary hypotheses, and ultimately provide insight into the structure and function microbes in their natural environments. The chapter will focus mainly on relatively closely-related (same genera or species) populations of 'wild' bacteria (i.e. outside of lab or microcosm settings). My goal is to provide an introduction for readers new to microbial evolutionary genomics, while raising outstanding questions in the field and synthesizing knowledge in a way that is useful to more experienced readers.

The chapter begins by asking the question, how do we define and identify ecologically distinct species of bacteria (Sect. 17.2)? It then describes different models of speciation, and the importance of natural selection in these models (Sect. 17.3). The major classes of tests for natural selection in genome sequences are briefly described (Sect. 17.4), and applied to a population of natural *Vibrio* genomes (Sect. 17.5), focusing on the McDonald-Kreitman test (Sect. 17.5.2) and the long-range haplotype test (Sect. 17.5.3). Other methods that can be applied to detect selection in rarely recombining bacteria, including time course studies and convergent evolution, and in the 'flexible' (horizontally transferred) component of the genome, are discussed briefly (Sect. 17.6). The chapter closes with an outlook (Sect. 17.7) on how new datasets and populations models are beginning to be incorporated into a better understanding of microbial evolution and ecology.

17.2 Recombination and the Bacterial Species Problem

Partitioning biological diversity into discrete units is challenging in general, and perhaps most challenging in microbes. To begin with, microbes are by definition microscopic and we can only categorize them into a limited number of morphological classes based on cell wall characteristics, shape and size, presence or absence of flagella, etc. They are much more diverse in terms of physiology and metabolic capability, leading to the problem of how to properly weight an abundance of traits into a

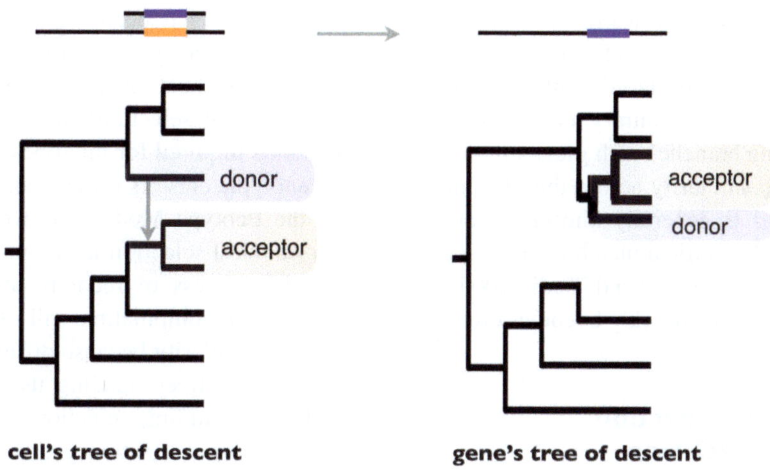

Fig. 17.1 Recombination can result in incongruence between gene-trees and cell-trees. A piece of DNA with flanking homology (*grey shading in upper panel*) is recombined from a donor into an acceptor genome, replacing the original allele (*orange*) with a new one (*purple*). The result is that the acceptor genome now has an identical allele as the donor, so the acceptor and donors branch closely together on the gene tree (*right*; branch lengths not to scale), whereas the acceptor and donor cells share a much more distant common ancestor

meaningful species classification scheme. The same problem arises when trying to classify multicellular organisms into species based on shared traits.

One solution to this problem is to privilege genetic data over other measurable traits. The reasoning behind this solution is that genetic similarity provides the best evidence that two individuals are similar *by descent*, as opposed to *by chance* or *by convergence*. I will discuss the concept of convergence in Sect. 17.6, but for now let's explore the idea of descent. When we talk about descent in bacteria, we could mean at least two different things: one is the bifurcating tree of cellular descent by clonal cell division; the other is the tree describing the replication of DNA molecules. When a cell divides in two, its genome replicates into two copies as well. The tree of cellular descent is identical to the tree of genomic descent. Now imagine that after the first cell division, one of the daughter cells encounters a molecule of DNA in its environment. The DNA – let's say it encodes an allele of a gene already present in the genome – enters the cell and replaces the original version of the gene by the mechanism of homologous recombination. The history of this gene is now different from the history of the cell. They are described by different trees. In the cell's tree, the two daughter cells branch together. In the gene's tree, the daughter that accepted the foreign DNA branches with the source of that DNA rather than with the other daughter (Fig. 17.1). I am intentionally using the word 'tree' instead of 'phylogeny' because the latter usually implies relationships between species, whereas my intention is to more generally describe patterns of descent. So which tree do we care about, the gene's tree of descent or the cell's?

Let's begin by examining the cell's tree. This tree describes the exponential process of binary cell division. The tree topology remains the same, no matter how many genes have been swapped for different alleles. In what I will call the purely *clonal* scenario, absolutely no genes have been swapped by recombination. In the extreme opposite of the clonal scenario, genes are exchanged at a rate that far outpaces cell division, so the tree for any given gene will have nothing to do with the cell's tree. In the clonal scenario, the gene's tree and the cell's tree are identical, so DNA sequence data from any (or every) gene in the genome can be used to infer, using a model of sequence evolution, the correct tree of cellular descent with reasonable statistical certainty. Now we have a trustworthy tree, but we are still left with the problem of defining species: where should we make a cut in

the tree to divide one species from another? Just like species definitions based on morphology or physiology, we are faced with a decision. Should we make an arbitrary cut in the tree, perhaps dividing branches with greater than 95 % DNA sequence similarity across the genome? A given threshold is generally chosen because it provides a good empirical match to other species definitions (Konstantinidis and Tiedje 2005), but this type of reasoning quickly becomes circular.

17.3 Natural Selection and Speciation

17.3.1 Models of Bacterial Speciation

Up until this point, I have focused on using genetic similarity to infer patterns of descent. But is this really what we want from a bacterial species concept? I argue that we should care more about the *process* that generates genetic similarity than the exact level of genetic similarity itself. The process is an *evolutionary process*, involving natural selection of the fittest within a diverse, replicating population. A good example of a process-based species concept is the Ecotype Model, developed by Cohan and others (see (Cohan and Perry 2007) for a comprehensive overview). In its simplest form, the Ecotype Model defines species as independent evolutionary units. If a mutation occurs in the genome of a member of one species, it only competes with members of the same species, all sharing the same ecological niche. If the mutation is adaptive, genomes containing the mutation will multiply more rapidly, or escape predation more effectively, than those without the mutation, eventually dominating the population in a *selective sweep*. Importantly, the selective sweep will have absolutely no effect on other populations that compete in different ecological niches. New species emerge when a member of an existing species gains a function (by mutation or recombination) that allows it to exploit a new ecological niche, founding a new evolutionarily independent population. The process described by the Ecotype Model generates *clusters of ecological and genetic similarity*. Although it has been suggested that clusters of genetic similarity could arise through neutral processes (by mutation and genetic drift alone, without natural selection), theory suggests that selection is required for microbes to differentiate into genotypic clusters (Polz et al. 2013).

In the Ecotype Model, the evolutionary process of natural selection is paramount. But what about the process of recombination? Taken to an extreme, recombination will obscure clusters of genetic similarity because different genes will have different trees, leading us to infer different clusters. Adding selection to this scenario of extreme recombination yields a model that I will call Gene Ecology. In this model, genes, not species, inhabit ecological niches and are the targets of natural selection. Species only exist insofar as genes have to work together in order to reproduce themselves within genomes. This model may be extreme – ignoring epistasis and gradual coevolution among genes – but it can be a useful tool for understanding the distribution of different genes in different environments (Coleman and Chisholm 2010; Delong 2006; Mandel et al. 2009).

One way of moderating the extreme gene-centrism of Gene Ecology is to introduce elements of Mayr's Biological Species concept (de Queiroz 2005; Mayr 1942). This concept defines species based on reproductive isolation, so strictly speaking, it does not apply to asexually-reproducing bacteria. In sexual reproduction, genes are recombined every generation. Reproductive isolation therefore results in separate gene pools. In bacteria, reproduction is decoupled from recombination, and genes from very distantly related bacteria can be exchanged by recombination (Koonin et al. 2002). Therefore, we can never expect bacterial species to have completely isolated gene pools. But we needn't discard the Biological Species concept entirely. In bacteria that recombine frequently, different genes could be selected in different niches, independently of the cell or genome that they (transiently) inhabit. This begins to resemble Gene Ecology. But if there is preferential recombination among cells in the same niche (due to physical proximity, or

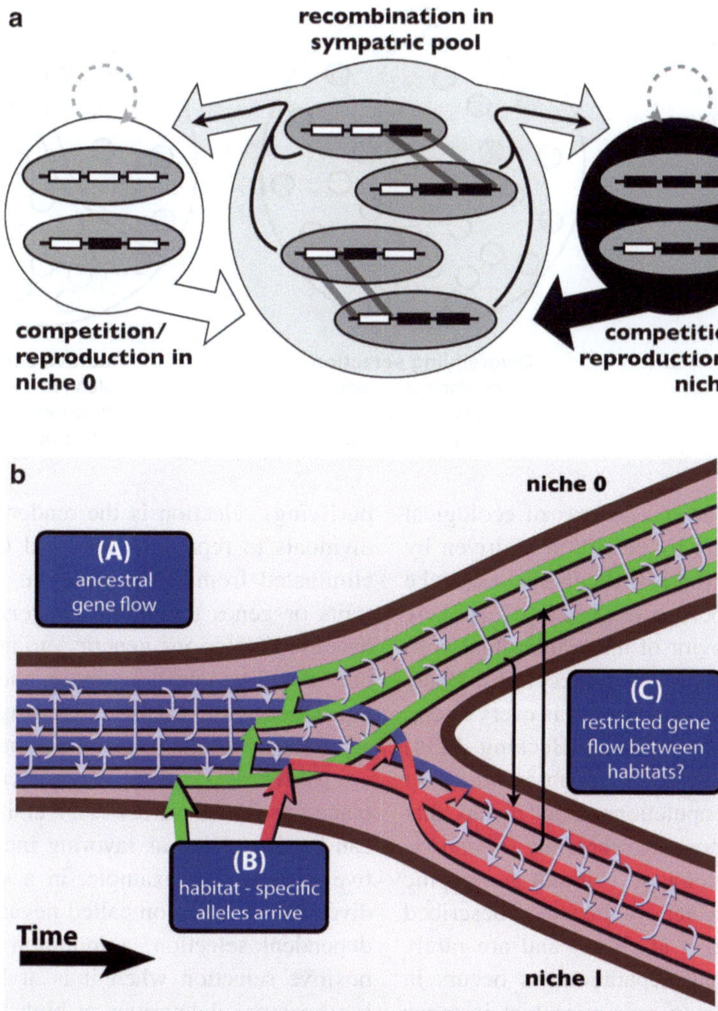

Fig. 17.2 A model of ecological differentiation for sympatric recombining bacteria. (**a**) A sympatric model (Modified from Friedman et al. 2013) in which microbial cells (*dark grey ovals*) compete in either of two niches. Cells containing mostly *black* alleles are best adapted to niche 1; *white* alleles to niche 0. Gene conversion of homologous loci (*diagonal lines*) take place in a sympatric, mixed pool of genotypes from both niches, and cells return to the niche to which their genotype is best adapted (e.g. in this 3-locus example, genotypes with mostly white alleles go to niche 0; those with mostly black alleles go to niche 1). Some degree of allopatry could be added to the model by increasing the rate of recombination within niches (*dashed circular arrows*). (**b**) The resulting temporal dynamics of such a model, supported by data from *V. cyclitrophicus* populations adapted to large- or small-particle niches in the marine water column (Modified from Shapiro et al. 2012. Reprinted with permission from John Kaufmann and from © AAAS 2012. All Rights Reserved)). *Thin gray* or *black arrows* represent recombination within or between ecologically associated populations. *Thick red* or *green colored arrows* represent acquisition of adaptive alleles for different habitats/niches

increased efficiency of recombination due to DNA sequence similarity, or both), a hybrid of Gene Ecology and Biological Species might apply. This is the sort of hybrid model that I proposed for a pair of closely-related, recombining populations of marine *Vibrio* bacteria, described in Sect. 17.5.1 and illustrated in Fig. 17.2.

I hope that you now have some appreciation for the connections between speciation, recombination and selection. From here on, I won't focus any further on the bacterial species problem

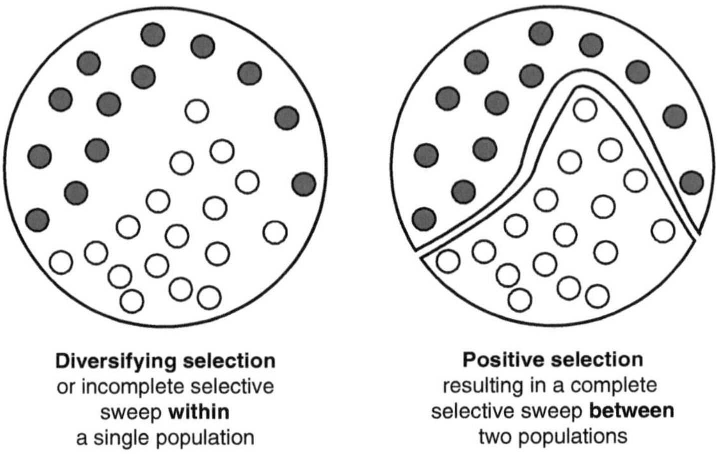

Fig. 17.3 The concepts of positive and diversifying selection depend on population boundaries. The *right* and *left* panels differ only in the definition of boundaries between populations. *Small circles* represent individual sampled bacteria with different alleles (*red filled or empty circles*) at a polymorphic locus

per se, but instead on the process of ecological differentiation, which by definition is driven by selection for adaptation to different niches. I take this 'adaptationist' perspective because it is likely to describe the behavior of many microbial populations on earth. With some exceptions, Baas-Becking's theory that "everything is everywhere; the environment selects" (Baas-Becking 1934) has been largely supported by observations of natural microbial populations. This means that microbial populations are generally *sympatric* (part of the "same country," without geographic structure, e.g. freshwater cyanobacteria described in van Gremberghe et al. 2011) and are rarely separated by physical separation, as occurs in *allopatric* speciation (a rare microbial instance of which is presented by hotspring archaea; see Whitaker et al. 2003). In reality, most microbes probably fall on the spectrum between absolute sympatry and allopatry. The point is that physical separation is much less important for microbes than for most species of plants and animals. As a result, natural selection should be the most important contributor to ecological differentiation and speciation (Fig. 17.2).

17.3.2 Forces of Natural Selection

Natural selection can operate in a variety of ways, but I find it useful to consider three major forms of selection: negative, positive, and diversifying selection. Negative (sometimes called purifying) selection is the tendency of unfit individuals to reproduce less and therefore to be eliminated from the population. This results in traits or genes remaining *conserved* over time, because deleterious genetic variants are weeded out. Positive selection favors the survival and reproduction of variants conferring a competitive advantage over the rest of the population. During a *selective sweep*, positively selected variants replace unselected variants. Diversifying selection can be thought of as favoring incomplete selective sweeps. For example, in a special case of diversifying selection called negative frequency-dependent selection, a mutation is favored by positive selection when it is at low frequency, but becomes deleterious at high frequency. The mutation never sweeps the entire population, but fluctuates around an intermediate frequency. Depending on how boundaries between populations are drawn, diversifying and positive selection can be hard to distinguish (Fig. 17.3).

17.4 Signatures of Selection and Adaptive Divergence

The goal of microbial ecological and evolutionary genomics is to use genetic sequences sampled from microbial populations to learn how these populations adapt to different niches. To solve this reverse ecology problem, we need to identify signatures of selection and niche adaptation in microbial genomes. A whole battery of sequence-

based statistical tests for selection have been developed, but because most of them were designed for sexual populations we must be careful which tests we choose to apply to asexual microbes (Shapiro et al. 2009). The basic premise of these tests is to define patterns of genetic variation that are shaped by selection, and distinguish them from the *neutral* patterns expected by random mutation and genetic drift.

One of the most popular tests for selection involves comparing the relative rates of nonsynonymous (amino acid-changing) to synonymous mutations, often called the *dN/dS* ratio. The key assumption is that nonsynonymous changes (measured by *dN*) affect protein structure, change the phenotype, and are thus subject to natural selection. Synonymous changes have no effect on protein structure, and are thus subject to less natural selection and reflect mostly random mutation and genetic drift. In fact, synonymous mutations may also be under selection for RNA stability, translational efficiency, etc., e.g. (Gingold and Pilpel 2011; Raghavan et al. 2012), but the *dN/dS* test assumes that selection is generally stronger on nonsynonymous mutations. Imagine that we have sequenced orthologous protein-coding genes from two species, aligned the two sequences and counted nucleotide differences between them. We can then count the differences as synonymous or nonsynonymous according to the genetic code, and normalize the counts by the number of synonymous or nonsynonymous sites, respectively, to obtain *dN* and *dS*. Averaged across the entire gene, $dN/dS \approx 1$ suggests very little selection at the protein level (characteristic of pseudogenes), $dN/dS > 1$ suggests very strong positive selection to fix different amino acids between species, and $dN/dS < 1$ suggests negative selection for a conserved protein structure in both species. I have deliberately chosen one of the simplest possible applications of *dN/dS* in order to illustrate the principle, but applications of *dN/dS* can be tailored to consider more than two species, or to consider separately individual codons within a gene (Yang 2008; Yang and Nielsen 2002).

A powerful extension of *dN/dS* called the McDonald-Kreitman (MK) test (McDonald and Kreitman 1991) employs protein-coding sequences sampled both within and between species. We have already discussed the difficulties in defining species boundaries in bacteria, but the MK test can still be useful. If we are satisfied with a species concept based on adaptive divergence (i.e. ecological differentiation driven by positive selection), the MK test can be flipped on its head: rather than testing for positive selection between a priori defined species, we are instead testing whether a sample of gene sequences come from the same or different species (Simmons et al. 2008; Vos 2011). I will return to the 'flipped' MK in Sect. 17.5.2, but will first describe the original version of the test.

The key assumption of the MK test is that in the absence of selection, the *dN/dS* ratio should remain constant over time and, thus, be the same for fixed substitutions (between species) as for segregating polymorphism (within species). The MK test normalizes *DN/DS* (measured between a pair of species) by *PN/PS*, the equivalent measure *within* one or both of the species. In this case *D* and *P* refer to divergence and polymorphism, and *DN, DS, PN,* and *PS* are the absolute counts (rather than rates per nonsynonymous or synonymous site) of each category of mutation in the gene of interest. The Fixation Index (FI) is defined as $(DN/DS)/(PN/PS)$, with FI >1 suggesting positive selection between species and FI <1 suggesting negative selection. The MK test is preferable to simply computing *dN/dS* because an average $dN/dS > 1$ across an entire gene is very unlikely, even if a few individual amino acids are genuinely under positive selection. By normalizing by *PN/PS*, the MK test is more sensitive. Second, $dN/dS > 1$ may occur due to a *relaxation of negative selection* rather than positive selection, whereas FI >1 is much more likely to indicate positive selection only.

Tests for selection need not be based on the genetic code, like *dN/dS* and the MK test. Another group of tests that I will collectively call *allele-frequency* and *haplotype-frequency* tests look for mutations (or clusters of mutations linked together as alleles or haplotypes) that have risen to an unexpectedly high frequency

in the population, suggesting positive or diversifying selection. Allele-frequency tests, such as Tajima's *D* (Tajima 1989) or Fay and Wu's *H* (Fay and Wu 2000), calculate mutation frequencies within a gene or region of interest, under the assumption of no recombination within it. Haplotype-frequency tests, including the long-range haplotype (LRH) test and its variations, explicitly consider recombination as a sort of 'clock' (Sabeti et al. 2002; Voight et al. 2006). When a new mutation occurs in the genome, it is necessarily linked to other mutations on the chromosome. This haplotype of mutations is initially long, spanning the entire chromosome. In sexual population, recombination occurs with some frequency every generation by crossing-over of homologous chromosomes. This results in the slow erosion of the haplotype, from the edges of the chromosome toward the new mutation. As a result, older mutations will be part of shorter haplotypes than newer mutations. If they are neutral to fitness, new mutations should not rise very quickly, or at all, to high frequency in the population. But if they are subject to positive selection, they are more likely to increase in frequency. If the increase in frequency is fast relative to the recombination 'clock,' selected mutations will tend to be observed at high frequency on unexpectedly long haplotype backgrounds.

The LRH test was designed with sexual populations in mind, and is not expected to work in bacteria – at least not in its original form. While many bacteria are capable of homologous recombination, they do so by gene conversion rather than crossing over. Instead of eroding linear haplotypes from the edges inward, gene conversion generates a characteristic 'patchwork' pattern known as the *clonal frame* (Milkman and Bridges 1990). The clonal frame refers to the chromosome background, with its own clonal ancestry and tree topology (presumably congruent with the tree of cellular descent), which is interrupted by short recombinant blocks, usually a few kilobases (kb) that have been introduced by gene conversion. These recombinant blocks have different evolutionary histories than the clonal frame. In the clonal frame model, because gene conversion events are of fairly uniform size, there should be little association between haplotype length and frequency in positively selected regions of the genome. Therefore, the original formulation of the LRH test is not strictly applicable to bacteria.

17.5 Testing for Selection in Bacterial Genomes

In the last section, I touched on some of the issues involved in applying tests for selection to recombining bacterial genomes. On the one hand, if bacteria are perfectly clonal (no recombination), every gene in the genome will be linked in the same clonal frame. When an adaptive mutation occurs in a particular genome in the population, the resulting selective sweep will bring to high frequency not only the adaptive mutation, but any other neutral or slightly deleterious mutations that happen to be 'hitchhiking' in the same genome (Hanage et al. 2006; Shapiro et al. 2009). Selective sweeps therefore purge population diversity genomewide, and it becomes difficult, based on any of the tests described above, to distinguish adaptive mutations from hitchhikers (Fig. 17.4). On the other hand, if bacteria recombine frequently or promiscuously, care must be taken to ensure that recombination does not obscure or lead to false signals of selection. Recombination has the potential to introduce adaptive alleles (by homologous recombination), or entirely new genes or operons (by illegitimate recombination, often mediated by phage, plasmids or integrative conjugative elements). We could in principle design tests to look for recombination events that are adaptive, based on a consistent association with a niche or phenotype of interest, unexpectedly high population frequency, or recombination across species boundaries. For example, in an analysis of *Streptococcus* genomes, we found that genes recombined between recognized species tended to have higher *dN/dS* than genes recombined within species, suggesting that recombination across species boundaries requires positive selection (Shapiro et al. 2009).

Fig. 17.4 In clonal populations, selected mutations (*red circle*) can be confused with neutral mutations (*blue diamonds*) in the genome (*horizontal line*)

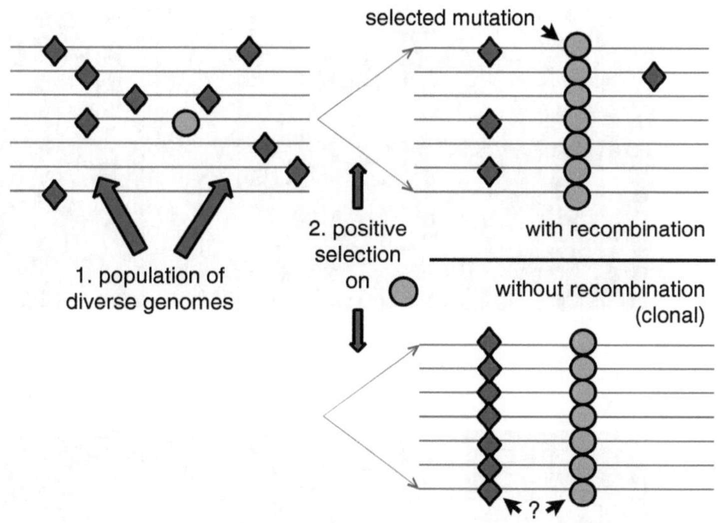

I will now walk through a workflow for detecting regions of the genome under selection in populations of bacteria. I will use the example of marine *Vibrio cyclitrophicus*, which my colleagues and I have studied extensively (Hunt et al. 2008; Shapiro et al. 2012; Szabó et al. 2013), highlighting aspects of the analyses that can be generalized to other data, and focusing on the interplay between selection and ecological differentiation.

17.5.1 Ecological Differentiation Among Marine *Vibrio*

In 2006, we sampled coastal seawater off the coast of Massachusetts, and ran it through a series of progressively finer filters. We then isolated *Vibrio* from each of the filters on *Vibrio*-selective media. I will focus on two groups of isolates: those from the largest filter, which catches mainly large particles (>63 μm, consisting primarily of zooplankton) and those from an intermediate filter, which catches small particles that are still larger than a typical *Vibrio* cell (~1 μm in diameter). The large and small particles are proxies for different microhabitats in the water column, and thus constitute a potential axis of ecological differentiation.

We sequenced 20 whole genomes from two closely-related clusters of *V. cyclitrophicus* that appeared to have undergone a recent habitat switch, finding that just a few loci in the genome appear to have driven the switch (Fig. 17.2b). We inferred that the ecological switch had been relatively recent because all the isolates had identical 16S sequences, and only differed by about ten mutations in the faster-evolving *hsp60* gene. Genomewide, 725 single nucleotide polymorphisms (SNPs) clearly partitioned the large – and small-particle isolates into two distinct groups. Surprisingly, these 725 'ecoSNPs' were not distributed evenly across the genome, but were clustered in only 11 regions, the three densest of which contained >80 % of ecoSNPs (Shapiro et al. 2012). Outside of these regions, SNPs tended to conflict with the partitioning of large- and small-particle isolates. The extent of recombination and conflicting phylogenetic signal is displayed in a STARRInIGHTS plot (Fig. 17.5a and b), showing many small, sometimes barely visible 'constellations' of support (in white) for many different phylogenetic partitions, none of which accounts for much of the genome. These conflicting phylogenetic signals suggested high rates of homologous recombination since the divergence of these isolates, with recombination breakpoints inferred to have occurred about once per kilobase. Together, this suggests that large- and small-particle populations actually constituted a single homogeneously recombining

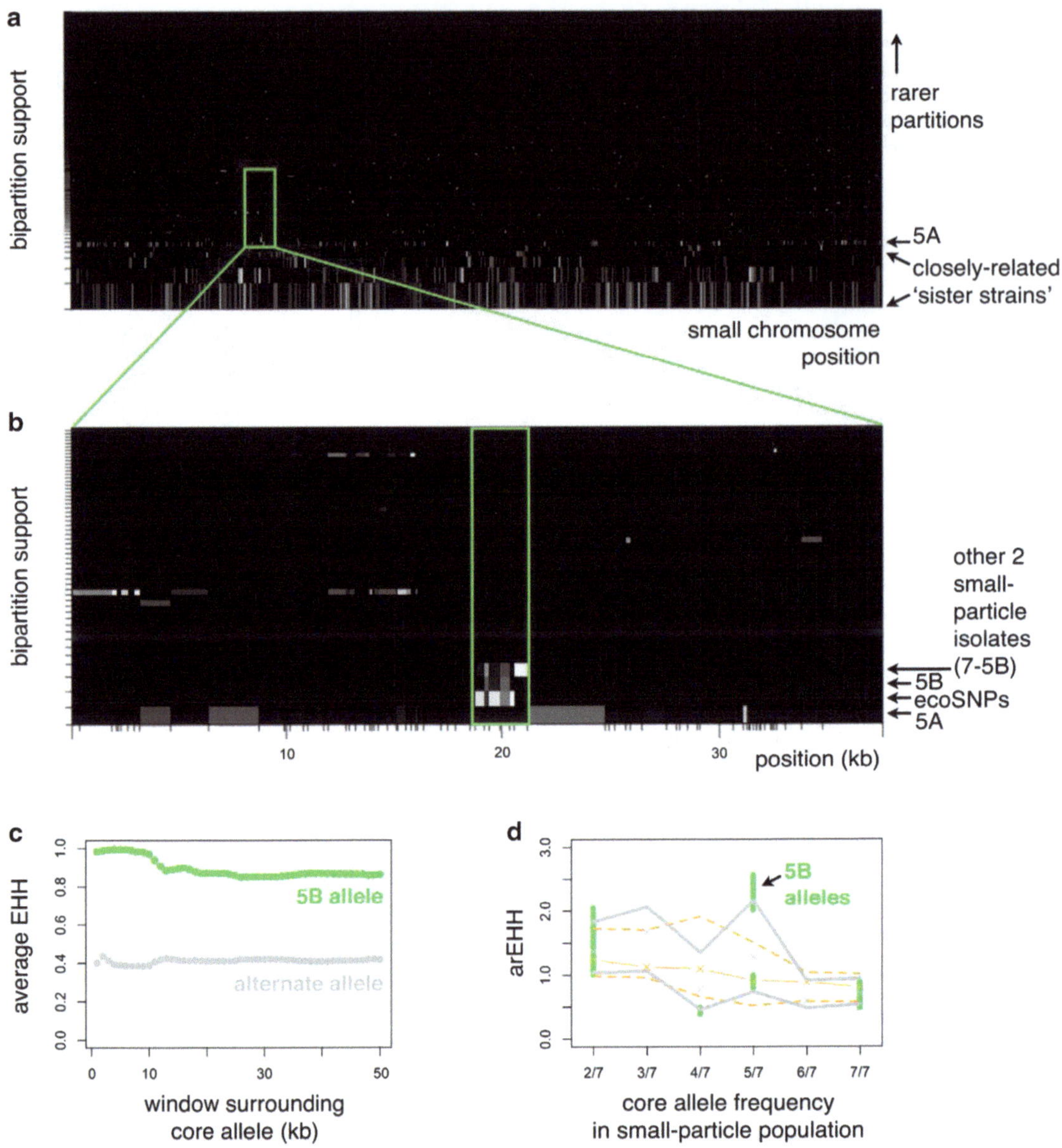

Fig. 17.5 Recombination and selection at the *RpoS2/RTX* locus in *V. cyclitrophicus* genomes. (**a**) Recombination blocks supporting different phylogenetic partitions across the small chromosome. Strain-based Tree Analysis and Recombinant Region Inference In Genomes from High-Throughput Sequencing projects (STARRInIGHTS; http://almlab.mit.edu/starrinights.html) was used to infer breakpoints between recombination blocks across the chromosome (x-axis). *Brighter white* indicates higher numbers of SNPs within a block supporting a particular partition of the 20 genomes. Partitions are ranked on the y-axis in increasing order of their prevalence genomewide. The row width is also proportional to genomewide prevalence of each partition. (**b**) Detail of 37.5 kb surrounding the *RpoS2/RTX* region (*green box*). *Small tick marks* on the x-axis indicate recombination breakpoints. (**c**) Decay of linkage (average extended haplotype homozygosity) with distance around a representative SNP in the *RpoS2/RTX* region. The 5B-supporting variant (*green*) is surrounded by a longer linked haplotype than the alternate allele (*grey*; present in the other two small-particle genomes and the large-particle outgroup). (**d**) SNPs within the ~2 kb *RpoS2/RTX* region (*green points*) at frequency 5/7, supporting the 5B partition, have high average relative EHH (arEHH) compared to neutral simulations (*dashed orange lines*) and other sites on the small chromosome (*grey lines*). Lines denote upper and lower 95 % confidence bounds and x denotes the median arEHH

population for most of their history (Fig. 17.2b). The ecoSNP regions are the exception, and we reasoned that they might contain alleles conferring adaptation to different microhabitats, driving ecological differentiation. Certain genes in the 'flexible' genome, differing in their pattern of presence and absence across the 20 sampled genomes, are probably also involved in ecological differentiation, and I will discuss them briefly in Sect. 17.6.

Before formally testing the ecoSNP regions for evidence of divergent positive selection between habitats, I will briefly discuss the implications of these regions having been acquired by recombination from very distant relatives of the 20 sequenced isolates – which could be considered evidence for selection in and of itself (Shapiro et al. 2009). Acquisition by recombination is by no means a necessary characteristic of positively selected loci, but it certainly adds a layer of evidence. There are two main reasons why we suspect the ecoSNP regions to have been acquired by recombination. First, they constitute only a small fraction of a genome that mostly rejects the ecoSNP phylogeny, making it highly unlikely that they are part of the clonal frame. Second, most genes in the ecoSNP regions have very high rates of synonymous substitutions (*dS*) between habitats; several times higher than anywhere else in the genome. Such high *dS* is best explained by recombination with relatives beyond the 20 genomes considered here. One consequence of such high *dS* is that, despite relatively high nonsynonymous divergence (*dN*), traditional tests for positive selection at the protein level (such as *dN/dS* and the MK test) suffer a substantial loss of power. Potentially due to this power loss, none of the genes in the three densest ecoSNP regions have FI significantly greater than 1 (Shapiro et al. 2012).

17.5.2 Insights from the MK Test

As alluded to in Sect. 17.4, the MK test can be 'flipped' in order to test whether two populations constitute distinct species (Shapiro et al. 2009; Vos 2011). Using a species concept based on adaptive divergence, Vos proposed that if the genomewide FI is significantly greater than 1 between populations, then these populations can be considered separate species (Vos 2011). Computing FI genomewide can be done by pooling genes, but combining genes with different histories of recombination, and different levels of polymorphism and divergence, can lead to biased estimates of FI. To control for this, the observed genomewide FI can be compared to the expected neutral distribution of FI, obtained by summing *DN*, *DS*, *PN* and *PS* across a set of bootstrapped contingency tables with marginal sums equal to those at each individual gene (Shapiro et al. 2007). By repeating this bootstrap resampling procedure 1,000 times, I was able to obtain an empirical *p*-value for the deviation of the observed from the expected FI.

Using all 4,491 aligned core *V. cyclitrophicus* genes, the genomewide FI is significantly greater than expected – but only when *PN* and *PS* are estimated from the large-particle population, not from the small-particle population (Table 17.1). Even though *PN/PS* is similar in both populations, the overall level of polymorphism is much lower (about half) in the small-particle population, which might explain the ambiguous results. In general, both *DN/DS* and *PN/PS* decrease as genes with progressively higher divergence (*DN* + *DS*) between habitats are included. If we accept that divergence measures evolutionary time, then this is consistent with purifying selection purging deleterious nonsynonymous mutations over time, both within and between populations. However, there is an exception to this trend: the highest *PN/PS* is actually observed in the seven most highly divergent genes, in the small-particle population (Table 17.1). This suggests diversifying selection among small-particle strains might be acting to increase *PN/PS*, specifically among the most divergent genes. Meanwhile, in the large-particle population, *PN/PS* is low among the most divergent genes, resulting in much stronger evidence for speciation in these genes ($FI = 1.93$, $p = 0.008$) than elsewhere in the genome. Overall, this reinforces that a few highly divergent genes seem to be driving ecological differentiation. However, it also reinforces

Table 17.1 MK test applied to core genes in *V. cyclitrophicus* ecological populations

divergence btw. habitats	inclusion criterion	# genes included	*DN*	*DS*	*DN/DS*	*PN*	*PS*	*PN/PS*	FI(obs)		FI(exp)*	*p*(obs>exp)**
						small-particle population polymorphism						
low	0≤(*DN*+*DS*)<5	4,475	16	48	0.33	7,473	34,349	0.22	1.53	≈	1.62	0.542
medium	5≤(*DN*+*DS*)<10	9	13	46	0.28	21	190	0.11	2.56	≈	2.15	0.093
high	10≤(*DN*+*DS*)<inf	7	73	334	0.22	25	110	0.23	0.96	≈	1.18	0.800
all	none	4,491 (all)	102	428	0.24	7,519	34,649	0.22	1.10	≈	1.13	0.773
		RpoS2	23	76	0.30	15	41	0.37	0.83		Fisher test *p* = 0.698	
						large-particle population polymorphism						
						PN	*PS*	*PN/PS*	FI(obs)		FI(exp)*	*p*(obs>exp)**
low		same as above				15,204	69,747	0.22	1.53	>	0.95	0.029
medium		same as above				63	420	0.15	1.88	>	1.12	0.025
high		same as above				26	230	0.11	1.93	>	1.31	0.008
all						15,293	70,397	0.22	1.10	>	0.88	0.001
		RpoS2 same as above				1	1	1	0.303		Fisher test *p* = 0.421	

[a]Mean FI in 1,000 bootstrap resamplings
[b]Based on 1,000 bootstrap resamplings

how different genes in the genome speciate at different rates (Retchless and Lawrence 2010), making it difficult to decide on a firm threshold between species.

Let's consider one of the ecoSNP regions as a candidate driver of ecological differentiation between large- and small-particle habitats: the single densest ecoSNP region, located on the smaller of the two chromosomes, encodes an RTX toxin and RpoS2, an RNA polymerase sigma factor involved in stress response. The *RTX* gene has sequence similarity to an excreted cytotoxic protein in *V. cholerae* (Lin et al. 1999). *RTX* is highly divergent between habitats, with ten fixed nonsynonymous changes and significant domain reorganization: the gene is split into three aligned coding regions, with other domains uniquely present in either small- or large-particle genomes only. The sigma factor appears to be a *Vibrio*-specific second copy of RpoS (hence the "RpoS2" designation), the first copy of which is located on the large chromosome. The *RpoS2* gene contains 23 fixed nonsynonymous differences (*DN*) between small- and large-particle isolates – the highest *DN* in the genome – three of which occur in predicted DNA binding domains (Lee and Gralla 2002). An additional two DNA binding residues differ between RpoS2 and the canonical RpoS, but are identical in large- and small-particle genomes. These observations suggest, first, that RpoS2 may target different DNA binding sites than the canonical stress-response sigma factor, and second, that RpoS2 may have experienced functional modifications between small- and large-particle isolates – potentially modulating major differences in gene expression between habitats. And yet, the MK test does not support this evidence of positive selection between habitats.

For the moment, let's assume that selection is real, and the MK test simply lacked power to detect it. There are at least two reasons why this could happen. First, in addition to the 23 fixed nonsynonymous differences, *RpoS2* also contains $DS = 76$ fixed synonymous differences (Table 17.1). This is consistent with *RpoS2* having diversified for a long time outside the populations considered here, and different alleles being acquired recently in different habitats (Fig. 17.2b). If all 99 substitutions were acquired simultaneously by recombination, even if some of the nonsynonymous substitutions were adaptive, the signal of positive selection might be obscured by high *DS*. Second, *RpoS2* contains a lot of polymorphism, particularly within the small-particle population, with a *PN/PS* ratio slightly higher than the *DN/DS* ratio (Table 17.1). This suggests that *RpoS2* might be under diversifying selection within the small-particle population, resulting in FI <1.

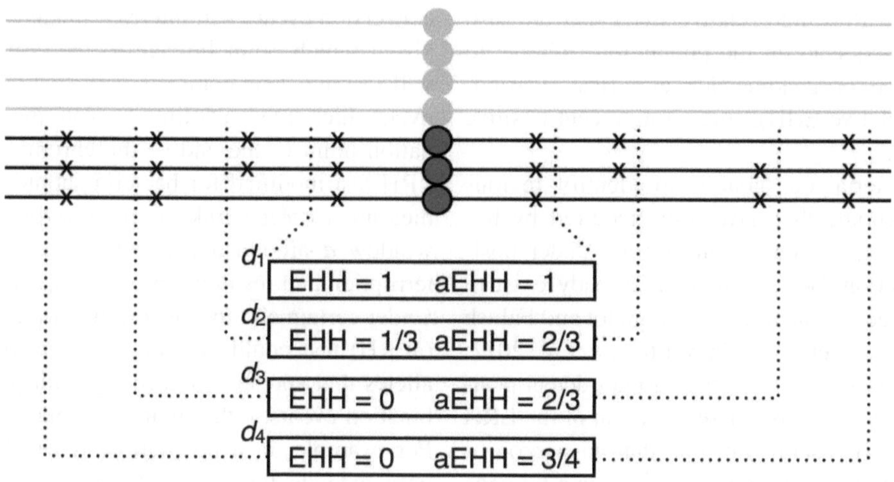

Fig. 17.6 Illustration of 'classical' and 'average' long-range haplotype (LRH) tests. The extended haplotype homozygosity (EHH) is defined in a window *d* around a core site (*circles*) as the probability that two randomly chosen chromosomes with the same allele at the core site (*black horizontal lines*) are also identical at *all* polymorphic sites within *d*. For example, in window d_1, all polymorphic sites have identical alleles (all have the 'x' allele), yielding EHH = 1. The average EHH (aEHH) is also 1. In window d_2, EHH and aEHH differ. This is because there are two identical haplotypes and one unique haplotype in d_2, yielding EHH = 1/3. However, aEHH is averaged over four polymorphic sites, with homozygosities of 1/3, 1, 1, and 1/3 (from *left to right*), for an average equal to 2/3. Note that EHH eventually decays to 0 with increasing distance from the core site, whereas aEHH can fluctuate up and down between 0 and 1. To compute the average *relative* EHH (arEHH), the aEHH values for the first allele at the core site (shown) would be normalized by aEHH in haplotypes around the other allele (*grey circles*; surrounding polymorphism not shown)

Because *DN/DS* = 0.30 is quite high (compared to *DN/DS* = 0.22 across other highly-divergent genes; see Table 17.1), within-population diversifying selection might be more important than elevated *DS* in explaining why the MK test does not detect significant positive selection between habitats.

17.5.3 Long-Range Haplotype Testing in Bacteria

I do not mean to push the hypothesis of positive selection on *RpoS2* too hard, merely to highlight some of the difficulties in applying and interpreting tests for selection. In the same spirit, let's investigate another test for selection: a variant of the LRH test. I explained in Sect. 17.4 that we don't expect the classical LRH test to work in microbes because of its assumptions about recombination. Yet a simple modification of the LRH test might be used to detect selection in bacterial genomes. In its original form (Sabeti et al. 2002), the LRH was implemented as follows. First, a 'core' nucleotide site is defined, with at least two polymorphic variants at the site. The extended haplotype homozygosity (EHH) is defined in a window *d* around the core site as the probability that two randomly chosen chromosomes from the population are identical (homozygous) at *all* single nucleotide polymorphisms (SNPs) in *d* (Fig. 17.6). EHH ranges from 0 (all extended haplotypes are different; rapid breakdown of linkage due to recombination) to 1 (all extended haplotypes are identical; perfect linkage, uninterrupted by recombination). Relative EHH is defined separately for each nucleotide variant at a core site as the EHH for the variant divided by the average EHH of other variants at the same locus, in order to control for the local recombination rate. Relative EHH ranges from 0 to infinity, with higher values indicating a variant on an unusually long haplotype, relative to other variants at the same locus. The LRH test then identifies

variants with a high relative EHH that are also at high frequency in the population (relative to a modeled or genomewide empirical distribution of relative EHH), suggesting recent positive selection.

In bacteria, our goal is to identify regions of the genome that have been acquired by homologous gene conversion *and* are under positive selection. Several methods already exist to identify recombinant regions (Didelot and Falush 2007; Didelot et al. 2010; Marttinen et al. 2012; Shapiro et al. 2012), but they do not detect positive selection; this is where a version of the LRH test could be useful. Imagine that an operon of size 10 kb arrives in a bacterial genome by gene conversion, replacing the existing allele. The new allele comes from a distantly related lineage, but confers a selective advantage to its recipient. The allele can now spread in the populations by two mechanisms: gene conversion into other genomes, or clonal expansion of the genome that originally acquired the new allele. The predominant mechanism will depend on the strength of selection relative to the rate of recombination, and we can reasonably expect a combination of mechanisms to operate in bacteria with moderate to high recombination rates, like *Vibrio* and *E. coli* (Schubert et al. 2009; Touchon et al. 2009).

First consider a clonal expansion following acquisition of the new allele. The new allele will increase in frequency on a haplotype background consisting of the entire chromosome (the clonal frame). If this selected genome sweeps to fixation without any further recombination in either the selected genome or other genomes in the population, both selected and unselected alleles will have identical haplotype lengths, and any version of the LRH test will be uninformative (and the selected variant will be indistinguishable from other mutations linked in the clonal frame; see Fig. 17.4). If, however, some (neutral) recombination events occur within the population before and during the clonal expansion, we expect the selected allele to be in better *average* linkage with the clonal frame than the unselected allele at the same locus. It will be in better *average* linkage to sites anywhere on the chromosome, because gene conversion events will interrupt linkage for a few kb, at which point linkage to the clonal frame will resume. This contrasts with sexual crossing over, which interrupts linkage from the recombination point to the end of the chromosome. An LRH test modified for bacteria should therefore measure *average* linkage (homozygosity) in a window *d* around a core site, to allow for interruption and resumption of linkage (Fig. 17.6). Under certain circumstances, this 'average' LRH (aLRH) test could also be sensitive to selected alleles that spread primarily by multiple recombination events, rather than by clonal expansion. For example, in a population in which linkage only extends for ~1 kb (a realistic figure for *V. cyclitrophicus* and other populations such as *Leptospirillum* (Simmons et al. 2008) and *E. coli* (Mau et al. 2006) due to frequent, potentially overlapping gene conversions), an adaptive allele of ~10 kb that spreads rapidly by recombination of roughly the same block of DNA, might stand out as a long haplotype at high frequency.

I will now demonstrate how the aLRH test might work in practice, using the small chromosome of *V. cyclitrophicus* as an example. I do not intend this as a rigorous benchmarking of the test, but rather as an illustration of how such a test could be useful, and of its shortcomings.

The STARRInIGHTS plot, an illustration of recombination 'blocks' of the genome containing SNPs supporting different binary partitions of the isolates, shows that the *RpoS2/RTX* locus contains a dense cluster of ecoSNPs, supporting the partition between the 7 small- and 13 large-particle isolates (Fig. 17.5a and b). The partitions are ranked on the *y*-axis according to the prevalence of SNPs supporting them genomewide. The partition shown just below the ecoSNP partition, for example, is supported by 796 sites genomewide, slightly more than the 725 ecoSNPs. This partition – let's call it 5A – groups together a clade of 5 small-particle isolates, and is almost entirely restricted to the small chromosome (790/796 SNPs), strongly suggesting that the 5A strains share a closely related copy of the small chromosome. However, the *RpoS2/RTX* locus supports a grouping of a *different* five small-particle isolates – let's call it 5B – that is

phylogenetically inconsistent with 5A. SNPs supporting 5B are common on the large chromosome (630 SNPs), and rare on the small chromosome. The only 52 5B-supporting SNPs on the small chromosome occur in the *RpoS2/RTX* region. This suggests that a new *RpoS2/RTX* allele spread by recombination through the 5B isolates, which all share a chromosomal background supporting the 5A phylogeny. Does this suggest positive selection acting to favor a new allele in a cryptic, ecologically distinct 5B population, or perhaps diversifying selection acting within the small-particle population?

To help evaluate these scenarios, I applied the aLRH test to the small chromosome of seven small-particle isolates and two large-particle isolates (as an outgroup). For each phylogenetically informative SNP (with a minor allele present in two or more isolates), I calculated the average homozygosity (average EHH) within windows of up to 50 kb centered around the SNP. The choice of window size is somewhat arbitrary, but was chosen to be sensitive to relatively long stretches of recombined DNA (a few genes or an operon), without picking up linkage across the clonal frame (spanning the whole chromosome, on the order of hundreds to thousands of kb). The average EHH surrounding a representative 5B-supporting allele in the *RpoS2/RTX* region stays near the maximum value of 1 within a window of ~10 kb, whereas the alternative allele (present in the two non-5B small-particle isolates and the large-particle isolates) has a stable average EHH of ~0.4 (Fig. 17.5c). We can then calculate the average relative EHH (arEHH) for the 5B variant as arEHH $\approx 1/0.4 \approx 2.5$ within a ~10 kb window. It turns out that this 5B-SNP, and others in the ~2 kb *RpoS/RTX* region, have exceptionally high values of arEHH, meaning that they have unusually long linked haplotypes for their frequency (Fig. 17.5d). Remarkably, it is mostly 5B-SNPs and *not* ecoSNPs (fixed in all seven small-particle isolates) that are responsible for the high arEHH in the region. This is a clear shortcoming of the aLRH test: the ecoSNPs are likely under selection, but do not occur on very long haplotypes, so they are missed.

In parallel to the 'real' small chromosome data, I analyzed a 'matched' dataset (with the same rate of polymorphism) from a coalescent model (Hudson 2002) simulating a population evolving neutrally and recombining by gene conversion (with population recombination rate $\rho = 1$ gene conversion of tract length 500 bp per generation; ρ as high as 10 and tract length of 5 kb also fit the real data reasonably well). Instead of the characteristic exponential decay of relative EHH with SNP frequency observed in humans and other sexual populations (Sabeti et al. 2002), both simulated and real microbial populations show a relatively uniform distribution of arEHH across SNP frequencies (Fig. 17.5d). The slight drop in arEHH of high-frequency SNPs might be due to small clusters of ecoSNPs recombined into different clonal frames, resulting in rapid decay of linkage across the recombination breakpoint. An alternative, but not exclusive explanation is that near-identical pairs of 'sister strains,' almost certainly sharing a clonal frame, are responsible for the slightly higher arEHH of SNPs at frequency 2/7. Whatever the explanation, factors such as population subdivision and uneven gene flow, are likely contributors to the difference between the real and simulated data. One clear example of this is the spike in arEHH at frequency 5/7 in the real data, corresponding to support for the 5A partition across the entire small chromosome (likely a feature of this chromosome's clonal frame phylogeny). And yet the 5B-supporting SNPs in the *RpoS2/RTX* region stand out as having an *even higher* arEHH than 5A-SNPs (shown as green points at frequency 5/7 in Fig. 17.5d). This certainly suggests that a relatively large segment of DNA has been recombined into the 5A isolates. Whether this is due to recent positive selection on a cryptic sub-population or diversifying selection within the small-particle population is unknown, but the aLRH analysis does show that the *RpoS2/RTX* region stands out from other loci in the genome, and from neutral simulations. It also demonstrates how the aLRH test can be used to explain the results of the MK test, and to more fully explore the complex layers of selective pressures on this locus.

17.5.4 Conclusions from *Vibrio* Ecological Genomics

What conclusions can we draw about natural selection acting on these *V. cyclitrophicus* populations, and in particular on ecologically differentiated regions of the genome such as *RpoS2/RTX*? First, these regions were probably acquired by recombination from more distantly related populations. The stable maintenance (of different allelic versions in different ecological populations) of foreign pieces of DNA – which we would generally expect to be maladaptive – immediately suggests divergent positive selection between microhabitats. However, because the evolutionary history of ecoSNP regions such as *RpoS2/RTX* is so different from the history of the ecological populations themselves, standard tests for selection give ambiguous results. Second, natural selection may be operating on different levels of organization: both between and within populations. The *RpoS2* gene, for example, is highly differentiated between large- and small-particle populations, but is also highly polymorphic within the small-particle population. This could suggest diversifying selection within the small-particle population, or divergent positive selection between two small-particle cryptic sub-populations adapted to unobserved niches. Without further knowledge of these unobserved niches, or evidence of reduced gene flow between putative sub-populations, it is difficult to distinguish between diversifying and positive selection scenarios (Fig. 17.3). Third – and related to the second point – selective processes are inextricably linked to ecological differentiation and speciation: the processes that define and bound populations. In the *V. cyclitrophicus* study, we were able to define populations based on both ecological (particle size preference) and biological (preference for recombination within rather than between populations) criteria. This allowed us to identify as candidates for selection genes and alleles that are highly divergent between populations. However, this does not guarantee that the populations we defined are by any means 'the most important,' and other important axes of ecological differentiation and selection may well exist. Fourth, 'flexible' genes, acquired by illegitimate recombination, may also be subject to habitat-specific selection and are likely involved in ecological differentiation. In most comparative studies of microbial genomes, it is almost implicitly assumed that this is the case (Haegeman and Weitz 2012). However, beyond looking for stable and significant associations between gene presence/absence and a particular niche or phenotype of interest, formal tests for selection on the flexible genome are not well developed. However, neutral models of genome evolution are starting to be explored (Haegeman and Weitz 2012), and tests for selection based on deviation from such models will hopefully follow.

Finally, the population genomic tests for selection I have described will never constitute unassailable proof of positive selection. They can, however, generate strong and specific hypotheses about the mechanisms of ecological differentiation. The adaptive value of putative selected loci can be tested in competition experiments. For example, isogenic *Vibrios* with different *RpoS2* alleles could be competed in microcosms containing different mixtures of large- and small-particles obtained from filtered, sterilized seawater. Data can also be compared across studies to discern trends. In *E. coli* experimental evolution studies, *RpoS* was frequently mutated or duplicated in replicate populations subjected to heat stress (Riehle et al. 2001; Tenaillon et al. 2012). This reinforces that *RpoS* genes may be frequent targets of selection in novel environments.

How general are these conclusions, and to what extent can the analyses I've described in *Vibrio* be applied to other populations of microbes? One study of hotspring *Sulfolobus* archaea showed a similar pattern of ecological differentiation with gradual reduction in gene flow between populations (Cadillo-Quiroz et al. 2012). However, the population-specific regions (equivalent to ecoSNP regions) were much larger (several hundred kb) and widely distributed across the genome, making it difficult to pinpoint genes under divergent positive selection between habitats.

The *Sulfolobus* study is also a truer example of reverse ecology than the *Vibrio*. This is because the two *Sulfolobus* populations were not predefined based on known ecology, but were inferred *de novo* based on comparative genomics. They were subsequently found to have distinct growth characteristics, but more work will be required to understand the nature of their niche partitioning, and the genetic changes that drive it.

The MK test failed to identify any genes under selection between *Sulfolobus* populations, either because many genes are involved in ecological differentiation, each under only weak selection, or because of lack of recombination within the extended 'continents' of population-specific differentiation. While the MK test lacked power to pinpoint selected genes within large linked continents, a variant of the LRH test might be successful in genomes with longer stretches of linkage, as appears to be more the case in the *Sulfolobus* than the *Vibrio* genomes.

In a study of vaccine evasion by serotype-switching in highly recombining *Streptococcus pneumoniae*, the recombination events leading to serotype switches were generally much larger (~20–40 kb) than other recombinations in the genome (Croucher et al. 2011). This suggests that serotype-switch alleles may have spread rapidly in the population due to selection (probably some form of frequency-dependent selection) before the long stretches of linkage could be eroded by recombination. *Streptococcus* could therefore be another good candidate to apply a variant of the LRH test for selection. Generally speaking, the balance between selection and recombination within and between populations will dictate the sorts of tests for selection that can be applied (Fraser et al. 2009; Shapiro et al. 2009). In the next section, I will describe tests that perform well regardless of the recombination rate.

17.6 Convergence and Evolution in Real Time

So far, I have been considering 'cross-sectional' samples of genomes at a single moment in time. It is remarkable how much can be inferred about the evolutionary history of a population based on this kind of data, but microbes give us the opportunity to go further. In many natural microbial populations, large population sizes and high genetic diversity mean that the exact same mutation is likely to arise independently in different genomes. This allows a form of pseudo-replication, even in natural settings, that provides the basis for a class of tests for selection based on convergent evolution. Moreover, many bacteria replicate very rapidly, allowing us to track them in real time, and effectively watch evolutionary processes in action. The best examples of positive or diversifying selection in action come from viruses (HIV-1 in particular), and some of the techniques developed in their study (e.g. Pybus and Rambaut 2009) will prove very useful as time-course sequencing of natural bacterial populations becomes more feasible. For example, *dN* and *dS* can be measured precisely over time in replicate populations in order to estimate the strength of selection on nonsynonymous mutations (Neher and Leitner 2010).

In some instances, evolution in 'stasis' is just as interesting as evolution in action. In 2009, we isolated *Vibrio* from the same coastal location, using the same sampling strategy as 2006 in order to test whether the ecological associations we initially observed were stable over time (Szabó et al. 2013). Although some of the populations observed in 2006 were not observed at all in 2009 (perhaps due to association with cryptic niches, or stochastic extinctions), most of those that were observed again also had the same habitat association as in 2006. This suggests that particle-associated microniches are, to a large extent, stable over time. In particular, the small- and large-particle associated *V. cyclitrophicus* populations were re-identified in 2009, again based on shared habitat association and similarity in the *hsp60* phylogenetic marker gene. We did not sequence whole genomes of the 2009 isolates, but instead asked specifically whether the habitat-specific flexible genes were still associated with the same habitats in 2009. This type of stable association of flexible genes is not necessarily expected, because flexible genes are exchanged frequently by recombination in these populations,

resulting in even the most closely-related pairs of 'sister' strains (almost certainly sharing a recent clonal ancestor) differing by several flexible genes (Shapiro et al. 2012). In fact, all five flexible genes tested in a PCR-based screen were consistently habitat-associated in 2009.

These genes, involved in mannose-sensitive hemagglutinin (MSH) pilus biosynthesis, intercellular adhesion, and surface carbohydrate biosynthesis, are generally present in large- but not small-particle associated isolates. Given the expected high flexible genome turnover, it is difficult to explain the stable habitat-association of these genes without invoking selection. The involvement of the MSH genes in attachment to chitinous surfaces (Frischkorn et al. 2013; Meibom et al. 2005) also suggests they may be positively selected in large-particle strains, which likely rely on attachment to chitinous copepods or diatoms. The stable habitat association of these genes strongly suggests, but of course does not prove, that they are under habitat-specific positive selection. Alternatively, the association could be maintained by a preference for habitat-specific recombination. Even though we did observe such a preference, it was only discernible in a small fraction of the genome (Shapiro et al. 2012). Given the high turnover of flexible genes, it is hard to imagine how recombination alone could result in the stable gene-habitat association without invoking some form of selection. Overall, this is consistent with findings from other bacteria showing that flexible gene content tends to be structure by horizontal transfer within ecological niches with similar selective pressures (Boucher et al. 2011; Smillie et al. 2011).

As discussed extensively throughout this chapter, *dN/dS*, the MK test, allele-frequency- and haplotype-based tests for selection will often lack power or fail in highly clonal (rarely recombining) populations. An attractive alternative is a class of tests for selection based on convergent evolution, the independent fixation of the same mutation in different clonal backgrounds. Given a sufficient sample of related genomes, convergence has the potential to pinpoint positive selection to a single mutation. Alternatively, if several different mutations in the same gene confer the same selected phenotype, then convergence at the gene level would provide a sensitive test. For example, in a pioneering application of convergence to study pathogenic *E. coli*, Sokurenko and colleagues found a significant excess of nonsynonymous mutations in the adhesin gene *FimH*, specifically in lineages (tips of the phylogenetic tree) that independently transitioned from a gut commensal to a uropathogenic phenotype (Sokurenko 2004). Both *dN/dS* and allele-frequency tests lacked power to detect selection in *FimH*, whereas convergence provided strong evidence for selection at the level of the entire protein, and at individual amino acid sites. Sokurenko and colleagues began with a specific hypothesis about selection on *FimH*, but convergence can also be applied genomewide. For example, a comparative genomic study of *Salmonella* Typhi identified convergent mutations in *gyrA* as likely targets of selection for antibiotic resistance (Holt et al. 2008). Depending on the mutation rate and evolutionary time scales considered, a certain level of 'background' convergence is expected even in the absence of selection (Rokas and Carroll 2008). Provided that care is taken to control for this, convergence will provide a powerful framework for detecting selection in a wide variety of microbial genomes, ranging from highly clonal to highly recombining.

17.7 Outlook

In this chapter, I have drawn from selected examples to illustrate how one should go about assessing evidence for selection and ecological differentiation among microbial genomes. I introduced the major classes of tests for selection, and attempted to highlight their strengths and weaknesses, and cases in which special care is warranted in interpreting them. In particular, haplotype-based tests and tests for 'unexpected' recombination across great phylogenetic distances, although not yet fully developed for use in bacteria, have great potential for inferring selection in recombining populations (Shapiro et al. 2009). The MK

test provides a useful framework for assessing adaptive divergence genomewide, but often lacks power when applied to loci recently acquired by recombination with distance relatives. Convergence tests and time-course studies provide powerful tools for both recombining and clonal populations, provided that sufficient evolutionary time has transpired for convergent mutations to occur, or to observe changes in allele frequencies over time. There is currently no 'standardized' way to detect selection in microbial genomes; the current 'best practice' is to apply the tests that are justified given the recombination rate, and to interpret test results with a critical eye.

I focused on comparisons among closely-related whole genome sequences, but of course there are other rich sources of data that can be used to test for selection. The most obvious of these is metagenomic shotgun sequencing data, which yields allele-frequency information, but very limited information on how sequences of DNA are linked together into genomes. This accentuates the joint problem of detecting selection within or between species, while simultaneously defining the boundaries between species. However, when combined with ecological metadata and/or time course sampling, metagenomics can be a powerful tool for identifying genes, alleles and populations associated with particular phenotypes or ecological niches (Delong 2006; Denef et al. 2010b). Associations between protein expression and environmental metadata can be revealing as well. For example, proteins involved in cobalamin biosynthesis, phosphate uptake and motility were more highly expressed in late- than early-colonizing strains of *Leptospirillum* in an acid mine community, suggesting particular adaptations to the nutrient-limited environment experienced by late-colonizers (Denef et al. 2010a). Many of these differentially expressed proteins also had high *dN/dS* between early- and late-colonizers, further suggesting niche partitioning by positive selection.

As well as developing new and more refined tests for selection in microbial genomes and exploiting new datasets, we are also improving our basic understanding of the boundaries between microbial populations. In our current understanding, 'globally adaptive' genes may sweep through multiple sub-populations at once (Majewski and Cohan 1999), while sub-populations are maintained by local adaptation and preferences for within-population recombination. However, new findings are complicating models based on recombination and selection alone. For example, microbial populations might also be defined based on cooperation within but not between populations (Cordero et al. 2012). In such cases, it might be appropriate to consider selection at the level of the population, not just the individual or the gene. While genes are generally exchanged freely as 'public goods' (McInerney et al. 2011), they can sometimes become private to specific populations due to ecological boundaries, or barriers to recombination. An emerging challenge in microbial evolutionary genomics will therefore be to devise tests for selection that account for, and test the generality of, these new population models. Two papers have recently applied phylogenetic convergence tests for selection, discussed in Sect. 17.6, to identify genotype-phenotype associations. The tests were applied genomewide to identify variants associated with host-specificity (Sheppard et al. 2013) and antibiotic resistance (Farhat et al. 2013), respectively.

References

Baas-Becking LGM (1934) Geobiologie of Inleiding Tot de Milieukunde. W.P. Van Stockum & Zoon, The Hague

Boucher Y, Cordero OX, Takemura A et al (2011) Local mobile gene pools rapidly cross species boundaries to create endemicity within global *Vibrio cholerae* populations. mBio 2:e00335–10

Cadillo-Quiroz H, Didelot X, Held NL et al (2012) Patterns of gene flow define species of thermophilic archaea. PLoS Biol 10:e1001265

Cohan FM, Perry EB (2007) A systematics for discovering the fundamental units of bacterial diversity. Curr Biol 17:R373–R386

Coleman ML, Chisholm SW (2010) Ecosystem-specific selection pressures revealed through comparative population genomics. Proc Natl Acad Sci USA 107:18634–18639

Cordero OX, Wildschutte H, Kirkup B et al (2012) Ecological populations of bacteria act as socially cohesive units of antibiotic production and resistance. Science 337:1228–1231

Croucher NJ, Harris SR, Fraser C et al (2011) Rapid pneumococcal evolution in response to clinical interventions. Science 331:430–434

de Queiroz K (2005) Ernst Mayr and the modern concept of species. Proc Natl Acad Sci USA 102(Suppl 1):6600–6607

Delong EF (2006) Community genomics among stratified microbial assemblages in the ocean's interior. Science 311:496–503

Denef VJ, Kalnejais LH, Mueller RS et al (2010a) Proteogenomic basis for ecological divergence of closely related bacteria in natural acidophilic microbial communities. Proc Natl Acad Sci USA 107:2383–2390

Denef VJ, Mueller RS, Banfield JF (2010b) AMD biofilms: using model communities to study microbial evolution and ecological complexity in nature. ISME J 4:599–610

Didelot X, Falush D (2007) Inference of bacterial microevolution using multilocus sequence data. Genetics 175:1251–1266

Didelot X, Lawson D, Darling A, Falush D (2010) Inference of homologous recombination in bacteria using whole-genome sequences. Genetics 186:1435–1449

Doolittle WF, Papke RT (2006) Genomics and the bacterial species problem. Genome Biol 7:116

Farhat MR, Shapiro BJ, Kieser KJ, Sultana R, Jacobson KR, Victor TC, Warren RM, Streicher EM, Calver A, Sloutsky A, Kaur D, Posey JE, Plikaytis B, Oggioni MR, Gardy JL, Johnston JC, Rodrigues M, Tang PK, Kato-Maeda M, Borowsky ML, Muddukrishna B, Kreiswirth BN, Kurepina N, Galagan J, Gagneux S, Birren B, Rubin EJ, Lander ES, Sabeti PC, Murray M (2013) Genomic analysis identifies targets of convergent positive selection in drug-resistant *Mycobacterium tuberculosis*. Nat Genet 45(10):1183–1189

Fay JC, Wu CI (2000) Hitchhiking under positive Darwinian selection. Genetics 155:1405–1413

Fraser C, Alm EJ, Polz MF et al (2009) The bacterial species challenge: making sense of genetic and ecological diversity. Science 323:741–746

Friedman J, Alm EJ, Shapiro BJ (2013) Sympatric speciation: when is it possible in bacteria? PLoS ONE 8:e53539

Frischkorn KR, Stojanovski A, Paranjpye R (2013) *Vibrio parahaemolyticus* type IV pili mediate interactions with diatom-derived chitin and point to an unexplored mechanism of environmental persistence. Environ Microbiol 15:1416–1427

Gingold H, Pilpel Y (2011) Determinants of translation efficiency and accuracy. Mol Syst Biol 7:1–13

Haegeman B, Weitz JS (2012) A neutral theory of genome evolution and the frequency distribution of genes. BMC Genomics 13:196–196

Hanage WP, Spratt BG, Turner KM, Fraser C (2006) Modelling bacterial speciation. Philos Trans R Soc Lond B Biol Sci 361:2039–2044

Holt KE, Parkhill J, Mazzoni CJ et al (2008) High-throughput sequencing provides insights into genome variation and evolution in *Salmonella typhi*. Nat Genet 40:987–993

Hudson RR (2002) Generating samples under a Wright-Fisher neutral model of genetic variation. Bioinformatics 18:337–338

Hunt DE, David LA, Gevers D et al (2008) Resource partitioning and sympatric differentiation among closely related bacterioplankton. Science 320:1081–1085

Konstantinidis KT, Tiedje JM (2005) Towards a genome-based taxonomy for prokaryotes. J Bacteriol 187:6258–6264

Koonin EV, Makarova KS, Wolf YI, Aravind L (2002) Horizontal gene transfer and its role in the evolution of prokaryotes. In: Syvanen M, Kado CI (eds) Horizontal gene transfer, 2nd edn. Academic, London, pp 277–304

Lee SJ, Gralla JD (2002) Promoter use by sigma 38 (rpoS) RNA polymerase. Amino acid clusters for DNA binding and isomerization. J Biol Chem 277: 47420–47427

Li YF, Costello JC, Holloway AK, Hahn MW (2008) "Reverse ecology" and the power of population genomics. Evolution 62:2984–2994

Lin W, Fullner KJ, Clayton R et al (1999) Identification of a *Vibrio cholerae* RTX toxin gene cluster that is tightly linked to the cholera toxin prophage. Proc Natl Acad Sci USA 96:1071–1076

Majewski J, Cohan FM (1999) Adapt globally, act locally: the effect of selective sweeps on bacterial sequence diversity. Genetics 152:1459–1474

Mandel MJ, Wollenberg MS, Stabb EV et al (2009) A single regulatory gene is sufficient to alter bacterial host range. Nature 457:215–218

Marttinen P, Hanage WP, Croucher NJ et al (2012) Detection of recombination events in bacterial genomes from large population samples. Nucleic Acids Res 40:e6

Mau B, Glasner JD, Darling AE, Perna NT (2006) Genome-wide detection and analysis of homologous recombination among sequenced strains of *Escherichia coli*. Genome Biol 7:R44

Mayr E (1942) Systematics and the origin of species. Columbia University Press, New York

McDonald JH, Kreitman M (1991) Adaptive protein evolution at the Adh locus in Drosophila. Nature 351:652–654

McInerney JO, Pisani D, Bapteste E, O'Connell MJ (2011) The public goods hypothesis for the evolution of life on earth. Biol Direct 6:41

Meibom KL, Blokesch M, Dolganov NA et al (2005) Chitin induces natural competence in *Vibrio cholerae*. Science 310:1824–1827

Milkman R, Bridges MM (1990) Molecular evolution of the *Escherichia coli* chromosome. III. Clonal frames. Genetics 126:505–517

Neher RA, Leitner T (2010) Recombination rate and selection strength in HIV intra-patient evolution. PLoS Comput Biol 6:e1000660

Polz MF, Alm EJ, Hanage WP (2013) Horizontal gene transfer and the evolution of bacterial and archaeal population structure. Trends Genet 29:170

Pybus OG, Rambaut A (2009) Evolutionary analysis of the dynamics of viral infectious disease. Nat Rev Genet 10:540–550

Raghavan R, Kelkar YD, Ochman H (2012) A selective force favoring increased G+C content in bacterial genes. Proc Natl Acad Sci 109:14504–14507

Retchless AC, Lawrence JG (2010) Phylogenetic incongruence arising from fragmented speciation in enteric bacteria. Proc Natl Acad Sci USA 107:11453–11458

Riehle MM, Bennett AF, Long AD (2001) Genetic architecture of thermal adaptation in *Escherichia coli*. Proc Natl Acad Sci USA 98:525–530

Rokas A, Carroll SB (2008) Frequent and widespread parallel evolution of protein sequences. Mol Biol Evol 25:1943–1953

Sabeti PC, Reich DE, Higgins JM et al (2002) Detecting recent positive selection in the human genome from haplotype structure. Nature 419:832–837

Schubert S, Darlu P, Clermont O et al (2009) Role of intraspecies recombination in the spread of pathogenicity islands within the *Escherichia coli* species. PLoS Pathog 5:e1000257

Shapiro JA, Huang W, Zhang C et al (2007) Adaptive genic evolution in the Drosophila genomes. Proc Natl Acad Sci USA 104:2271–2276

Shapiro BJ, David LA, Friedman J, Alm EJ (2009) Looking for Darwin's footprints in the microbial world. Trends Microbiol 17:196–204

Shapiro BJ, Friedman J, Cordero OX et al (2012) Population genomics of early events in the ecological differentiation of bacteria. Science 336:48–51

Sheppard SK, Didelot X, Meric G, Torralbo A, Jolley KA et al (2013) Genome-wide association study identifies vitamin B5 biosynthesis as a host specificity factor in Campylobacter. Proc Natl Acad Sci USA 110: 11923–11927

Simmons SL, DiBartolo G, Denef VJ et al (2008) Population genomic analysis of strain variation in *Leptospirillum* group II bacteria involved in acid mine drainage formation. Plos Biol 6:1427–1442

Smillie CS, Smith MB, Friedman J et al (2011) Ecology drives a global network of gene exchange connecting the human microbiome. Nature 480: 241–244

Sokurenko EV (2004) Selection footprint in the FimH adhesin shows pathoadaptive niche differentiation in *Escherichia coli*. Mol Biol Evol 21: 1373–1383

Szabó G, Preheim SP, Kauffman KM et al (2013) Reproducibility of Vibrionaceae population structure in coastal bacterioplankton. ISME J 7:509–519

Tajima F (1989) Statistical method for testing the neutral mutation hypothesis by DNA polymorphism. Genetics 123:585–595

Tenaillon O, Rodriguez-Verdugo A, Gaut RL et al (2012) The molecular diversity of adaptive convergence. Science 335:457–461

Touchon M, Hoede C, Tenaillon O et al (2009) Organised genome dynamics in the *Escherichia coli* species results in highly diverse adaptive paths. PLoS Genet 5:e1000344

van Gremberghe I, Leliaert F, Mergeay J et al (2011) Lack of phylogeographic structure in the freshwater cyanobacterium *Microcystis aeruginosa* suggests global dispersal. PLoS ONE 6:e19561

Voight BF, Kudaravalli S, Wen X, Pritchard JK (2006) A map of recent positive selection in the human genome. Plos Biol 4:e72

Vos M (2011) A species concept for bacteria based on adaptive divergence. Trends Microbiol 19:1–7

Whitaker RJ, Banfield JF (2006) Population genomics in natural microbial communities. Trends Ecol Evol 21:508–516

Whitaker RJ, Grogan DW, Taylor JW (2003) Geographic barriers isolate endemic populations of hyperthermophilic archaea. Science 301:976–978

Yang Z (2008) Likelihood ratio tests for detecting positive selection and application to primate lysozyme evolution. Mol Biol Evol 15(5):568–573

Yang Z, Nielsen R (2002) Codon-substitution models for detecting molecular adaptation at individual sites along specific lineages. Mol Biol Evol 19: 908–917